Multimodal Scene Understanding

Multimodal Scene Understanding

Algorithms, Applications and Deep Learning

Edited by

Michael Ying Yang

Bodo Rosenhahn

Vittorio Murino

Academic Press is an imprint of Elsevier
125 London Wall, London EC2Y 5AS, United Kingdom
525 B Street, Suite 1650, San Diego, CA 92101, United States
50 Hampshire Street, 5th Floor, Cambridge, MA 02139, United States
The Boulevard, Langford Lane, Kidlington, Oxford OX5 1GB, United Kingdom

Notices

Knowledge and best practice in this field are constantly changing. As new research and experience broaden our understanding, changes in research methods, professional practices, or medical treatment may become necessary.

Practitioners and researchers must always rely on their own experience and knowledge in evaluating and using any information, methods, compounds, or experiments described herein. In using such information or methods they should be mindful of their own safety and the safety of others, including parties for whom they have a professional responsibility.

To the fullest extent of the law, neither the Publisher nor the authors, contributors, or editors, assume any liability for any injury and/or damage to persons or property as a matter of products liability, negligence or otherwise, or from any use or operation of any methods, products, instructions, or ideas contained in the material herein.

Library of Congress Cataloging-in-Publication Data
A catalog record for this book is available from the Library of Congress

British Library Cataloguing-in-Publication Data
A catalogue record for this book is available from the British Library

ISBN: 978-0-12-817358-9

Publisher: Mara Conner
Acquisition Editor: Tim Pitts
Editorial Project Manager: Joshua Mearns
Production Project Manager: Kamesh Ramajogi
Designer: Victoria Pearson

Typeset by VTeX

Contents

List of Contributors

Hanno Ackermann Institute for Information Processing, Leibniz University Hanover, Hanover, Germany

Yanlong Cao State Key Laboratory of Fluid Power and Mechatronic Systems, School of Mechanical Engineering, Zhejiang University, Hangzhou, China
Key Laboratory of Advanced Manufacturing Technology of Zhejiang Province, School of Mechanical Engineering, Zhejiang University, Hangzhou, China

Yanpeng Cao State Key Laboratory of Fluid Power and Mechatronic Systems, School of Mechanical Engineering, Zhejiang University, Hangzhou, China
Key Laboratory of Advanced Manufacturing Technology of Zhejiang Province, School of Mechanical Engineering, Zhejiang University, Hangzhou, China

Nesrine Chehata Univ. Paris-Est, LASTIG MATIS, IGN, ENSG, Saint-Mande, France
EA G&E Bordeaux INP, Université Bordeaux Montaigne, Pessac, France

Daniel Cremers Department of Informatics, Technical University of Munich, Munich, Germany
Artisense GmbH, Garching, Germany

Nuno C. Garcia Pattern Analysis & Computer Vision (PAVIS), Istituto Italiano di Tecnologia (IIT), Genova, Italy
Universita' degli Studi di Genova, Genova, Italy

Jaume Gibert Eurecat, Centre Tecnològic de Catalunya, Unitat de Tecnologies Audiovisuals, Barcelona, Spain

Lluis Gomez Computer Vision Center, Universitat Autònoma de Barcelona, Barcelona, Spain

Raul Gomez Eurecat, Centre Tecnològic de Catalunya, Unitat de Tecnologies Audiovisuals, Barcelona, Spain
Computer Vision Center, Universitat Autònoma de Barcelona, Barcelona, Spain

Valerie Gouet-Brunet Univ. Paris-Est, LASTIG MATIS, IGN, ENSG, Saint-Mande, France

Dayan Guan State Key Laboratory of Fluid Power and Mechatronic Systems, School of Mechanical Engineering, Zhejiang University, Hangzhou, China
Key Laboratory of Advanced Manufacturing Technology of Zhejiang Province, School of Mechanical Engineering, Zhejiang University, Hangzhou, China

Hai Huang Institute for Applied Computer Science, Bundeswehr University Munich, Neubiberg, Germany

Dorota Iwaszczuk Photogrammetry & Remote Sensing, Technical University of Munich, Munich, Germany

Asako Kanezaki National Institute of Advanced Industrial Science and Technology, Tokyo, Japan

Dimosthenis Karatzas Computer Vision Center, Universitat Autònoma de Barcelona, Barcelona, Spain

Vladimir V. Kniaz State Res. Institute of Aviation Systems (GosNIIAS), Moscow, Russia
Moscow Institute of Physics and Technology (MIPT), Moscow, Russia

Vladimir A. Knyaz State Res. Institute of Aviation Systems (GosNIIAS), Moscow, Russia
Moscow Institute of Physics and Technology (MIPT), Moscow, Russia

Zoltan Koppanyi Department of Civil, Envir. & Geod Eng., The Ohio State University, Columbus, OH, United States

Ryohei Kuga Graduate School of Information Science and Technology, Osaka University, Osaka, Japan

Andreas Kuhn Institute for Applied Computer Science, Bundeswehr University Munich, Neubiberg, Germany

Arnaud Le Bris Univ. Paris-Est, LASTIG MATIS, IGN, ENSG, Saint-Mande, France

Clément Mallet Univ. Paris-Est, LASTIG MATIS, IGN, ENSG, Saint-Mande, France

Yasuyuki Matsushita Graduate School of Information Science and Technology, Osaka University, Osaka, Japan

Helmut Mayer Institute for Applied Computer Science, Bundeswehr University Munich, Neubiberg, Germany

Mario Michelini Institute for Applied Computer Science, Bundeswehr University Munich, Neubiberg, Germany

Pietro Morerio Pattern Analysis & Computer Vision (PAVIS), Istituto Italiano di Tecnologia (IIT), Genova, Italy

Vittorio Murino Pattern Analysis & Computer Vision (PAVIS), Istituto Italiano di Tecnologia (IIT), Genova, Italy
Universita' degli Studi di Verona, Verona, Italy

Maroun Ojail CEA, LIST, Gif-sur-Yvette Cedex, France

Walid Ouerghemmi Univ. Paris-Est, LASTIG MATIS, IGN, ENSG, Saint-Mande, France
Aix-Marseille Université, CNRS ESPACE UMR 7300, Aix-en-Provence, France

Erwan Piriou CEA, LIST, Gif-sur-Yvette Cedex, France

Martyna Poreba Univ. Paris-Est, LASTIG MATIS, IGN, ENSG, Saint-Mande, France

Tristan Postadjian Univ. Paris-Est, LASTIG MATIS, IGN, ENSG, Saint-Mande, France

Anne Puissant CNRS UMR 7362 LIVE-Université de Strasbourg, Strasbourg, France

Christoph Reinders Institute for Information Processing, Leibniz University Hanover, Hanover, Germany

Bodo Rosenhahn Institute for Information Processing, Leibniz University Hanover, Hanover, Germany

Imane Salhi CEA, LIST, Gif-sur-Yvette Cedex, France
Univ. Paris-Est, LASTIG MATIS, IGN, ENSG, Saint-Mande, France

Can Jozef Saul Robert College, Istanbul, Turkey

Matthias Schmitz Institute for Applied Computer Science, Bundeswehr University Munich, Neubiberg, Germany

Yusuke Sugano Graduate School of Information Science and Technology, Osaka University, Osaka, Japan

Charles K. Toth Department of Civil, Envir. & Geod Eng., The Ohio State University, Columbus, OH, United States

Vladyslav Usenko Department of Informatics, Technical University of Munich, Munich, Germany

Lukas von Stumberg Department of Informatics, Technical University of Munich, Munich, Germany
Artisense GmbH, Garching, Germany

Cyril Wendl Univ. Paris-Est, LASTIG MATIS, IGN, ENSG, Saint-Mande, France
Student at Ecole Polytechnique Fédérale de Lausanne (EPFL), Lausanne, Switzerland

Jiangxin Yang State Key Laboratory of Fluid Power and Mechatronic Systems, School of Mechanical Engineering, Zhejiang University, Hangzhou, China
Key Laboratory of Advanced Manufacturing Technology of Zhejiang Province, School of Mechanical Engineering, Zhejiang University, Hangzhou, China

Michael Ying Yang Scene Understanding Group, University of Twente, Enschede, The Netherlands

Alper Yilmaz Department of Civil, Envir. & Geod Eng., The Ohio State University, Columbus, OH, United States

Bing Zha Department of Civil, Envir. & Geod Eng., The Ohio State University, Columbus, OH, United States

CHAPTER 1

Introduction to Multimodal Scene Understanding

Michael Ying Yang*, Bodo Rosenhahn†, Vittorio Murino‡

*University of Twente, Enschede, The Netherlands †Leibniz University Hannover, Hannover, Germany
‡Istituto Italiano di Tecnologia, Genova, Italy

Contents

1.1 Introduction

While humans constantly extract meaningful information from visual data almost effortlessly, it turns out that simple visual tasks such as recognizing, detecting and tracking objects, or, more difficult, understanding what is going on in the scene, are extremely challenging problems for machines. To design artificial vision systems that can reliably process information as humans do has many potential applications in fields such as robotics, medical imaging, surveillance, remote sensing, entertainment or sports science, to name a few. It is therefore our ultimate goal to be able to emulate the human visual system and processing capabilities with computational algorithms.

Computer vision has contributed to a broad range of tasks to the field of artificial intelligence, such as estimating physical properties from an image, e.g., depth and motion, as well as estimating semantic properties, e.g., labeling each pixel with a semantic class. A fundamental goal of computer vision is to discover the semantic information within a given scene, namely, *understanding* a scene, which is the basis for many applications: surveillance, autonomous driving, traffic safety, robot navigation, vision-guided mobile navigation systems, or activity recognition. Understanding a scene from an image or a video requires much more than recording and extracting some features. Apart from visual information, humans make use of further sensor data, e.g. from audio signals, or acceleration. The net goal is to find a mapping

Multimodal Scene Understanding
https://doi.org/10.1016/B978-0-12-817358-9.00007-X

to derive semantic information from sensor data, which is an extremely challenging task partially due to the ambiguities in the appearance of the data. These ambiguities may arise either due to the physical conditions such as the illumination and the pose of the scene components, or due to the intrinsic nature of the sensor data itself. Therefore, there is the need of capturing local, global or dynamic aspects of the acquired observations, which are to be utilized to interpret a scene. Besides, all information which is possible to extract from a scene must be considered in context in order to get a comprehensive representation, but this information, while it is easily captured by humans, is still difficult to extract by machines.

Using big data leads to a big step forward in many applications of computer vision. However, the majority of scene understanding tasks tackled so far involve visual modalities only. The main reason is the analogy to our human visual system, resulting in large multipurpose labeled image datasets. The unbalanced number of labeled samples available among different modalities result in a big gap in performance when algorithms are trained separately [1]. Recently, a few works have started to exploit the synchronization of multimodal streams to transfer semantic information from one modality to another, e.g. RGB/Lidar [2], RGB/depth [3,4], RGB/infrared [5,6], text/image [7], image/Inertial Measurement Units (IMU) data [8,9].

This book focuses on recent advances in algorithms and applications that involve multiple sources of information. Its aim is to generate momentum around this topic of growing interest, and to encourage interdisciplinary interactions and collaborations between computer vision, remote sensing, robotics and photogrammetry communities. The book will also be relevant to efforts on collecting and analyzing multisensory data corpora from different platforms, such as autonomous vehicles [10], surveillance cameras [11], unmanned aerial vehicles (UAVs) [12], airplanes [13] and satellites [14]. On the other side, it is undeniable that deep learning has transformed the field of computer vision, and now rivals human-level performance in tasks such as image recognition [15], object detection [16], and semantic segmentation [17]. In this context, there is a need for new discussions as regards the roles and approaches for multisensory and multimodal deep learning in the light of these new recognition frameworks.

In conclusion, the central aim of this book is to facilitate the exchange of ideas on how to develop algorithms and applications for multimodal scene understanding. The following are some of the scientific questions and challenges we hope to address:

- What are the general principles that help in the fusion of multimodal and multisensory data?
- How can multisensory information be used to enhance the performance of generic high-level vision tasks, such as object recognition, semantic segmentation, localization, and scene reconstruction, and empower new applications?
- What are the roles and approaches of multimodal deep learning?

To address these challenges, a number of peer-reviewed chapters from leading researchers in the fields of computer vision, remote sensing, and machine learning have been selected. These chapters provide an understanding of the state-of-the-art, open problems, and future directions related to multimodal scene understanding as a relevant scientific discipline.

The editors sincerely thank everyone who supported the process of preparing this book. In particular, we thank the authors, who are among the leading researchers in the field of multimodal scene understanding. Without their contributions in writing and peer-reviewing the chapters, this book would not have been possible. We are also thankful to Elsevier for the excellent support.

1.2 Organization of the Book

An overview of each of the book chapters is given in the following.

Chapter 2: Multimodal Deep Learning for Multisensory Data Fusion

This chapter investigates multimodal encoder–decoder networks to harness the multimodal nature of multitask scene recognition. In its position regarding the current state of the art, this work was distinguished by: (1) the use of the U-net architecture, (2) the application of translations between all modalities of the learning package and the use of monomodal data, which improves intra-modal self-encoding paths, (3) the independent mode of operation of the encoder–decoder, which is also useful in the case of missing modalities, and (4) the image-to-image translation application managed by more than two modalities. It also improves the multitasking reference network and automatic multimodal coding systems. The authors evaluate their method on two public datasets. The results of the tests illustrate the effectiveness of the proposed method in relation to other work.

Chapter 3: Multimodal Semantic Segmentation: Fusion of RGB and Depth Data in Convolutional Neural Networks

This chapter investigates the fusion of optical multispectral data (red-green-blue or near infrared-red-green) with 3D (and especially depth) information within a deep learning CNN framework. Two ways are proposed to use 3D information: either 3D information is directly introduced into the classification fusion as a depth measure or information about normals is estimated and provided as input to the fusion process. Several fusion solutions are considered and compared: (1) Early fusion: RGB and depth (or normals) are merged before being provided to the CNN. (2) RGB and depth (or normals) are simply concatenated and directly provided to common CNN architectures. (3) RGB and depth (or normals) are provided as two distinct inputs to a Siamese CNN dedicated to fusion. Such methods are tested on two benchmark datasets: an indoor terrestrial one (Stanford) and an aerial one (Vaihingen).

Chapter 4: Learning Convolutional Neural Networks for Object Detection with Very Little Training Data

This chapter addresses the problem of learning with very few labels. In recent years, convolutional neural networks have shown great success in various computer vision tasks, whenever they are trained on large datasets. The availability of sufficiently large labeled data, however, limits possible applications. The presented system for object detection is trained with very few training examples. To this end, the advantages of convolutional neural networks and random forests are combined to learn a patch-wise classifier. Then the random forest is mapped to a neural network and the classifier is transformed to a fully convolutional network. Thereby, the processing of full images is significantly accelerated and bounding boxes can be predicted. In comparison to the networks for object detection or algorithms for transfer learning, the required amount of labeled data is considerably reduced. Finally, the authors integrate GPS-data with visual images to localize the predictions on the map and multiple observations are merged to further improve the localization accuracy.

Chapter 5: Multimodal Fusion Architectures for Pedestrian Detection

In this chapter, a systematic evaluation of the performances of a number of multimodal feature fusion architectures is presented, in the attempt to identify the optimal solutions for pedestrian detection. Recently, multimodal pedestrian detection has received extensive attention since the fusion of complementary information captured by visible and infrared sensors enables robust human target detection under daytime and nighttime scenarios. Two important observations can be made: (1) it is useful to combine the most commonly used concatenation fusion scheme with a global scene-aware mechanism to learn both human-related features and correlation between visible and infrared feature maps; (2) the two-stream semantic segmentation without multimodal fusion provides the most effective scheme to infuse semantic information as supervision for learning human-related features. Based on these findings, a unified multimodal fusion framework for joint training of semantic segmentation and target detection is proposed, which achieves state-of-the-art multispectral pedestrian detection performance on the KAIST benchmark dataset.

Chapter 6: ThermalGAN: Multimodal Color-to-Thermal Image Translation for Person Re-Identification in Multispectral Dataset

This chapter deals with color-thermal cross-modality person re-identification (Re-Id). This topic is still challenging, in particular for video surveillance applications. In this context, it is demonstrated that conditional generative adversarial networks are effective for cross-modality prediction of a person appearance in thermal image conditioned by a probe color image. Discriminative features can be extracted from real and synthesized thermal images for effective matching of thermal signatures. The main observation is that thermal cameras coupled with generative adversarial network (GAN) Re-Id framework can significantly improve the Re-Id performance in low-light conditions. A ThermalGAN framework for cross-modality person Re-Id in the visible range and infrared images is so proposed. Furthermore, a large-scale multispectral ThermalWorld dataset is collected, acquired with FLIR ONE PRO cameras, usable both for Re-Id and visual objects in context recognition.

Chapter 7: A Review and Quantitative Evaluation of Direct Visual–Inertia Odometry

This chapter combines complementary features of visual and inertial sensors to solve direct sparse visual–inertial odometry problem in the field of simultaneous localization and mapping (SLAM). By introducing a novel optimization problem that minimizes camera geometry and motion sensor errors, the proposed algorithm estimates camera pose and sparse scene geometry precisely and robustly. As the initial scale can be very far from the optimum, a technique is proposed called dynamic marginalization, where multiple marginalization priors and constraints on the maximum scale difference are considered. Extensive quantitative evaluation on the EuRoC dataset demonstrates that the described visual–inertial odometry method outperforms other state-of-the-art methods, both the complete system as well as the IMU initialization procedure.

Chapter 8: Multimodal Localization for Embedded Systems: A Survey

This chapter presents a survey of systems, sensors, methods, and application domains of multimodal localization. The authors introduce the mechanisms of various sensors such as inertial measurement units (IMUs), global navigation satellite system (GNSS), RGB cameras (with global shutter and rolling shutter technology), IR and Event-based cameras, RGB-D cameras, and Lidar sensors. It leads the reader to other survey papers and thus covers the corresponding research areas exhaustively. Several types of sensor fusion methods are also illustrated. Moreover, various approaches and hardware configurations for specific applications (e.g. autonomous mobile robots) as well as real products (such as Microsoft Hololens and Magic Leap One) are described.

Chapter 9: Self-supervised Learning from Web Data for Multimodal Retrieval

This chapter addresses the problem of self-supervised learning from image and text data which is freely available from web and social media data. Thereby features of a convolutional neural network can be learned without requiring labeled data. Web and social media platforms provide a virtually unlimited amount of this multimodal data. This free available bunch of data is then exploited to learn a multimodal image and text embedding, aiming to leverage the semantic knowledge learned in the text domain and transfer it to a visual model for semantic image retrieval. A thorough analysis and performance comparisons of five different state-of-the-art text embeddings in three different benchmarks are reported.

Chapter 10: 3D Urban Scene Reconstruction and Interpretation from Multisensor Imagery

This chapter presents an approach for 3D urban scene reconstruction based on the fusion of airborne and terrestrial images. It is one step forward towards a complete and fully automatic pipeline for large-scale urban reconstruction. Fusion of images from different platforms (terrestrial, UAV) has been realized by means of pose estimation and 3D reconstruction of the observed scene. An automatic pipeline for level of detail 2 building model reconstruction is proposed, which combines a reliable scene and building decomposition with a subsequent primitive-based reconstruction and assembly. Level of detail 3 models are obtained by integrating the results of facade image interpretation with an adapted convolutional neural network (CNN), which employs the 3D point cloud as well as the terrestrial images.

Chapter 11: Decision Fusion of Remote Sensing Data for Land Cover Classification

This chapter presents a framework for land cover classification by late decision fusion of multimodal data. The data include imagery with different spatial as well as temporal resolution and spectral range. The main goal is to build a practical and flexible pipeline with proven techniques (i.e., CNN and random forest) for various data and appropriate fusion rules. The different remote sensing modalities are first classified independently. Class membership maps calculated for each of them are then merged at pixel level, using decision fusion rules, before the final label map is obtained from a global regularization. This global regularization aims at dealing with spatial uncertainties. It relies on a graphical model, involving a fit-to-data term related to merged class membership measures and an image-based contrast sensitive regularization term. Two use cases demonstrate the potential of the work and limitations of the proposed methods are discussed.

Chapter 12: Cross-modal Learning by Hallucinating Missing Modalities in RGB-D Vision

Diverse input data modalities can provide complementary cues for several tasks, usually leading to more robust algorithms and better performance. This chapter addresses the challenge of how to learn robust representations leveraging multimodal data in the training stage, while considering limitations at test time, such as noisy or missing modalities. In particular, the authors consider the case of learning representations from depth and RGB videos, while relying on RGB data only at test time. A new approach to training a hallucination network has been proposed that learns to distill depth features through multiplicative connections of spatiotemporal representations, leveraging soft labels and hard labels, as well as distance between feature maps. State-of-the-art results on the video action classification dataset are reported.

Note: The color figures will appear in color in all electronic versions of this book.

References

[1] A. Nagrani, S. Albanie, A. Zisserman, Seeing voices and hearing faces: cross-modal biometric matching, in: IEEE Conference on Computer Vision and Pattern Recognition, CVPR, 2018.

[2] M.Y. Yang, Y. Cao, J. McDonald, Fusion of camera images and laser scans for wide baseline 3D scene alignment in urban environments, ISPRS Journal of Photogrammetry and Remote Sensing 66 (6S) (2011) 52–61.

[3] A. Krull, E. Brachmann, F. Michel, M.Y. Yang, S. Gumhold, C. Rother, Learning analysis-by-synthesis for 6d pose estimation in rgb-d images, in: IEEE International Conference on Computer Vision, ICCV, 2015.

[4] O. Hosseini, O. Groth, A. Kirillov, M.Y. Yang, C. Rother, Analyzing modular cnn architectures for joint depth prediction and semantic segmentation, in: International Conference on Robotics and Automation, ICRA, 2017.

[5] M.Y. Yang, Y. Qiang, B. Rosenhahn, A global-to-local framework for infrared and visible image sequence registration, in: IEEE Winter Conference on Applications of Computer Vision, 2015.

[6] A. Wu, W.-S. Zheng, H.-X. Yu, S. Gong, J. Lai, Rgb-infrared cross-modality person re-identification, in: IEEE International Conference on Computer Vision, ICCV, 2017.

[7] D. Huk Park, L. Anne Hendricks, Z. Akata, A. Rohrbach, B. Schiele, T. Darrell, M. Rohrbach, Multimodal explanations: justifying decisions and pointing to the evidence, in: IEEE Conference on Computer Vision and Pattern Recognition, CVPR, 2018.

[8] C. Reinders, H. Ackermann, M.Y. Yang, B. Rosenhahn, Object recognition from very few training examples for enhancing bicycle maps, in: IEEE Intelligent Vehicles Symposium, IV, 2018, pp. 1–8.

[9] T. von Marcard, R. Henschel, M.J. Black, B. Rosenhahn, G. Pons-Moll, Recovering accurate 3d human pose in the wild using imus and a moving camera, in: European Conference on Computer Vision, ECCV, 2018, pp. 614–631.

[10] A. Geiger, P. Lenz, R. Urtasun, Are we ready for autonomous driving? The kitti vision benchmark suite, in: IEEE Conference on Computer Vision and Pattern Recognition, CVPR, 2012.

[11] S. Oh, A. Hoogs, A.G.A. Perera, N.P. Cuntoor, C. Chen, J.T. Lee, S. Mukherjee, J.K. Aggarwal, H. Lee, L.S. Davis, E. Swears, X. Wang, Q. Ji, K.K. Reddy, M. Shah, C. Vondrick, H. Pirsiavash, D. Ramanan, J. Yuen, A. Torralba, B. Song, A. Fong, A.K. Roy-Chowdhury, M. Desai, A large-scale benchmark dataset for event recognition in surveillance video, in: IEEE Conference on Computer Vision and Pattern Recognition, CVPR, 2011, pp. 3153–3160.

[12] F. Nex, M. Gerke, F. Remondino, H. Przybilla, M. Baumker, A. Zurhorst, Isprs benchmark for multi-platform photogrammetry, in: Annals of the Photogrammetry, Remote Sensing and Spatial Information Science, 2015, pp. 135–142.

[13] Z. Zhang, M. Gerke, G. Vosselman, M.Y. Yang, A patch-based method for the evaluation of dense image matching quality, International Journal of Applied Earth Observation and Geoinformation 70 (2018) 25–34.

[14] X. Han, X. Huang, J. Li, Y. Li, M.Y. Yang, J. Gong, The edge-preservation multi-classifier relearning framework for the classification of high-resolution remotely sensed imagery, ISPRS Journal of Photogrammetry and Remote Sensing 138 (2018) 57–73.

[15] A. Krizhevsky, I. Sutskever, G.E. Hinton, Imagenet classification with deep convolutional neural networks, in: Advances in Neural Information Processing Systems, NIPS, 2012, pp. 1097–1105.

[16] S. Ren, K. He, R. Girshick, J. Sun, Faster r-cnn: towards real-time object detection with region proposal networks, in: Advances in Neural Information Processing Systems, NIPS, 2015, pp. 91–99.

[17] J. Long, E. Shelhamer, T. Darrell, Fully convolutional networks for semantic segmentation, in: IEEE Conference on Computer Vision and Pattern Recognition, CVPR, 2015.

CHAPTER 2

Deep Learning for Multimodal Data Fusion

Asako Kanezaki*, Ryohei Kuga†, Yusuke Sugano†, Yasuyuki Matsushita†

*National Institute of Advanced Industrial Science and Technology, Tokyo, Japan †Graduate School of Information Science and Technology, Osaka University, Osaka, Japan

Contents

Multimodal Scene Understanding
https://doi.org/10.1016/B978-0-12-817358-9.00008-1

2.1 Introduction

Scene understanding is one of the most important tasks for various applications including robotics and autonomous driving and has been an active research area in computer vision for a long time. The goal of scene understanding can be divided into several different tasks, such as depth reconstruction and semantic segmentation. Traditionally, these different tasks have been studied independently, resulting in their own tailored methods. Recently, there has been a growing demand for a single unified framework to achieve multiple tasks at a time unlike previous approaches. By sharing a part of the learned estimator, such a multitask learning framework is expected to achieve better performance with a compact representation.

In most of the prior work, multitask learning is formulated with a motivation to train a shared feature representation among different tasks for efficient feature encoding [1–3]. Accordingly, in recent convolutional neural network (CNN)-based methods, multitask learning often employs an encoder–decoder network architecture [1,2,4]. If, for example, the target tasks are semantic segmentation and depth estimation from RGB images, multitask networks encode the input image to a shared low-dimensional feature representation and then estimate depth and semantic labels with two distinct decoder networks.

While such a shared encoder architecture can constrain the network to extract a common feature for different tasks, one limitation is that it cannot fully exploit the multimodal nature of the training dataset. The representation capability of the shared representation in the above example is not limited to image-to-label and image-to-depth conversion tasks, but it can also represent the common feature for *all* of the cross-modal conversion tasks such as depth-to-label as well as within-modal dimensionality reduction tasks such as image-to-image. By incorporating these additional conversion tasks during the training phase, the multitask network is expected to learn more efficient shared feature representation for the diverse target tasks.

In this chapter, we introduce a recent method named the multimodal encoder–decoder networks method [5] for multitask scene recognition. The model consists of encoders and decoders for each modality, and the whole network is trained in an end-to-end manner taking into account all conversion paths—both cross-modal encoder–decoder pairs and within-modal self-encoders. As illustrated in Fig. 2.1, all encoder–decoder pairs are connected via a single shared latent representation in the method. In addition, inspired by the U-net architecture [6, 7], the decoders for pixel-wise image conversion tasks such as semantic segmentation also take a shared skipped representation from all encoders. Since the whole network is jointly trained using multitask losses, these two shared representations are trained to extract the common feature representation among all modalities. Unlike multimodal auto-encoders [1], this method can further utilize auxiliary unpaired data to train self-encoding paths and consequently improve the cross-modal conversion performance. In the experiments using two

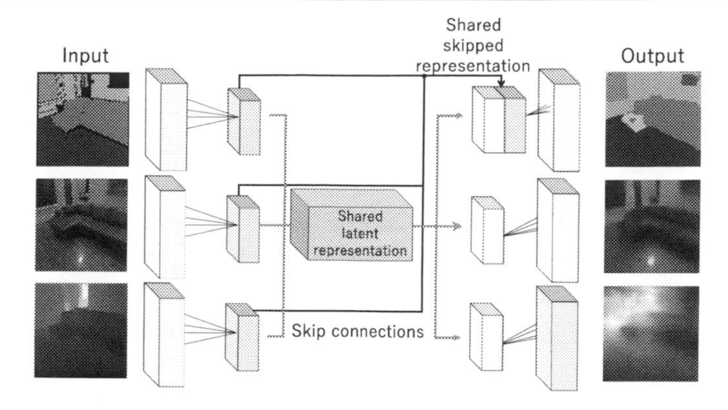

Figure 2.1 : Overview of the multimodal encoder–decoder networks. The model takes data in multiple modalities, such as RGB images, depth, and semantic labels, as input, and generates multimodal outputs in a multitask learning framework.

public datasets, we show that the multimodal encoder–decoder networks perform significantly better on cross-modal conversion tasks.

The remainder of this chapter is organized as follows. In Sect. 2.2, we summarize in an overview various methods on multimodal data fusion. Next, we describe the basics of multimodal deep learning techniques in Sect. 2.3 and the latest work based on those techniques in Sect. 2.4. We then introduce the details of multimodal encoder–decoder networks in Sect. 2.5. In Sect. 2.6, we show experimental results and discuss the performance of multimodal encoder–decoder networks on several benchmark datasets. Finally, we conclude this chapter in Sect. 2.7.

2.2 Related Work

Multitask learning is motivated by the finding that the feature representation for one particular task could be useful for the other tasks [8]. In prior work, multiple tasks, such as scene classification, semantic segmentation [9], character recognition [10] and depth estimation [11, 12], have been addressed with a single input of an RGB image, which is referred to as *single-modal multitask learning*. Hand et al. [13] demonstrated that multitask learning of gender and facial parts from one facial image leads to better accuracy than individual learning of each task. Hoffman et al. [14] proposed a modality hallucination architecture based on CNNs, which boosts the performance of RGB object detection using depth information only in the training phase. Teichmann et al. [15] presented neural networks for scene classification, object detection, segmentation of a street view image. Uhrig et al. [16] proposed an instance-level segmentation method via simultaneous estimation of semantic labels, depth, and instance center direction. Li et al. [17] proposed fully convolutional neural networks for segmentation and saliency tasks. In these previous approaches, the feature representation of a single input modality is shared in an intermediate layer for solving multiple tasks. In contrast, the multimodal encoder–decoder networks [5] described in Sect. 2.5 fully utilize the multimodal training data by learning cross-modal shared representations through joint multitask training.

There have been several prior attempts for utilizing multimodal inputs for deep neural networks. They proposed the use of multimodal input data, such as RGB and depth images [18], visual and textual features [19], audio and video [2], and multiple sensor data [20], for single-task neural networks. In contrast to such multimodal single-task learning methods, relatively few studies have been made on *multimodal* & *multitask* learning. Ehrlich et al. [21] presented a method to identify a person's gender and smiling based on two feature modalities extracted from face images. Cadena et al. [1] proposed neural networks based on auto-encoder for multitask estimation of semantic labels and depth.

Both of these single-task and multitask learning methods with multimodal data focused on obtaining better shared representation from multimodal data. Since straightforward concatenation of extracted features from different modalities often results in lower estimation accuracy, some prior methods tried to improve the shared representation by singular value decomposition [22], encoder–decoder [23], auto-encoder [2,1,24], and supervised mapping [25]. While the multimodal encoder–decoder networks are also based on the encoder–decoder approach, one employs the U-net architecture for further improving the learned shared representation, particularly in high-resolution convolutional layers.

Most of the prior works also assume that all modalities are available for the single-task or multitask in both training and test phases. One approach for dealing with the missing modal data is to perform zero-filling, which fills the missing elements in the input vector by zeros [2, 1]. Although these approaches allow the multimodal networks to handle missing modalities and cross-modal conversion tasks, it has not been fully discussed whether such a zero-filling approach can be also applied to recent CNN-based architectures. Sohn et al. [19] explicitly estimated missing modal data from available modal data by deep neural networks. In a difficult task, such as a semantic segmentation with many classes, the missing modal data is estimated inaccurately, which has a negative influence on performance of the whole network. Using the multimodal encoder–decoder networks, at the test phase encoder–decoder paths work individually even for missing modal data. Furthermore, it can perform conversions between all modalities in the training set, and it can utilize single-modal data to improve within-modal self-encoding paths during the training.

Recently, many image-to-image translation methods based on deep neural networks have been developed [7,26–32]. In contrast to that they address image-to-image translation on two different modalities, StarGAN [33] was recently proposed to efficiently learn the translation on

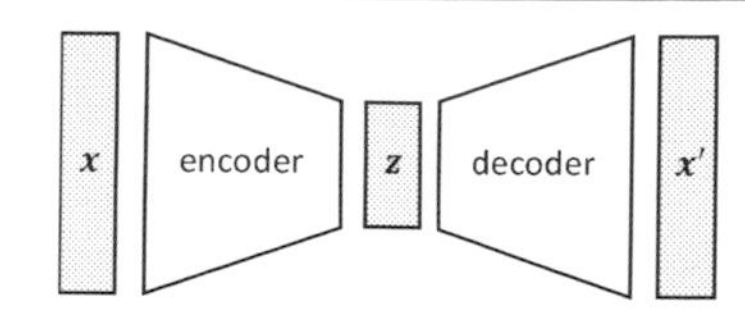

Figure 2.2 : Architecture of Auto-encoder.

more than two domains. The multimodal encoder–decoder networks is also applicable to the translation on more than two modalities. We describe the details of this work in Sect. 2.4 and the basic methods behind this work in Sect. 2.3.

2.3 Basics of Multimodal Deep Learning: VAEs and GANs

This section introduces the basics of multimodal deep learning for multimodal image translation. We first mention auto-encoder, which is the most basic neural network consisting of an encoder and a decoder. Then we introduce an important extension of auto-encoder named variational auto-encoder (VAE) [34,35]. VAEs consider a standard normal distribution for latent variables and thus they are useful for generative modeling. Next, we describe generative adversarial network (GAN) [36], which is the most well-known way of learning deep neural networks for multimodal data generation. Concepts of VAEs and GANs are combined in various ways to improve the distribution of latent space for image generation with, *e.g.*, VAE-GAN [37], adversarial auto-encoder (AAE) [38], and adversarial variational Bayes (AVB) [39], which are described later in this section. We also introduce the adversarially learned inference (ALI) [40] and the bidirectional GAN (BiGAN) [41], which combine the GAN framework and the inference of latent representations.

2.3.1 Auto-Encoder

An auto-encoder is a neural network that consists of an encoder network and a decoder network (Fig. 2.2). Letting $\mathbf{x} \in \mathbb{R}^d$ be an input variable, which is a concatenation of pixel values when managing images, an encoder maps it to a latent variable $\mathbf{z} \in \mathbb{R}^r$, where r is usually much smaller than d. A decoder maps $\mathbf{z}$ to the output $\mathbf{x}' \in \mathbb{R}^d$, which is the reconstruction of the input $\mathbf{x}$. The encoder and decoder are trained so as to minimize the reconstruction errors such as the following squared errors:

$$\mathcal{L}_{\mathrm{AE}} = \mathbb{E}_{\mathbf{x}}[\|\mathbf{x} - \mathbf{x}'\|^2]. \tag{2.1}$$

The purpose of an auto-encoder is typically dimensionality reduction, or in other words, unsupervised feature / representation learning. Recently, as well as the encoding process, more

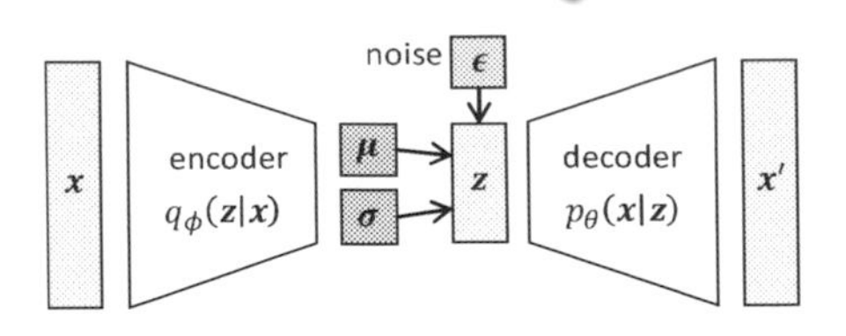

Figure 2.3 : Architecture of VAE [34,35].

attention has been given to the decoding process which has the ability of data generation from latent variables.

2.3.2 Variational Auto-Encoder (VAE)

The variational auto-encoder (VAE) [34,35] regards an auto-encoder as a generative model where the data is generated from some conditional distribution $p(\mathbf{x}|\mathbf{z})$. Letting ϕ and θ denote the parameters of the encoder and the decoder, respectively, the encoder is described as a recognition model $q_\phi(\mathbf{z}|\mathbf{x})$, whereas the decoder is described as an approximation to the true posterior $p_\theta(\mathbf{x}|\mathbf{z})$. The marginal likelihood of an individual data point $\mathbf{x}_i$ is written as follows:

$$\log p_\theta(\mathbf{x}_i) = D_{\mathrm{KL}}\left(q_\phi(\mathbf{z}|\mathbf{x}_i)||p_\theta(\mathbf{z}|\mathbf{x}_i)\right) + \mathcal{L}(\theta, \phi; \mathbf{x}_i). \tag{2.2}$$

Here, D_{KL} stands for the Kullback–Leibler divergence. The second term in this equation is called the (variational) lower bound on the marginal likelihood of data point i, which can be written as

$$\mathcal{L}(\theta, \phi; \mathbf{x}_i) = -D_{\mathrm{KL}}\left(q_\phi(\mathbf{z}|\mathbf{x}_i)||p_\theta(\mathbf{z})\right) + \mathbb{E}_{\mathbf{z}\sim q_\phi(\mathbf{z}|\mathbf{x}_i)}[\log p_\theta(\mathbf{x}_i|\mathbf{z})]. \tag{2.3}$$

In the training process, the parameters ϕ and θ are optimized so as to minimize the total loss $-\sum_i \mathcal{L}(\theta, \phi; \mathbf{x}_i)$, which can be written as

$$\mathcal{L}_{\mathrm{VAE}} = D_{\mathrm{KL}}\left(q_\phi(\mathbf{z}|\mathbf{x})||p_\theta(\mathbf{z})\right) - \mathbb{E}_{\mathbf{z}\sim q_\phi(\mathbf{z}|\mathbf{x})}[\log p_\theta(\mathbf{x}|\mathbf{z})]. \tag{2.4}$$

Unlike the original auto-encoder which does not consider the distribution of latent variables, VAE assumes the prior over the latent variables such as the centered isotropic multivariate Gaussian $p_\theta(\mathbf{z}) = \mathcal{N}(\mathbf{z}; \mathbf{0}, \mathbf{I})$. In this case, we can let the variational approximate posterior be a multivariate Gaussian with a diagonal covariance structure:

$$\log q_\phi(\mathbf{z}|\mathbf{x}_i) = \log \mathcal{N}(\mathbf{z}; \boldsymbol{\mu}_i, \boldsymbol{\sigma}_i^2\mathbf{I}), \tag{2.5}$$

where the mean $\boldsymbol{\mu}_i$ and the standard deviation $\boldsymbol{\sigma}_i$ of the approximate posterior are the outputs of the encoder (Fig. 2.3). In practice, the latent variable $\mathbf{z}_i$ for data point i is calculated as follows:

$$\mathbf{z}_i = \boldsymbol{\mu}_i + \boldsymbol{\sigma}_i \cdot \boldsymbol{\epsilon}, \quad \text{where} \quad \boldsymbol{\epsilon} \sim \mathcal{N}(\mathbf{0}, \mathbf{I}). \tag{2.6}$$

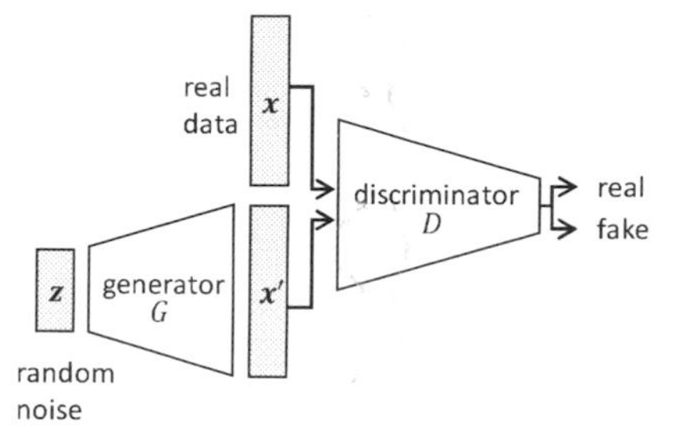

Figure 2.4 : Architecture of GAN [36].

2.3.3 Generative Adversarial Network (GAN)

The generative adversarial network (GAN) [36] is one of the most successful framework for data generation. It consists of two networks: a generator G and a discriminator D (shown in Fig. 2.4), which are jointly optimized in a competitive manner. Intuitively, the aim of the generator is to generate a sample $\mathbf{x}'$ from a random noise vector $\mathbf{z}$ that can fool the discriminator, *i.e.*, make it believe $\mathbf{x}'$ is real. The aim of the discriminator is to distinguish a fake sample $\mathbf{x}'$ from a real sample $\mathbf{x}$. They are simultaneously optimized via the following two-player minimax game:

$$\min_{G} \max_{D} \mathcal{L}_{\text{GAN}}, \quad \text{where} \quad \mathcal{L}_{\text{GAN}} = \mathbb{E}_{\mathbf{x}\sim p_{\text{data}}}[\log D(\mathbf{x})] + \mathbb{E}_{\mathbf{z}\sim p(\mathbf{z})}[\log(1 - D(G(\mathbf{z})))]. \quad (2.7)$$

From the perspective of a discriminator D, the objective function is a simple cross-entropy loss function for the binary categorization problem. A generator G is trained so as to minimize $\log(1 - D(G(\mathbf{z})))$, where the gradients of the parameters in G can be back-propagated through the outputs of (fixed) D. In spite of its simplicity, GAN is able to train a reasonable generator that can output realistic data samples.

Deep convolutional GAN (DCGAN) [42] was proposed to utilize GAN for generating natural images. The DCGAN generator consists of a series of four fractionally-strided convolutions that convert a 100-dimensional random vector (in uniform distribution) into a 64×64 pixel image. The main characteristics of the proposed CNN architecture are threefold. First, they used the all convolutional net [43] which replaces deterministic spatial pooling functions (such as maxpooling) with strided convolutions. Second, fully connected layers on top of convolutional features were eliminated. Finally, batch normalization [44], which normalize the input to each unit to have zero mean and unit variance, was used to stabilize learning.

DCGAN still has the limitation in resolution of generated images. StackGAN [45] was proposed to generate high-resolution (*e.g.*, 256×256) images by using two-stage GANs. Stage-I GAN generator in StackGAN generates low-resolution (64×64) images, which are input into

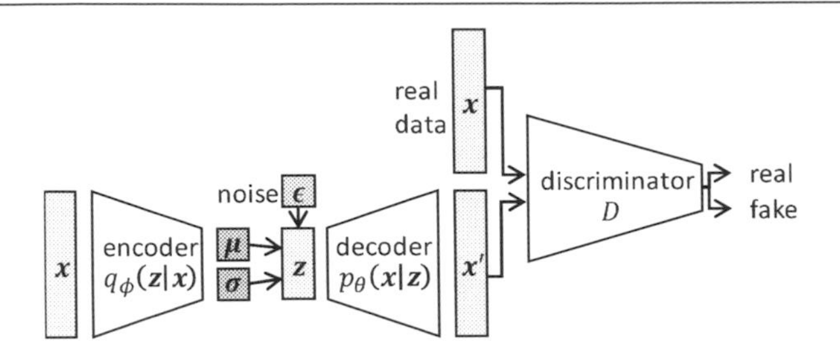

Figure 2.5 : Architecture of VAE-GAN [37].

Stage-II GAN generator that outputs high-resolution (256 × 256) images. Discriminators in Stage-I and Stage-II distinguish output images and real images of corresponding resolutions, respectively. More stable StackGAN++ [46] was later proposed, which consists of multiple generators and multiple discriminators arranged in a tree-like structure.

2.3.4 VAE-GAN

VAE-GAN [37] is a combination of VAE [34,35] (Sect. 2.3.2) and GAN [36] (Sect. 2.3.3) as shown in Fig. 2.5. The VAE-GAN model is trained with the following criterion:

$$\mathcal{L}_{\text{VAE-GAN}} = \mathcal{L}_{\text{VAE}^*} + \mathcal{L}_{\text{GAN}}, \tag{2.8}$$

$$\mathcal{L}_{\text{VAE}^*} = D_{\text{KL}}\left(q_\phi(\mathbf{z}|\mathbf{x})||p_\theta(\mathbf{z})\right) - \mathbb{E}_{\mathbf{z}\sim q_\phi(\mathbf{z}|\mathbf{x})}[\log p_\theta(D(\mathbf{x})|\mathbf{z})], \tag{2.9}$$

where $\mathcal{L}_{\text{GAN}}$ is the same as that in Eq. (2.7). Note that VAE-GAN replaces the element-wise error measures in VAE (*i.e.*, the second term in Eq. (2.4)) by the similarity measures trained through a GAN discriminator (*i.e.*, the second term in Eq. (2.9)). Here, a Gaussian observation model for $D(\mathbf{x})$ with mean $D(\mathbf{x}')$ and identity covariance is introduced:

$$p_\theta(D(\mathbf{x})|\mathbf{z}) = \mathcal{N}(D(\mathbf{x})|D(\mathbf{x}'), \mathbf{I}). \tag{2.10}$$

In this way, VAE-GAN improves the sharpness of output images, in comparison to VAE.

It has been said that GANs could suffer from mode collapse; the generator in GAN learns to produce samples with extremely low variety. VAE-GAN mitigates this problem by introducing the prior distribution for $\mathbf{z}$. The learned latent space is therefore continuous and is able to produce meaningful latent vectors, *e.g.*, visual attribute vectors corresponding to eyeglasses and smiles in face images.

2.3.5 Adversarial Auto-Encoder (AAE)

The adversarial auto-encoder (AAE) [38] was proposed to introduce the GAN framework to perform variational inference using auto-encoders. Similarly to VAE [34,35] (Sect. 2.3.2),

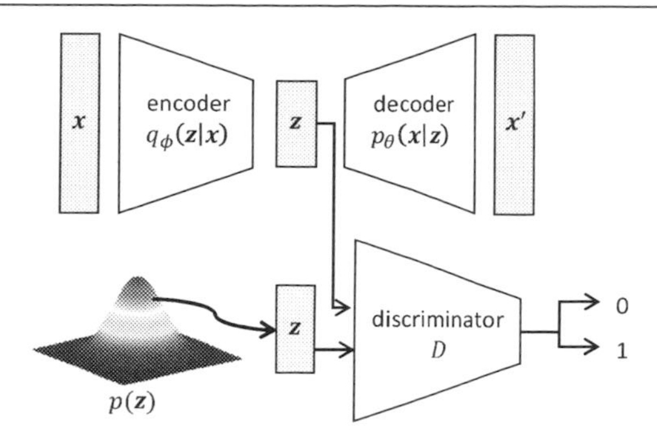

Figure 2.6 : Architecture of AAE [38].

AAE is a probabilistic autoencoder that assumes the prior over the latent variables. Let $p(\mathbf{z})$ be the prior distribution we want to impose, $q(\mathbf{z}|\mathbf{x})$ be an encoding distribution, $p(\mathbf{x}|\mathbf{z})$ be the decoding distribution, and $p_{\text{data}}(\mathbf{x})$ be the data distribution. An aggregated posterior distribution of $q(\mathbf{z})$ is defined as follows:

$$q(\mathbf{z}) = \int_{\mathbf{x}} q(\mathbf{z}|\mathbf{x}) p_{\text{data}}(\mathbf{x}) d\mathbf{x}. \tag{2.11}$$

The regularization of AAE is to match the aggregated posterior $q(\mathbf{z})$ to an arbitrary prior $p(\mathbf{z})$. While the auto-encoder attempts to minimize the reconstruction error, an adversarial network guides $q(\mathbf{z})$ to match $p(\mathbf{z})$ (Fig. 2.6). The cost function of AAE can be written as

$$\mathcal{L}_{\text{AAE}} = \mathcal{L}_{\text{AE}} + \mathbb{E}_{q_\phi(\mathbf{z}|\mathbf{x})}[\log D(\mathbf{z})] + \mathbb{E}_{\mathbf{z} \sim p(\mathbf{z})}[\log(1 - D(\mathbf{z}))], \tag{2.12}$$

where $\mathcal{L}_{\text{AE}}$ represents the reconstruction error defined in Eq. (2.1).

In contrast that VAE uses a KL divergence penalty to impose a prior distribution on latent variables, AAE uses an adversarial training procedure to encourage $q(\mathbf{z})$ to match to $p(\mathbf{z})$. An important difference between VAEs and AAEs lies in the way of calculating gradients. VAE approximates the gradients of the variational lower bound through the KL divergence by Monte Carlo sampling, which needs the access to the exact functional form of the prior distribution. On the other hand, AAEs only need to be able to sample from the prior distribution for inducing $q(\mathbf{z})$ to match $p(\mathbf{z})$. AAEs therefore can impose arbitrarily complicated distributions (*e.g.*, swiss roll distribution) as well as black-box distributions.

2.3.6 Adversarial Variational Bayes (AVB)

Adversarial variational Bayes (AVB) [39] also uses adversarial training for VAE, which enables the usage of arbitrarily complex inference models. As shown in Fig. 2.7, the inference

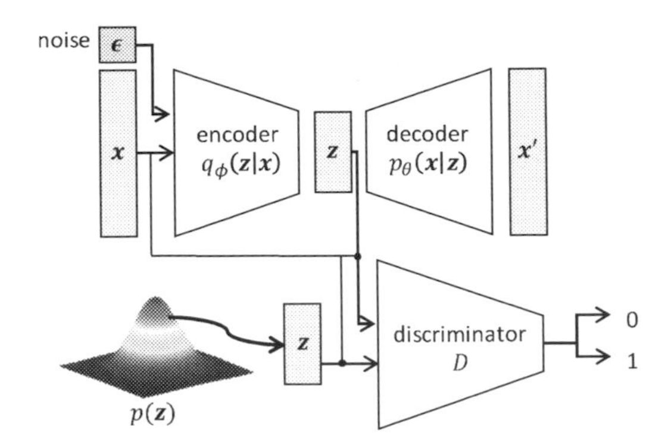

Figure 2.7 : Architecture of AVB [39].

model (*i.e.* encoder) in AVB takes the noise ϵ as additional input. For its derivation, the optimization problem of VAE in Eq. (2.4) is rewritten as follows:

$$\begin{aligned}&\max_\theta \max_\phi \mathbb{E}_{p_{\text{data}}(\mathbf{x})}[-D_{\text{KL}}\left(q_\phi(\mathbf{z}|\mathbf{x})||p(\mathbf{z})\right) + \mathbb{E}_{q_\phi(\mathbf{z}|\mathbf{x})} \log p_\theta(\mathbf{x}|\mathbf{z})]\\ &= \max_\theta \max_\phi \mathbb{E}_{p_{\text{data}}(\mathbf{x})}\mathbb{E}_{q_\phi(\mathbf{z}|\mathbf{x})}\left(\log p(\mathbf{z}) - \log q_\phi(\mathbf{z}|\mathbf{x}) + \log p_\theta(\mathbf{x}|\mathbf{z})\right). \end{aligned} \tag{2.13}$$

This can be optimized by stochastic gradient descent (SGD) using the reparameterization trick [34,35], when considering an explicit $q_\phi(\mathbf{z}|\mathbf{x})$ such as the centered isotropic multivariate Gaussian. For a black-box representation of $q_\phi(\mathbf{z}|\mathbf{x})$, the following objective function using the discriminator $D(\mathbf{x}, \mathbf{z})$ is considered:

$$\max_D \mathbb{E}_{p_{\text{data}}(\mathbf{x})}\mathbb{E}_{q_\phi(\mathbf{z}|\mathbf{x})} \log \sigma(D(\mathbf{x}, \mathbf{z})) + \mathbb{E}_{p_{\text{data}}(\mathbf{x})}\mathbb{E}_{p(\mathbf{z})} \log(1 - \sigma(D(\mathbf{x}, \mathbf{z}))), \tag{2.14}$$

where $\sigma(t) \equiv (1 + e^{-t})^{-1}$ denotes the sigmoid function. For $p_\theta(\mathbf{x}|\mathbf{z})$ and $q_\phi(\mathbf{z}|\mathbf{x})$ fixed, the optimal discriminator $D^*(\mathbf{x}, \mathbf{z})$ according to Eq. (2.14) is given by

$$D^*(\mathbf{x}, \mathbf{z}) = \log q_\phi(\mathbf{z}|\mathbf{x}) - \log p(\mathbf{z}). \tag{2.15}$$

The optimization objective in Eq. (2.13) is rewritten as

$$\max_\theta \max_\phi \mathbb{E}_{p_{\text{data}}(\mathbf{x})}\mathbb{E}_{q_\phi(\mathbf{z}|\mathbf{x})}\left(-D^*(\mathbf{x}, \mathbf{z}) + \log p_\theta(\mathbf{x}|\mathbf{z})\right). \tag{2.16}$$

Using the reparameterization trick [34,35], this can be rewritten with a suitable function $\mathbf{z}_\phi(\mathbf{x}, \epsilon)$ in the form

$$\max_\theta \max_\phi \mathbb{E}_{p_{\text{data}}(\mathbf{x})}\mathbb{E}_\epsilon\left(-D^*(\mathbf{x}, \mathbf{z}_\phi(\mathbf{x}, \epsilon)) + \log p_\theta(\mathbf{x}|\mathbf{z}_\phi(\mathbf{x}, \epsilon))\right). \tag{2.17}$$

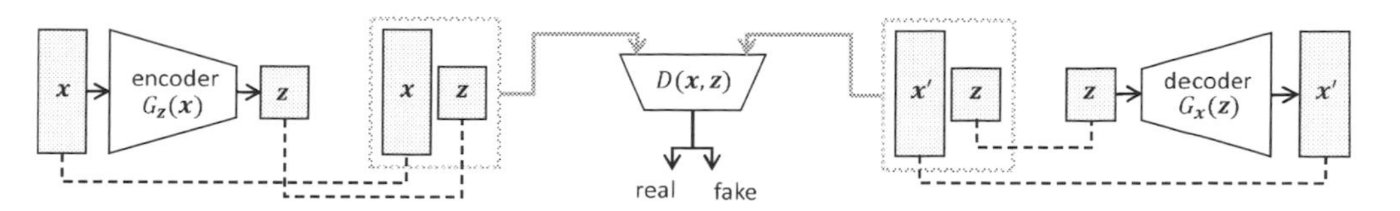

Figure 2.8 : Architecture of ALI [40] and BiGAN [41].

The gradients of Eq. (2.17) w.r.t. ϕ and θ as well as the gradient of Eq. (2.14) w.r.t. the parameters for D are computed to apply SGD updates. Note that AAE [38] (Sect. 2.3.5) can be regarded as an approximation to AVB, where $D(\mathbf{x}, \mathbf{z})$ is restricted to the class of functions that do not depend on $\mathbf{x}$ (*i.e.*, $D(\mathbf{x}, \mathbf{z}) \equiv D(\mathbf{z})$).

2.3.7 ALI and BiGAN

The adversarially learned inference (ALI) [40] and the Bidirectional GAN (BiGAN) [41], which were proposed at almost the same time, have the same model structure shown in Fig. 2.8. The model consists of two generators $G_\mathbf{z}(\mathbf{x})$ and $G_\mathbf{x}(\mathbf{z})$, which correspond to an encoder $q(\mathbf{z})$ and a decoder $p(\mathbf{x})$, respectively, and a discriminator $D(\mathbf{x}, \mathbf{z})$. In contrast that the original GAN [36] lacks the ability to infer $\mathbf{z}$ from $\mathbf{x}$, ALI and BiGAN can induce latent variable mapping, similarly to auto-encoder. Instead of the explicit reconstruction loop of auto-encoder, the GAN-like adversarial training framework is used for training the networks. Owing to this, trivial factors of variation in the input are omitted and the learned features become insensitive to these trivial factors of variation. The objective of ALI/BiGAN is to match the two joint distributions $q(\mathbf{x}, \mathbf{z}) = q(\mathbf{x})q(\mathbf{z}|\mathbf{x})$ and $p(\mathbf{x}, \mathbf{z}) = p(\mathbf{z})p(\mathbf{x}|\mathbf{z})$ by optimizing the following function:

$$\min_{G_\mathbf{x}, G_\mathbf{z}} \max_{D} \mathcal{L}_{\text{BiGAN}}, \quad \text{where } \mathcal{L}_{\text{BiGAN}} = \mathbb{E}_{q(\mathbf{x})}[\log D(\mathbf{x}, G_\mathbf{z}(\mathbf{x}))] + \mathbb{E}_{p(\mathbf{z})}[\log(1 - D(G_\mathbf{x}(\mathbf{z}), \mathbf{z}))]. \tag{2.18}$$

Here, a discriminator network learns to discriminate between a sample pair $(\mathbf{x}, \mathbf{z})$ drawn from $q(\mathbf{x}, \mathbf{z})$ and one from $p(\mathbf{x}, \mathbf{z})$. The concept of using bidirectional mapping is also important for image-to-image translation methods, which are described in the next section.

2.4 Multimodal Image-to-Image Translation Networks

Recent advances of deep neural networks enabled realistic translation of multimodal images. Basic ideas of such techniques are derived from VAEs and GANs. In this section, we introduce the latest works on this topic: image-to-image translation in two different

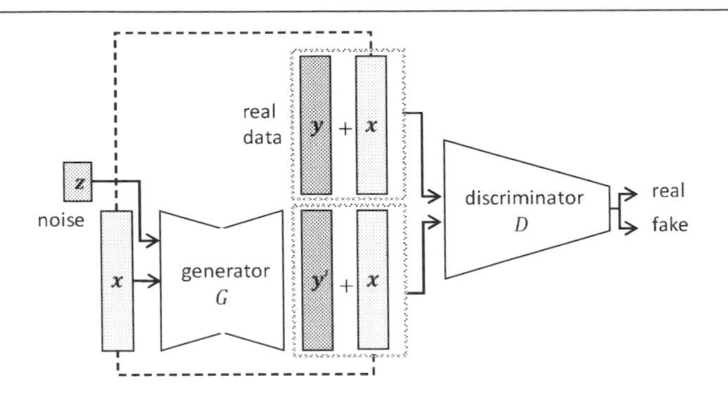

Figure 2.9 : Architecture of pix2pix [7].

domains/modals using deep neural networks. There are a few latest methods such as StarGAN [33] that is able to manage image-to-image translation among more than two domains, however, we do not address this topic in this chapter.

2.4.1 Pix2pix and Pix2pixHD

Pix2pix [7] is one of the earliest works on image-to-image translation using a conditional GAN (cGAN) framework. The purpose of this work is to learn a generator that takes an image $\mathbf{x}$ (*e.g.*, an edge map) as input and outputs an image $\mathbf{y}$ in a different modality (*e.g.*, a color photo). Whereas the generator in a GAN is only dependent on a random vector $\mathbf{z}$, the one in a cGAN also depends on another observed variable, which is the input image $\mathbf{x}$ in this case. In the proposed setting in [7], the discriminator also depends on $\mathbf{x}$. The architecture of pix2pix is shown in Fig. 2.9. The objective of the cGAN is written as follows:

$$\mathcal{L}_{\text{cGAN}}(G, D) = \mathbb{E}_{\mathbf{x},\mathbf{y}}[\log D(\mathbf{x}, \mathbf{y})] + \mathbb{E}_{\mathbf{x},\mathbf{z}}[\log(1 - D(\mathbf{x}, G(\mathbf{x}, \mathbf{z})))]. \tag{2.19}$$

The above objective is then mixed with ℓ_1 distance between the ground truth and the generated image:

$$\mathcal{L}_{\ell_1}(G) = \mathbb{E}_{\mathbf{x},\mathbf{y},\mathbf{z}}[\|\mathbf{y} - G(\mathbf{x}, \mathbf{z})\|_1]. \tag{2.20}$$

Note that images generated using ℓ_1 distance tend to be less blurring than ℓ_2 distance. The final objective of pix2pix is

$$\min_G \max_D \mathcal{L}_{\text{pix2pix}}(G, D), \quad \text{where} \quad \mathcal{L}_{\text{pix2pix}}(G, D) = \mathcal{L}_{\text{cGAN}}(G, D) + \lambda \mathcal{L}_{\ell_1}(G), \tag{2.21}$$

where λ controls the importance of the two terms.

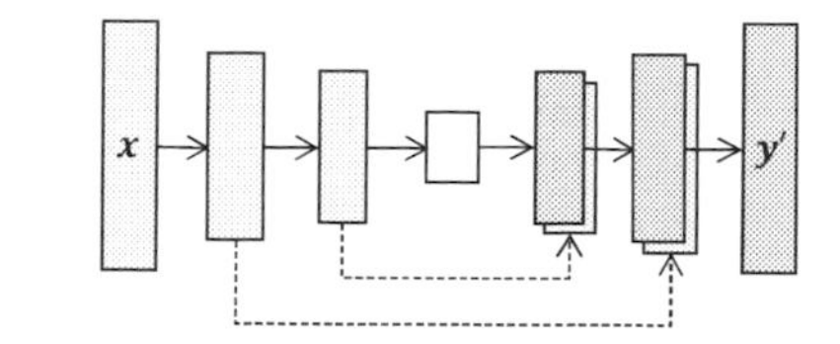

Figure 2.10 : Architecture of U-Net [6].

The generator of pix2pix takes the U-Net architecture [6], which is an encoder–decoder network with skip connections between intermediate layers (see Fig. 2.10). Because the U-Net transfers low-level information to high-level layers, it improves the quality (*e.g.*, the edge sharpness) of output images.

More recent work named pix2pixHD [32] generates 2048×1024 high-resolution images from semantic label maps using the pix2pix framework. They improved the pix2pix framework by using a coarse-to-fine generator, a multiscale discriminator architecture, and a robust adversarial learning objective function. The generator G consists of two sub-networks $\{G_1, G_2\}$, where G_1 generates 1024×512 images and G_2 generates 2048×1024 images using the outputs of G_1. Multiscale discriminators $\{D_1, D_2, D_3\}$, who have an identical network structure but operate at different image scales, are used in a multitask learning setting. The objective of pix2pixHD is

$$\min_{G}\left(\left(\max_{D_1, D_2, D_3} \sum_{k=1,2,3} \mathcal{L}_{\text{cGAN}}(G, D_k)\right) + \lambda \sum_{k=1,2,3} \mathcal{L}_{\text{FM}}(G, D_k)\right), \tag{2.22}$$

where $\mathcal{L}_{\text{FM}}(G, D_k)$ represents the newly proposed *feature matching loss*, which is the summation of ℓ_1 distance between the discriminator's ith-layer outputs of the real and the synthesized images.

2.4.2 CycleGAN, DiscoGAN, and DualGAN

CycleGAN [26] is a newly designed image-to-image translation framework. Suppose that $\mathbf{x}$ and $\mathbf{y}$ denote samples in a source domain X and a target domain Y, respectively. Previous approaches such as pix2pix [7] (described in Sect. 2.4.1) only lean a mapping $G_Y : X \to Y$, whereas CycleGAN also learns a mapping $G_X : Y \to X$. Here, two generators $\{G_\mathbf{y}, G_\mathbf{x}\}$ and two discriminators $\{D_\mathbf{y}, D_\mathbf{x}\}$ are jointly optimized (Fig. 2.11A). $G_\mathbf{y}$ translates $\mathbf{x}$ to $\mathbf{y}'$, which is then fed into $G_\mathbf{x}$ to generate $\mathbf{x}'$ (Fig. 2.11B). In the same manner, $G_\mathbf{x}$ translates $\mathbf{y}$ to $\mathbf{x}'$, which is then fed into $G_\mathbf{y}$ to generate $\mathbf{y}'$ (Fig. 2.11C). The ℓ_1 distances between the input and the output are summed up to calculate the following *cycle-consistency loss*:

$$\mathcal{L}_{\text{cyc}}(G_\mathbf{x}, G_\mathbf{y}) = \mathbb{E}_{\mathbf{x}\sim p_{\text{data}}(\mathbf{x})}[\|\mathbf{x} - G_\mathbf{x}(G_\mathbf{y}(\mathbf{x}))\|_1] + \mathbb{E}_{\mathbf{y}\sim p_{\text{data}}(\mathbf{y})}[\|\mathbf{y} - G_\mathbf{y}(G_\mathbf{x}(\mathbf{y}))\|_1]. \tag{2.23}$$

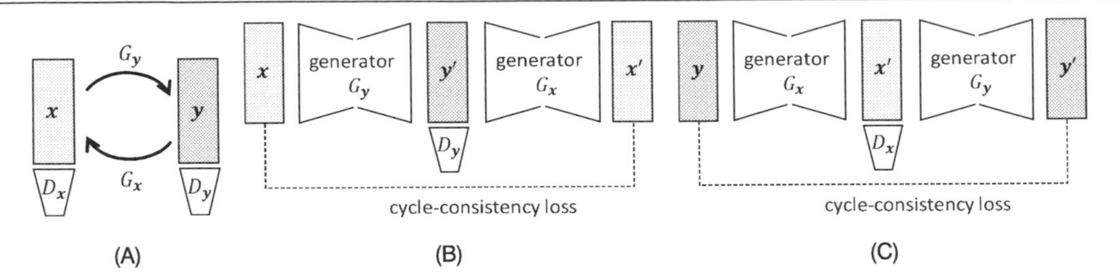

Figure 2.11 : Architecture of CycleGAN [26], DiscoGAN [27], and DualGAN [28]. (A) two generators $\{G_y, G_x\}$ and two discriminators $\{D_y, D_x\}$ are jointly optimized. (B) forward cycle-consistency and (C) backward cycle-consistency.

The most important factor of the *cycle-consistency loss* is that paired training data is unnecessary. In contrast that Eq. (2.20) in pix2pix loss requires **x** and **y** corresponding to the same sample, Eq. (2.23) only needs to compare the input and the output in the respective domains. The full objective of CycleGAN is

$$\mathcal{L}_{\text{cycleGAN}}(G_{\mathbf{x}}, G_{\mathbf{y}}, D_{\mathbf{x}}, D_{\mathbf{y}}) = \mathbb{E}_{\mathbf{y}\sim p_{\text{data}}(\mathbf{y})}[\log D_{\mathbf{y}}(\mathbf{y})] + \mathbb{E}_{\mathbf{x}\sim p_{\text{data}}(\mathbf{x})}[\log(1 - D_{\mathbf{y}}(G_{\mathbf{y}}(\mathbf{x})))] + \mathbb{E}_{\mathbf{x}\sim p_{\text{data}}(\mathbf{x})}[\log D_{\mathbf{x}}(\mathbf{x})] + \mathbb{E}_{\mathbf{y}\sim p_{\text{data}}(\mathbf{y})}[\log(1 - D_{\mathbf{x}}(G_{\mathbf{x}}(\mathbf{y})))] + \lambda\mathcal{L}_{\text{cyc}}(G_{\mathbf{x}}, G_{\mathbf{y}}). \tag{2.24}$$

DiscoGAN [27] and DualGAN [28] were proposed in the same period that CycleGAN was proposed. The architectures of those methods are identical to that of CycleGAN (Fig. 2.11). DiscoGAN uses ℓ_2 loss instead of ℓ_1 loss for the cycle-consistency loss. DualGAN replaced the sigmoid cross-entropy loss used in the original GAN by Wasserstein GAN (WGAN) [47].

2.4.3 CoGAN

A coupled generative adversarial network (CoGAN) [29] was proposed to learn a joint distribution of multidomain images. CoGAN consists of a pair of GANs, each of which synthesizes an image in one domain. These two GANs share a subset of parameters as shown in Fig. 2.12. The proposed model is based on the idea that a pair of corresponding images in two domains share the same high-level concepts. The objective of CoGAN is a simple combination of the two GANs:

$$\mathcal{L}_{\text{CoGAN}}(G_1, G_2, D_1, D_2) = \mathbb{E}_{\mathbf{x}_2\sim p_{\text{data}}(\mathbf{x}_2)}[\log D_2(\mathbf{x}_2)] + \mathbb{E}_{\mathbf{z}\sim p(\mathbf{z})}[\log(1 - D_2(G_2(\mathbf{z})))] + \mathbb{E}_{\mathbf{x}_1\sim p_{\text{data}}(\mathbf{x}_1)}[\log D_1(\mathbf{x}_1)] + \mathbb{E}_{\mathbf{z}\sim p(\mathbf{z})}[\log(1 - D_1(G_1(\mathbf{z})))]. \tag{2.25}$$

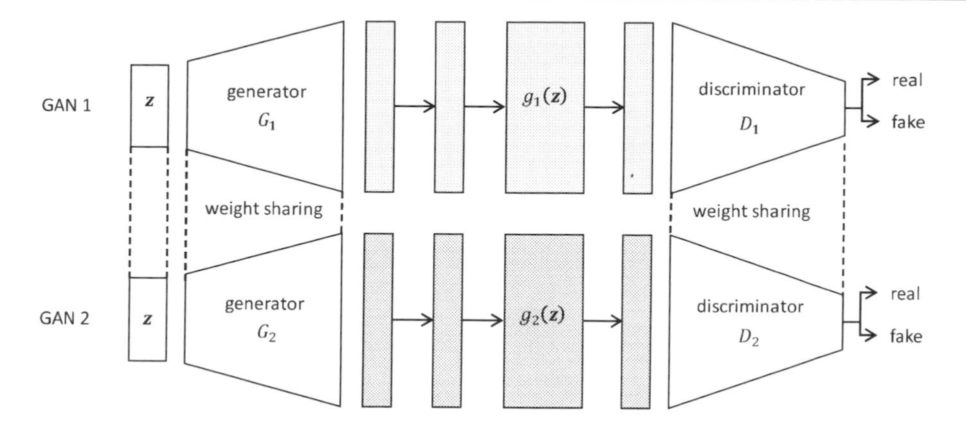

Figure 2.12 : Architecture of CoGAN [29].

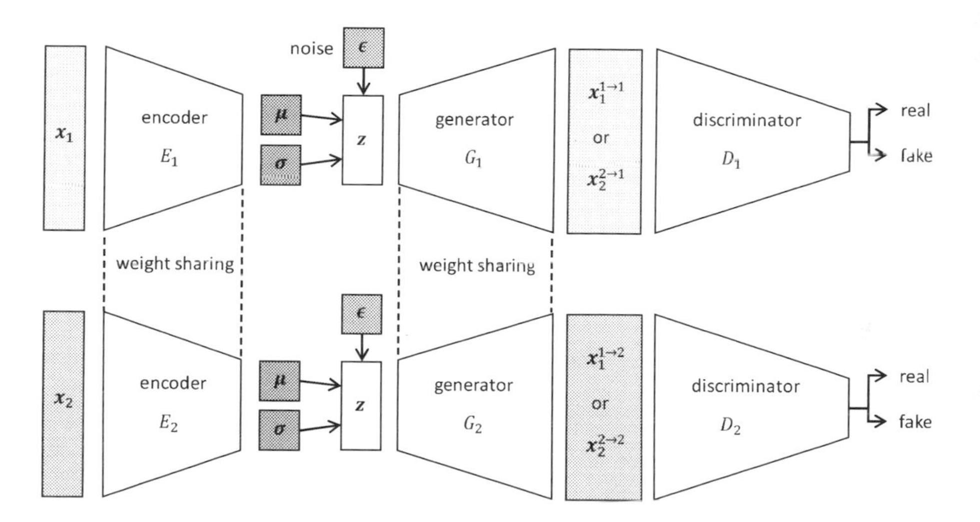

Figure 2.13 : Architecture of UNIT [30].

Similarly to the GAN frameworks introduced in Sect. 2.4.2, CoGAN does not require pairs of corresponding images as supervision.

2.4.4 UNIT

Unsupervised Image-to-Image Translation (UNIT) Networks [30] combined the VAE-GAN [37] model (Sect. 2.3.4) and the CoGAN [29] model (Sect. 2.4.3) as shown in Fig. 2.13. UNIT also does not require pairs of corresponding images as supervision. The sets of networks $\{E_1, G_1, D_1\}$ and $\{E_2, G_2, D_2\}$ correspond to VAE-GANs, whereas the set of networks

$\{G_1, G_2, D_1, D_2\}$ correspond to CoGAN. Because the two VAEs share the weights of the last few layers of encoders as well as the first few layers of decoders (*i.e.*, generators), an image of one domain $\mathbf{x}_1$ is able to be translated to an image in the other domain $\mathbf{x}_2$ through $\{E_1, G_2\}$, and vice versa. Note that the weight-sharing constraint alone does not guarantee that corresponding images in two domains will have the same latent code $\mathbf{z}$. It was, however, shown that, by E_1 and E_2, a pair of corresponding images in the two domains can be mapped to a common latent code, which is then mapped to a pair of corresponding images in the two domains by G_1 and G_2. This benefits from the cycle-consistency constraint described below.

The learning problems of the VAEs and GANs for the image reconstruction streams, the image translation streams, and the cycle-reconstruction streams are jointly solved:

$$\min_{E_1,E_2,G_1,G_2} \max_{D_1,D_2} \mathcal{L}_{\text{UNIT}}, \tag{2.26}$$

where

$$\begin{aligned}\mathcal{L}_{\text{UNIT}} = \lambda_1\mathcal{L}_{\text{VAE}}(E_1, G_1) + \lambda_2\mathcal{L}_{\text{GAN}}(E_2, G_1, D_1) + \lambda_3\mathcal{L}_{\text{CC}}(E_1, G_1, E_2, G_2) + \\ \lambda_1\mathcal{L}_{\text{VAE}}(E_2, G_2) + \lambda_2\mathcal{L}_{\text{GAN}}(E_1, G_2, D_2) + \lambda_3\mathcal{L}_{\text{CC}}(E_2, G_2, E_1, G_1).\end{aligned} \tag{2.27}$$

Here, $\mathcal{L}_{\text{VAE}}(\cdot)$ and $\mathcal{L}_{\text{GAN}}(\cdot)$ are the same as those in Eq. (2.4) and Eq. (2.7), respectively. $\mathcal{L}_{\text{CC}}(\cdot)$ represents the cycle-consistency constraint given by the following VAE-like objective function:

$$\begin{aligned}\mathcal{L}_{\text{CC}}(E_a, G_a, E_b, G_b) = D_{\text{KL}}\left(q_a(\mathbf{z}_a|\mathbf{x}_a)||p_\theta(\mathbf{z})\right) + D_{\text{KL}}\left(q_b(\mathbf{z}_b|\mathbf{x}_a^{a\to b})||p_\theta(\mathbf{z})\right) - \\ \mathbb{E}_{\mathbf{z}_b\sim q_b(\mathbf{z}_b|\mathbf{x}_a^{a\to b})}[\log p_{G_a}(\mathbf{x}_a|\mathbf{z}_b)].\end{aligned} \tag{2.28}$$

The third term in the above function ensures a twice translated resembles the input one, whereas the KL terms penalize the latent variable $\mathbf{z}$ deviating from the prior distribution $p_\theta(\mathbf{z}) \equiv \mathcal{N}(\mathbf{z}|\mathbf{0}, \mathbf{I})$ in the cycle-reconstruction stream.

2.4.5 Triangle GAN

A triangle GAN (Δ-GAN) [31] is developed for semi-supervised cross-domain distribution matching. This framework requires only a few *paired* samples in two different domains as supervision. Δ-GAN consists of two generators $\{G_{\mathbf{x}}, G_{\mathbf{y}}\}$ and two discriminators $\{D_1, D_2\}$, as shown in Fig. 2.14. Suppose that $\mathbf{x}'$ and $\mathbf{y}'$ represent the translated image from $\mathbf{y}$ and $\mathbf{x}$ using $G_{\mathbf{x}}$ and $G_{\mathbf{y}}$, respectively. The fake data pair $(\mathbf{x}', \mathbf{y})$ is sampled from the joint distribution $p_{\mathbf{x}}(\mathbf{x}, \mathbf{y}) = p_{\mathbf{x}}(\mathbf{x}|\mathbf{y})p(\mathbf{y})$, and vice versa. The objective of Δ-GAN is to match the three joint distribution $p(\mathbf{x}, \mathbf{y})$, $p_{\mathbf{x}}(\mathbf{x}, \mathbf{y})$, and $p_{\mathbf{y}}(\mathbf{x}, \mathbf{y})$. If this is achieved, the learned bidirectional mapping $p_{\mathbf{x}}(\mathbf{x}|\mathbf{y})$ and $p_{\mathbf{y}}(\mathbf{y}|\mathbf{x})$ are guaranteed to generate fake data pairs $(\mathbf{x}', \mathbf{y})$ and $(\mathbf{x}, \mathbf{y}')$ that are indistinguishable from the true data pair $(\mathbf{x}, \mathbf{y})$.

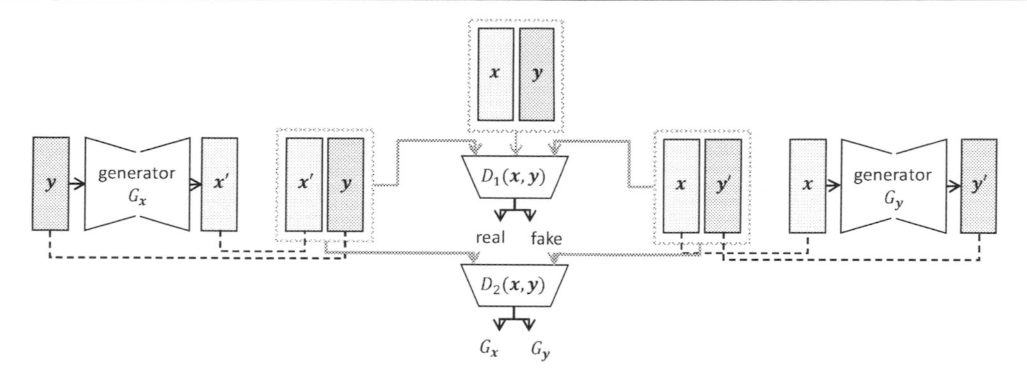

Figure 2.14 : Architecture of Δ-GAN [31].

The objective function of Δ-GAN is given by

$$\min_{G_\mathbf{x},G_\mathbf{y}}\max_{D_1,D_2} \mathbb{E}_{(\mathbf{x},\mathbf{y})\sim p(\mathbf{x},\mathbf{y})}[\log D_1(\mathbf{x},\mathbf{y})] + \mathbb{E}_{\mathbf{y}\sim p(\mathbf{y}),\mathbf{x}'\sim p_\mathbf{x}(\mathbf{x}|\mathbf{y})}[\log\left((1-D_1(\mathbf{x}',\mathbf{y}))\cdot D_2(\mathbf{x}',\mathbf{y})\right)]$$
$$+\mathbb{E}_{\mathbf{x}\sim p(\mathbf{x}),\mathbf{y}'\sim p_\mathbf{y}(\mathbf{y}|\mathbf{x})}[\log\left((1-D_1(\mathbf{x},\mathbf{y}'))\cdot(1-D_2(\mathbf{x},\mathbf{y}'))\right)]. \tag{2.29}$$

The discriminator D_1 distinguishes whether a sample pair is from the true data distribution $p(\mathbf{x},\mathbf{y})$ or not. If this is not from $p(\mathbf{x},\mathbf{y})$, D_2 is used to distinguish whether a sample pair is from $p_\mathbf{x}(\mathbf{x},\mathbf{y})$ or $p_\mathbf{y}(\mathbf{x},\mathbf{y})$.

Δ-GAN can be regarded as a combination of cGAN and BiGAN [41]/ALI [40] (Sect. 2.3.7). Eq. (2.29) can be rewritten as

$$\min_{G_\mathbf{x},G_\mathbf{y}}\max_{D_1,D_2} \mathcal{L}_{\Delta\text{-GAN}}, \tag{2.30}$$

where

$$\begin{aligned}
\mathcal{L}_{\Delta\text{-GAN}} &= \mathcal{L}_{\text{cGAN}} + \mathcal{L}_{\text{BiGAN}},\\
\mathcal{L}_{\text{cGAN}} &= \mathbb{E}_{p(\mathbf{x},\mathbf{y})}[\log D_1(\mathbf{x},\mathbf{y})] + \mathbb{E}_{p_\mathbf{x}(\mathbf{x}',\mathbf{y})}[\log(1-D_1(\mathbf{x}',\mathbf{y}))]\\
&\quad + \mathbb{E}_{p_\mathbf{y}(\mathbf{x},\mathbf{y}')}[\log(1-D_1(\mathbf{x},\mathbf{y}'))],\\
\mathcal{L}_{\text{BiGAN}} &= \mathbb{E}_{p_\mathbf{x}(\mathbf{x}',\mathbf{y})}[\log D_2(\mathbf{x}',\mathbf{y})] + \mathbb{E}_{p_\mathbf{y}(\mathbf{x},\mathbf{y}')}[\log(1-D_2(\mathbf{x},\mathbf{y}'))].
\end{aligned}$$

2.5 Multimodal Encoder–Decoder Networks

The architecture of the multimodal encoder–decoder networks [5] is illustrated in Fig. 2.15. To exploit the commonality among different tasks, all encoder / decoder pairs are connected with each other via the shared latent representation. In addition, if the decoding task

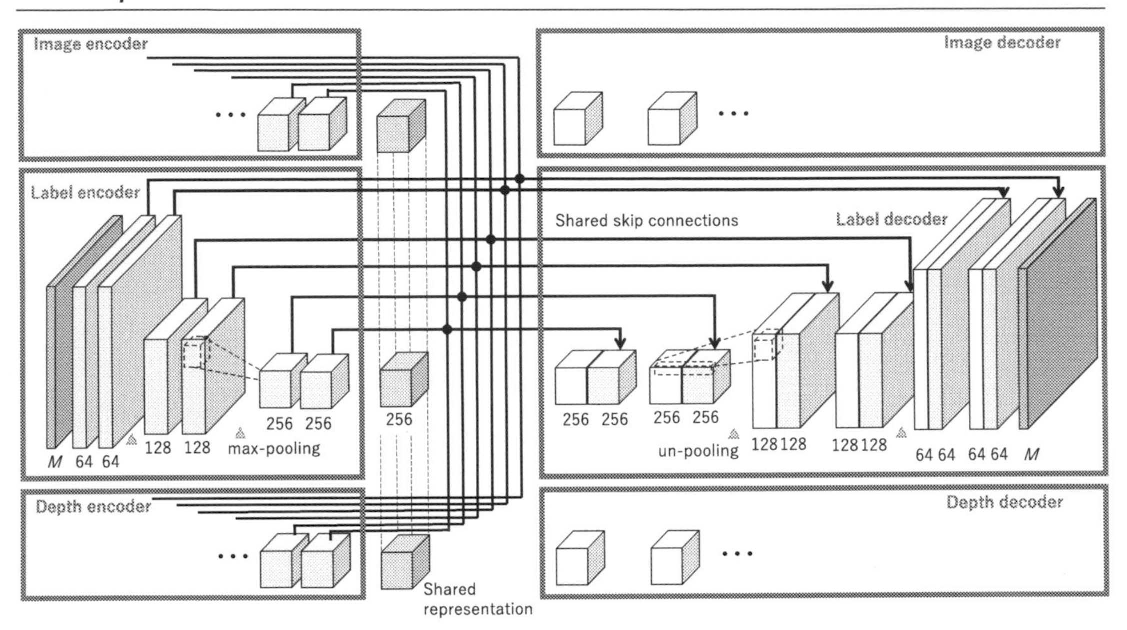

Figure 2.15 : Model architecture of the multimodal encoder–decoder networks. This model consists of encoder–decoder networks with the shared latent representation. Depending on the task, a decoder also employs the U-net architecture and is connected with all encoders via shared skip connections. The network consists of *Conv+Norm+ReLU* modules except for the final layer, which is equivalent to *Conv*. For experiments described in Sect. 2.6, we use kernel size 3 x 3 with stride 1 and padding size 1 for all convolution layers, and kernel size 2 x 2 and stride 2 for max-pooling.

is expected to benefit from high-resolution representations, the decoder is further connected with all encoders via shared skip connections in a similar manner to the U-net architecture [6]. Given one input modality, the encoder generates a single representation, which is then decoded through different decoders into all available modalities. The whole network is trained by taking into account all combinations of the conversion tasks among different modalities.

In what follows, we discuss the details of the architecture of the multimodal encoder–decoder networks for the task of depth and semantic label estimation from RGB images assuming a training dataset consisting of three modalities: RGB images, depth maps, and semantic labels. In this example, semantic segmentation is the task where the advantage of the skip connections has been already shown [7], while such high-resolution representations are not always beneficial to depth and RGB image decoding tasks. It is also worth noting that the task and the number of modalities are not limited to this particular example. More encoders and decoders can be added to the model, and the decoder can be trained with different tasks and loss functions within this framework.

2.5.1 Model Architecture

As illustrated in Fig. 2.15, each convolution layer (*Conv*) in the encoder is followed by a batch-normalization layer (*Norm*) and activation function (*ReLU*). Two max-pooling operations are placed in the middle of seven *Conv+Norm+ReLU* components, which makes the dimension of the latent representation be 1/16 of the input. Similarly, the decoder network consists of seven *Conv+Norm+ReLU* components except for the final layer, while max-pooling operations are replaced by un-pooling operations for expanding a feature map. The max-pooling operation pools a feature map by taking its maximum values, and the un-pooling operation restores the pooled feature into an un-pooled feature map using switches where the locations of the maxima are recorded. The final output of the decoder is then rescaled to the original input size. In case for a classification task, the rescaled output may be further fed into a softmax layer to yield a class probability distribution.

As discussed earlier, all encoder / decoder pairs are connected with the shared latent representation. Let $\mathbf{x}_r$, $\mathbf{x}_s$, and $\mathbf{x}_d$ be the input modalities for each encoder; RGB image, semantic label, and depth map, and $\mathbf{E}_r$, $\mathbf{E}_s$, and $\mathbf{E}_d$ be the corresponding encoder functions, then the outputs from each encoder $\mathbf{r}$, which means latent representation, is defined by $\mathbf{r} \in \{\mathbf{E}_r(\mathbf{x}_r), \mathbf{E}_s(\mathbf{x}_s), \mathbf{E}_d(\mathbf{x}_d)\}$. Here, the dimension of $\mathbf{r}$ is $C \times H \times W$, where C, H, and W are the number of channels, height, and width, respectively. Because the intermediate output $\mathbf{r}$ from all the encoders form the same shape $C \times H \times W$ by the convolution and pooling operations at all the encoder functions $\mathbf{E} \in \{\mathbf{E}_r, \mathbf{E}_s, \mathbf{E}_s\}$, we can obtain the output $\mathbf{y} \in \{\mathbf{D}_r(\mathbf{r}), \mathbf{D}_s(\mathbf{r}), \mathbf{D}_d(\mathbf{r})\}$ from any $\mathbf{r}$, where $\mathbf{D}_r$, $\mathbf{D}_s$, and $\mathbf{D}_d$ are decoder functions. The latent representations between encoders and decoders are not distinguished among different modalities, *i.e.*, the latent representation encoded by an encoder is fed into all decoders, and at the same time, each decoder has to be able to decode latent representation from any of the encoders. In other words, latent representation is shared for by all the combinations of encoder / decoder pairs.

For semantic segmentation, the U-net architecture with skip paths are also employed to propagate intermediate low-level features from encoders to decoders. Low-level feature maps in the encoder are concatenated with feature maps generated from latent representations and then convolved in order to mix the features. Since we use 3×3 convolution kernels with 2×2 max-pooling operators for the encoder and 3×3 convolution kernels with 2×2 un-pooling operators for the decoder, the encoder and decoder networks are symmetric (U-shape). The introduced model has skip paths among all combinations of the encoders and decoders, and also shares the low-level features in a similar manner to the latent representation.

2.5.2 Multitask Training

In the training phase, a batch of training data is passed through all forwarding paths for calculating losses. For example, given a batch of paired RGB image, depth, and semantic label images, three decoding losses from the RGB image / depth / label decoders are first computed for the RGB image encoder. The same procedure is then repeated for depth and label encoders, and the global loss is defined as the sum of all decoding losses from nine encoder–decoder pairs, *i.e.*, three input modalities for three tasks. The gradients for the whole network are computed based on the global loss by back-propagation. If the training batch contains unpaired data, only within-modal self-encoding losses are computed. In the following, we describe details of the cost functions defined for decoders of RGB images, semantic labels, and depth maps.

RGB images. For RGB image decoding, we define the loss $\mathcal{L}_{\mathbf{r}}$ as the ℓ_1 distance of RGB values as

$$\mathcal{L}_I = \frac{1}{N} \sum_{x \in P} \|\mathbf{r}(x) - \mathbf{f}_r(x)\|_1, \tag{2.31}$$

where $\mathbf{r}(x) \in \mathbb{R}^3_+$ and $\mathbf{f}_r(x) \in \mathbb{R}^3$ are the ground truth and predicted RGB values, respectively. If the goal of the network is realistic RGB image generation from depth and label images, the RGB image decoding loss may be further extended to DCGAN-based architectures [42].

Semantic labels. In this chapter, we define a label image as a map in which each pixel has a one-hot-vector that represents the class that the pixel belongs to. The number of the input and output channels is thus equivalent to the number of classes. We define the loss function of semantic label decoding by the pixel-level cross-entropy. Let K be the number of classes, the softmax function is then written as

$$p^{(k)}(x) = \frac{\exp\left(f_s^{(k)}(x)\right)}{\sum_{i=1}^{K} \exp\left(f_s^{(i)}(x)\right)}, \tag{2.32}$$

where $f_s^{(k)}(x) \in \mathbb{R}$ indicates the value at the location x in the kth channel ($k \in \{1 \dots, K\}$) of the tensor given by the final layer output. Letting P be the whole set of pixels in the output and N be the number of the pixels, the loss function $\mathcal{L}_s$ is defined as

$$\mathcal{L}_s = \frac{1}{N} \sum_{x \in P} \sum_{k=1}^{K} t_k(x) \log p^{(k)}(x), \tag{2.33}$$

where $t_k(x) \in \{0, 1\}$ is the kth channel ground-truth label, which is one if the pixel belongs to the kth class, and zero, otherwise.

Depth maps. For the depth decoder, we use the distance between the ground truth and predicted depth maps. The loss function $\mathcal{L}_d$ is defined as

$$\mathcal{L}_d = \frac{1}{N} \sum_{x \in P} |d(x) - f_d(x)|, \tag{2.34}$$

where $d(x) \in \mathbb{R}_+$ and $f_d(x) \in \mathbb{R}$ are the ground truth and predicted depth values, respectively. We normalize the depth values so that they range in $[0, 255]$ by linear interpolation. In the evaluation step, we revert the normalized depth map into the original scale.

2.5.3 Implementation Details

In the multimodal encoder–decoder networks, learnable parameters are initialized by a normal distribution. We set the learning rate to 0.001 and the momentum to 0.9 for all layers with weight decay 0.0005. The input image size is fixed to 96×96. We use both paired and unpaired data as training data, which are randomly mixed in every epoch. When a triplet of the RGB image, semantic label, and depth map of a scene is available, training begins with the RGB image as input and the corresponding semantic label, depth map, and the RGB image itself as output. In the next step, the semantic label is used as input, and the others as well as the semantic label are used as output for computing individual losses. These steps are repeated on all combinations of the modalities and input / output. A loss is calculated on each combination and used for updating parameters. In case the triplet of the training data is unavailable (for example, when the dataset contains extra unlabeled RGB images), the network is updated only with the reconstruction loss in a similar manner to auto-encoders. We train the network by the mini-batch training method with a batch including at least one labeled RGB image paired with either semantic labels or depth maps if available, because it is empirically found that such paired data contribute to convergence of training.

2.6 Experiments

In this section, we evaluate the multimodal encoder–decoder networks [5] for semantic segmentation and depth estimation using two public datasets: NYUDv2 [48] and Cityscape [49]. The baseline model is the single-task encoder–decoder networks (*enc-dec*) and single-modal (RGB image) multitask encoder–decoder networks (*enc-decs*) that have the same architecture as the multimodal encoder–decoder networks. We also compare the multimodal encoder–decoder networks to multimodal auto-encoders (*MAEs*) [1], which concatenates latent representations of auto-encoders for different modalities. Since the shared representation in MAE is the concatenation of latent representations in all the modalities, it is required to explicitly input zero-filled pseudo signals to estimate the missing modalities. Also, MAE uses fully connected layers instead of convolutional layers, so that input images are flattened when fed into the first layer.

For semantic segmentation, we use the mean intersection over union (MIOU) scores for the evaluation. IOU is defined as

$$IOU = \frac{TP}{TP + FP + FN}, \tag{2.35}$$

where TP, FP, and FN are the numbers of true positive, false positive, and false negative pixels, respectively, determined over the whole test set. MIOU is the mean of the IOU on all classes.

For depth estimation, we use several evaluation measures commonly used in prior work [50, 11,12]:

- Root mean squared error: $\sqrt{\frac{1}{N}\sum_{x \in P}(d(x) - \hat{d}(x))^2}$
- Average relative error: $\frac{1}{N}\sum_{x \in P}\frac{|d(x) - \hat{d}(x)|}{d(x)}$
- Average log 10 error: $\frac{1}{N}\sum_{x \in P}\left|\log_{10}\frac{d(x)}{\hat{d}(x)}\right|$
- Accuracy with threshold: Percentage of $x \in P$ s.t. $\max\left(\frac{d(x)}{\hat{d}(x)}, \frac{\hat{d}(x)}{d(x)}\right) < \delta$,

where $d(x)$ and $\hat{d}(x)$ are the ground-truth depth and predicted depth at the pixel x, P is the whole set of pixels in an image, and N is the number of the pixels in P.

2.6.1 Results on NYUDv2 Dataset

NYUDv2 dataset has 1448 images annotated with semantic labels and measured depth values. This dataset contains 868 images for training and a set of 580 images for testing. We divided test data into 290 and 290 for validation and testing, respectively, and used the validation data for early stopping. The dataset also has extra 407,024 unpaired RGB images, and we randomly selected 10,459 images as unpaired training data, while *other* class was not considered in both of training and evaluation. For semantic segmentation, following prior work [51, 52,12], we evaluate the performance on estimating 12 classes out of all available classes. We trained and evaluated the multimodal encoder–decoder networks on two different image resolutions 96×96 and 320×240 pixels.

Table 2.1 shows depth estimation and semantic segmentation results of all methods. The first six rows show results on 96×96 input images, and each row corresponds to MAE [1], single-task encoder–decoder with and without skip connections, single-modal multitask

Table 2.1: Performance comparison on the NYUDv2 dataset. The first six rows show results on 96 x 96 input images, and each row corresponds to MAE [1], single-task encoder–decoder with and without U-net architecture, single-modal multitask encoder–decoder, the multimodal encoder–decoder networks with and without extra RGB training data. The next seven rows show results on 320 x 240 input images in comparison with baseline depth estimation methods [53–55,11,50].

		Depth Estimation						Semantic Segmentation
		Error			Accuracy			
		Rel	log10	RMSE	$\delta < 1.25$	$\delta < 1.25^2$	$\delta < 1.25^3$	MIOU
96 × 96	MAE [1]	1.147	0.290	2.311	0.098	0.293	0.491	0.018
	enc-dec (U)	–	–	–	–	–	–	0.357
	enc-dec	0.340	0.149	1.216	0.396	0.699	0.732	–
	enc-decs	0.321	0.150	1.201	0.398	0.687	0.718	0.352
	multimodal enc-decs	0.296	0.120	1.046	0.450	0.775	**0.810**	0.411
	multimodal enc-decs (+extra)	**0.283**	**0.119**	**1.042**	**0.461**	**0.778**	**0.810**	**0.420**
320 × 240	Make3d [53]	0.349	–	1.214	0.447	0.745	0.897	–
	DepthTransfer [54]	0.350	0.131	1.200	–	–	–	–
	Discrete-continuous CRF [55]	0.335	0.127	1.060	–	–	–	–
	Eigen et al. [11]	**0.215**	–	0.907	0.601	**0.887**	**0.971**	–
	Liu et al. [50]	0.230	0.095	0.824	**0.614**	0.883	**0.971**	–
	multimodal enc-decs	0.228	0.088	0.823	0.576	0.849	0.867	0.543
	multimodal enc-decs (+extra)	0.221	**0.087**	**0.819**	0.579	0.853	0.872	**0.548**

encoder–decoder, the multimodal encoder–decoder networks without and with extra RGB training data. The first six columns show performance metrics for depth estimation, and the last column shows semantic segmentation performances. The performance of the multimodal encoder–decoder networks was better than the single-task network (enc-dec) and single-modal multitask encoder–decoder network (enc-decs) on all metrics even without the extra data, showing the effectiveness of the multimodal architecture. The performance was further improved with the extra data, and achieved the best performance in all evaluation metrics. It shows the benefit of using unpaired training data and multiple modalities to learn more effective representations.

In addition, next seven rows show results on 320 × 240 input images in comparison with baseline depth estimation methods including Make3d [53], DepthTransfer [54], Discrete-continuous CRF [55], Eigen et al. [11], and Liu et al. [50]. The performance was improved when trained on higher resolution images, in terms of both depth estimation and semantic segmentation. The multimodal encoder–decoder networks achieved a better performance than baseline methods, and comparable to methods requiring CRF-based optimization [55] and a large amount of labeled training data [11] even without the extra data.

More detailed results on semantic segmentation on 96 × 96 images are shown in Table 2.2. Each column shows class-specific IOU scores for all models. The multimodal encoder–

Table 2.2: Detailed IOU on the NYUDv2 dataset. Each column shows class-specific IOU scores for all models.

	MAE	enc-dec (U)	enc-decs	multimodal enc-decs	multimodal enc-decs (+extra)
book	0.002	0.055	0.071	**0.096**	0.072
cabinet	0.033	0.371	0.382	0.480	**0.507**
ceiling	0.000	0.472	0.414	0.529	**0.534**
floor	0.101	0.648	0.659	0.704	**0.736**
table	0.020	0.197	0.222	0.237	**0.299**
wall	0.023	0.711	0.706	0.745	**0.749**
window	0.022	0.334	**0.363**	0.321	0.320
picture	0.005	0.361	0.336	0.414	**0.422**
blinds	0.001	0.274	0.234	0.303	**0.304**
sofa	0.004	0.302	0.300	0.365	**0.375**
bed	0.006	0.370	0.320	**0.455**	0.413
tv	0.000	0.192	0.220	0.285	**0.307**
mean	0.018	0.357	0.352	0.411	**0.420**

decoder networks with extra training data outperforms the baseline models with 10 out of the 12 classes and achieved 0.063 points improvement in MIOU.

2.6.2 Results on Cityscape Dataset

The Cityscape dataset consists of 2975 images for training and 500 images for validation, which are provided together with semantic labels and disparity. We divide the validation data into 250 and 250 for validation and testing, respectively, and used the validation data for early stopping. This dataset has 19,998 additional RGB images without annotations, and we also used them as extra training data. There are semantic labels of 19 class objects and a single background (unannotated) class. We used the 19 classes (excluding the background class) for evaluation. For depth estimation, we used the disparity maps provided together with the dataset as the ground truth. Since there were missing disparity values in the raw data unlike NYUDv2, we adopted the image inpainting method [57] to interpolate disparity maps for both training and testing. We used image resolutions 96×96 and 512×256, while the multimodal encoder–decoder networks was trained on half-split 256×256 images in the 512×256 case.

The results are shown in Table 2.3, and the detailed comparison on semantic segmentation using 96×96 images are summarized in Table 2.4. The first six rows in Table 2.3 show a comparison between different architectures using 96×96 images. The multimodal encoder–decoder networks achieved improvement over both of the MAE [1] and the baseline networks in most of the target classes. While the multimodal encoder–decoder networks without extra data did not improve MIOU, it resulted in 0.043 points improvement with extra data. The multimodal encoder–decoder networks also achieved the best performance on the depth estimation task, and the performance gain from extra data illustrates the generalization capability

Table 2.3: Performance comparison on the Cityscape dataset.

		Depth Estimation						Semantic Segmentation
		Error			Accuracy			
		Rel	**log10**	**RMSE**	$\delta < 1.25$	$\delta < 1.25^2$	$\delta < 1.25^3$	**MIOU**
96×96	MAE [1]	3.675	0.441	34.583	0.213	0.395	0.471	0.099
	enc-dec (U)	–	–	–	–	–	–	0.346
	enc-dec	0.380	0.125	8.983	0.602	0.780	0.870	–
	enc-decs	0.365	0.117	8.863	0.625	0.798	0.880	0.356
	multimodal enc-decs	0.387	0.115	8.267	0.631	0.803	0.887	0.346
	multimodal enc-decs (+extra)	**0.290**	**0.100**	**7.759**	**0.667**	**0.837**	**0.908**	**0.389**
512×256	Segnet [56]	–	–	–	–	–	–	0.561
	multimodal enc-decs	**0.201**	**0.076**	5.528	0.759	**0.908**	**0.949**	0.575
	multimodal enc-decs (+extra)	0.217	0.080	**5.475**	**0.765**	**0.908**	**0.949**	**0.604**

Table 2.4: Detailed IOU on the Cityscape dataset.

	MAE	**enc-dec**	**enc-decs**	**multimodal enc-decs**	**multimodal enc-decs (+extra)**
road	0.688	0.931	0.936	0.925	**0.950**
side walk	0.159	0.556	0.551	0.529	**0.640**
build ing	0.372	0.757	0.769	0.770	**0.793**
wall	0.022	0.125	0.128	0.053	**0.172**
fence	0.000	0.054	0.051	0.036	**0.062**
pole	0.000	0.230	0.220	0.225	**0.280**
traffic light	0.000	0.100	0.074	0.049	**0.109**
traffic sigh	0.000	0.164	0.203	0.189	**0.231**
vegetation	0.200	0.802	0.805	0.805	**0.826**
terrain	0.000	0.430	0.446	0.445	**0.498**
sky	0.295	0.869	0.887	0.867	**0.890**
person	0.000	0.309	0.318	0.325	**0.365**
rider	0.000	0.040	**0.058**	0.007	0.036
car	0.137	0.724	0.743	0.720	**0.788**
truck	0.000	0.062	0.051	**0.075**	0.035
bus	0.000	0.096	0.152	0.153	**0.251**
train	0.000	0.006	0.077	**0.133**	0.032
motor cycle	0.000	0.048	0.056	0.043	**0.108**
bicycle	0.000	0.270	0.241	0.218	**0.329**
mean	0.099	0.346	0.356	0.346	**0.389**

of the described training strategy. The next three rows show results using 512×256, and the multimodal encoder–decoder networks achieved a better performance than the baseline method [56] on semantic segmentation.

2.6.3 Auxiliary Tasks

Although the main goal of the described approach is semantic segmentation and depth estimation from RGB images, in Fig. 2.16 we show other cross-modal conversion pairs, *i.e.*,

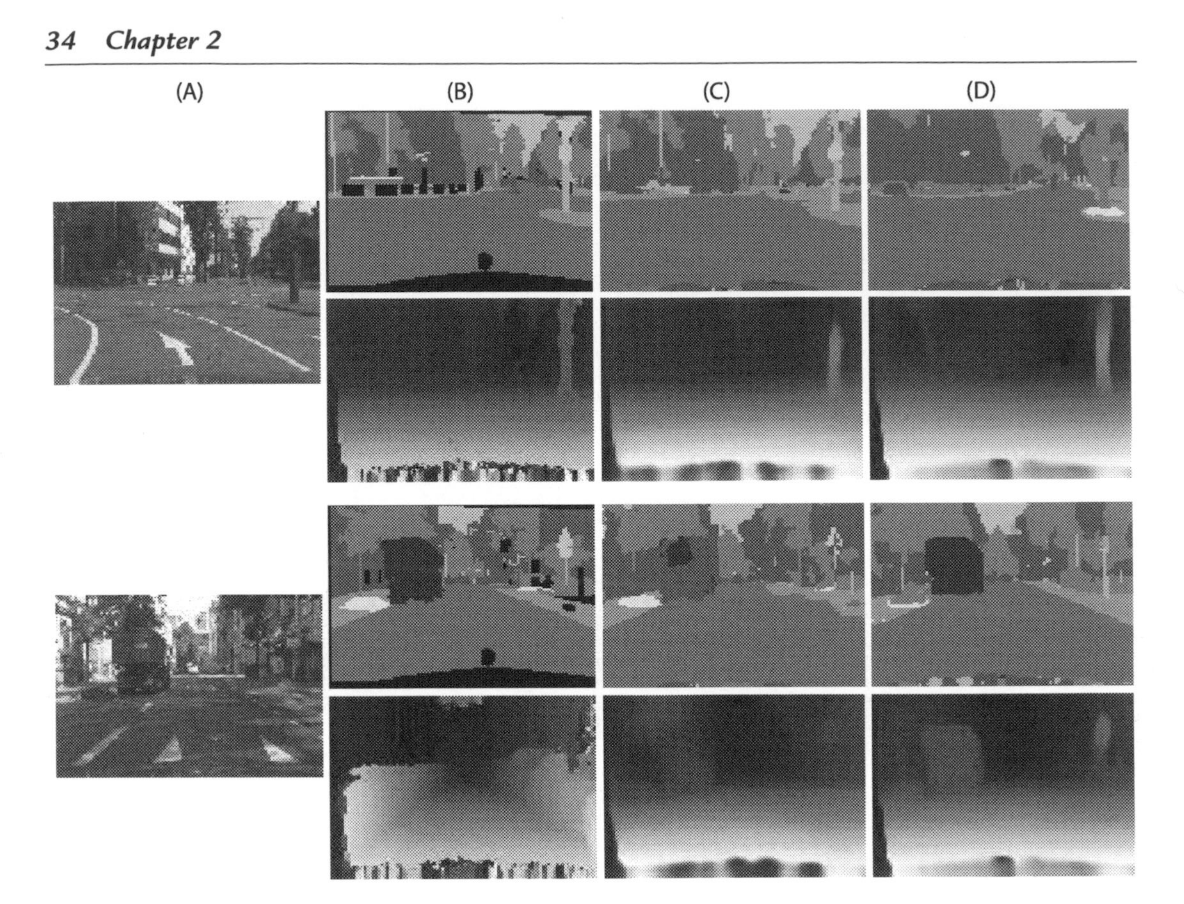

Figure 2.16 : Example output images from the multimodal encoder–decoder networks on the Cityscape dataset. From left to right, each column corresponds to (A) input RGB image, (B) the ground-truth semantic label image (top) and depth map (bottom), (C) estimated label and depth from RGB image, (D) estimated label from depth image (top) and estimated depth from label image (bottom).

semantic segmentation from depth images and depth estimation from semantic labels on cityscape dataset. From left to right, each column corresponds to ground truth (A) RGB image, (B) upper for semantic label image lower for depth map and (C) upper for image-to-label lower for image-to-depth, (D) upper for depth-to-label lower for label-to-depth. The ground-truth depth maps are ones after inpainting. As can be seen, the multimodal encoder–decoder networks could also reasonably perform these auxiliary tasks.

More detailed examples and evaluations on NYUDv2 dataset is shown in Fig. 2.17 and Table 2.5. The left side of Fig. 2.17 shows examples of output images corresponding to all of the above-mentioned tasks. From top to bottom on the left side, each row corresponds to the ground truth, (A) RGB image, (B) semantic label image, estimated semantic labels from

Figure 2.17 : Example outputs from the multimodal encoder–decoder networks on the NYUDv2 dataset. From top to bottom on the left side, each row corresponds to (A) the input RGB image, (B) the ground-truth semantic label image, (C) estimation by enc-dec, (D) image-to-label estimation by the multimodal encoder–decoder networks, (E) depth-to-label estimation by the multimodal encoder–decoder networks, (F) estimated depth map by the multimodal encoder–decoder networks (normalized to $[0, 255]$ for visualization), (G) estimation by enc-dec, (H) image-to-depth estimation by the multimodal encoder–decoder networks, and (I) label-to-depth estimation by the multimodal encoder–decoder networks. In addition, the right side shows image decoding tasks, where each block corresponds to (A) the ground-truth RGB image, (B) label-to-image estimate, and (C) depth-to-image estimate.

Table 2.5: Comparison of auxiliary task performances on the NYUDv2.

	Depth Estimation						Semantic Segmentation
	Error			Accuracy			
	Rel	log10	RMSE	$\delta < 1.25$	$\delta < 1.25^2$	$\delta < 1.25^3$	MIOU
image-to-depth	0.283	0.119	1.042	0.461	0.778	0.810	-
label-to-depth	0.258	0.128	1.114	0.452	0.741	0.779	-
image-to-label	-	-	-	-	-	-	0.420
depth-to-label	-	-	-	-	-	-	0.476

(C) the baseline enc-dec model, (D) image-to-label, (E) depth-to-label conversion paths of the multimodal encoder–decoder networks, (F) depth map (normalized to [0, 255] for visualization) and estimated depth maps from (G) enc-dec, (H) image-to-depth, (I) label-to-depth. Interestingly, these auxiliary tasks achieved better performances than the RGB input cases. A clearer object boundary in the label and depth images is one of the potential reasons of the performance improvement. In addition, the right side of Fig. 2.17 shows image decoding tasks and each block corresponds to (A) the ground-truth RGB image, (B) semantic label, (C) depth map, (D) label-to-image, and (E) depth-to-image. Although the multimodal encoder–decoder networks could not correctly reconstruct the input color, object shapes can be seen even with the simple image reconstruction loss.

2.7 Conclusion

In this chapter, we introduced several state-of-the-art approaches on deep learning for multimodal data fusion as well as basic techniques behind those works. In particular, we described a new approach named multimodal encoder–decoder networks for efficient multitask learning with a shared feature representation. In the multimodal encoder–decoder networks, encoders and decoders are connected via the shared latent representation and shared skipped representations. Experiments showed the potential of shared representations from different modalities to improve the multitask performance.

One of the most important issues in future work is to investigate the effectiveness of the multimodal encoder–decoder networks on different tasks such as image captioning and DCGAN-based image translation. More detailed investigation on learned shared representations during multitask training is another important future direction to understand why and how the multimodal encoder–decoder architecture addresses the multimodal conversion tasks.

References

[1] C. Cadena, A. Dick, I.D. Reid, Multi-modal auto-encoders as joint estimators for robotics scene understanding, in: Proceedings of Robotics: Science and Systems, 2016.

[2] J. Ngiam, A. Khosla, M. Kim, J. Nam, H. Lee, A.Y. Ng, Multimodal deep learning, in: Proceedings of International Conference on Machine Learning, ICML, 2011.

[3] N. Srivastava, R.R. Salakhutdinov, Multimodal learning with deep Boltzmann machines, in: Proceedings of Advances in Neural Information Processing Systems, NIPS, 2012.

[4] A. Kendall, Y. Gal, R. Cipolla, Multi-task learning using uncertainty to weigh losses for scene geometry and semantics, arXiv preprint, arXiv:1705.07115.

[5] R. Kuga, A. Kanezaki, M. Samejima, Y. Sugano, Y. Matsushita, Multi-task learning using multi-modal encoderdecoder networks with shared skip connections, in: Proceedings of International Conference on Computer Vision Workshops, ICCV, 2017.

[6] O. Ronneberger, P. Fischer, T. Brox, U-net: convolutional networks for biomedical image segmentation, in: Proceedings of International Conference on Medical Image Computing and Computer-Assisted Intervention, 2015.

[7] P. Isola, J.-Y. Zhu, T. Zhou, A.A. Efros, Image-to-image translation with conditional adversarial networks, in: Proceedings of IEEE Conference on Computer Vision and Pattern Recognition, CVPR, 2017, pp. 5967–5976.

[8] R. Caruana, Multitask learning, Machine Learning 28 (1) (1997) 41–75.

[9] Y. Liao, S. Kodagoda, Y. Wang, L. Shi, Y. Liu, Understand scene categories by objects: a semantic regularized scene classifier using convolutional neural networks, in: Proceedings of IEEE International Conference on Robotics and Automation, ICRA, 2016.

[10] Y. Yang, T. Hospedales, Deep multi-task representation learning: a tensor factorisation approach, in: Proceedings of International Conference on Learning Representations, ICLR, 2017.

[11] D. Eigen, C. Puhrsch, R. Fergus, Depth map prediction from a single image using a multi-scale deep network, in: Proceedings of Advances in Neural Information Processing Systems, NIPS, 2014.

[12] D. Eigen, R. Fergus, Predicting depth, surface normals and semantic labels with a common multi-scale convolutional architecture, in: Proceedings of International Conference on Computer Vision, ICCV, 2015.

[13] E.M. Hand, R. Chellappa, Attributes for improved attributes: a multi-task network for attribute classification, in: Proceedings of AAAI Conference on Artificial Intelligence, 2017.

[14] J. Hoffman, S. Gupta, T. Darrell, Learning with side information through modality hallucination, in: Proceedings of IEEE Conference on Computer Vision and Pattern Recognition, CVPR, 2016.

[15] M. Teichmann, M. Weber, M. Zoellner, R. Cipolla, R. Urtasun, Multinet: real-time joint semantic reasoning for autonomous driving, in: Proceedings of 2018 IEEE Intelligent Vehicles Symposium, IV, 2018.

[16] J. Uhrig, M. Cordts, U. Franke, T. Brox, Pixel-level encoding and depth layering for instance-level semantic labeling, in: Proceedings of German Conference on Pattern Recognition, 2016.

[17] X. Li, L. Zhao, L. Wei, M.H. Yang, F. Wu, Y. Zhuang, H. Ling, J. Wang, DeepSaliency: multi-task deep neural network model for salient object detection, IEEE Transactions on Image Processing 25 (8) (2016) 3919–3930.

[18] A. Eitel, J.T. Springenberg, L. Spinello, M. Riedmiller, W. Burgard, Multimodal deep learning for robust RGB-D object recognition, in: Proceedings of IEEE/RSJ International Conference on Intelligent Robots and Systems, IROS, 2015.

[19] K. Sohn, W. Shang, H. Lee, Improved multimodal deep learning with variation of information, in: Proceedings of Advances in Neural Information Processing Systems, NIPS, 2014, pp. 2141–2149.

[20] V. Radu, N.D. Lane, S. Bhattacharya, C. Mascolo, M.K. Marina, F. Kawsar, Towards multimodal deep learning for activity recognition on mobile devices, in: Proceedings of the 2016 ACM International Joint Conference on Pervasive and Ubiquitous Computing: Adjunct, 2016, pp. 185–188.

[21] M. Ehrlich, T.J. Shields, T. Almaev, M.R. Amer, Facial attributes classification using multi-task representation learning, in: Proceedings of the IEEE Conference on Computer Vision and Pattern Recognition Workshops, 2016.

[22] E. Bruni, N.K. Tran, M. Baroni, Multimodal distributional semantics, Journal of Artificial Intelligence Research 49 (1) (2014) 1–47.

[23] I.V. Serban, A.G. Ororbia II, J. Pineau, A.C. Courville, Multi-modal variational encoder–decoders, arXiv preprint, arXiv:1612.00377.

[24] C. Silberer, V. Ferrari, M. Lapata, Visually grounded meaning representations, IEEE Transactions on Pattern Analysis and Machine Intelligence 39 (11) (2017) 2284–2297.

[25] G. Collell, T. Zhang, M. Moens, Imagined visual representations as multimodal embeddings, in: Proceedings of AAAI Conference on Artificial Intelligence, 2017.

[26] J.-Y. Zhu, T. Park, P. Isola, A.A. Efros, Unpaired image-to-image translation using cycle-consistent adversarial networks, in: Proceedings of International Conference on Computer Vision, ICCV, 2017.

[27] T. Kim, M. Cha, H. Kim, J.K. Lee, J. Kim, Learning to discover cross-domain relations with generative adversarial networks, in: Proceedings of International Conference on Machine Learning, ICML, 2017.

[28] Z. Yi, H. Zhang, P. Tan, M. Gong, Dualgan: unsupervised dual learning for image-to-image translation, in: Proceedings of International Conference on Computer Vision, ICCV, 2017.

[29] M.-Y. Liu, O. Tuzel, Coupled generative adversarial networks, in: Proceedings of Advances in Neural Information Processing Systems, NIPS, 2016.

[30] M.-Y. Liu, T. Breuel, J. Kautz, Unsupervised image-to-image translation networks, in: Proceedings of Advances in Neural Information Processing Systems, NIPS, 2017.

[31] Z. Gan, L. Chen, W. Wang, Y. Pu, Y. Zhang, H. Liu, C. Li, L. Carin, Triangle generative adversarial networks, in: Proceedings of Advances in Neural Information Processing Systems, NIPS, 2017.

[32] T.-C. Wang, M.-Y. Liu, J.-Y. Zhu, A. Tao, J. Kautz, B. Catanzaro, High-resolution image synthesis and semantic manipulation with conditional gans, in: Proceedings of IEEE Conference on Computer Vision and Pattern Recognition, CVPR, 2018.

[33] Y. Choi, M. Choi, M. Kim, J.-W. Ha, S. Kim, J. Choo, Stargan: unified generative adversarial networks for multi-domain image-to-image translation, in: Proceedings of IEEE Conference on Computer Vision and Pattern Recognition, CVPR, 2018.

[34] D.P. Kingma, M. Welling, Auto-encoding variational Bayes, in: Proceedings of International Conference on Learning Representations, ICLR, 2014.

[35] D.J. Rezende, S. Mohamed, D. Wierstra, Stochastic backpropagation and approximate inference in deep generative models, in: Proceedings of International Conference on Machine Learning, ICML, 2014.

[36] I. Goodfellow, J. Pouget-Abadie, M. Mirza, B. Xu, D. Warde-Farley, S. Ozair, A. Courville, Y. Bengio, Generative adversarial nets, in: Proceedings of Advances in Neural Information Processing Systems, NIPS, 2014.

[37] A.B.L. Larsen, S.K. Sønderby, H. Larochelle, O. Winther, Autoencoding beyond pixels using a learned similarity metric, in: Proceedings of International Conference on Machine Learning, ICML, 2016.

[38] A. Makhzani, J. Shlens, N. Jaitly, I. Goodfellow, Adversarial autoencoders, in: Proceedings of International Conference on Learning Representations, ICLR, 2016.

[39] L. Mescheder, S. Nowozin, A. Geiger, Adversarial variational Bayes: unifying variational autoencoders and generative adversarial networks, in: Proceedings of International Conference on Machine Learning, ICML, 2017.

[40] V. Dumoulin, I. Belghazi, B. Poole, O. Mastropietro, A. Lamb, M. Arjovsky, A. Courville, Adversarially learned inference, in: Proceedings of International Conference on Learning Representations, ICLR, 2017.

[41] J. Donahue, P. Krähenbühl, T. Darrell, Adversarial feature learning, in: Proceedings of International Conference on Learning Representations, ICLR, 2017.

[42] A. Radford, L. Metz, S. Chintala, Unsupervised representation learning with deep convolutional generative adversarial networks, in: Proceedings of International Conference on Learning Representations, ICLR, 2016.

[43] J. Springenberg, A. Dosovitskiy, T. Brox, M. Riedmiller, Striving for simplicity: the all convolutional net, in: Proceedings of International Conference on Learning Representations (Workshop Track), ICLR, 2015.

[44] S. Ioffe, C. Szegedy, Batch normalization: accelerating deep network training by reducing internal covariate shift, in: Proceedings of International Conference on Machine Learning, ICML, 2015.

[45] H. Zhang, T. Xu, H. Li Stackgan, Text to photo-realistic image synthesis with stacked generative adversarial networks, in: Proceedings of International Conference on Computer Vision, ICCV, 2017.

[46] H. Zhang, T. Xu, H. Li, S. Zhang, X. Wang, X. Huang, D. Metaxas, Stackgan++: realistic image synthesis with stacked generative adversarial networks, IEEE Transactions on Pattern Analysis and Machine Intelligence (2019), https://doi.org/10.1109/TPAMI.2018.2856256.

[47] M. Arjovsky, S. Chintala, L. Bottou, Wasserstein generative adversarial networks, in: Proceedings of International Conference on Machine Learning, ICML, 2017.

[48] N. Silberman, D. Hoiem, P. Kohli, R. Fergus, Indoor segmentation and support inference from rgbd images, in: Proceedings of European Conference on Computer Vision, ECCV, 2012, pp. 746–760.

[49] M. Cordts, M. Omran, S. Ramos, T. Rehfeld, M. Enzweiler, R. Benenson, U. Franke, S. Roth, B. Schiele, The cityscapes dataset for semantic urban scene understanding, in: Proceedings of IEEE Conference on Computer Vision and Pattern Recognition, CVPR, 2016.

[50] F. Liu, C. Shen, G. Lin, Deep convolutional neural fields for depth estimation from a single image, in: Proceedings of IEEE Conference on Computer Vision and Pattern Recognition, CVPR, 2015.

[51] A. Hermans, G. Floros, B. Leibe, Dense 3D semantic mapping of indoor scenes from RGB-D images, in: Proceedings of IEEE International Conference on Robotics and Automation, ICRA, 2014.

[52] A. Wang, J. Lu, G. Wang, J. Cai, T.-J. Cham, Multi-modal unsupervised feature learning for RGB-D scene labeling, in: Proceedings of European Conference on Computer Vision, ECCV, 2014.

[53] A. Saxena, M. Sun, A.Y. Ng, Make3d: learning 3d scene structure from a single still image, IEEE Transactions on Pattern Analysis and Machine Intelligence 31 (5) (2009) 824–840.

[54] K. Karsch, C. Liu, S.B. Kang, Depth transfer: depth extraction from video using non-parametric sampling, IEEE Transactions on Pattern Analysis and Machine Intelligence 36 (11) (2014) 2144–2158.

[55] M. Liu, M. Salzmann, X. He, Discrete-continuous depth estimation from a single image, in: Proceedings of IEEE Conference on Computer Vision and Pattern Recognition, CVPR, 2014, pp. 716–723.

[56] V. Badrinarayanan, A. Kendall, R. Cipolla, Segnet: a deep convolutional encoder–decoder architecture for image segmentation, IEEE Transactions on Pattern Analysis and Machine Intelligence 39 (12) (2017) 2481–2495.

[57] A. Telea, An image inpainting technique based on the fast marching method, Journal of Graphics Tools 9 (1) (2004) 23–34.

CHAPTER 3

Multimodal Semantic Segmentation: Fusion of RGB and Depth Data in Convolutional Neural Networks

Zoltan Koppanyi*, Dorota Iwaszczuk†, Bing Zha*, Can Jozef Saul‡, Charles K. Toth*, Alper Yilmaz*

**Department of Civil, Envir. & Geod Eng., The Ohio State University, Columbus, OH, United States †Photogrammetry & Remote Sensing, Technical University of Munich, Munich, Germany ‡Robert College, Istanbul, Turkey*

Contents

3.1 Introduction

Semantic segmentation of an image is an easy task for humans; for instance, if asked, a child can easily delineate an object in an image. It is trivial for us, but not for computers as the

Multimodal Scene Understanding
https://doi.org/10.1016/B978-0-12-817358-9.00009-3

high-level cognitive process of semantically segmenting an image cannot be transformed into exact mathematical expressions. Additionally, there is no strict definition of object categories or "boundaries". To tackle these "soft" problems, various machine learning algorithms were developed for semantic segmentation in the past decades. Early studies used "descriptors" for representing object categories. Recent advances in neural networks allow for training deep learning models which "learn" these "descriptors". These types of networks are called convolutional neural networks (CNN or ConvNet).

This chapter addresses the problem of semantic labeling using deep CNNs. We use the terms of "semantic labeling", "semantic scene understanding", "semantic image understanding" and "segmentation and labeling" interchangeably, which is considered as segmentation of an image into object categories by labeling pixels semantically. Clearly, this process involves two steps: first, objects with their boundaries in the image are localized, and then, in the second step, they are labeled. As we will see later, these two steps can be simultaneously solved with a CNN.

Both the earlier and more recent studies on semantic labeling use the three-channel red, green, blue (RGB) images. Aside from the common adoption of the RGB channels, one can expect that additional scene observations, such as specular bands or 3D information, can improve image segmentation and labeling. Note that any 3D data acquired from LiDAR or other sensor can be converted to a 2.5D representation, which is often a depth image. Depth images extend the RGB images by incorporating 3D information. This information is helpful to distinguish objects with different 3D shapes. However, finding potential ways to combine or fuse these modalities in the CNN framework remains an open question.

This chapter presents the concepts and CNN architectures that tackle this problem. In addition to an overview of the RGB and depth fusion networks and their comparison, we report two main contributions. First, we investigate various color space transformations for RGB and depth fusion. The idea behind this approach is to eliminate the redundancy of the RGB space by converting the image to either normalized-RGB space or HSI space, and then substituting one of the channels with depth. Second, we report results that indicate that using normals instead of depth can significantly improve the generalization ability of the networks in semantic segmentation.

The rest of the chapter is organized as follows. First, we give an overview of existing CNN structures that fuse RGB and depth images. We present the basic ideas of building up a network from a single image classifier to a Siamese network architecture, and then three approaches, namely, fusion through color space transformations, networks with four-channel inputs, and Siamese networks for RGB and depth fusion, are discussed. The "Methods" section provides the description of experiments conducted to investigate the three approaches, including datasets used, the data splitting and preprocessing, color space transformations,

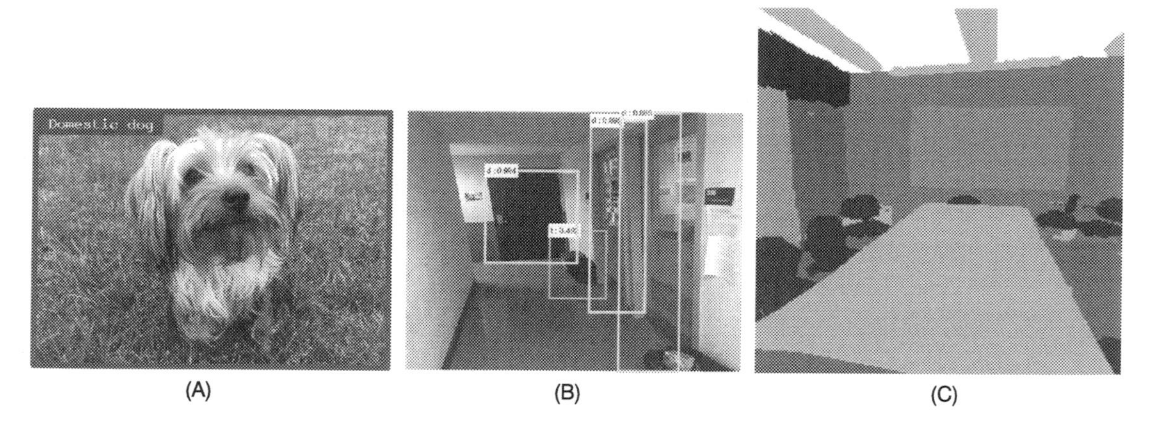

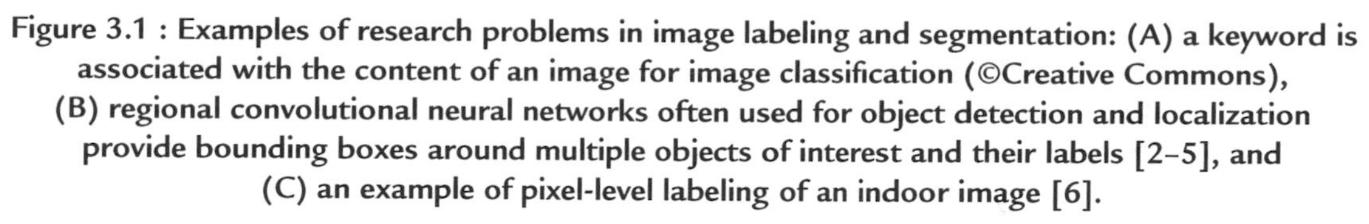

Figure 3.1 : Examples of research problems in image labeling and segmentation: (A) a keyword is associated with the content of an image for image classification (©Creative Commons), (B) regional convolutional neural networks often used for object detection and localization provide bounding boxes around multiple objects of interest and their labels [2–5], and (C) an example of pixel-level labeling of an indoor image [6].

and the parametrization of the investigated network architectures. The investigation is carried out on the Stanford indoor dataset and the ISPRS aerial dataset. Both datasets provide depth images, however, these images are derived differently for each dataset; in particular, flash LiDAR sensor is used to acquire the Stanford dataset which directly provides depth images. For the ISPRS dataset we apply dense reconstruction to generate the depth images. In the "Results and Discussion" section, we present our results, comparison of the approaches, and discussion on our findings. The chapter ends with a conclusion and possible future research directions.

3.2 Overview

CNNs are used for problems where the input can be defined as array-like data, such as images, and there is spatial relationship between the array elements. Three research problems are considered to be the main application of CNNs for image classification.

First of these research problems is the labeling of an entire image based on its content; see Fig. 3.1A. This problem is referred to as the image classification, and it is well suited for interpreting the content of an entire image, and thus, can be used for image indexing and image retrieval. We will see later that networks, which solve single image classification problems, are the basis of other networks that are designed for solving more complex segmentation problems. The second research problem is to detect and localize objects of interests by providing bounding boxes around them, and then labeling the boxes [1]; see Fig. 3.1B. It is noteworthy that these two research areas have been very active due to the ImageNet Large Scale Visual Recognition Challenge (ILSVRC) [1]. Finally, the third research problem, which is the focus of this chapter, is pixel-level labeling of images, where the task is to annotate all pixels of an image; see Fig. 3.1C.

Pixel-level labeling, and object detection and localization problems overlap, because annotation of all pixels eventually leads to identifying regions where the objects are located [7]. The choice of which approach should be used, however, depends on the application. Networks developed for object detection, such as the regional convolutional networks [2–5], might be more suitable for recognizing multiple objects in an image, such as chair, table, trash bin, and provide less accurate boundaries or contours of the objects. In contrast, pixel-level labeling allows for identifying larger image regions, such as ceilings and floors, and provides more accurate object boundaries [7]. In general, pixel-level labeling is better suited for remote sensing problems, such as land use classification [8–11]. Note that some remote sensing applications may require object detection, such as detecting vehicles [12], aircraft [13], or ships [14] from aerial photos.

This chapter discusses networks that solve pixel-level labeling tasks using multiple data sources, such as RGB and depth information. The most common architectures for pixel-level labeling consist of an encoder that transforms an image into a lower dimensional feature space, and a decoder that converts this feature space back to the image space [6]. The encoder part of this network architecture is often a variant of an image classification network. For this reason, in the next subsection, we briefly present some notable CNNs for image classification tasks. This discussion will be followed with an overview of encoder–decoder network structures for pixel-level labeling. Finally, we discuss the latest network architectures for fusing RGB and depth data. At the end of this section, we present the most notable datasets that might be used for developing deep models for RGB and depth fusion. Detailed discussion on various layer types, such as convolution, pooling, unpooling, ReLU layers, is not the intent of this chapter, and thus, it is not included in the remainder of the text. The reader can find detailed information on these subjects in [15, p. 321].

3.2.1 Image Classification and the VGG Network

Network architectures developed for data fusion heavily rely on the results achieved in image classification using CNNs. Image classification is a very active research field, and recent developments can be tracked from the results reported at the annual ILSVCR competition [1]. Most notable networks from this competition are LeNet-5 [16], AlexNet [17], VGG [18], GoogLeNet [19], DenseNet [20] and ResNet [21].

Most commonly used image classifier that is also adopted in many data fusion architectures is the VGG developed by the Visual Geometry Group at University of Oxford. VGG is composed of blocks of 2 to 3 convolution-ReLU (Rectified Linear Units) pairs and a max-pooling layer at the end of each block; see Fig. 3.2A. The depth or feature channel size of the blocks increases with the depth of the network. Finally, three fully connected layers and a soft-max function map the features into label probabilities at the tail of the VGG network. Several variations of the VGG network exist, which differ in the number of blocks. For example, VGG-16 network indicates that the network consists of 13 layers of convolutional-ReLU pairs as well as three fully connected layers resulting in 16 blocks; see Fig. 3.2A. The latest implementations add batch normalization and drop-out layers to the originally reported VGG network in order to improve training and the generalization of the network [22].

In VGG and other networks, the convolution is implemented as shared convolution, not as a locally connected layer [15, p. 341] to handle fewer parameters, speed up the training, and use less memory. Shared weights within one channel assumes that the same features appear at various locations of the input images. This assumption is generally valid for remote sensing and indoor images. It is worth to mention that, since the ceiling or floor regions are located at the top and bottom of the images, using locally connected layer might provide an interesting insight and research direction for segmentation of indoor images.

The channels in the convolutional layers can be interpreted as filters, highlighting patterns in an image that the network has learned from the training data. In the first convolution layer, these patterns, referred to as features, are associated with basic/primitive image content; for instance, corners, edges, etc. Deeper in the network, the features describe complex relationships of these basic image content; for more details see [15, p. 360].

It is noteworthy that VGG networks are not considered as the best network architecture for image classification problems due to their accuracy, the number of parameters learned, and the inference time. The reader can find comparisons of various CNNs in [23]. The latest deep models that outperform VGG contain significantly more convolutional layers. To train these deeper networks is harder due to the propagation of potential numerical errors stemming from very low residuals in the training step. For this reason, ResNet utilizes residual connections that allow for training a 1000-layer deep network; see [21]. While this field is rapidly evolving, current data fusion networks typically adopt the VGG network due to its simple architecture.

3.2.2 Architectures for Pixel-level Labeling

As opposed to image classification, pixel-level labeling requires annotating the pixels of the entire image. Consequently, the output is an array similar to the size of the input. This can be

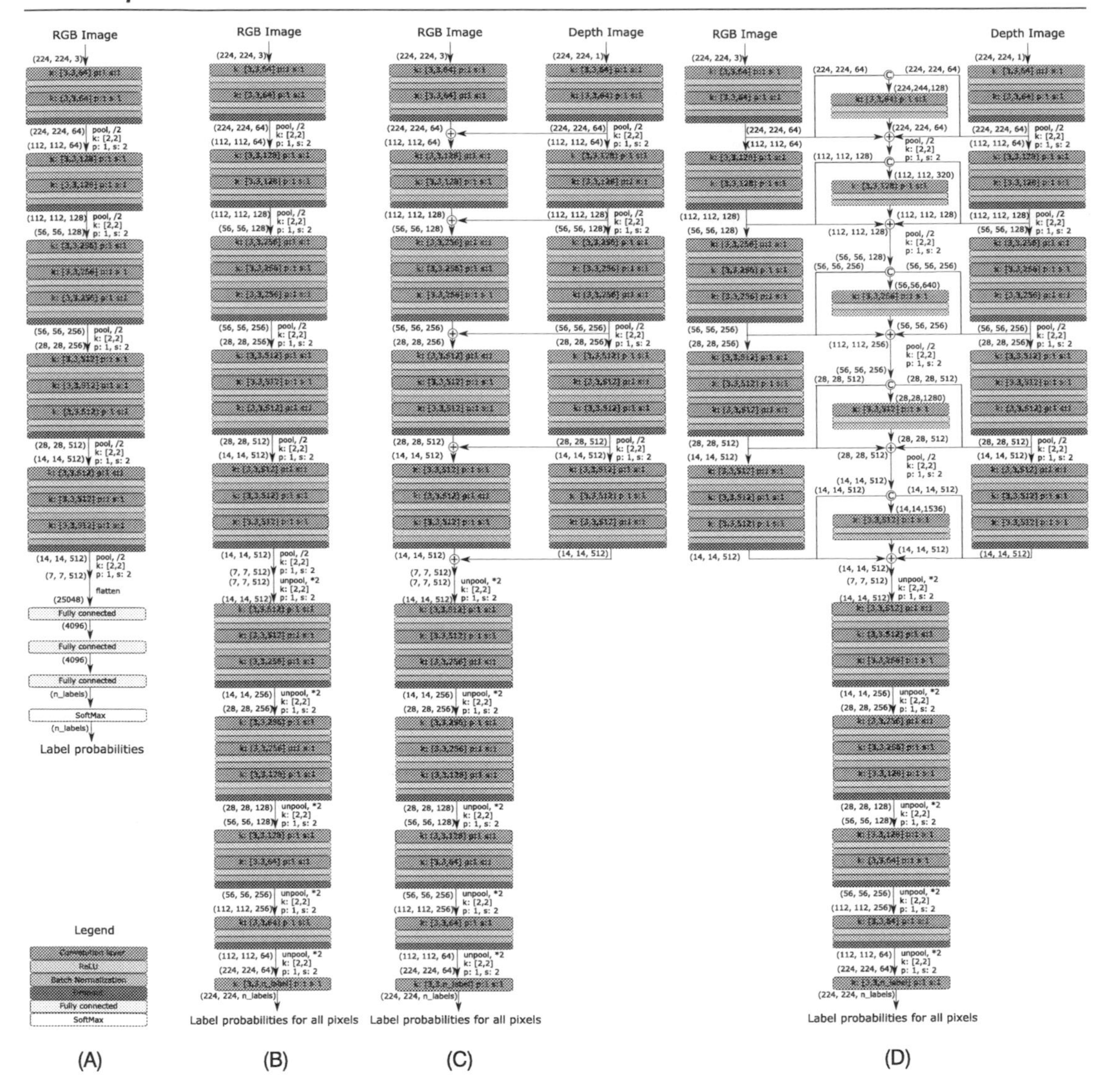

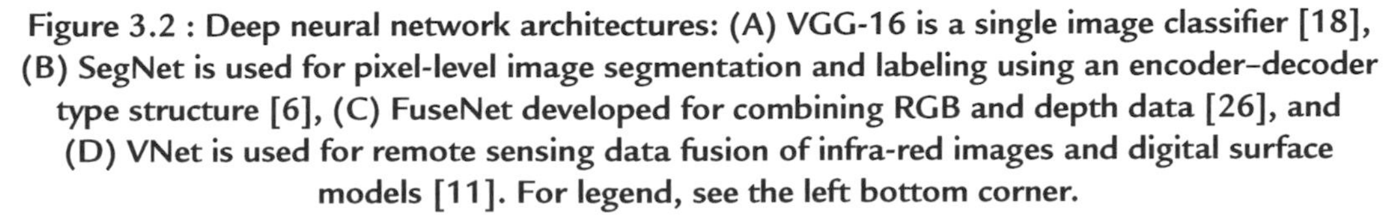

Figure 3.2 : Deep neural network architectures: (A) VGG-16 is a single image classifier [18], (B) SegNet is used for pixel-level image segmentation and labeling using an encoder–decoder type structure [6], (C) FuseNet developed for combining RGB and depth data [26], and (D) VNet is used for remote sensing data fusion of infra-red images and digital surface models [11]. For legend, see the left bottom corner.

achieved by utilizing a single image classifier network and by discarding the classifier tail of VGG, *i.e.* removing the fully connected layer and soft-max, and instead, utilizing a series of unpooling operations along with additional convolutions. An unpooling operation allows for increasing the width and height of the convolutional layer and decreases the number of channels. This results in generating an output at the end of the network that has the original image size; see Fig. 3.2B. This concept is referred to as encoder–decoder network, such as SegNet [6]. SegNet adopts a VGG network as encoder, and mirrors the encoder for the decoder, except the pooling layers are replaced with unpooling layers; see Fig. 3.2B. An advantage of utilizing an image classifier is that the weights trained on image classification datasets can be used for the encoder. This can be considered a benefit as the image classification datasets are typically larger, such that the weights learned using these datasets are likely to be more accurate. Adopting these weights as initial weights in the encoder part of the network is referred to as transfer learning. Here, we mention another notable network, called U-Net, which has a similar encoder–decoder structure but the encoder and decoder features are connected forming a U-shaped network topography [24].

The main disadvantage of encoder–decoder networks is the pooling-unpooling strategy which introduces errors at segment boundaries [6]. Extracting accurate boundaries is generally important for remote sensing applications, such as delineating small patches corresponding to buildings, trees or cars. SegNet addresses this issue by tracking the indices of max-pooling, and uses these indices during unpooling to maintain boundaries. DeepLab, a recent pixel-level labeling network, tackles the boundary problem by using atrous spatial pyramid pooling and a conditional random field [25]. Additionally, the latest DeepLab version integrates ResNet into its architecture, and thus benefits from a more advanced network structure.

3.2.3 Architectures for RGB and Depth Fusion

It is reasonable to expect that using additional information about the scene, such as depth, would improve the segmentation and labeling tasks. In this chapter, our focus is on end-to-end deep learning models that fuse RGB and depth data using CNNs; hence we investigate three different approaches:

- **Approach 1**: Depth image can be incorporated in the network input by modifying the color space of the RGB image. In this case, one of the three image channels in the modified color space is replaced with the one-channel depth image. The main idea behind this modification is that the RGB color space carries redundant information, hence removing the redundancy does not influence scene segmentation. In our experiments, we exploit this redundancy by using "normalized" RGB space, where one channel is redundant, and therefore, can be replaced with scene depth or normal which augments the geometric information present in the image. Another color representation, namely HSI, is also investigated, where the HS channels contain the geometric scene information, and the intensity channel contains "scene" illumination. Considering that the illumination component is not descriptive of scene geometry, it can be replaced with depth or normal information. Using color transformations from RGB to the adopted color spaces allow training the existing neural networks, such as SegNet, DeepLab or U-Net [24], without any modifications. This integration has the advantage of using same number of weights as it would be used for RGB images. In addition, weights learned on large datasets for image classification can be directly used conjecturing that transfer learning for this problem holds. Here, we investigate several color space modifications, such as when the last channel of the HSI representation is replaced by depth values or surface normals.

- **Approach 2**: The three-channel RGB image can be stacked together with the depth image resulting in a four-channel input. In this case, the network structure requires small changes at the head of the network and the number of learnable parameters is almost the same, which is an advantage. However, it is not clear whether the depth information can be successfully incorporated using this approach due to both the different nature of and the correlation between data sources. We investigate the SegNet network with four-channel input, where the fourth channel is the one-channel depth or the surface normal image.
- **Approach 3**: Siamese network structures are designed to extract features from two data sources independently, and then the features of independent network branches can be combined at the end of certain convolutional blocks; see Figs. 3.2C and 3.2D. Clearly, these networks have more parameters to be estimated, but the data structures treat the data sources differently. We investigate two Siamese network structures, namely FuseNet [26] and VNet [11]. FuseNet utilizes two parallel VGG-16 networks as encoders for RGB and depth, see Fig. 3.2C, where RGB encoder–decoder part is the same as SegNet, and the depth branch is identical to a VGG with one-channel input without the classifier head. The outputs of the two networks are combined with a simple addition of two tensors from the RGB and depth branches at the end of each convolutional block. VNet is a more complex network in terms of structure; see Fig. 3.2D. While it still has two separate VGG-16 branches for RGB and depth, there is a middle branch which integrates the features generated at the end of each convolutional block of the two VGG-16 networks. This middle branch concatenate the tensors along the feature (depth) channel; doubling the length of the feature channel. The next convolutional layer reduces the length of the feature channels, and thus enables adding the three tensors, *i.e.* the two tensors of the VGG-16 convolutional blocks and the middle tensor in the next step. The decoder operates on the middle sub-network. Clearly, the VNet has more parameters than FuseNet. Our implementation of the network architecture presented in Fig. 3.2D is based on [11].

Furthermore, instead of depth representation, one can also derive 3D normal vector (needle) images from the depth images and use them as input. Note that surface normal vector representation of the 3D space might be more general in terms of describing objects, since it does not depend on camera pose. It is clear that the depth values change when the camera pose

Table 3.1: Overview of datasets for RGB and depth fusion; datasets include annotated images; the size of the dataset is the number of annotated images.

Dataset	Size	Description
NYUv2 [29]	1.5k	Indoor scenes, captured by Kinect v1
SUN-RGBD [30]	10k	Indoor scenes, 47 indoor and 800 object categories, captured by Kinect v2 and Intel RealSense, includes other datasets, such as NYUv2.
SceneNet RGB-D [30]	>1M	Synthetic, photorealistic indoor videos from 15k trajectories
Stanford 2D-3D-Semantics Dataset [27]	70k	6000 m^2 indoor area; dataset includes depths, surface normals, semantic annotations, global and metadata from a 360° camera, depth obtained by structured-light sensor
ISPRS 2D Semantic Labeling [28]	33+38	High resolution aerial images; urban scene of a village (Vaihingen) and city (Potsdam) captured from aircraft; three-channel images (infra-red, blue, green), DSM derived with dense reconstruction, and LiDAR point cloud

changes, but the normals are fixed to the local plane of the object; hence, they are view independent. This property is expected to allow for more general learning of various objects. 3D normal vectors can be represented as a three-channel image that contains the X, Y, Z components of the vector at each pixel. We also derive a one-channel input by labeling the normal vectors based on their quantized directions. This one-channel normal label image allows for simpler representation, and thus requires fewer network parameters. These two normal representations are investigated for all fusion approaches presented above.

3.2.4 Datasets and Benchmarks

Successful training of deep learning methods requires large set of labeled segments. Generating accurate labels are labor intensive, and therefore, open datasets and benchmarks are important for developing and testing new network architectures. Table 3.1 lists the most notable benchmark datasets that contain both RGB and depth data. Currently, the Stanford 2D-3D-Semantics Dataset [27] is the largest database of images captured from real scenes along with depth data as well as ground truth labels. In the remote sensing domain, the ISPRS 2D Semantic Labeling benchmark [28] is the most recent overhead imagery dataset that contains high resolution orthophotos, images with labels of six land cover categories, digital surface models (DSM) and point clouds captured in urban areas.

3.3 Methods

This section presents our experimental setup, used for investigating the semantic segmentation of indoor and airborne images. First, we review the Stanford 2D-3D-Semantics and ISPRS 2D

Table 3.2: Overview of the datasets used for training, validation and testing.

Dataset	Training [# of images]	Validation [# of images]	Testing [# of images]
Stanford 2D-3D-S	9294 (90% of A1)	1033 (10% of A1)	3704 (100% of A3)
ISPRS Vaihingen	8255 (65%)	1886 (15%)	2523 (20%)

Semantic Labeling datasets, and then the data split for training and testing applied in our investigation, including data preprocessing, and the derivation of 3D normals. Next, we present our normal labeling concept that converts the 3D normal vectors into labels. The section continues by introducing the color space transformations for RGB and depth fusion. Finally, the hyper-parameters used during the network training are presented.

3.3.1 Datasets and Data Splitting

The Stanford 2D-3D-Semantics Dataset (2D-3D-S) [27] contains RGB and depth images of 6 different indoor areas. The dataset were mainly captured in office rooms and hallways, and small part of it in a lobby and auditorium. The dataset is acquired by a Matterport Camera system, which combines cameras and three structured-light sensors to capture 360-degree RGB and depth images. The sensor provides reconstructed 3D textured meshes and raw RGB-D images. The dataset contains annotated pixel-level labels for each image. The objects are categorized into 13 classes: ceiling, floor, wall, column, beam, window, door, table, chair, bookcase, sofa, board, and clutter. The sofa class is underrepresented in the dataset, and thus categorized as clutter in our tests.

Due to the large size of the dataset, we use Area 1 of the Stanford dataset for training, and Area 3 for testing. This strategy allows for estimating the generalization error of the networks, because the training and testing are performed in two independent scenes. The training dataset is divided into 90% training and 10% validation sets, respectively, resulting in 9294 and 1033 images. The validation set is used for evaluating the network performance at each epoch of the training. The test data, *i.e.* Area 3, contains 3704 images. An overview of the training, validation and test datasets can be found in Table 3.2.

The ISPRS 2D Semantic Labeling Dataset is an airborne image collection, consisting of high resolution true orthophotos and corresponding digital surface models (DSMs) derived with dense image matching. The images are annotated manually into six land use and object categories, *i.e.* impervious surfaces, building, low vegetation, tree, car and clutter/background. The dataset acquired in a city (Potsdam) and a village (Vaihingen). Here, the Vaihingen dataset is investigated, which contains 33 high resolution orthophoto mosaics. The images have a ground sampling distance (GSD) of 9 cm, and are false-colored using the near-infrared, red and green spectral bands (NRG). Table 3.2 lists the data used for training, validation, and testing.

3.3.2 Preprocessing of the Stanford Dataset

This subsection presents the methods used for image resizing, depth filtering and normalization. Sample images can be seen in the first two columns of Fig. 3.3.

Resizing. Original image size of the dataset is 1080×1080 pixels. Deep learning models use significantly smaller image size due to memory and computation considerations. We should note that larger image size does not necessarily lead to more accurate result. We adopt the strategies developed by other authors [26,6], and resize the images to 224×224 pixels. Sample RGB images can be seen in the first row of Fig. 3.3.

Depth filtering. Depth images in the dataset contain missing pixels. Since missing pixels can significantly influence the training and testing process, depth interpolation is performed by applying image inpainting strategies [31,32]. This approach, first, calculates initial guesses of the missing depth value using local statistical analysis, and then one iteratively estimates the missing depth values using discrete cosine transform. Sample depth images can be seen in the second row of Fig. 3.3.

Data normalization. RGB values are normalized according to the input layer of the neural networks. Range of depth values depends on the scene and the maximum operating range of the sensor. For indoor scenes, depth values are typically up to few meters. Therefore, we use 0 and 10 m range to normalize the data. Values larger than 10 m are truncated to this maximum value.

Calculation of normal vectors. Surface normal vectors are provided with the dataset. Examples of 3D normal images can be seen in the third row of Fig. 3.3.

3.3.3 Preprocessing of the ISPRS Dataset

This subsection presents the preprocessing steps for the ISPRS dataset. These steps include resizing and cropping of the orthophotos and the calculation of the normal vectors. Sample images can be seen in the last two columns of Fig. 3.3.

Resizing and cropping. The orthophoto mosaics from the ISPRS dataset are roughly 2000×2500 pixels. In order to reduce the images to the desired 224×224 pixel size, first, the images are cropped into 448×448 tiles, providing an about 60% overlap. Then the cropped areas are

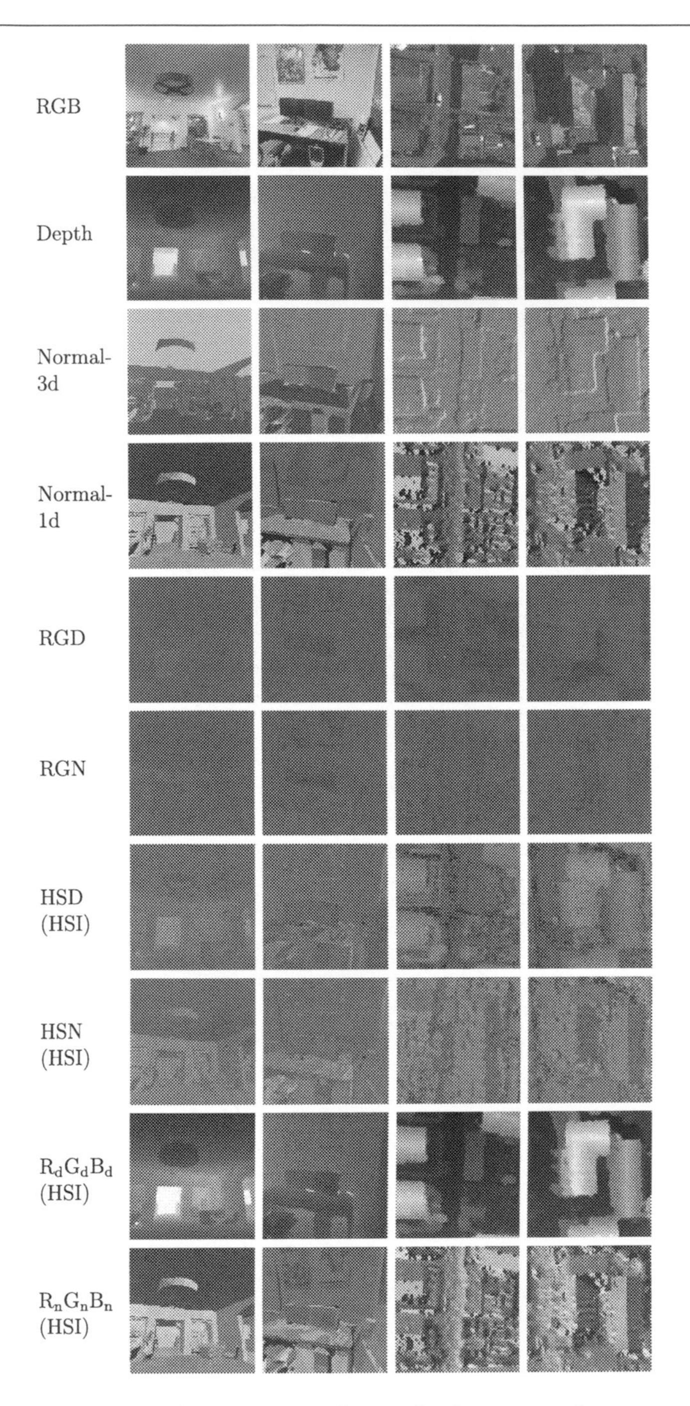

Figure 3.3 : Different data modality representations. The first two columns are samples from the Stanford dataset, and the last two columns are samples from the ISPRS dataset. Each row presents a different modality.

resized to 224 × 224 using nearest neighborhood interpolation. The final dataset consists of 12,664 images. Sample images can be seen in the first row of Fig. 3.3.

Depth filtering. The data are already filtered for outliers, and therefore, no data preprocessing is needed. Sample depth images can be seen in the second row of Fig. 3.3.

Calculation of normal vectors. Surface normal data is directly not available from the ISPRS dataset. Therefore, surface normals are derived from the DSM using least squares plane fitting. Given a point, the computation of a normal at that point is performed as follows. First, the eight neighbor grid points ($\mathbf{z}_i$, $i = 1, ..., 8$) are obtained around the point ($\mathbf{z}_c$), and then their differences from the point are calculated, *i.e.* $\overline{\mathbf{z}}_i = \mathbf{z}_i - \mathbf{z}_c$. Then the points are filtered based on the vertical differences and a predefined threshold. Given $\overline{\mathbf{z}}_i$, $i = 1, ..., k$, $k \leq 8$, we formulate a constrained optimization problem:

$$\hat{\mathbf{n}} = \underset{||\mathbf{n}||=1}{\operatorname{argmin}}\, \mathbf{n}^\top (\sum_{i=1}^{k} \overline{\mathbf{z}}_i \overline{\mathbf{z}}_i^\top)\mathbf{n}, \tag{3.1}$$

where $\mathbf{n}$ is the surface normal, $\hat{\mathbf{n}}$ is its estimate. This problem can be solved with singular value decomposition; the $\hat{\mathbf{n}}$ solution is the eigenvector of $\sum_{i=1}^{k} \overline{\mathbf{z}}_i \overline{\mathbf{z}}_i^\top$ corresponding to the smallest eigenvalue. Sample 3D normal images can be seen in the third row of Fig. 3.3.

3.3.4 One-channel Normal Label Representation

The one-channel surface normals are derived from the 3D normals by mapping the three-channel normal vectors into a unit sphere. This mapping can be simply done by transforming the normal unit vectors into spherical coordinates, *i.e.* $\theta = \tan^{-1}(\frac{n_y}{n_x})$ and $\phi = \cos^{-1}(n_z)$, since $r = ||\mathbf{n}|| = \sqrt{n_x^2 + n_y^2 + n_z^2} = 1$. At this point, the angle values have to be associated with integer numbers (labels) in order to represent the normal values in 8-bit image format, *i.e.* $(\theta, \phi) \rightarrow [0, ..., 255]$. Here, we opted to divide the sphere at every 18 degrees, which results in 200 values. These values are saved as gray image. This one-channel representation was calculated for both the Stanford and ISPRS datasets. Sample one-channel normal images can be seen in the fourth row of Fig. 3.3.

3.3.5 Color Spaces for RGB and Depth Fusion

We investigate two possibilities to fuse RGB and depth information through color space transformations. The goal is to keep the three-channel structure of RGB images, but substitute one of the channels with either the depth or normal labels.

Normalized RGB color space. Let r, g and b be the normalized RGB values, such that

$$r = \frac{R}{R+G+B}; g = \frac{G}{R+G+B}; b = \frac{B}{R+G+B}. \tag{3.2}$$

In this color space representation, one of the channels is redundant due to the fact that it can be recovered from the other two channels using the condition $r + g + b = 1$. Therefore, for example, we can replace the blue channel with depth D or normals (one-channel representation) N, normalized to [0, 1]. This results in fused images rgD and rgN. Sample rgD and rgN images can be seen in the fifth and sixth rows of Fig. 3.3. Note that in terms of energy, blue represents about 10% of the visible spectrum in average.

HSI color space. The second color space fusion is based on hue, saturation and intensity (HSI) representation. In this representation, intensity I is replaced with depth D or normals N, which results in HSD and HSN representation, respectively, where

$$H = \begin{cases} \cos^{-1} \frac{0.5[(r-g)+(r-b)]}{\sqrt{(r-g)^2+(r-g)(g-b)}}, \\ 2\pi - \cos^{-1} \frac{0.5[(r-g)+(r-b)]}{\sqrt{(r-g)^2+(r-g)(g-b)}}, \end{cases} \tag{3.3}$$

$$S = 1 - 3\min(r, g, b). \tag{3.4}$$

Sample HSI and HSN images can be seen in the seventh and eighth rows of Fig. 3.3.

Finally, we investigate a representation which is a transformation of the HSD or HSN representation back to the RGB space. Therefore, we calculate

$$(R_d, G_d, B_d) = \begin{cases} (b, c, a), h = H, \text{ for } H < \frac{2}{3}\pi, \\ (a, b, c), h = H - \frac{2}{3}\pi, \text{for } \frac{2}{3}\pi \leq H < \frac{4}{3}\pi, \\ (c, a, b), h = H - \frac{4}{3}\pi, \text{ for } \frac{4}{3}\pi \leq H < 2\pi, \end{cases} \tag{3.5}$$

where

$$a = D(1 - S), \tag{3.6}$$

$$b = D(1 + \frac{S\cos(h)}{\cos(\frac{\pi}{3} - h)}), \tag{3.7}$$

$$c = 3D - (a + b), \tag{3.8}$$

for color space modification using depth D or (R_n, G_n, B_n) calculated using (3.5), where

$$a = N(1 - S), \tag{3.9}$$

$$b = N(1 + \frac{S\cos(h)}{\cos(\frac{\pi}{3} - h)}), \tag{3.10}$$

$$c = 3N - (a + b), \tag{3.11}$$

for color space modification using normals N. Sample HSI and HSN images can be seen in the last two rows row of Fig. 3.3.

3.3.6 Hyper-parameters and Training

Due to the similarity in the network structures of investigated networks, the same hyper-parameters are used in the experiments for all three networks. We use pretrained weights for the VGG-16 part of the networks, which are trained on the large-scale ImageNet dataset [1]. When the network has an additional one-channel branch, such as the depth branch of FuseNet, or it has an additional input channel, *i.e.* four-channel network, the three-channel pretrained weights of the input convolution layer of VGG-16 are averaged and used for initializing the extra channel. A stochastic gradient descent (SGD) solver and cross-entropy loss are used for training. The entire training dataset is passed through the network at each epoch. Within an epoch, the mini batch size is set to 4 and the batch was randomly shuffled at each iteration. The momentum and weight decay was set to 0.9 and 0.0005, respectively. The learning rate was 0.005 and decreased by 10% at every 30 epochs. The best model is chosen according to the performance on the validation set measured after each epoch. We run 300 epochs for all tests, but the best loss is typically achieved between 100–150 epochs.

3.4 Results and Discussion

The predication images calculated from the test set are evaluated based on the elements of the confusion matrix. The TP_i, FP_i and FN_i, respectively, denote the total number of true positive, false positive and false negative pixels of the ith label, where $i = 1, ..., K$, K is the number of label classes. We use three standard metrics to measure the overall performance of the networks. Global accuracy represents the percentage of the correctly classified all pixels and is defined as

$$\text{GlobAcc} = \frac{1}{N}\sum_{i=1}^{K}\text{TP}_i \tag{3.12}$$

and N denotes the total number of annotated pixels. The mean accuracy measures the average accuracy over all classes:

$$\text{MeanAcc} = \frac{1}{K}\sum_{i=1}^{K}\frac{\text{TP}_i}{\text{TP}_i + \text{FP}_i}. \tag{3.13}$$

Table 3.3: Results on Stanford indoor dataset; Training: 90% of Area 1, 9294 images; Validation: 10% of Area 1, 1033 images; Testing: 100% of Area 3, 3704 images. *Network abbreviations: the name of the networks consists of two parts: the first is the base network, *i.e.* SegNet, FuseNet and VNet, see Fig. 3.2, following the input type. Note that, in some cases, the networks are slightly changed to be able to accept the different input size. For instance, RGBD and RGBN represent four-channel inputs. RGB+D and RGB+N indicates that the network has separate branches for depth or normal, respectively. $X_nY_nZ_n$ is three-channel input of 3D surface normals.

Network*	GlobalAcc	MeanAcc	Mean IoU
Segnet RGB	76.4%	56.0%	70.8%
Segnet Depth Only	71.0%	50.8%	67.5%
Segnet HSD	73.4%	53.6%	74.2%
Segnet HSN	77.4%	57.1%	71.2%
Segnet $R_dG_dB_{d(HSI)}$	73.5%	49.7%	64.1%
Segnet $R_nG_nB_{n(HSI)}$	74.6%	51.4%	67.2%
Segnet rgD	76.1%	55.6%	71.3%
Segnet rgN	76.3%	53.4%	69.2%
Segnet RGBD	75.4%	53.6%	68.5%
Segnet RGBN	76.7%	57.4%	74.4%
Segnet RGBDN	68.7%	45.2%	58.9%
FuseNet RGB+D	76.0%	54.9%	73.7%
FuseNet RGB+N	**81.5**%	**62.4**%	76.1%
FuseNet RGB+$X_nY_nZ_n$	81.4%	62.4%	**78.1**%
VNet RGB+D	78.3%	54.2%	68.5%
VNet RGB+N	80.0%	56.1%	70.4%

Finally, intersection over union (IoU) is defined as the average value of the intersection of the prediction and ground truth over the union of them:

$$\mathrm{IoU} = \frac{1}{K}\sum_{i=1}^{K}\frac{\mathrm{TP}_i}{\mathrm{TP}_i + \mathrm{FP}_i + \mathrm{FN}_i}. \tag{3.14}$$

3.4.1 Results and Discussion on the Stanford Dataset

Results for the various network and input structures are presented in Table 3.3. Fig. 3.4 shows the accuracies of each class for different networks. The baseline solution, the SegNet with RGB inputs, achieves 76.4% global accuracy in our tests. If only depth images are used in SegNet, then the accuracy drops by 5%. Fusing RGB and depth through color space transformations results in noticeable decrease ($\sim$ 1%–3%) in the network performance.

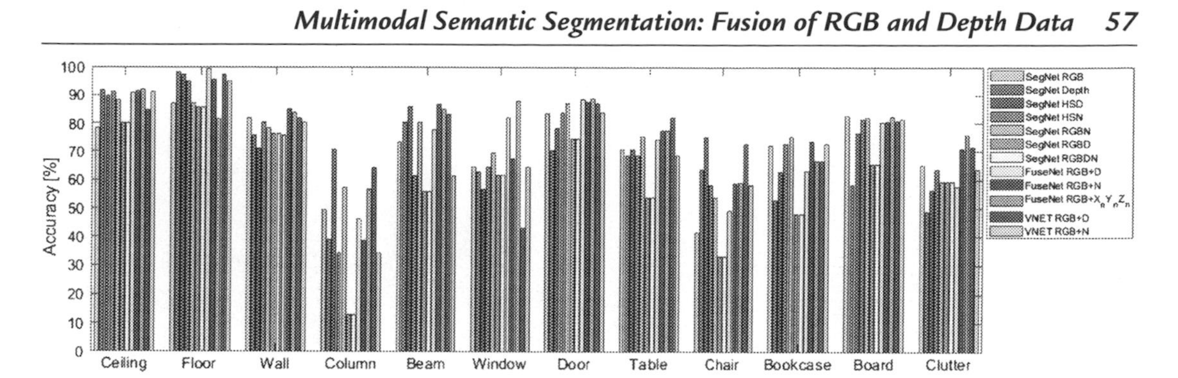

Figure 3.4 : Per-class accuracies for the Stanford dataset.

The four-channel variants of the SegNet, *i.e.* SegNet RGBD and SegNet RGBN, produce similar results to the SegNet RGB. In these cases, the information from the depth data clearly makes no contribution to the segmentation process; the network uses only the pretrained weights of SegNet RGB, and the weights associated with the depth are neglected.

The FuseNet architecture with depth input, *i.e.* FuseNet RGB+D, produces slightly worse results than the baseline solution in our tests; it particularly achieves 76.0% global accuracy, which is 0.4% less than the accuracy produced by SegNet, which can be considered insignificant. The mean IoU in this case is better by $\sim$ 3% than the baseline. Note that this 76% accuracy is similar to the result reported in [26] on the SUN RGB-D benchmark dataset. In our tests, we found that FuseNet with normals, *i.e.* FuseNet RGB+N, provides the best results in terms of global and mean accuracy. The $\sim$ 5% margin between either the SegNet RGB or FuseNet RGB+D and the RGB+N is significant. The 3D normal representation provides the same global accuracy as the FuseNet RGB+N, while it outperforms by 2% margin in the mean IoU.

Discussion. Results suggest that depth does not improve the overall accuracy for the adopted dataset. Using normals instead of depth, however, improves almost all types of RGB and depth fusion approaches with a significant margin. Since the test dataset is a different area (Area 3) than the training (Area 1), these results indicate that the normals improve the generalization ability of the network. In other words, normals are better general representation of indoor scenes than depth.

In order to support this hypothesis, we also evaluated the networks using data from the Area 1 only. No samples were taken from Area 3 neither for the training nor for the testing. These results are presented in Table 3.4. In this training scenario, the images from Area 1 are divided into 10% (1033 images) training and 90% (9294 images) validation sets. We used this extreme split in order to rely on the pretrained SegNet weights; consequently, the achieved accuracies presented in Table 3.4 are significantly smaller than in the first test scenario.

Table 3.4: Results for Area 1 only; Training: 10% of Area 1, 1033 images; Testing and validation: 90% of Area 1, 9294 images.

Channels	GlobalAcc	MeanAcc	Mean IoU
SegNet RGB	65.0%	61.8%	45.2%
SegNet rgD	58.5%	53.1%	36.9%
SegNet rgN	58.7%	53.1%	36.5%
SegNet RGBD	68.6%	58.8%	44.9%
SegNet RGBN	68.8%	59.1%	45.4%
FuseNet RGB+D	**71.9**%	**73.0**%	**48.8**%
FuseNet RGB+N	69.1%	58.0%	43.6%

Table 3.4 shows no or a slight performance gap between depth and normal data for all approaches. FuseNet shows the best results by effectively combining RGB and depth; FuseNet outperforms SegNet with 6% margin as opposed to the results presented in Table 3.3; note that similar improvements were reported by previous studies [26]. It is important to note that FuseNet is able to combine RGB and depth with very low number of training samples, which prompts the question whether the network actually learns the depth features, or just the addition operator applied between the network branches is an efficient, but deterministic way to share depth information with the VGG-16 network.

The per-class accuracies presented in Fig. 3.4 show that each network has a different accuracy per each class. FuseNet RGB+D provides the best overall accuracy metrics, but other networks, such as SegNet HSD, performs better in majority of classes, such as column, beam, chair. This indicates that an ensemble classifier using these networks might outperform any individual network.

3.4.2 Results and Discussion on the ISPRS Dataset

Results for the ISPRS Vaihingen dataset are listed in Table 3.5, the baseline solution is the three-channel SegNet NRG network, which achieves 85.1% global accuracy. Note that SegNet NRG is the same network as SegNet RGB which was discussed in the previous section; we use NRG term to emphasize that the ISPRS images are false color infra-red, red, green images. As expected, the accuracy drops down by about 10% using only depth data. All network solutions produce similar results close to the baseline solution. SegNet NRGD, NRG+N and VNet NRG+N slightly outperform the SegNet NRG in all three metrics; though, the difference is not significant. The best results are provided by the VNet NRG+N achieving 85.6% global, 50.8% mean accuracy and 60.4% mean IoU. Per-class accuracies presented in Fig. 3.5. Here, we see that all networks have a very similar performance in all classes. A slight difference can be seen in the clutter class, where SegNet NRGD outperforms the other networks.

Table 3.5: Results for ISPRS Vaihingen dataset; Training: 65%, 8255 images; Validation: 15%, 1886 images; Testing: 20%, 2523 images. *Network abbreviations: the name of the networks in the first column consists of two parts: the first is the base network, *i.e.* SegNet, FuseNet and VNet, see Fig. 3.2, following the input type. NRG denotes the infra-red, red and green bands, which is the image format of the ISPRS dataset. Note that, in some cases, the first layer of the networks is slightly changed to be able to accept the different input size. For instance, NRGD and NRGN represent four-channel inputs. NRGB+D and NRGB+N indicates that the network has separate branches for depth or normal, respectively. $X_nY_nZ_n$ is three-channel input of 3D surface normals.

Networks*	GlobalAcc	MeanAcc	Mean IoU
Segnet NRG	85.1%	49.7%	59.3%
Segnet Depth Only	75.9%	38.9%	49.9%
Segnet HSD	84.7%	46.6%	56.4%
Segnet HSN	82.8%	45.2%	54.8%
Segnet NRGD	74.2%	39.8%	50.9%
Segnet NRGN	85.2%	48.5%	58.4%
Segnet NRGDN	84.8%	46.7%	57.1%
FuseNet NRG+D	84.4%	47.3%	57.4%
FuseNet NRG+N	85.5%	48.1%	57.9%
FuseNet NRG+$X_nY_nZ_n$	85.1%	47.2%	57.3%
VNet NRG+D	85.6%	48.2%	58.4%
VNet NRG+N	**85.6%**	**50.8%**	**60.4%**

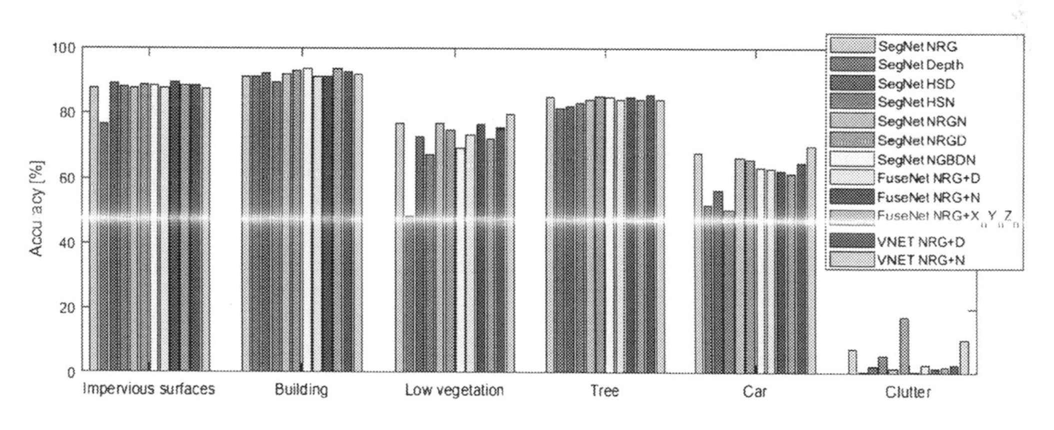

Figure 3.5 : Per-class accuracies for the ISPRS dataset.

Discussion. Overall, almost all accuracies presented in Table 3.5 are within the error margin and are around 85%. Using normals have insignificant improvement due to the fact that almost all surface normals point in the same direction (upward).

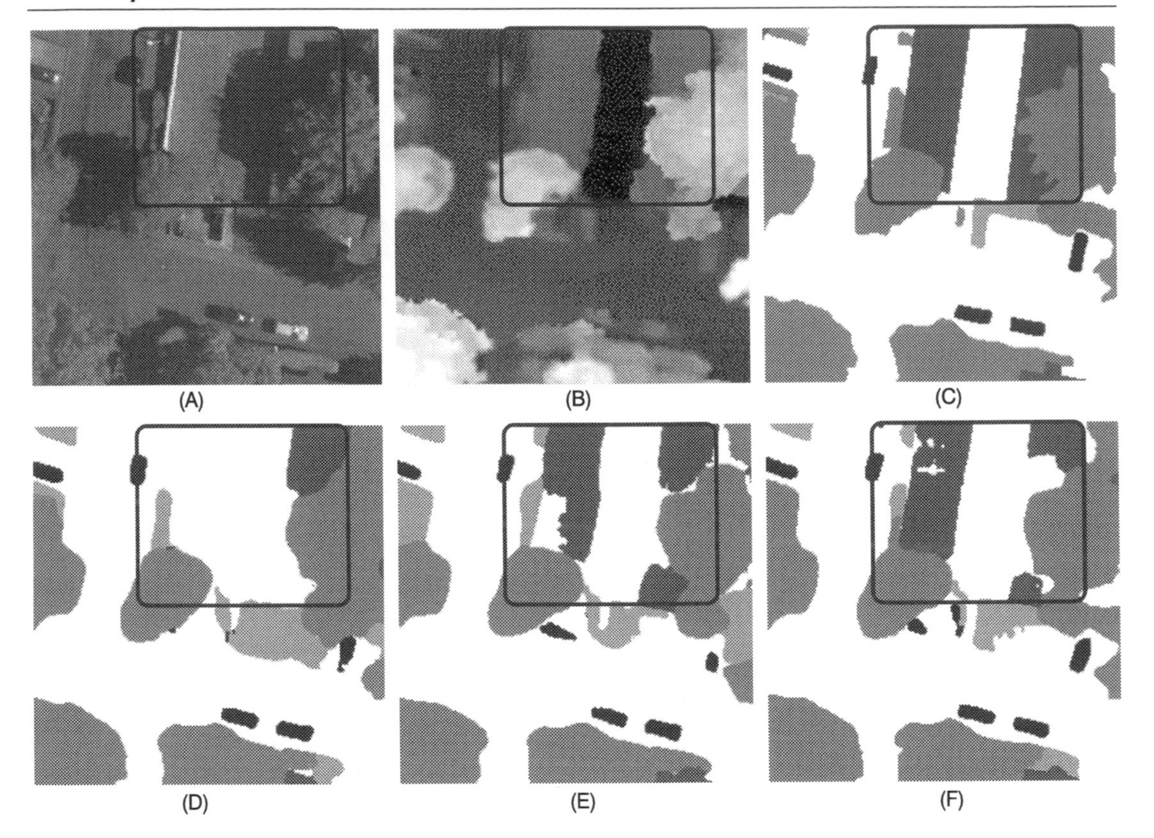

Figure 3.6 : (A) NRG image, (B) depth, (C) ground truth, (D) SegNet NRG, (E) FuseNet NRG+D, (F) FuseNet NRG+N; Label colors: white – impervious surface; red – building; light green – low vegetation, dark green – tree, blue – car, black – background.

Table 3.5 shows very slight improvement of VNet over FuseNet, while FuseNet NRG+D does not achieve the accuracy of the baseline SegNet architecture. It is noteworthy that the reported global accuracies on the ISPRS Vaihingen dataset using either SegNet or FuseNet are around 91% [11]. The main reason of this relatively large performance gap between Table 3.5 and [11] is potentially due both to the difference in evaluation technique and the random selection of images in the training process.

Although SegNet NRG has similar results to VNet or FuseNet, there are examples, where data fusion using depth outperforms SegNet; see an example in Fig. 3.6. The red rectangle shows a section, where SegNet NRG does not detect a building due to the color of the building roof is very similar to the road; see Figs. 3.6D and 3.6C for ground truth. Unlike SegNet, FuseNet NRG+D partially segment that building, see Fig. 3.6E, due to the depth difference between the roof and road surface in the depth image presented in Fig. 3.6B. Also,

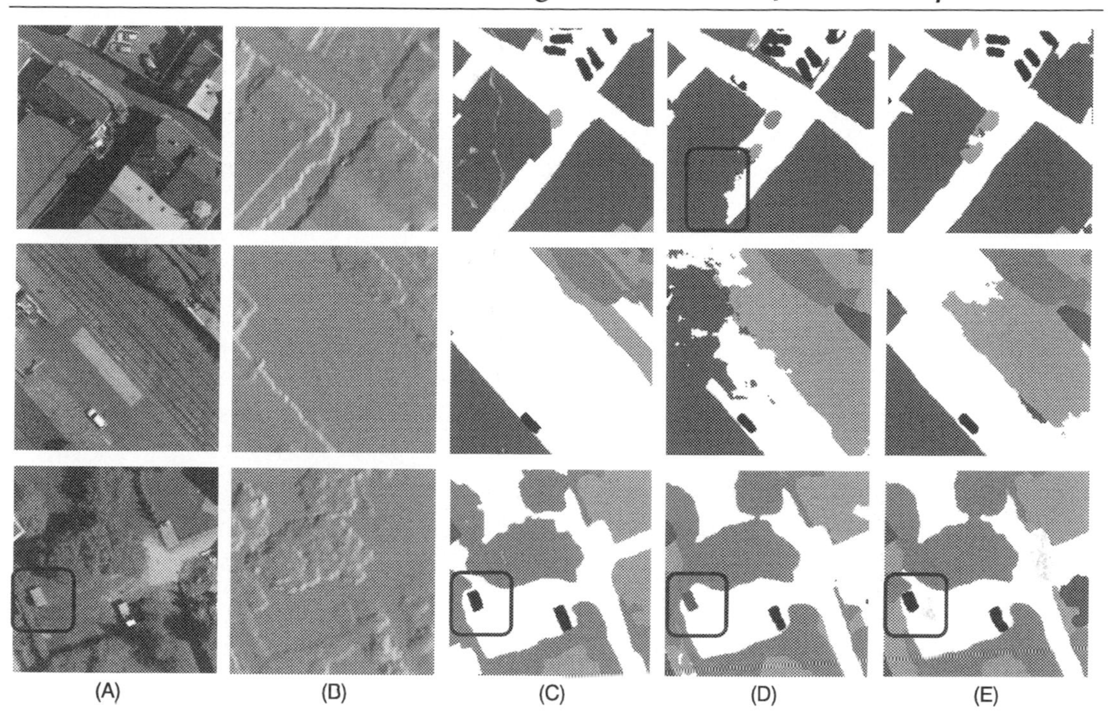

Figure 3.7 : (A) NRG image, (B) 3D normals, (C) ground truth labels, (D) SegNet NRG, (E) FuseNet NRG+N; Label colors: white – impervious surface; red – building; light green – low vegetation, dark green – tree, blue – car, black – background.

note that the depth image is not sharp around the building contours, and the prediction image of FuseNet NRG+D follows this irregularity; see Fig. 3.6B and 3.6E. This indicates that the quality of the depth image is important to obtain reliable segment contours.

Using normals instead of depth as input also does not improve the global accuracy significantly. Most object categories are horizontal, such as grass or a road, and therefore, normals point upwards containing no distinctive information of the object class itself. In fact, after analyzing the confusion matrices, we found that FuseNet NRG+N slightly decreased (0.7%) the false positives of the building class and improved the true positives for the impervious surface class with the same ratio. This indicates that the network better differentiate between buildings and roads. The reason for this that some roofs in the dataset are not flat, such as gable or hipped roof. In these cases, normals have a distinguishable direction that guides the network to correctly identify some buildings. Two examples can be seen in the first two rows of Fig. 3.7. The figures demonstrate that the FuseNet NRG+N has sharper object edges than a single SegNet NRG classifier. Finally, the last row in Fig. 3.7 presents a case when the NRG based network does not classify correctly an object based on the ground truth; see the object boxed by the red rectangle. The ground truth and FuseNet NRG+N mark that object as car, and SegNet as building. This example demonstrates that normal or depth data has semantical impact on interpreting a patch.

3.5 Conclusion

The chapter reviewed the current end-to-end learning approaches to combining RGB and depth for image segmentation and labeling. We conducted tests with SegNet, FuseNet and VNet architectures using various inputs including RGB, depth, normal labels, HSD, HSN and 3D surface normals. Networks were trained on indoor and remote sensing images. We reported similar results as other authors in terms of global accuracy; around 78% on the Stanford and 85% on ISPRS dataset. For indoor scenes, when the training and testing datasets were captured from different areas, we noticed 1–5% improvement with using normal labels instead of depth in almost all fusion approaches. This indicates that using surface normals is a more general representation of the scene that can be exploited in neural network trained for semantic segmentation of indoor images. Best global accuracy achieved by FuseNet RGB+N was 81.5% global accuracy on the Stanford indoor dataset. Similar performance gaps between networks using depth or normal is not achieved for aerial images due to differences between indoor and aerial image scenes.

In summary, the accuracy provided by a simpler SegNet RGB network is close to both FuseNet and VNet indicating that there is still room for improvements, especially for remote sensing data. Future models might consider more recent networks, such as residual connections used in ResNet, or conditional random field implemented in DeepLab. It is also expected that future networks become deeper. Directly incorporating some depth knowledge into the network; for instance, directed dropouts of weights in the convolution layer, might be an interesting research direction. For high resolution images, image segmentation at various resolution (image pyramid) might increase the global consistency of the inference.

Augmenting Siamese networks by adding of tensors, such as in FuseNet or VNet, clearly improves results and is a simple way to fuse RGB and depth data. These networks outperform the original SegNet network in indoor datasets. In the future, more powerful neural network architectures are expected to be developed and thus can be adopted for these two modalities.

References

[1] O. Russakovsky, J. Deng, H. Su, J. Krause, S. Satheesh, S. Ma, Z. Huang, A. Karpathy, A. Khosla, M. Bernstein, A.C. Berg, L. Fei-Fei, ImageNet large scale visual recognition challenge, International Journal of Computer Vision 115 (3) (2015) 211–252, https://doi.org/10.1007/s11263-015-0816-y.

[2] R. Girshick, J. Donahue, T. Darrell, J. Malik, Rich feature hierarchies for accurate object detection and semantic segmentation, in: Proceedings of the 2014 IEEE Conference on Computer Vision and Pattern Recognition, CVPR '14, IEEE Computer Society, Washington, DC, USA, 2014, pp. 580–587.

[3] R.B. Girshick, Fast R-CNN, CoRR, arXiv:1504.08083, arXiv:1504.08083, http://arxiv.org/abs/1504.08083.

[4] S. Ren, K. He, R. Girshick, J. Sun, Faster r-cnn: towards real-time object detection with region proposal networks, IEEE Transactions on Pattern Analysis and Machine Intelligence 39 (6) (2017) 1137–1149, https://doi.org/10.1109/TPAMI.2016.2577031.

[5] K. He, G. Gkioxari, P. Dollár, R.B. Girshick, Mask R-CNN, CoRR, arXiv:1703.06870, arXiv:1703.06870, http://arxiv.org/abs/1703.06870.

[6] V. Badrinarayanan, A. Kendall, R. Cipolla, Segnet: a deep convolutional encoder–decoder architecture for image segmentation, IEEE Transactions on Pattern Analysis and Machine Intelligence 39 (12) (2017) 2481–2495, https://doi.org/10.1109/TPAMI.2016.2644615.

[7] P.H.O. Pinheiro, R. Collobert, From image-level to pixel-level labeling with convolutional networks, in: 2015 IEEE Conference on Computer Vision and Pattern Recognition, CVPR, 2015, pp. 1713–1721.

[8] C. Yang, F. Rottensteiner, C. Heipke, 23 apr 2018 classification of land cover and land use based on convolutional neural networks, ISPRS Annals of the Photogrammetry, Remote Sensing and Spatial Information Sciences IV (2018) 251–258, https://doi.org/10.5194/isprs-annals-IV-3-251-2018.

[9] W. Yao, P. Poleswki, P. Krzystek, Classification of urban aerial data based on pixel labelling with deep convolutional neural networks and logistic regression, ISPRS Annals of the Photogrammetry, Remote Sensing and Spatial Information Sciences XLI (2016) 405–410, https://doi.org/10.5194/isprs-archives-XLI-B7-405-2016.

[10] X. Wei, K. Fu, X. Gao, M. Yan, X. Sun, K. Chen, H. Sun, Semantic pixel labelling in remote sensing images using a deep convolutional encoder–decoder model, Remote Sensing Letters 9 (3) (2018) 199–208, https://doi.org/10.1080/2150704X.2017.1410291.

[11] N. Audebert, B.L. Saux, S. Lefevre, Beyond RGB: very high resolution urban remote sensing with multimodal deep networks, ISPRS Journal of Photogrammetry and Remote Sensing 140 (2018) 20–32, https://doi.org/10.1016/j.isprsjprs.2017.11.011.

[12] C. Yang, W. Li, Z. Lin, Vehicle object detection in remote sensing imagery based on multi-perspective convolutional neural network, ISPRS International Journal of Geo-Information 7 (7) (2018) 249, https://doi.org/10.3390/ijgi7070249, http://www.mdpi.com/2220-9964/7/7/249.

[13] Y. Cao, X. Niu, Y. Dou, Region-based convolutional neural networks for object detection in very high resolution remote sensing images, in: 2016 12th International Conference on Natural Computation, Fuzzy Systems and Knowledge Discovery, ICNC-FSKD, 2016, pp. 548–554.

[14] Y. Ren, C. Zhu, S. Xiao, Small object detection in optical remote sensing images via modified faster r-cnn, Applied Sciences 8 (5) (2018) 813, https://doi.org/10.3390/app8050813, http://www.mdpi.com/2076-3417/8/5/813.

[15] I. Goodfellow, Y. Bengio, A. Courville, Deep Learning, MIT Press, 2016.

[16] Y. Lecun, L. Bottou, Y. Bengio, P. Haffner, Gradient-based learning applied to document recognition, Proceedings of the IEEE 86 (11) (1998) 2278–2324, https://doi.org/10.1109/5.726791.

[17] A. Krizhevsky, I. Sutskever, G.E. Hinton, Imagenet classification with deep convolutional neural networks, in: Proceedings of the 25th International Conference on Neural Information Processing Systems, vol. 1, NIPS'12, Curran Associates Inc., USA, 2012, pp. 1097–1105, http://dl.acm.org/citation.cfm?id=2999134.2999257.

[18] K. Simonyan, A. Zisserman, Very deep convolutional networks for large-scale image recognition, CoRR, arXiv:1409.1556, arXiv:1409.1556, http://arxiv.org/abs/1409.1556.

[19] C. Szegedy, W. Liu, Y. Jia, P. Sermanet, S. Reed, D. Anguelov, D. Erhan, V. Vanhoucke, A. Rabinovich, Going deeper with convolutions, in: 2015 IEEE Conference on Computer Vision and Pattern Recognition, CVPR, 2015, pp. 1–9.

[20] G. Huang, Z. Liu, K.Q. Weinberger, Densely connected convolutional networks, CoRR, arXiv:1608.06993, arXiv:1608.06993, http://arxiv.org/abs/1608.06993.

[21] K. He, X. Zhang, S. Ren, J. Sun, Deep residual learning for image recognition, arXiv preprint, arXiv:1512.03385.

[22] M. Simon, E. Rodner, J. Denzler, Imagenet pre-trained models with batch normalization, CoRR, arXiv:1612.01452, arXiv:1612.01452, http://arxiv.org/abs/1612.01452.

[23] A. Canziani, A. Paszke, E. Culurciello, An analysis of deep neural network models for practical applications, CoRR, arXiv:1605.07678, arXiv:1605.07678, http://arxiv.org/abs/1605.07678.

[24] O. Ronneberger, P. Fischer, T. Brox, U-net: convolutional networks for biomedical image segmentation, CoRR, arXiv:1505.04597, arXiv:1505.04597, http://arxiv.org/abs/1505.04597.

[25] L. Chen, G. Papandreou, I. Kokkinos, K. Murphy, A.L. Yuille, Deeplab: semantic image segmentation with deep convolutional nets, atrous convolution, and fully connected crfs, IEEE Transactions on Pattern Analysis and Machine Intelligence 40 (4) (2018) 834–848, https://doi.org/10.1109/TPAMI.2017.2699184.

[26] C. Hazirbas, L. Ma, C. Domokos, D. Cremers, Fusenet: incorporating depth into semantic segmentation via fusion-based cnn architecture, in: Asian Conference on Computer Vision, 2016.

[27] I. Armeni, A. Sax, A.R. Zamir, S. Savarese, Joint 2D-3D-semantic data for indoor scene understanding, arXiv e-prints, arXiv:1702.01105.

[28] ISPRS 2D semantic labeling – vaihingen data, http://www2.isprs.org/commissions/comm3/wg4/2d-sem-label-vaihingen.html. (Accessed 24 August 2018).

[29] P.K. Nathan Silberman, Derek Hoiem, R. Fergus, Indoor segmentation and support inference from rgbd images, in: ECCV, 2012.

[30] S. Song, S.P. Lichtenberg, J. Xiao, Sun rgb-d: a rgb-d scene understanding benchmark suite, in: 2015 IEEE Conference on Computer Vision and Pattern Recognition, CVPR, 2015, pp. 567–576.

[31] D. Garcia, Robust smoothing of gridded data in one and higher dimensions with missing values, Computational Statistics & Data Analysis 54 (4) (2010) 1167–1178.

[32] G. Wang, D. Garcia, Y. Liu, R. De Jeu, A.J. Dolman, A three-dimensional gap filling method for large geophysical datasets: application to global satellite soil moisture observations, Environmental Modelling & Software 30 (2012) 139–142.

CHAPTER 4

Learning Convolutional Neural Networks for Object Detection with Very Little Training Data

Christoph Reinders*, **Hanno Ackermann***, **Michael Ying Yang**†, **Bodo Rosenhahn***

*Institute for Information Processing, Leibniz University Hanover, Hanover, Germany †Scene Understanding Group, University of Twente, Enschede, The Netherlands

Contents

Multimodal Scene Understanding
https://doi.org/10.1016/B978-0-12-817358-9.00010-X

4.1 Introduction

Cycling as a mode of transport has attracted growing interest. Cities are transforming urban transportation to improve their infrastructure. Amsterdam and Copenhagen, for example, are pioneers for cycling-friendly cities. While current development shows more and more infrastructure improvements, road conditions can vary greatly. Cyclists are frequently confronted with challenges such as absence of bicycle lanes, being overlooked by cars, or bad roads. Arising safety concerns represent a barrier for using bicycles. Thus, recommending fast and safe routes for cyclists has great potential in terms of environmental and mobility aspects. This, in turn, requires detailed information about roads and traffic regulations.

For cars, precise information has become available. Google, for example, started the Google Street View project in which data is captured by many cars. These are equipped with stereo cameras which already offer a good 3D estimation in a certain range, lidar, and other sensors. Additionally the cars provide computational power as well as power supply. In research, popular datasets like GTSRB [1], KITTI [2], and Cityscapes [3] have been published.

In recent years, users are increasingly involved in the data collection. Crowdsourcing data enables to create large amount of real-world datasets. For example, the smart phone app Waze collects data such as GPS-position and speed from multiple users to predict traffic jams. OpenStreetMap aims to build a freely available map of the world to which users can easily contribute.

Machine learning techniques have shown great success for analyzing this data. Most supervised methods especially convolutional neural networks, however, require large datasets of labeled data. While large datasets have been published regarding cars, for cyclists very few labeled data are available although appearance, point of view, and positioning of even relevant objects differ. Unfortunately, labeling data is costly and requires a huge amount of work.

Our aim is to collect information which is of interest to cyclists. Analyzing street data for cyclists cannot be straightforwardly done by using data captured for cars due to different perspectives, different street signs, and routes prohibited for cars but not for bicycles, as shown

Figure 4.1 : Real-world data has great potential to provide traffic information that is of interest to cyclists. For example, roads that are prohibited for cars but free for cyclists (left), bicycle lines in parks (middle), or bicycle boulevards which are optimized for cyclists (right). All three examples are recognized by our system.

in Fig. 4.1. For collecting real-world data, we involve users by using smart phones that are attached to their bicycles. Compared to other systems, like for example Google Street View, our recording system consists of a single consumer camera and can only rely on a limited power supply and little computational power. On the other hand, our system has very low hardware costs and is highly scalable so that crowdsourcing becomes possible.

Although capturing data becomes easy with this system, generating labels is still very expensive. Thus, in this chapter we further address the problem of learning with extremely little labeled data to recognize traffic signs relevant for cyclists. We combine multiple machine learning techniques to create a system for object detection. Convolutional neural networks (CNNs) have shown to learn strong feature representations. On the other hand, random forests (RFs) achieve very good results in regression and classification tasks even when little labeled data is available. To combine both advantages we generate a feature extractor using a CNN and train a random forest based on the features. We map the random forest to a neural network and transform the full pipeline into a fully convolutional network. Thus, due to the shared features, processing full images is significantly accelerated. The resulting probability map is used to perform object detection. In a next step, we integrate information of a GPS-sensor to localize the detections on the map.

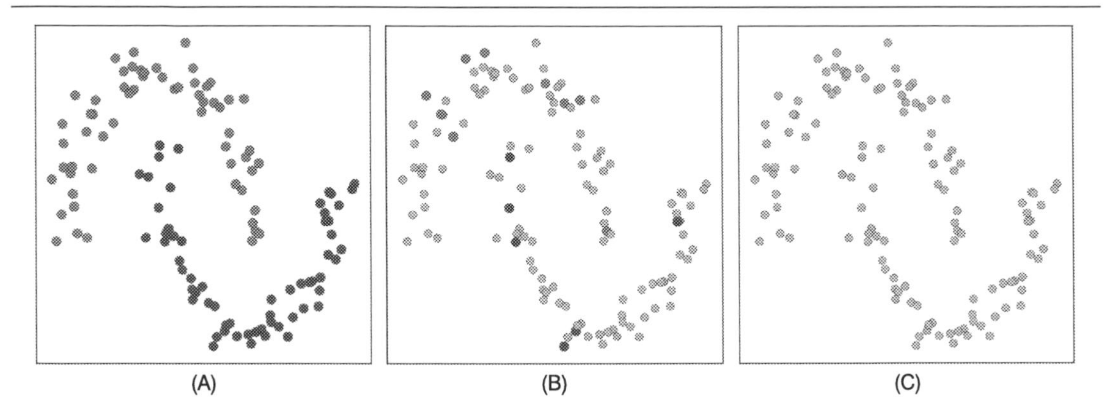

Figure 4.2 : Supervised learning methods (A) are trained on input–target pairs to classify the data points into the classes red and blue. For semi-supervised learning methods (B) only a few labeled data points are given additional to a large amount of unlabeled data points. Unsupervised learning methods (C) process the training data without any labels and try to find structures in the data.

This chapter further extends our previous work [4]. We increased the traffic sign dataset for training and testing from 297 to 524 examples. Additionally, the feature generating CNN is improved and a larger network is used. Finally, the GPS-sensors from the smart phones have shown to be not very precise. To improve the localization accuracy, we added a clustering process which identifies and merges multiple observations of the same traffic sign.

4.2 Fundamentals

In this section, the fundamental concepts used throughout this chapter are presented. At first, a short overview over different types of learning is given. In the second section, the origins of neural networks and convolutional neural networks are presented as well as a brief introduction to the so-called *back-propagation* algorithm for training neural networks. In the last section, random forests which consist of multiple decision trees are explained.

4.2.1 Types of Learning

Machine learning algorithms can be broadly divided into supervised learning, semi-supervised learning, and unsupervised learning [5, pp. 102–105]. The division depends on the training data that is provided during the learning process. In this section the different types of learning algorithms are briefly presented. An example of training data for each method is illustrated in Fig. 4.2.

Supervised learning. In supervised learning, labeled data is provided to the learning process meaning that each example is annotated with the desired target output. The dataset for training consists of N input–target pairs,

$$X = \{(x^{(1)}, y^{(1)}), (x^{(2)}, y^{(2)}), \ldots, (x^{(N)}, y^{(N)})\}.$$

Each training pair consists of input data $x^{(i)}$ together with the corresponding target $y^{(i)}$. The goal of the algorithm is to learn a mapping from the input data to the target value so that the target y^* for some unseen data x^* can be predicted. Supervised learning can be thought of as a teacher that evaluates the performance and identifies errors during training to improve the algorithm.

Common tasks in supervised learning are classification and regression. In classification, the target is a discrete value which represents a category such as "red" and "blue" (see Fig. 4.2A). Another example is the classification of images to predict the object that is shown in the image such as for instance "car", "plane", or "bicycle". In regression, the target is a real value such as "dollars" or "length". A regression task is for instance the prediction of house prices based on data of the properties such as location or size.

Popular examples of supervised learning methods are random forests, support vector machines, and neural networks.

Semi-supervised learning. Semi-supervised learning is a mixture between supervised learning and unsupervised learning. Typically, supervised learning methods require large amounts of labeled data. Whereas the collection of data is often cheap, labeling data can usually only be achieved at enormous costs because experts have to annotate the data manually. Semi-supervised learning combines supervised learning and the usage of unlabeled data. For that, the dataset consists of N labeled examples,

$$X_l = \{(x^{(1)}, y^{(1)}), (x^{(2)}, y^{(2)}), \ldots, (x^{(N)}, y^{(N)})\},$$

together with M unlabeled examples,

$$X_u = \{x^{(N+1)}, x^{(N+2)}, \ldots, x^{(N+M)}\}.$$

Usually, the number of labeled examples N is much smaller than the number of unlabeled examples M. Semi-supervised learning is illustrated in Fig. 4.2B for the classification of data points.

Unsupervised learning. In unsupervised learning, unlabeled data is provided to the learning process. The aim of the learning algorithm is to find structures or relationships in the data without having information about the target. The training set consists of N training samples,

$$X = \{x^{(1)}, x^{(2)}, \ldots, x^{(N)}\}.$$

Popular unsupervised learning methods are for example k-means and mean shift. For unsupervised learning, usually, a distribution of the data has to be defined. K-means assumes that the data is convex and isotropic and requires a predefined number of classes. For mean shift a kernel along with a bandwidth is defined such as for example a Gaussian kernel. Typical tasks in unsupervised learning are density estimation, clustering, and dimensionality reduction. An example of a clustering task is shown in Fig. 4.2C.

4.2.2 Convolutional Neural Networks

Convolutional neural networks have shown great success in recent years. Especially since Alex Krizhevsky, Ilya Sutskever, and Geoffrey Hinton won the ImageNet Large-Scale Visual Recognition Challenge in 2012 the topic of research has attracted much attention [6]. In this section an overview of convolutional neural networks is given. First, the origins of neural networks are described and afterwards the learning process *back-propagation* is explained. Finally, convolutional neural networks are presented.

4.2.2.1 Artificial neuron

Neural networks have been biologically inspired by the brain. The brain is a very efficient information-processing system which consists of single units, so-called *neurons*, which are highly connected with each other [7, pp. 27–30].

The first model of an artificial neuron was presented by Warren McCulloch and Walter Pitts in 1943 [8] and was called the *McCulloch–Pitts model*. The scientists drew a connection between the biological neuron and logical gates and developed a simplified mathematical description of the biological model. In 1957, Frank Rosenblatt developed the so-called *perceptron* [9], which is based on the work of McCulloch and Pitts. The model has been further generalized to a so-called *artificial neuron*. An artificial neuron has N inputs and one output a, as illustrated in Fig. 4.3. Each input is multiplied by a weight w_i. Afterwards all weighted inputs are added up together with a bias b, which gives the weighted sum z:

$$z = \sum_{i=1}^{N} x_i \cdot w_i + b. \tag{4.1}$$

Finally, the output is calculated using an activation function $\phi(\cdot)$:

$$a = \phi(z). \tag{4.2}$$

Early models such as the perceptron often used a step function $\mathrm{g}(x)$ as activation function where g is defined as

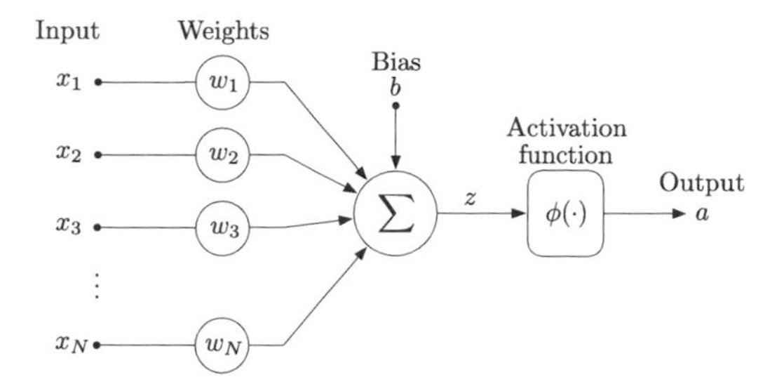

Figure 4.3 : Model of an artificial neuron. The inputs are weighted and added together with the bias. Afterwards, the weighted sum is passed to an activation function to calculate the output.

$$g(x) = \begin{cases} 1 & x \geq 0, \\ 0 & \text{otherwise.} \end{cases} \tag{4.3}$$

Another activation function that is commonly used is the sigmoid function,

$$\text{sig}(x) = \frac{1}{1+e^{-x}}. \tag{4.4}$$

Different from the step function, the sigmoid function is differentiable. This is an important property for training as shown later in this chapter. In recent years, another activation function, called the *rectified linear unit (ReLU)* became popular, which has been proposed by Krizhevsky et al. [6]. ReLU outputs 0 if the input value x is smaller than zero and otherwise the input value x:

$$\text{ReLU}(x) = \max(0, x). \tag{4.5}$$

Compared with the sigmoid function, ReLU is non-saturating for positive inputs and has shown to accelerate the training process.

This model of an artificial neuron can be used to classify input data into two classes. However, this only works for data which is linearly separable. To calculate more complex functions, multiple neurons are connected as presented in the next section.

4.2.2.2 Artificial neural network

Neural networks are arranged in layers. Each network has an input layer, one or more hidden layers, and an output layer. An example of a two-hidden-layer network is presented in Fig. 4.4. The design of the input and output layer is often straightforward compared to the design of the hidden layers. The input layer is the first layer in the network. Its so-called input neurons store the input data and perform no computation. If the input is for example an image of size 32×32 pixels, the number of input neurons is $1024 = 32 \cdot 32$. The output layer is the

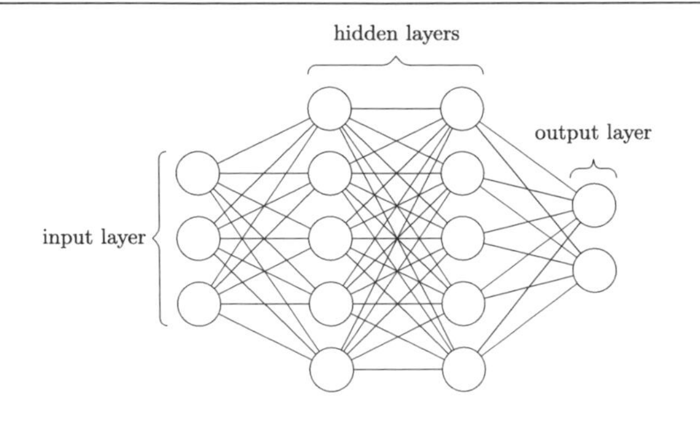

Figure 4.4 : A multilayer perceptron with two hidden layers.

last layer in the network and contains the output neurons. For example, if a network classifies an image into one out of ten classes, the output layer contains one neuron for each class, indicating the probability that the input belongs to this class.

Formally, a neural network has L layers where the first layer is the input layer and the Lth layer is the output layer. Neural networks with multiple layers are also called *multilayer perceptrons*. They are one of the simplest types of feed-forward neural networks, which means that the connected neurons build an acyclic directed graph.

Each layer l has n_l neurons and a neuron is connected to all neurons in the previous layer. The weight between the kth neuron in the $(l-1)$th layer and the jth neuron in the lth layer is denoted by $w^l_{k,j}$. Similarly, the bias of the jth neuron in the lth layer is denoted by b^l_j. Similar to Eq. (4.1) the weighted sum z^l_j of a neuron is calculated by

$$z^l_j = \sum_{k=1}^{n_{l-1}} a^{l-1}_k \cdot w^l_{k,j} + b^l_j. \tag{4.6}$$

Afterwards the activation function is used to calculate the output a^l_j of a neuron

$$a^l_j = \phi(z^l_j). \tag{4.7}$$

To simplify the notation, the formulas can be written in matrix form. For each layer, a bias vector b^l, a weighted sum vector z^l, and an output vector a^l are defined:

$$b^l = \begin{bmatrix} b^l_1 & b^l_2 & \dots & b^l_{n_l} \end{bmatrix}^{\mathrm{T}}, \tag{4.8}$$

$$z^l = \begin{bmatrix} z^l_1 & z^l_2 & \dots & z^l_{n_l} \end{bmatrix}^{\mathrm{T}}, \tag{4.9}$$

$$a^l = \begin{bmatrix} a^l_1 & a^l_2 & \dots & a^l_{n_l} \end{bmatrix}^{\mathrm{T}}. \tag{4.10}$$

The weights for each layer can be expressed by a matrix $w^l \in \mathbb{R}^{n_l \times n_{l-1}}$:

$$w^l = \begin{bmatrix} w^l_{1,1} & \dots & w^l_{n_{l-1},1} \\ \vdots & \ddots & \vdots \\ w^l_{1,n_l} & \dots & w^l_{n_{l-1},n_l} \end{bmatrix}. \tag{4.11}$$

The matrix has n_l rows and n_{l-1} columns which corresponds to the number of neurons in layer l and layer $(l-1)$. The kth row of w^l contains all weights of the kth neuron in layer l connecting the neuron to all neurons in layer $(l-1)$. Using the matrix form, the weighted sum for each layer can be calculated by multiplying the weight matrix with the output of the previous layer and adding the biases:

$$z^l = w^l a^{l-1} + b^l. \tag{4.12}$$

Applying the activation to each element of the weighted sum vector, the output of each layer is calculated by

$$a^l = \phi(z^l). \tag{4.13}$$

A neural network takes some input data x and calculates the output of the network $\mathrm{N}(x)$, which is defined as the output of the last layer $\mathrm{N}(x) = a^L(x)$. This is done by processing the network layer by layer. First of all the input data is passed to the input neurons $a^1(x) = x$. Afterwards all layers are processed by calculating the weighted sum and the output. Finally, the output of the network $a^L(x)$ is calculated.

4.2.2.3 Training

The goal of the training process is to automatically learn the weights and biases. Finding the optimal configuration can be challenging especially in larger networks which have thousands or millions of parameters. Whereas the McCulloch–Pitts model used fixed weights, Rosenblatt proposed a learning rule for adjusting the weights of a perceptron. In the 1970s and 1980s a much faster algorithm called *back-propagation* has been developed. Various researchers have worked towards similar ideas, including Werbos [10] and Parker [11]. The back-propagation algorithm has been popularized by a work of Rumelhart, Hinton, and Williams in 1986 [12].

Cost function. In order to learn the weights and biases, a training set of input vectors $x^{(i)}$ together with a corresponding set of target vectors $y^{(i)}$ are provided. During a *forward pass* the output $a^L(x)$ of a network is calculated. To quantify the error made by a network a loss function C is introduced. A loss function that is often used is the *Euclidean loss*. It is defined as the sum over the squared differences between the target vector and the output vector

$$\mathrm{C} = \frac{1}{2}\left\| y^{(i)} - a^L(x^{(i)}) \right\|^2 = \frac{1}{2}\sum_j \left(y_j^{(i)} - a_j^L(x^{(i)}) \right)^2. \tag{4.14}$$

Back-propagation. The objective of the training process is to adjust the weights and biases so that the loss is minimized. To understand how changes of the weights and biases change the cost function, let Δz_j^l be a small change that is added to the weighted sum of the jth neuron in the lth layer. Instead of $\phi(z_j^l)$ the neuron outputs $\phi(z_j^l + \Delta z_j^l)$, which is propagated through the network and leads to an overall change of the cost function by $\frac{\partial \mathrm{C}}{\partial z_j^l}\Delta z_j^l$. If the value of $\partial \mathrm{C}/\partial z_j^l$ is large, Δz_j^l can be chosen to have opposite sign so that the loss is reduced. If the value of $\partial \mathrm{C}/\partial z_j^l$ is small, Δz_j^l cannot improve the loss and the neuron is assumed to be nearly optimal. Let δ_j^l be defined as the error by the jth neuron in the lth layer,

$$\delta_j^l = \frac{\partial \mathrm{C}}{\partial z_j^l}. \tag{4.15}$$

Back-propagation is based on four fundamental equations. First of all the error of each neuron in the last layer is calculated:

$$\delta^L = \nabla_a \mathrm{C} \odot \phi'(z^L). \tag{4.16}$$

Afterwards the error is propagated backwards through the network so that step by step the error of the neurons in the $(l+1)$th layer is propagated to the neurons in the lth layer:

$$\delta^l = ((w^{l+1})^{\mathrm{T}} \delta^{l+1}) \odot \phi'(z^l). \tag{4.17}$$

The error can be used to calculate the partial derivatives $\partial \mathrm{C}/\partial w_{k,j}^l$ and $\partial \mathrm{C}/\partial b_j^l$ as follows:

$$\frac{\partial \mathrm{C}}{\partial w_{k,j}^l} = \delta_j^l a_k^{l-1}, \tag{4.18}$$

$$\frac{\partial \mathrm{C}}{\partial b_j^l} = \delta_j^l, \tag{4.19}$$

which indicate how a change of a weight or bias influences the loss function. Finally, the weights and biases are adjusted by subtracting the corresponding partial derivative scaled with a learning rate α from the current value,

$$\hat{w}_{k,j}^l = w_{k,j}^l - \alpha \cdot \frac{\partial \mathrm{C}}{\partial w_{k,j}^l}, \tag{4.20}$$

$$\hat{b}^l_j = b^l_j - \alpha \cdot \frac{\partial C}{\partial b^l_j}. \tag{4.21}$$

As a result, the weights and biases are optimized iteratively to minimize the loss function.

4.2.2.4 Convolutional neural networks

Neural networks as explained in the last section are also called *fully-connected neural networks*. Every neuron is connected to every neuron in the previous layer. For a color image of size 32×32 pixels with three color channels, this means that each neuron in the first hidden layer has $3072 = 32 \cdot 32 \cdot 3$ weights. Although at first glance acceptable, for large images the number of variables is extremely large. For example, a color image of size 200×200 pixels with three color channels already requires $120\,000 = 200 \cdot 200 \cdot 3$ weights for each neuron in the first hidden layer. Additionally, fully-connected neural networks are also not taking the spatial structure into account. Neurons that are far away from another are treated equally to neurons that are close together.

Convolutional neural networks are designed to take advantage of the two-dimensional structure of an input image. Instead of learning fully-connected neurons, filters are learned that are convolved over the input image. The idea has been inspired by the visual cortex. Hubel and Wiesel [13] showed in 1962 that some neural cells are sensitive to small regions in the visual field and respond to specific features. In 1998, convolutional neural networks were popularized by Lecun, Botto, Bengio, and Haffner [14].

The first difference between regular neural networks and convolutional neural networks is the arrangement of the data. In order to take into account that images or other volumetric data are processed, the data in convolutional neural networks is arranged in 3D volumes. Each data blob in a layer has three dimensions: height, width, and depth. Each layer receives an input 3D volume and transforms it to an output 3D volume.

Convolutional layer. Whereas fully-connected neurons are connect to every neuron in the previous layer, neurons in convolutional layers will be connected to only a small region of the input data. This region is called *local receptive field* for a hidden neuron. Instead of applying different weights and biases for every local receptive field, the neurons are using shared weights and biases. The idea behind this is that a feature that is learned at one position might also be useful at a different position. Additionally, the number of parameters will decrease significantly. The shared weights and biases are defined as *filter*.

The size of the filters is usually small in height and width whereas the depth always equals the depth of the input volume. For example, a filter of size $3 \times 5 \times 5$ is often used in the first hidden layer, i.e. 3 pixels in depth, because the input image has three color channels, 5 pixels in height, and 5 pixels in width. In general, a filter has depth C_{in}, height K_{h}, and width K_{w}, where C_{in} is the depth of the input volume. This results in $C_{\text{in}} \cdot K_{\text{h}} \cdot K_{\text{w}}$ weights plus one bias per filter.

Convolutional layers consist of a set of K filters. Each filter is represented as a three-dimensional matrix W_i of size $C_{\text{in}} \times K_{\text{h}} \times K_{\text{w}}$ and a bias b_i. To calculate the weighted sum Z of a layer, each filter is slided across the width and height of the input volume. Therefore the dot product between the weights of the filter W_i and the input volume I at any position is computed:

$$Z[i, y, x] = \sum_{c=0}^{C_{\text{in}}-1} \sum_{l=0}^{K_{\text{h}}-1} \sum_{m=0}^{K_{\text{w}}-1} W_i[c, l, m] I[c, y+l, x+m] + b_i \tag{4.22}$$

where i denotes the filter index and x, y the spatial position. Each filter produces a two-dimensional feature map. These feature maps are stacked along the depth and generate a three-dimensional volume Z. Afterwards the activation output A is calculated by applying the activation function to each element of the matrix Z,

$$A[i, y, x] = \phi(Z[i, y, x]). \tag{4.23}$$

Each convolutional layer transforms an input volume of size $C_{\text{in}} \times H_{\text{in}} \times W_{\text{in}}$ to an output volume of size $C_{\text{out}} \times H_{\text{out}} \times W_{\text{out}}$. Additional to the number of filters K and the filter size $K_{\text{h}} \times K_{\text{w}}$, further parameters of a convolutional layer are the *stride* S and *padding* P. The stride S defines the number of pixels a filter is moved when sliding a filter over the input volume. While $S = 1$ refers to the standard convolution, $S > 1$ will skip some pixels and lead to a smaller output size. The padding P adds additionally P pixels to the border of the input volume. For example, it can be used to create an output volume that has the same size as the input volume. In general, because the feature maps are stacked, the output depth C_{out} equals the number of filters K that are used. The width and height of the output volume can be calculated by

$$W_{\text{out}} = (W_{\text{in}} - K_{\text{w}} + 2P)/S + 1, \tag{4.24}$$

$$H_{\text{out}} = (H_{\text{in}} - K_{\text{h}} + 2P)/S + 1. \tag{4.25}$$

Pooling layer. Another layer that is commonly used is the pooling layer. Usually, it is inserted directly after a convolutional layer. Pooling operates independently on every depth slice and applies a filter of size $K \times K$ to summarize the information. Similar to convolutional layers, a stride S controls the amount of pixels the filter is moved. Most frequently *max pooling* is used. As shown in Fig. 4.5, max pooling takes the maximum over each region and outputs

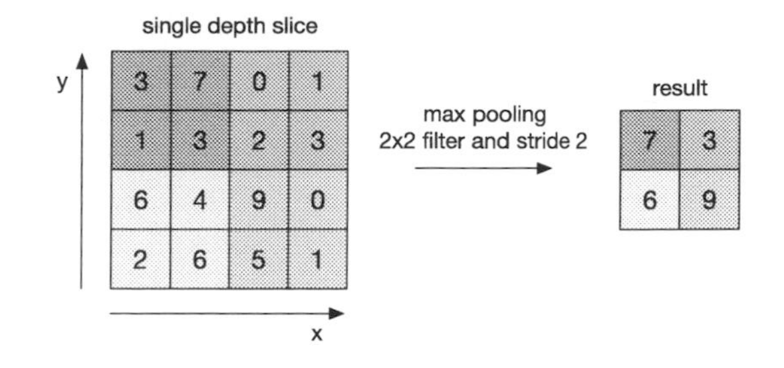

Figure 4.5 : Max pooling layer. Each depth slice is processed by taking the maximum of each 2 x 2 window.

a smaller feature map. For example, a filter size of 2×2 is commonly used which reduces the amount of information by a factor of 4. A pooling layer keeps the information that a feature has found but leaves out the exact position. By summarizing the features the spatial size is decreased. Thus, fewer parameters are needed in later layers which reduces the amount of memory that is required and the computation time. Other types of pooling functions are *average pooling* and *L2-norm pooling*.

Dropout layer. A common problem when using neuron networks is overfitting. Because of too many parameters, a network might learn to memorize the training data and does not generalize to unseen data. *Dropout* is a regularization technique to reduce overfitting [15]. A *dropout-ratio* p is defined and during training each neuron is temporarily removed with a probability p and kept with a probability $p - 1$, respectively. An example is illustrated in Fig. 4.6. This process is repeated so that in each iteration different networks are used. In general, dropout has shown to improve the performance of networks.

4.2.3 Random Forests

The random forest algorithm is a supervised learning algorithm for classification and regression. A random forest is an ensemble method that consists of multiple decision trees. The first work goes back to Ho [16] in 1995 who introduced random decision forests. Breiman [17] further developed the idea and presented the random forest algorithm in 2001.

4.2.3.1 Decision tree

A decision tree consists of *split nodes* $\mathcal{N}^{\text{Split}}$ and *leaf nodes* $\mathcal{N}^{\text{Leaf}}$. An example is illustrated in Fig. 4.7. Each split node $s \in \mathcal{N}^{\text{Split}}$ performs a split decision and routes a data sample x to the left child node $\text{cl}(s)$ or to the right child node $\text{cr}(s)$. When using axis-aligned split decisions the split rule is based on a single split feature $\text{f}(s)$ and a threshold value $\theta(s)$:

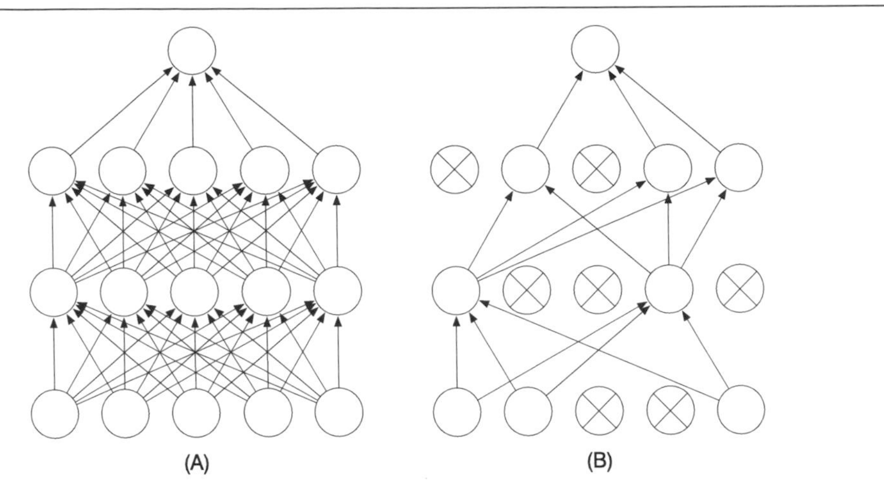

Figure 4.6 : Dropout temporarily removes neurons so that with each iteration different network structures are trained. (A) Standard neural network. (B) After applying dropout.

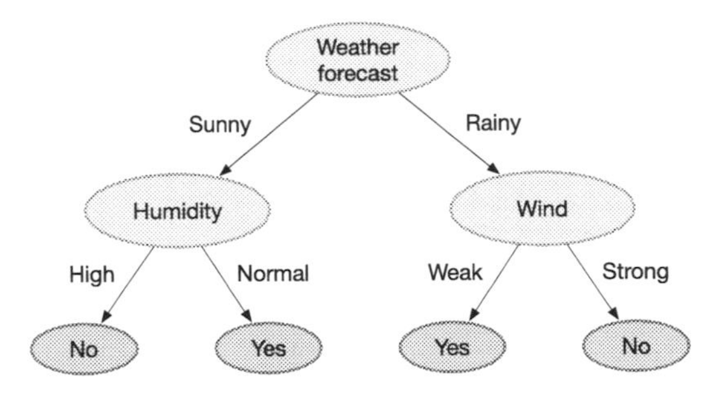

Figure 4.7 : An example of a decision tree that predicts whether or not to go hiking today. Split nodes (green) evaluate the data and route to the next node. Leaf nodes (blue) contain the possible outputs of the decision tree. Starting at the root node, the data is routed through the tree based on the split rules. Finally, a leaf node is reached which contains the decision output. In this example the output is "yes" or "no".

$$x \in \mathrm{cl}(s) \iff x_{\mathrm{f}(s)} < \theta(s), \tag{4.26}$$

$$x \in \mathrm{cr}(s) \iff x_{\mathrm{f}(s)} \geq \theta(s). \tag{4.27}$$

The data sample x is routed to the left child node if the value of feature $f(s)$ of x is smaller than a threshold $\theta(s)$ and to the right child node otherwise. All leaf nodes $l \in \mathcal{N}^{\mathrm{Leaf}}$ store votes for the classes $y^l = (y_1^l, \ldots, y_C^l)$, where C is the number of classes.

Decision trees are grown using training data. Starting at the root node, the data is recursively split into subsets. In each step the best split is determined based on a criterion. Commonly used criteria are *Gini index* and *entropy*:

$$\textit{Gini index:}\ \mathrm{G}(E) = 1 - \sum_{j=1}^{C} p_j^2, \tag{4.28}$$

$$\textit{entropy:}\ \mathrm{H}(E) = -\sum_{j=1}^{C} p_j \log p_j. \tag{4.29}$$

The algorithm for constructing a decision tree works as follows:

1. Randomly sample n training samples with replacement from the training dataset.
2. Create a root node and assign the sampled data to it.
3. Repeat the following steps for each node until all nodes consists of a single sample or samples of the same class:
   a. Randomly select m variables out of M possible variables.
   b. Pick the best split feature and threshold according to a criterion, for example Gini index or entropy.
   c. Split the node into two child nodes and pass the corresponding subsets.

For images, the raw image data is usually not used directly as input for the decision trees. Instead, features such as for instance HOG features or SIFT features are calculated for a full image or image patch. This additional step represents an essential difference between decision trees and convolutional neural networks. Convolutional neural networks in comparison are able to learn features automatically.

4.2.3.2 Random forest

The prediction with decision trees is very fast and operates on high-dimensional data. On the other hand, a single decision tree has overfitting problems as a tree grows deeper and deeper until the data is separated. This will reduce the training error but potentially results in a larger test error.

Random forests address this issue by constructing multiple decision trees. Each decision tree uses a randomly selected subset of training data and features. The output is calculated by averaging the individual decision tree predictions. As a result, random forests are still fast and additionally very robust to overfitting. An example of a decision tree and a random forest is shown in Fig. 4.8.

4.3 Related Work

In recent years, convolutional neural networks have become the dominant approach for many vision-based tasks such as object detection and scene analysis [18,19]. Girshick et al. [20]

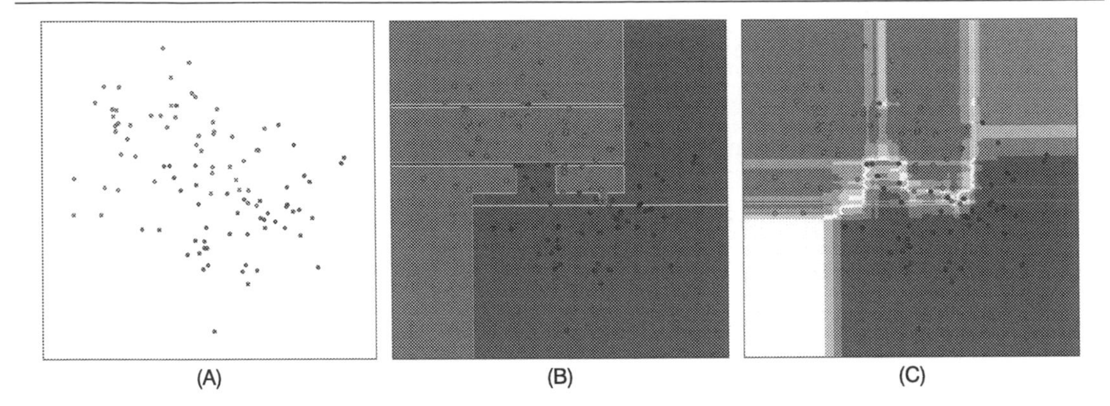

Figure 4.8 : A decision tree (B) and a random forest (C) are trained to classify the data points in (A). The decision boundaries are shown in (B) and (C). The dark red and blue colors indicate areas which are clearly classified as red and blue points, respectively. The random forest additionally models the uncertainty which is indicated by a color between red and blue.

proposed a multistage pipeline called regions with convolutional neural networks (R-CNNs) for the classification of region proposals to detect objects. It achieves good results but the pipeline is less efficient because features of each region proposal need be computed repeatedly. In the SPP-net [21], this problem has been addressed by introducing a pooling strategy to calculate the feature map only once and generate features in arbitrary regions. Fast R-CNN [22] further improves the speed and accuracy by combining multiple stages. A drawback of these algorithms is their large dependence on the region proposal method. Faster R-CNN [23] combines the region proposal mechanism and a CNN classifier within a single network by introducing a region proposal network. Due to shared convolutions, region proposals are generated at nearly no extra cost. Other networks such as SSD [24], YOLO [25], and RetinaNet [26] directly regress bounding boxes without generating object proposals in an end-to-end network. These one-stage detectors are extremely fast but come with some compromise in detection accuracy in general. Overall, the one-stage and two-stage convolutional neural networks for object detection perform very well. However, they typically consist of millions of variables and for estimating those, a large amount of labeled data is required for training.

Feature learning and transferring techniques have been applied to reduce the required amount of labeled data [27]. The problem of insufficient training data has also been addressed by other work such as in [28] and [29]. Moysset et al. [28] proposed a new model that predicts the bounding boxes directly. Wagner et al. [29] compared unsupervised feature learning methods and demonstrated performance boosts by pre-training. Although transfer learning techniques are applied, the networks still have a large number of variables for fine-tuning.

A different approach is the combination of random forests and neural networks. Deep neural decision forests [30] unifies both in a single system that is trained end-to-end. Sethi [31] and Welbl [32] presented a mapping of random forests to neural networks. The mapping can be used for several applications. Massiceti et al. [33] demonstrated the application for camera localization. Richmond et al. [34] explored the mapping of stacked RFs to CNNs and an approximate mapping back to perform semantic segmentation.

4.4 Traffic Sign Detection

In this section, we present a system for detecting traffic signs. To overcome the problem of lack of data, we first build a classifier that predicts the class probabilities of a single image patch. This is done in two steps. First, we train a CNN on a different dataset where a large amount of data is available. Afterwards we use the generated features, extract the feature vectors, and train a random forest. The resulting classifier can be used to perform patch-wise prediction and to build a probability map for a given full image. Subsequently, all traffic signs are extracted and the detection system outputs the class and the corresponding bounding box.

Finally, the processing of full images is accelerated. By mapping the random forest to a neural network, it becomes possible to combine feature generation and classification. Afterwards we transform the neural network to a fully convolutional network.

4.4.1 Feature Learning

We learn features by training a convolutional neural network CNN^{F}. The patch size is 32×32. We adopt the network architecture of Springenberg et al. [35], which yields good results on datasets like CIFAR-10, CIFAR-100, and ImageNet. The model *ALL-CONV-C* performed best and is used in this work. The network has a simple regular structure consisting of convolution layers only. Instead of pooling layers convolutional layers with stride of two are used. Additionally, the fully-connected layers that are usually at the end of a convolutional neural network, are replaced by 1×1 convolutional layers followed by an average pooling layer. Thus the output of the last layer has a spatial size of 1×1 and a depth C, where C equals the number of classes.

Because we have only very little labeled data available, we train the network on the larger dataset GTSRB [1]. After training, the resulting network CNN^{F} can be used to generate feature vectors by passing an input image to the network and performing a forward pass. The feature vectors can be extracted from the last convolutional layer or the last convolutional layer before the 1×1 convolutional layers, respectively. In our network this corresponds to the seventh convolutional layer, denoted by $\text{CNN}^{\text{F}}_{relu7}(x)$.

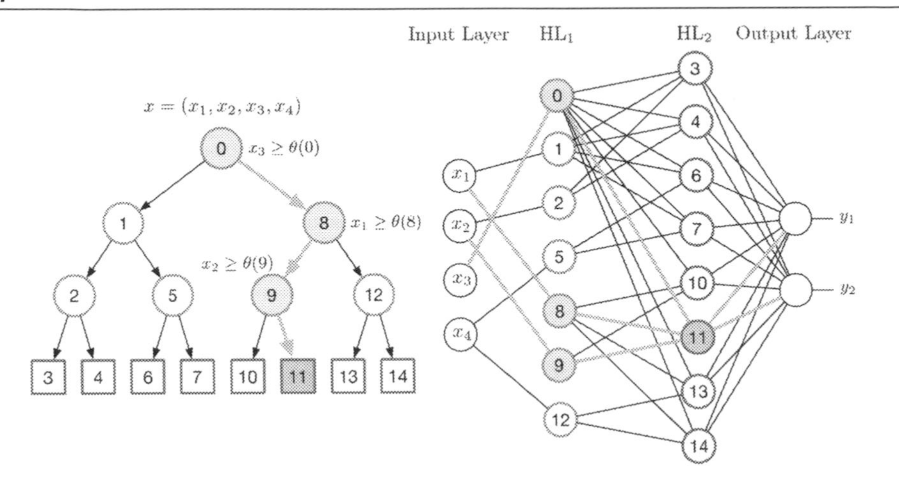

Figure 4.9 : A decision tree (left) and the mapped neural network (right). Each *split node* in the tree – indicated as circle – creates a neuron in the first hidden layer which evaluates the split rule. Each *leaf node*—indicated as rectangle—creates a neuron in the second hidden layer which determines the leaf membership. For example, a routing to leaf node 11 involves the split nodes $(0, 8, 9)$. The relevant connections for the corresponding calculation in the neural network are highlighted.

4.4.2 Random Forest Classification

Usually, neural networks perform very good in classification. However, if the data is limited, the large amount of parameters to be trained causes overfitting. Random forests [17] have shown to be robust classifiers even if few data are available. A random forest consists of multiple decision trees. Each decision tree uses a randomly selected subset of features and training data. The output is calculated by averaging the individual decision tree predictions.

After creating a feature generator, we calculate the feature vector $f^{(i)} = \mathrm{CNN}^{\mathrm{F}}_{relu7}(x^{(i)})$ for every input vector $x^{(i)}$. Based on the feature vectors we train a random forest that predicts the target values $y^{(i)}$. By combining the feature generator $\mathrm{CNN}^{\mathrm{F}}$ and the random forest, we construct a classifier that predicts the class probabilities for an image patch. This classifier can be used to process a full input image patch-wisely. Calculating the class probabilities for each image patch produces an output probability map.

4.4.3 RF to NN Mapping

Here, we present a method for mapping random forests to two-hidden-layer neural networks introduced by Sethi [31] and Welbl [32]. The mapping is illustrated in Fig. 4.9. Decision trees have been introduced in Sect. 4.2.3.1. A decision tree consists of *split nodes* $\mathcal{N}^{\mathrm{Split}}$ and *leaf nodes* $\mathcal{N}^{\text{Leaf}}$. Each split node $s \in \mathcal{N}^{\text{Split}}$ evaluates a split decision and routes a data sample x to the left child node $\text{cl}(s)$ or to the right child node $\text{cr}(s)$ based on a split feature $f(s)$ and a threshold value $\theta(s)$:

$$x \in \text{cl}(s) \iff x_{f(s)} < \theta(s), \tag{4.30}$$

$$x \in \text{cr}(s) \iff x_{f(s)} \geq \theta(s). \tag{4.31}$$

The data sample x is routed to the left child node if the value of feature $f(s)$ of x is smaller than a threshold $\theta(s)$ and to the right child node otherwise. All leaf nodes $l \in \mathcal{N}^{\text{Leaf}}$ store votes for the classes $y^l = (y_1^l, \ldots, y_C^l)$, where C is the number of classes. For each leaf a unique path $\text{P}(l) = (s_0, \ldots, s_d)$ from root node s_0 to leaf l over a sequence of split nodes $\{s_i\}_{i=0}^d$ exists, with $l \subseteq s_d \subseteq \cdots \subseteq s_0$. By evaluating the split rules for each split node along the path $\text{P}(l)$ the leaf membership can be expressed as

$$x \in l \iff \forall s \in \text{P}(l) : \begin{cases} x_{f(s)} < \theta(s) & \text{if } l \in \text{cl}(s), \\ x_{f(s)} \geq \theta(s) & \text{if } l \in \text{cr}(s). \end{cases} \tag{4.32}$$

First hidden layer. The first hidden layer computes all split decisions. It is constructed by creating one neuron $\text{H}_1(s)$ per split node evaluating the split decision $x_{f(s)} \geq \theta(s)$. The activation output of the neuron should approximate the following function:

$$\text{a}(\text{H}_1(s)) = \begin{cases} -1, & \text{if } x_{f(s)} < \theta(s), \\ +1, & \text{if } x_{f(s)} \geq \theta(s), \end{cases} \tag{4.33}$$

where -1 encodes a routing to the left child node and $+1$ a routing to the right child node. Therefore the $f(s)$th neuron of the input layer is connected to $\text{H}_1(s)$ with weight $w_{f(s),\text{H}_1(s)} = \text{c}_{\text{split}}$, where c_{split} is a constant. The bias of $\text{H}_1(s)$ is set to $b_{\text{H}_1(s)} = -\text{c}_{\text{split}} \cdot \theta(s)$. All other weights and biases are zero. As a result, the neuron $\text{H}_1(s)$ calculates the weighted sum,

$$z_{\text{H}_1(s)} = \text{c}_{\text{split}} \cdot x_{f(s)} - \text{c}_{\text{split}} \cdot \theta(s), \tag{4.34}$$

which is smaller than zero when $x_{f(s)} < \theta(s)$ is fulfilled and greater than or equal to zero otherwise. The activation function $\text{a}(\cdot) = \tanh(\cdot)$ is used which maps the weighted sum to a value between -1 and $+1$ according to the routing. The constant c_{split} controls the sharpness of the transition from -1 to $+1$.

Second hidden layer. The second hidden layer combines the split decisions from layer H_1 to indicate the leaf membership $x \in l$. One leaf neuron $\text{H}_2(l)$ is created per leaf node. It is connected to all split neurons $\text{H}_1(s)$ along the path $s \in \text{P}(l)$ as follows:

$$w_{\text{H}_1(s),\text{H}_2(l)} = \begin{cases} -\text{c}_{\text{leaf}} & \text{if } l \in \text{cl}(s), \\ +\text{c}_{\text{leaf}} & \text{if } l \in \text{cr}(s), \end{cases} \tag{4.35}$$

where c_{leaf} is a constant. The weights are sign matched according to the routing directions, i.e. negative when l is in the left subtree from s and positive otherwise. Thus, the activation of $H_2(l)$ is maximized when all split decisions routing to l are satisfied. All other weights are zero. To encode the leaf to which a data sample x is routed, the bias is set to

$$b_{H_2(l)} = -c_{leaf} \cdot (|P(l)| - 1), \tag{4.36}$$

so that the weighted sum of neuron $H_2(l)$ will be greater than zero when all split decisions along the path are satisfied and less than zero otherwise. By using the activation function $a(\cdot) = sigmoid(\cdot)$, the active neuron $H_2(l)$ with $x \in l$ will map close to 1 and all other neurons close to 0. Similar to c_{split}, a large value for c_{leaf} approximates a step function, whereas smaller values can relax the tree hardness.

Output layer. The output layer contains one neuron $H_3(c)$ for each class and is fully-connected to the previous layer H_2. Each neuron $H_2(l)$ indicates whether $x \in l$. The corresponding leaf node l in the decision tree stores the class votes y_c^l for each class c. To transfer the voting system, the weights are set proportional to the class votes:

$$w_{H_2(l),H_3(c)} = c_{output} \cdot y_c^l, \tag{4.37}$$

where c_{output} is a scaling constant to normalize the votes as explained in the following section. All biases are set to zero.

Random forest. Extending the mapping to random forests with T decision trees is simply done by mapping each decision tree and concatenating the neurons of the constructed neural networks for each layer. The neurons for each class in the output layer are created only once. They are fully-connected to the previous layer and by setting the constant c_{output} to $1/T$ the outputs of all trees are averaged. We denote the resulting neural network as NN^{RF}. It should be noted that the memory size of the mapped neural network grows linearly with the total number of split and leaf nodes. A possible network splitting strategy for very large random forests has been presented by Massiceti et al. [33].

4.4.4 Fully Convolutional Network

Mapping the random forest to a neural network allows one to join the feature generator and the classifier. For that we remove the classification layers from CNN^F, i.e. all layers after *relu7*, and append all layers from NN^{RF}. The constructed network CNN^{F+RF} processes an image patch and outputs the class probabilities. The convolutional neural network CNN^{F+RF} is converted to a fully convolutional network CNN^{FCN} by converting the fully-connected layers into convolutional layers, similar to [36]. The fully convolutional network operates on input images of any size and produces corresponding (possibly scaled) output maps. Compared with patch-wise processing, the classifier is naturally slided over the image evaluating the class probabilities at any position. At the same time the features are shared so that features in overlapping patches can be reused. This decreases the amount of computation and significantly accelerates the processing of full images.

4.4.5 Bounding Box Prediction

The constructed fully convolutional network processes a color image $I \in \mathbb{R}^{W \times H \times 3}$ of size $W \times H$ with three color channels and produces an output $O = \text{CNN}^{\text{FCN}}(I)$ with $O \in \mathbb{R}^{W' \times H' \times C}$. The output consists of C-dimensional vectors at any position which indicate the probabilities for each class. Due to stride and padding parameters, the size of the output map can be decreased. To detect objects of different sizes, we process the input image in multiple scales $S = \{s_1, \ldots, s_m\}$.

We extract potential object bounding boxes by identifying all positions in the output maps where the probability is greater than a minimal threshold $t_{\text{min}} = 0.2$. We describe a bounding box by

$$b = (b_{\text{x}}, b_{\text{y}}, b_{\text{w}}, b_{\text{h}}, b_{\text{c}}, b_{\text{s}}), \tag{4.38}$$

where $(b_{\text{x}}, b_{\text{y}})$ is the position of the center, $b_{\text{w}} \times b_{\text{h}}$ the size, b_{c} the class, and b_{s} the score. The bounding box size corresponds to the field of view which is equal to the size of a single image patch. All values are scaled according to the scale factor. The score b_{s} is equal to the probability in the output map.

For determining the final bounding boxes, we process the following three steps. First, we apply non-maximum suppression on the set of bounding boxes for each class to make the system more robust and accelerate the next steps. For that we iteratively select the bounding box with the maximum score and remove all overlapping bounding boxes. Second, traffic signs are special classes since the subject of one traffic sign can be included similarly in another traffic sign as illustrated in Fig. 4.10. We utilize this information by defining a list of parts that can occur in each class. A part is found when a bounding box b' with the corresponding class and an intersection over union (IoU) greater than 0.2 exists. If this is the case, we increase the score by $b'_{\text{s}} \cdot 0.2/P$, where P is the number of parts. Third, we perform non-maximum suppression on the set of all bounding boxes by iteratively selecting the bounding box with the maximum score and removing all bounding boxes with IoU > 0.5. The final predictions are determined by selecting all bounding boxes that have a score b_{s} greater than or equal to a threshold t_c for the corresponding class.

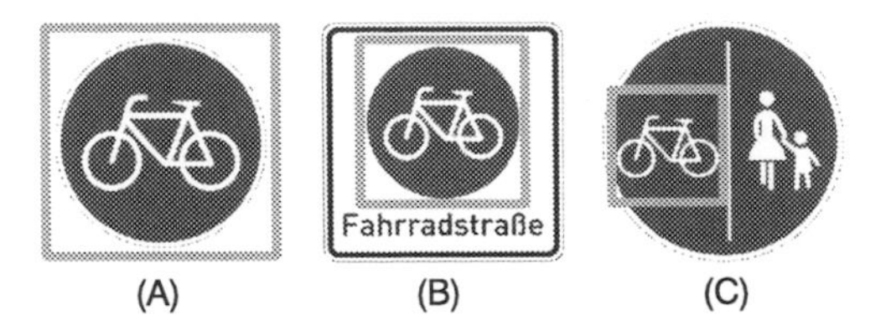

Figure 4.10 : The subject from class *237* (A) occurs similarly in class *244.1* (B) and class *241* (C). Due to very few training examples and the consequent low variability, parts of traffic signs are recognized. We utilize this information and integrate the recognition of parts into the bounding box prediction.

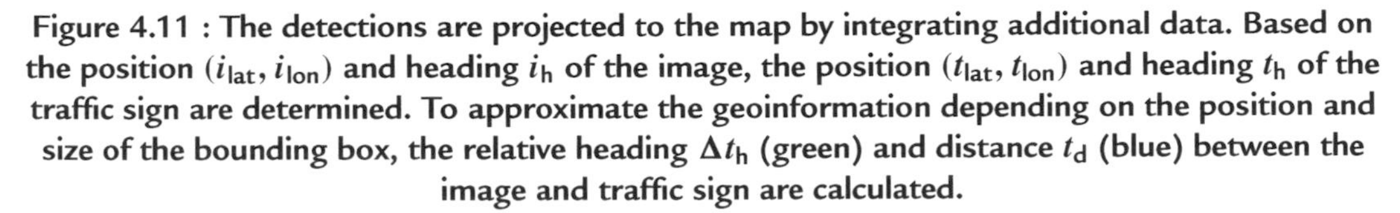

Figure 4.11 : The detections are projected to the map by integrating additional data. Based on the position (i_{lat}, i_{lon}) and heading i_h of the image, the position (t_{lat}, t_{lon}) and heading t_h of the traffic sign are determined. To approximate the geoinformation depending on the position and size of the bounding box, the relative heading Δt_h (green) and distance t_d (blue) between the image and traffic sign are calculated.

4.5 Localization

In this process we integrate additional data from other sensors to determine the position and heading of the traffic signs. For localization of the traffic signs, we use the GPS-position $(i_{\text{lat}}, i_{\text{lon}})$ and heading i_{h} of the images. The heading is the direction to which a vehicle is pointing. The data is included in our dataset, which is described in detail in Sect. 4.7. As illustrated in Fig. 4.11, we transform each bounding box

$$b = (b_{\text{x}}, b_{\text{y}}, b_{\text{w}}, b_{\text{h}}, b_{\text{c}}, b_{\text{s}}) \tag{4.39}$$

to a traffic sign

$$t = (t_{\text{lat}}, t_{\text{lon}}, t_{\text{h}}, t_{\text{c}}), \tag{4.40}$$

where $(t_{\mathrm{lat}}, t_{\mathrm{lon}})$ is the position, t_{h} the heading, and t_{c} the class. Since the position and viewing direction of the image are known, we approximate the traffic sign position and heading by calculating the relative heading Δt_{h} and distance t_{d}.

The relative heading is based on the horizontal position b_{x} of the bounding box in the image. A traffic sign which is located directly in the center of the image has the same heading than the image. A traffic sign on the left or right border has a relative heading of half of the angle of view. To determine the relative heading, we calculate the horizontal offset to the center of the image normalized by the image width i_{w}. Additionally, we multiply the value by the estimated angle of view α_{aov}. Thereby, the relative heading is calculated by

$$\Delta t_{\mathrm{h}} = \alpha_{\mathrm{aov}} \cdot \left(\frac{b_{\mathrm{x}}}{i_{\mathrm{w}}} - 0.5 \right). \tag{4.41}$$

The distance t_{d} between the position of the image and the position of the traffic sign is approximated by estimating the depth of the bounding box in the image. Traffic signs have a defined size $t_w \times t_{ht}$, where t_w is the width and t_{ht} the height. Since an approximate depth estimation is sufficient, we use the information about the size and assume a simple pinhole camera model. Given the focal length f and the sensor width s_{w} of the camera obtained from the data sheet and a bounding box with width b_{w}, we calculate the approximated distance by

$$t_{\mathrm{d}} = f \cdot \frac{t_w \cdot i_{\mathrm{w}}}{b_{\mathrm{w}} \cdot s_{\mathrm{w}}}. \tag{4.42}$$

Lastly, a traffic sign $t = (t_{\mathrm{lat}}, t_{\mathrm{lon}}, t_{\mathrm{h}}, t_{\mathrm{c}})$ is generated. The class t_{c} equals the bounding box class and the heading is calculated by adding the relative heading to the heading of the image $t_{\mathrm{h}} = i_{\mathrm{h}} + \Delta t_{\mathrm{h}}$. The traffic sign position $(t_{\mathrm{lat}}, t_{\mathrm{lon}})$ is determined by moving the position of the image by t_{d} in the direction t_{h}.

4.6 Clustering

Traffic signs can be observed multiple times in different images, cf. Fig. 4.12. In this section, an approach for merging multiple observations of the same traffic sign is presented which makes the localization more robust and improves the localization accuracy. For that, the generated geoinformation is used and the unsupervised clustering algorithm mean shift [37] is applied. A mean shift locates maxima of a density function and operates without predefining the number of clusters. Supervised learning algorithms are not applicable because no labels exist.

The mean shift is applied to the set of traffic signs $T_c = \{t^{(1)}, t^{(2)}, \ldots, t^{(N)}\}$ for each class c separately. As the traffic sign class is the same within the set, it can be omitted for clustering

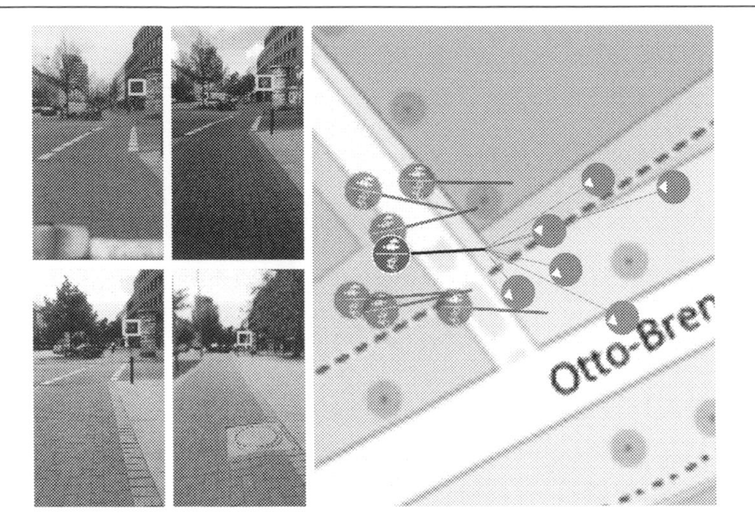

Figure 4.12 : Multiple observations of the same traffic sign are grouped based on their position and heading (left: detections in images, right: positions on map). Because neither the labels nor the number of clusters are known, mean shift clustering is used for processing the detected traffic signs.

so that each traffic sign has a three-dimensional data vector $t^{(i)} = (t_{\text{lat}}^{(i)}, t_{\text{lon}}^{(i)}, t_{\text{h}}^{(i)})$ consisting of latitude, longitude, and heading. The mean shift algorithm is extended by introducing a general function $\mathrm{D}(\cdot)$ to calculate the difference between two data points. This enables the processing of non-linear values and residue class groups such as for example angles. For the application in this work, the following difference function is used to calculate the difference between two traffic signs $t^{(a)}$ and $t^{(b)}$:

$$\mathrm{D}(t^{(a)}, t^{(b)}) = \left(t_{\text{lat}}^{(a)} - t_{\text{lat}}^{(b)},\, t_{\text{lon}}^{(a)} - t_{\text{lon}}^{(b)},\, \mathrm{D_h}(t_{\text{h}}^{(a)}, t_{\text{h}}^{(b)})\right). \tag{4.43}$$

Latitude and longitude are subtracted, whereas the difference between the headings is defined by $\mathrm{D_h}(\cdot)$. The function $\mathrm{D_h}(\cdot)$ subtracts two headings α and β which lie within the interval $[-180, 180]$ and this ensures that the difference lies within the same interval:

$$\mathrm{D_h}(\alpha, \beta) = \begin{cases} \alpha - \beta - 360 & \text{if } \alpha - \beta > +180, \\ \alpha - \beta + 360 & \text{if } \alpha - \beta < -180, \\ \alpha - \beta & \text{otherwise.} \end{cases} \tag{4.44}$$

The mean shift algorithm is generalized by integrating the difference function $\mathrm{D}(\cdot)$ into the function for calculating the mean shift $\mathrm{m}(y^t)$:

Figure 4.13 : Examples from the captured dataset. For instance, separate bicycle lines for cyclists (A) and (B), roads that are prohibited for cars but free for cyclists (C), or roads that allow for cycling in both directions (D).

$$\mathrm{m}\left(y^t\right) = \frac{\sum_{i=1}^{N} \mathrm{K}\left(\left\|\frac{\mathrm{D}(t^{(i)}, y^t)}{b}\right\|^2\right) \mathrm{D}\left(t^{(i)}, y^t\right)}{\sum_{i=1}^{N} \mathrm{K}\left(\left\|\frac{\mathrm{D}(t^{(i)}, y^t)}{b}\right\|^2\right)}, \tag{4.45}$$

where $\mathrm{K}(\cdot)$ is the kernel, b the bandwidth, and y^t the cluster position in iteration t. In this work, a multivariate Gaussian kernel $\mathrm{K}(x) = \exp\left(-\frac{1}{2}x\right)$ is used with different bandwidths for each dimension $b = (b_{\mathrm{lat}}, b_{\mathrm{lon}}, b_{\mathrm{h}})$. The bandwidths b_{lat} and b_{lon} are set to 0.00015 which equals approximately 10 meters for our geographic location and b_{h} to 30 degree.

The mean shift iteratively updates the cluster position by calculating the weighted mean shift. In each iteration the cluster position is translated by $\mathrm{m}(y^t)$ so that the updated cluster position y^{t+1} is calculated by $y^{t+1} = y^t + \mathrm{m}(y^t)$. A cluster is initialized at each data point. As a result, multiple instances of the same traffic sign are merged based on their position and heading.

4.7 Dataset

To collect data in real-world environments, smart phones are used for data recording because they can be readily attached to bicycles. Many people own a smart phone so that a large number of users can be involved. The recorded dataset consists of more than 60 000 images. Some examples are shown in Fig. 4.13.

4.7.1 Data Capturing

We developed an app for data recording which can be installed onto the smart phone. Using a bicycle mount, the smart phone is attached to the bike oriented in the direction of travel. While cycling, the app captures images and data from multiple sensors. Images of size 1080×1920 pixels are taken with a rate of one image per second. Sensor data is recorded from the built-in accelerometer, gyroscope, and magnetometer with a rate of ten data points per second. Furthermore, geoinformation is added using GPS. The data is recorded as often as the GPS-data is updated.

4.7.2 Filtering

After finishing a tour, the images are filtered to reduce the amount of data. Especially monotonous routes, e.g. in rural areas, produce many similar images. However, the rate with which images are captured cannot be reduced because this increases the risk of missing interesting situations.

We therefore introduce an adaptive filtering of the images. The objective is to keep images of potentially interesting situations that help to analyze traffic situations, but to remove redundant images. For instance, interesting situations could be changes in direction, traffic jams, bad road conditions, or obstructions like construction works or other road users. For filtering, we integrate motion information and apply a twofold filtering strategy based on decreases in speed and acceleration:

i. **Decreases in speed** indicate situations where the cyclist has to slow down because of potential traffic obstructions such as for example traffic jams, construction works, or other road users. Speed is provided by the GPS-data. We apply a derivative filter to detect decreases in speed. As filter, we use a derivative of Gaussian filter with a bandwidth, i.e. standard deviation, of $2\ \frac{\text{km}}{\text{h}^2}$.
ii. **Acceleration** is used to analyze the road conditions and to detect for example bumps. It is specified per axis. Each data point consists of a three-dimensional vector. We calculate the Euclidean norm of the vector and apply two smoothing filters with different time spans: One with a large and one with a short time span. Thus, we filter the noisy acceleration data and detect the situations in which the short-term average acceleration relative to the long-term average acceleration exceeds a threshold of k. For smoothing, we use Gaussian filters with bandwidths of 1.5 g and 10 g, with standard gravitational acceleration $g = 9.81\ \frac{\text{m}}{\text{s}^2}$, and set $k = 2.8$.

We filter the images by removing images if none of the two criteria indicate an interesting situation. The filtering process reduces the amount of data by a factor of 5 on average. Subsequently, the data is transferred to a server.

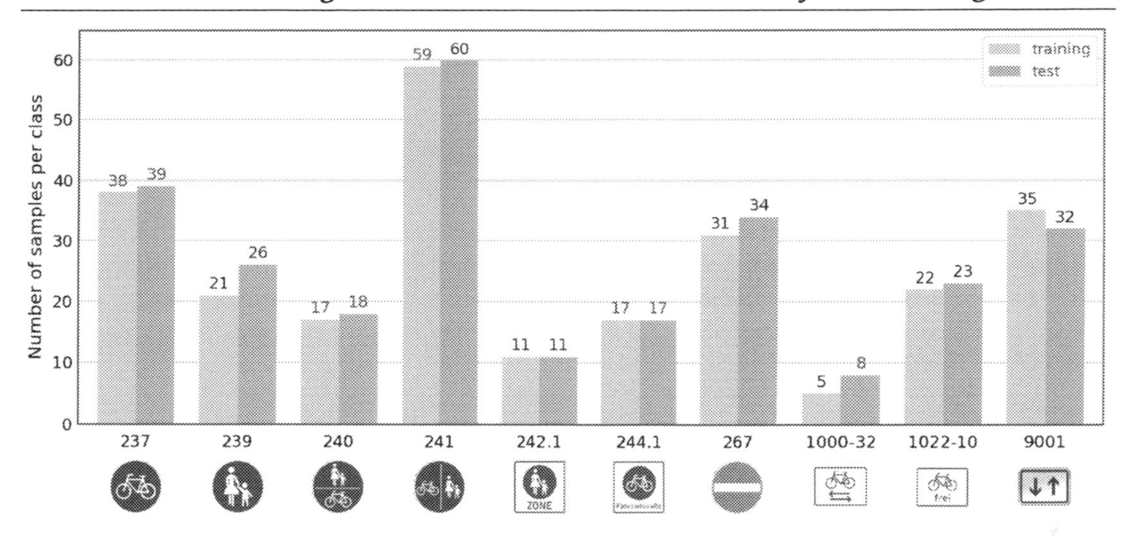

Figure 4.14 : Number of training and test samples in each class. On average only 26 samples are available per class for each set.

4.8 Experiments

Experiments are conducted to demonstrate the performance of the recognition system. Due to the limited amount of labeled data, the pipeline is trained on patches and then extended to perform object detection. First, results are presented on the classification of a single patch. Afterwards, the recognition performance is illustrated. The comparison of patch-wise processing and fully convolutional processing of full images is shown in the end. Random forests are trained and tested on an Intel(R) Core(TM) i7-7820X CPU @ 3.60GHz, and neural networks on a NVIDIA GeForce GTX 1080 Ti using *Tensorflow* [38] and *Keras* [39]. The proposed system is programmed in Python.

4.8.1 Training and Test Data

Ten different traffic signs that are interesting for cyclists are selected. Because the signs differ from traffic signs for cars, the availability of labeled data is very limited. Some classes come with little labeled data but for some classes no labeled data is available. To have ground truth data of our classes for training and testing, we manually annotated 524 bounding boxes of traffic signs in the images. The data is split into a training set and a test set using a split ratio of 50/50. In Fig. 4.14, the number of samples per class is shown. The training data consists of 256 samples for all 10 classes which corresponds to less than 26 samples per class on average. Please note that some traffic signs are very rare. Class *1000-32*, for example, has only five examples for training. Additionally, 2 000 background examples are randomly sampled for training and testing. The splitting is repeated multiple times and the results are averaged.

4.8.2 Classification

The first experiment evaluates the performance of the classification on patches. The evaluation is performed in two steps. First, the training for learning features is examined and, secondly, the classification on the target task.

For feature learning, the GTSRB [1] dataset is used since it is similar to our task and has a large amount of labeled data. The dataset consists of 39 209 examples for training and 12 630 examples for testing over 43 classes. After training, the convolutional neural network CNN^{F} achieves an accuracy of 97.0% on the test set.

In the next step, the learned features are used to generate a feature vector of each training example of our dataset, and then to train a random forest. For evaluation, the test data is processed similarly. A feature vector is generated for each example from the test set using the learned feature generator CNN^{F} and classified by the random forest subsequently.

Since the class distribution is imbalanced, we report the overall accuracy and the mean accuracy. The mean accuracy calculates the precision for each class independently and averages the results. The random forest classification achieves an accuracy of 96.8% and a mean accuracy of 94.8% on the test set. The confusion matrix is shown in Fig. 4.15. Six classes are classified without errors. All other classes, except from the background class, only contain a single, two, or three misclassified examples. Class *1000-32* which consist of five examples has larger error. Additionally, some background examples are classified as traffic signs and vice versa. Please refer to Fig. 4.15 for more information about the traffic signs the classes correspond to.

4.8.3 Object Detection

The next experiment is conducted to demonstrate the recognition performance of the proposed system. The task is to detect the position, size, and type of all traffic signs in an image. The images have a high diversity with respect to different perspectives, different lighting conditions, and motion blur.

The recognition system is constructed by extending the CNN for patch-wise classification to a fully convolutional network so that fast processing of full images is enabled. A filtering strategy is applied subsequently to predict bounding boxes. No additional training data is required

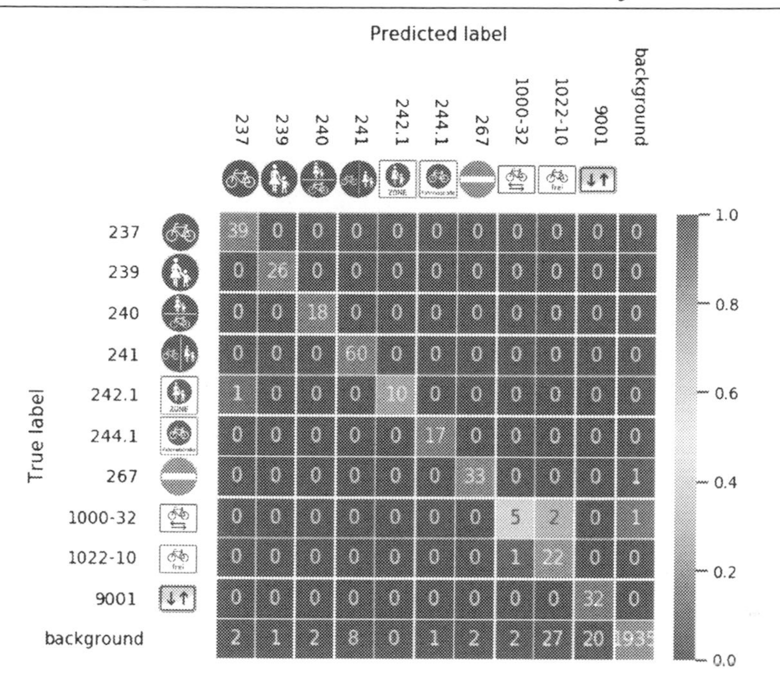

Figure 4.15 : Confusion matrix showing the performance of the classifier on the test set. The absolute number of samples are shown in the matrix.

during this process so that only 256 examples over 10 classes are used for training the recognition system. We process the images in eight different scales. Starting with the scale $s_0 = 1$, the image size is decreased from scale to scale by a factor of 1.3.

To evaluate the recognition performance, we process all images in the test set and match the predicted bounding boxes with the ground truth data. Each estimated bounding box is assigned to the ground truth bounding box with the highest overlap. The overlap is measured using the IoU and only overlaps with an IoU > 0.5 are considered.

All bounding boxes come with a score and the class specific threshold t_c determines if a bounding box is accepted or rejected as described in Sect. 4.4.5. For each class, the threshold t_c is varied and precision and recall are calculated. Precision measures the accuracy of the predictions i.e. how many of the predicted objects are correct. It is defined as

$$\text{Precision} = \frac{\text{TP}}{\text{TP} + \text{FP}}, \tag{4.46}$$

where TP is the number of true positives and FP the number of false positives. Recall measures the amount of objects that are found relative to the total number of objects that are

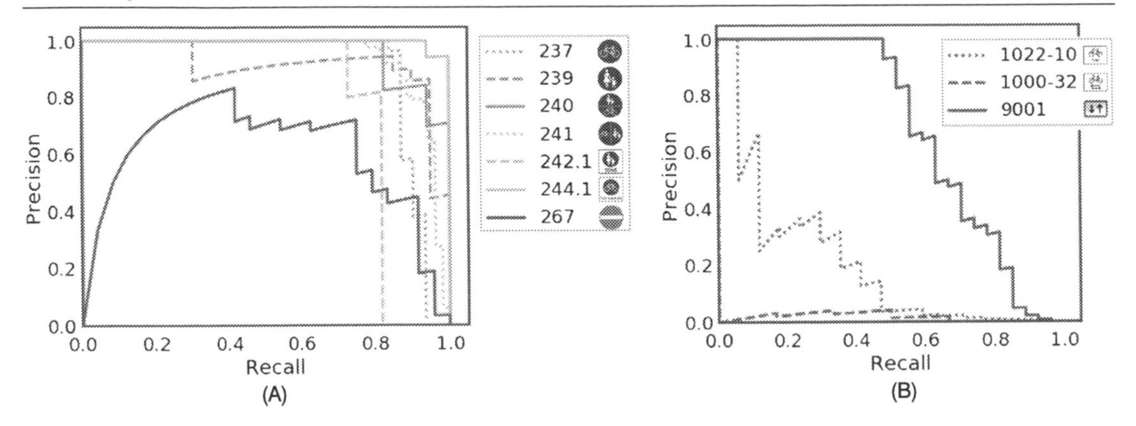

Figure 4.16 : Precision-recall curves for evaluating the recognition performance. (A) Standard traffic signs. (B) Info signs. The shape of the curves is erratic because few labeled data is available for training and testing.

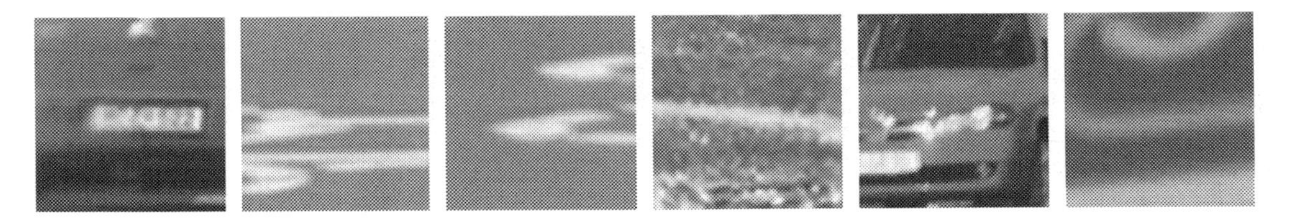

Figure 4.17 : Selected failure cases for class *267*.

available and is defined as follows:

$$\text{Recall} = \frac{\text{TP}}{\text{TP} + \text{FN}}, \tag{4.47}$$

where TP is the number of true positives and FN the number of false negatives. The resulting precision-recall curves are shown in Fig. 4.16. To facilitate understanding these results, two graphs are shown. In the first, the precision-recall curves of a group of standard traffic signs are plotted. The results are good. Some classes are detected almost perfectly. In the second graph, the precision-recall curves of a different group of traffic signs are plotted. These signs are much more difficult to recognize as they are black and white and do not have a conspicuous color. The performance of each class correlates with the number of examples that are available for training. Class *9001* with 35 training examples performs best, class *1022-10* with 22 training examples second best, and class *1000-32* with only 5 training examples worst. In Fig. 4.17 failure cases for class *267* are shown. Patches with similar appearance are extracted due to the limited variability with few training samples and missing semantic information since the broader context is not seen from the patch-wise classifier. To summarize the performance on each class the average precision (AP) is calculated. The results are presented in

Table 4.1: Average precision of each class on the test dataset.

class	237	239	240	241	242.1	244.1	267	1000-32	1022-10	9001
AP	0.901	0.930	0.964	0.944	0.801	0.996	0.677	0.023	0.215	0.679

Table 4.1. In total, the recognition system achieves a good mean average precision (mAP) of 0.713.

In the last step, the final bounding box predictions are determined. The threshold t_c of each class is selected by calculating the F1 score,

$$F1 = 2 \cdot \frac{\text{precision} \cdot \text{recall}}{\text{precision} + \text{recall}}, \tag{4.48}$$

for each precision-recall pair and choosing the threshold with the maximum F1 score. Some qualitative results are presented in Fig. 4.18. In each column, examples of a particular class are chosen at random. Examples that are recognized correctly are shown in the first three rows, examples which are recognized as traffic sign but in fact belong to the background or a different class are shown in the next two rows. These bounding box patches can have a similar color or structure. Examples that are not recognized are shown in the last two rows on the bottom. Some of these examples are twisted or covered by stickers.

4.8.4 Computation Time

In the third experiment we evaluate the computation time. Random forests are fast at test time for the classification of a single feature vector. When processing a full image, the random forest is applied to every patch in the feature maps. For an image of size 1080×1920 the feature maps are produced relatively fast using CNN^{F} and have a size of 268×478 so that 124 399 patches have to be classified to build the output probability map. The images are processed in eight different scales. All together, we measured an average processing time of more than 10 hours for a single image. Although the computation time could be reduced by using a more efficient language than Python, the time to access the memory represents a bottleneck due to a large overhead for accessing and preprocessing each patch.

For processing all in one pipeline, we constructed the fully convolutional network CNN^{FCN}. The network combines feature generation and classification and processes full images in one pass. The time for processing one image in eight different scales is only 4.91 seconds on average. Compared with the patch-wise processing using random forest, using the fully convolutional network reduces the processing time significantly.

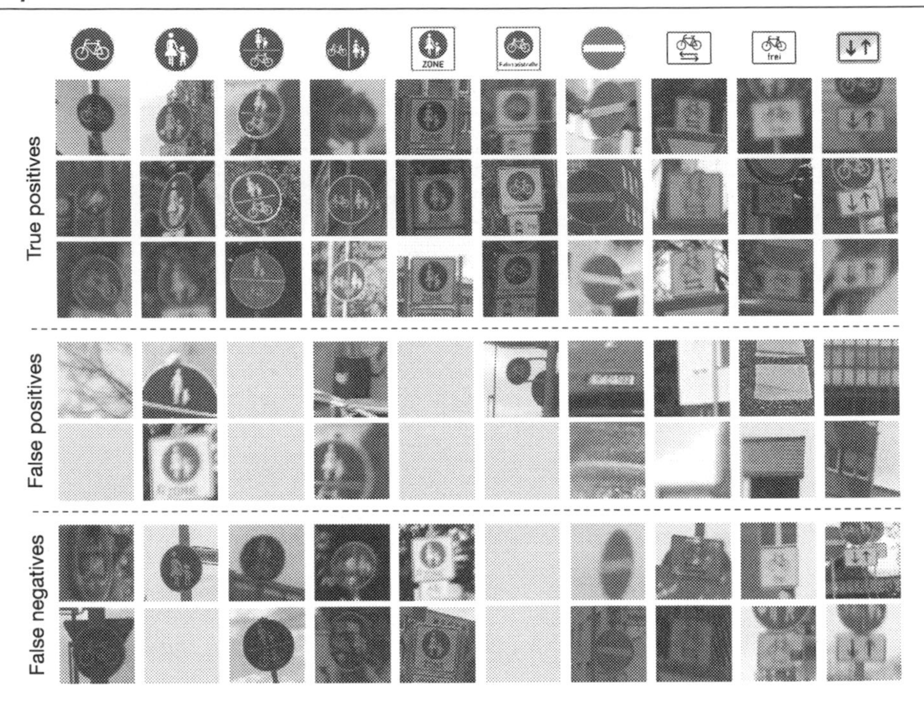

Figure 4.18 : Recognition results for randomly chosen examples of the test set. In each column, the ground truth traffic sign is shown on top along with correctly recognized traffic signs (first three rows), false positives (next two rows), and false negatives (last two rows on the bottom). Some classes have less than two false positives or false negatives, respectively, however.

4.8.5 Precision of Localizations

The last experiment is designed to demonstrate the localization performance. The localization maps the predicted bounding boxes in the image to positions on the map. Position and heading of a traffic sign are calculated based on the geoinformation of the image and the position and size of the bounding boxes.

For evaluation, we generate ground truth data by manually labeling all traffic signs on the map that are used in our dataset. In the next step, correctly detected traffic signs are matched with the ground truth data. The distance between two GPS-positions is calculated using the *haversine formula* [40]. The maximal possible difference of the heading is 90° because larger differences would show a traffic sign from the side or from the back. Each traffic sign is assigned to the ground truth traffic sign that has the minimum distance and a heading difference within the possible viewing area of 90°. The median of the localization error, i.e. the distance between the estimated position of the traffic sign and its ground truth position, is 6.79 m. Since the recorded GPS-data also includes the inaccuracies of each GPS-position, we can

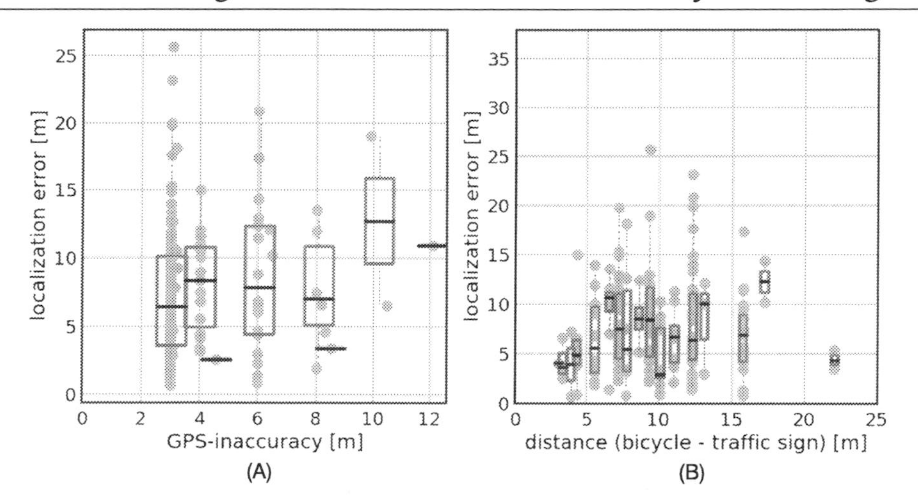

Figure 4.19 : The distance error with respect to GPS-inaccuracy (A) and distance between the recording device and the traffic sign (B). The black lines indicate the medians, the upper and bottom ends of the blue boxes the first and third quantile.

remove traffic signs which are estimated by more inaccurate GPS-positions. If traffic signs with a GPS-inaccuracy larger than the average of 3.79 m are removed, then the median of the localization error decreases to 6.44 m.

The errors of the localizations (y-axis) with respect to the GPS-inaccuracies (x-axis) are plotted in Fig. 4.19A. The orange dots indicate estimated positions of traffic signs. The black lines indicate the medians, the upper and bottom ends of the blue boxes the first and third quantiles. It can be seen that the localization error does not depend on the precision of the GPS-position as it does not increase with the latter. The localization errors (y-axis) with respect to the distances between the positions of the traffic signs and the GPS-positions (x-axis) are shown in Fig. 4.19B. It can be seen that the errors depend on the distance between traffic sign and bicycle as they increase with these distances. This can be explained by the fact that the original inaccuracies of the GPS-position are extrapolated, i.e. the larger the distances, the more the GPS-inaccuracies perturb the localizations.

Since smart phones are used as recording devices, the precision of the GPS-coordinates is lower than those used in GPS-sensors integrated in cars or in high-end devices. As the inaccuracies of the GPS-positions have a large influence on the localizations, we identify multiple observations of the same traffic sign in a clustering process to reduce the localization error. The performance is evaluated as before. The overall median of the localization error improves to 5.75 m. When measuring the performance for all traffic signs which have multiple observations the median of the localization error even decreases to 5.13 m.

4.9 Conclusion

We presented a system for object detection that is trained with very little labeled data. CNNs have shown great results in feature learning and random forests are able to build a robust classifier even if very few training examples are available. We combined the advantages of CNNs and random forests to construct a fully convolutional network for predicting bounding boxes. The system is built in three steps. First, we learned features using a CNN and trained a random forest to perform patch-wise classification. Second, the random forest is mapped to a neural network. Lastly, we transform the pipeline to a fully convolutional network to accelerate the processing of full images. Whereas deep learning typically depends on the availability of large datasets, the proposed system significantly reduces the required amount of labeled data.

The proposed system was evaluated on crowdsourced data with the aim of collecting traffic information for cyclists. We used our system to recognize traffic signs that are relevant for cyclists. The system is trained with only 26 examples per class on average. Furthermore, we showed how additional sensor information can be used to locate traffic signs on the map.

Acknowledgment

This research was supported by German Research Foundation DFG within Priority Research Program 1894 Volunteered Geographic Information: Interpretation, Visualization and Social Computing.

References

[1] J. Stallkamp, M. Schlipsing, J. Salmen, C. Igel, Man vs. computer: benchmarking machine learning algorithms for traffic sign recognition, Neural Networks 32 (2012) 323–332, https://doi.org/10.1016/j.neunet.2012.02.016.
[2] A. Geiger, P. Lenz, C. Stiller, R. Urtasun, Vision meets robotics: the KITTI dataset, The International Journal of Robotics Research 32 (11) (2013) 1231–1237.
[3] M. Cordts, M. Omran, S. Ramos, T. Rehfeld, M. Enzweiler, R. Benenson, U. Franke, S. Roth, B. Schiele, The Cityscapes dataset for semantic urban scene understanding, in: CVPR, 2016.
[4] C. Reinders, H. Ackermann, M.Y. Yang, B. Rosenhahn, Object recognition from very few training examples for enhancing bicycle maps, in: 2018 IEEE Intelligent Vehicles Symposium, IV, 2018.
[5] I. Goodfellow, Y. Bengio, A. Courville, Deep Learning, MIT Press, 2016.
[6] A. Krizhevsky, I. Sutskever, G.E. Hinton, Imagenet classification with deep convolutional neural networks, in: NIPS, 2012.
[7] J. Zurada, Introduction to Artificial Neural Systems, Jaico Publishing House, 1994, https://books.google.de/books?id=cMdjAAAACAAJ.
[8] W.S. McCulloch, W. Pitts, A logical calculus of the ideas immanent in nervous activity, The Bulletin of Mathematical Biophysics 5 (4) (1943) 115–133, https://doi.org/10.1007/BF02478259.
[9] F. Rosenblatt, The Perceptron, a Perceiving and Recognizing Automaton Project Para, Report, Cornell Aeronautical Laboratory, Cornell Aeronautical Laboratory, 1957, https://books.google.de/books?id=P_XGPgAACAAJ.

[10] P.J. Werbos, Beyond Regression: New Tools for Prediction and Analysis in the Behavioral Sciences, Ph.D. thesis, Harvard University, 1974.

[11] D. Parker, Learning Logic: Casting the Cortex of the Human Brain in Silicon, Technical report, Center for Computational Research in Economics and Management Science, Massachusetts Institute of Technology, Center for Computational Research in Economics and Management Science, 1985, https://books.google.de/books?id=2kS9GwAACAAJ.

[12] D.E. Rumelhart, J.L. McClelland, Parallel Distributed Processing: Explorations in the Microstructure of Cognition, vol. 1: Foundations, MIT Press, 1986.

[13] D. Hubel, T. Wiesel, Receptive fields, binocular interaction, and functional architecture in the cat's visual cortex, Journal of Physiology 160 (1962) 106–154.

[14] Y. Lecun, L. Bottou, Y. Bengio, P. Haffner, Gradient-based learning applied to document recognition, Proceedings of the IEEE 86 (11) (1998) 2278–2324, https://doi.org/10.1109/5.726791.

[15] N. Srivastava, G. Hinton, A. Krizhevsky, I. Sutskever, R. Salakhutdinov, Dropout: a simple way to prevent neural networks from overfitting, Journal of Machine Learning Research 15 (1) (2014) 1929–1958, http://dl.acm.org/citation.cfm?id=2627435.2670313.

[16] T.K. Ho, Random decision forests, in: Proceedings of the Third International Conference on Document Analysis and Recognition, vol. 1, ICDAR '95, IEEE Computer Society, Washington, DC, USA, 1995, p. 278, http://dl.acm.org/citation.cfm?id=844379.844681.

[17] L. Breiman, Random forests, Machine Learning 45 (1) (2001) 5–32, https://doi.org/10.1023/A:1010933404324.

[18] F. Kluger, H. Ackermann, M.Y. Yang, B. Rosenhahn, Deep learning for vanishing point detection using an inverse gnomonic projection, in: GCPR, 2017.

[19] M.Y. Yang, W. Liao, H. Ackermann, B. Rosenhahn, On support relations and semantic scene graphs, ISPRS Journal of Photogrammetry and Remote Sensing 131 (2017) 15–25.

[20] R. Girshick, J. Donahue, T. Darrell, J. Malik, Rich feature hierarchies for accurate object detection and semantic segmentation, in: Computer Vision and Pattern Recognition, 2014.

[21] K. He, X. Zhang, S. Ren, J. Sun, Spatial pyramid pooling in deep convolutional networks for visual recognition, in: European Conference on Computer Vision, 2014.

[22] R. Girshick, Fast R-CNN, in: ICCV, 2015.

[23] S. Ren, K. He, R. Girshick, J. Sun, Faster R-CNN: towards real-time object detection with region proposal networks, in: NIPS, 2015.

[24] W. Liu, D. Anguelov, D. Erhan, C. Szegedy, S. Reed, C.-Y. Fu, A.C. Berg, SSD: single shot multibox detector, in: ECCV, 2016.

[25] J. Redmon, A. Farhadi, Yolo9000: better, faster, stronger, arXiv preprint, arXiv:1612.08242.

[26] T. Lin, P. Goyal, R.B. Girshick, K. He, P. Dollár, Focal loss for dense object detection, CoRR, arXiv:1708.02002, arXiv:1708.02002, http://arxiv.org/abs/1708.02002.

[27] M. Oquab, L. Bottou, I. Laptev, J. Sivic, Learning and transferring mid-level image representations using convolutional neural networks, in: CVPR, 2014, pp. 1717–1724.

[28] B. Moysset, C. Kermorvant, C. Wolf, Learning to detect and localize many objects from few examples, CoRR, arXiv:1611.05664.

[29] R. Wagner, M. Thom, R. Schweiger, G. Palm, A. Rothermel, Learning convolutional neural networks from few samples, in: IJCNN, 2013, pp. 1–7.

[30] P. Kontschieder, M. Fiterau, A. Criminisi, S.R. Bulò, Deep neural decision forests, in: ICCV, 2015, pp. 1467–1475.

[31] I.K. Sethi, Entropy nets: from decision trees to neural networks, Proceedings of the IEEE 78 (10) (1990) 1605–1613, https://doi.org/10.1109/5.58346.

[32] J. Welbl, Casting random forests as artificial neural networks (and profiting from it), in: GCPR, Springer, Cham, 2014, pp. 765–771.

[33] D. Massiceti, A. Krull, E. Brachmann, C. Rother, P.H.S. Torr, Random forests versus neural networks – what's best for camera localization?, in: ICRA, IEEE, 2017, pp. 5118–5125.

[34] D.L. Richmond, D. Kainmueller, M.Y. Yang, E.W. Myers, C. Rother, Relating cascaded random forests to deep convolutional neural networks for semantic segmentation, CoRR, arXiv:1507.07583.
[35] J.T. Springenberg, A. Dosovitskiy, T. Brox, M. Riedmiller, Striving for simplicity: the all convolutional net, in: ICLR, 2015, pp. 1–14, arXiv:1412.6806.
[36] E. Shelhamer, J. Long, T. Darrell, Fully convolutional networks for semantic segmentation, IEEE Transactions on Pattern Analysis and Machine Intelligence 39 (4) (2017) 640–651, https://doi.org/10.1109/TPAMI.2016.2572683.
[37] D. Comaniciu, P. Meer, Mean Shift: a robust approach toward feature space analysis, IEEE Transactions on Pattern Analysis and Machine Intelligence 24 (2002) 603–619.
[38] M. Abadi, A. Agarwal, P. Barham, E. Brevdo, Z. Chen, C. Citro, G.S. Corrado, A. Davis, J. Dean, M. Devin, S. Ghemawat, I. Goodfellow, A. Harp, G. Irving, M. Isard, Y. Jia, R. Jozefowicz, L. Kaiser, M. Kudlur, J. Levenberg, D. Mané, R. Monga, S. Moore, D. Murray, C. Olah, M. Schuster, J. Shlens, B. Steiner, I. Sutskever, K. Talwar, P. Tucker, V. Vanhoucke, V. Vasudevan, F. Viégas, O. Vinyals, P. Warden, M. Wattenberg, M. Wicke, Y. Yu, X. Zheng, TensorFlow: large-scale machine learning on heterogeneous systems, software available from tensorflow.org, 2015, https://www.tensorflow.org/.
[39] F. Chollet, et al., Keras, https://keras.io, 2015.
[40] J. Inman, Navigation and Nautical Astronomy, for the Use of British Seamen, F. & J. Rivington, 1849.

CHAPTER 5

Multimodal Fusion Architectures for Pedestrian Detection

Dayan Guan[*,†], Jiangxin Yang[*,†], Yanlong Cao[*,†], Michael Ying Yang[‡], Yanpeng Cao[*,†]

[*]State Key Laboratory of Fluid Power and Mechatronic Systems, School of Mechanical Engineering, Zhejiang University, Hangzhou, China [†]Key Laboratory of Advanced Manufacturing Technology of Zhejiang Province, School of Mechanical Engineering, Zhejiang University, Hangzhou, China [‡]Scene Understanding Group, University of Twente, Enschede, The Netherlands

Contents

5.1 Introduction

In recent years, pedestrian detection has received wide attention in the computer vision community [42,9,13,16,15,6]. Given images captured in various real-world environment, pedes-

Multimodal Scene Understanding
https://doi.org/10.1016/B978-0-12-817358-9.00011-1

Figure 5.1 : Detections generated by a well-trained visible pedestrian detector [60]. (A) Detections generated using an image from the public Caltech testing dataset [12]; (B) detections generated using a visible image from the public KAIST testing dataset [26] captured during daytime; (C) detections generated using a visible image from the public KAIST testing dataset [26] captured during nighttime in good illumination condition; (D) detections generated using a visible image from the public KAIST testing dataset [26] captured during nighttime in bad illumination condition. Please note that green bounding boxes (BBs) represent true positives, red BBs represent false positives, and red BBs in dashed line represent false negatives.

trian detection solution is needed to generate bounding boxes to identify individual pedestrian instances accurately. It supplies a vital functionality to assist a number of human-centric applications, such as video surveillance [54,2,36], urban scene analysis [17,7,63] and autonomous driving [56,35,61].

Although some improvements have been achieved in the past years, it remains a challenging task to develop a robust pedestrian detector for real applications. Most of the detectors existed are trained using visible images only, thus their performances are unstable in various illumination environments as illustrated in Fig. 5.1. In order to overcome the limitation,

Figure 5.2 : Examples of pedestrian samples in the multimodal (visible and thermal) images captured in daytime and nighttime scenes [26]. It should be noted that multimodal data can supply complementary information about the target which could be effectively integrated to obtain more robust detection results.

multimodal pedestrian detection solutions are studied by many researchers to facilitate robust pedestrian detection for around-the-clock application [33,31,51,41,26,20]. The underlying reason is that multimodal images can supply complementary information about the target as shown in Fig. 5.2, therefore more robust detection results can be generated by the fusion of multimodal data. Designing an effect fusion architecture which can adaptively integrate multimodal features is critical to improving the detection results.

In Fig. 5.3, we illustrate the workflow of our proposed multimodal fusion framework for joint training of segmentation supervision and pedestrian detection. It contains three modules including feature learning/fusion, pedestrian detection, and segmentation supervision. The feature learning/fusion extracts features in individual channels (visible and thermal) and then integrate them to obtain multimodal feature maps. The pedestrian detection module generates predictions of the targets (confidence scores and bounding boxes) utilizing the generated multimodal feature maps. The segmentation supervision module improves the distinctiveness of features in individual channels (visible and thermal) through the supervision learning of segmentation masks. Pedestrian detection and segmentation supervision are trained end-to-end using a multitask loss function.

Based on this baseline framework, we organize experiments to explore a number of multimodal fusion models to identify the most effective scheme for the joint learning task of

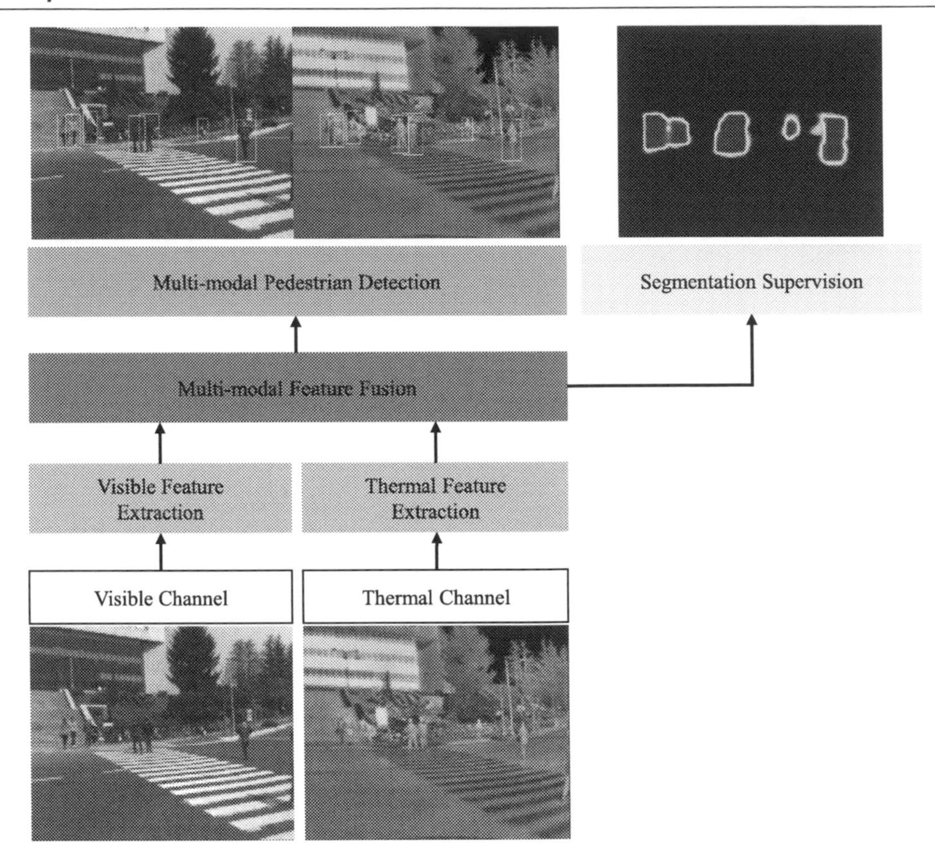

Figure 5.3 : The architecture of our proposed multimodal fusion framework for joint training of pedestrian detection and segmentation supervision.

pedestrian detection and segmentation supervision. For multimodal pedestrian detection, we evaluate three different multimodal feature fusion schemes (concatenation, maximization, and summation) and two deep neural network (DNN) models with and without a scene-aware mechanism. It can be observed that integrating the concatenation fusion scheme with a global scene-aware mechanism leads to better learning of both human-related features and correlation between visible and thermal feature maps. Moreover, we explore four different multimodal segmentation supervision infusion architectures (fused segmentation supervision with/without a scene-aware mechanism and two-stream semantic segmentation with/without a scene-aware mechanism). We found that the two-stream segmentation supervision with a scene-aware mechanism can better infuse semantic information to supervise the training of visible and thermal features. Our proposed method outperforms state-of-the-art multimodal pedestrian detectors and achieves higher detection accuracy using less runtime on public KAIST benchmark. In summary, our **contributions** are as follows:

1. We evaluate a number of feature fusion strategies for multimodal pedestrian detection. Compared with other basic fusion algorithms, integrating the concatenation with a global scene-aware mechanism leads to better learning of both human-related features and correlation between visible and thermal feature maps.
2. We experimentally reveal that the two-stream segmentation supervision infusion architecture, which individually infuses visible/thermal semantic information into their corresponding feature maps without multimodal fusion, provides the most effective scheme to make use of semantic information as supervision for visual feature learning.
3. We present a unified framework for joint training of multimodal segmentation supervision and target detection using a multitask loss function. Our method achieves the most accurate results with the least runtime compared with the current state-of-the-art multimodal pedestrian detectors [26,28,30].

This chapter further extends our previous work [22,23] in two aspects. First, we systematically evaluate three basic multimodal feature fusion schemes (concatenation, maximization, and summation) and two fusion models with and without a scene-aware mechanism. Our experimental results reveal that the basic concatenation fusion scheme combined with a global scene-aware mechanism performs better than other alternatives. Second, we utilize the original KAIST testing dataset and the improved one provided by Liu et al. [37] for quantitative evaluation. In both datasets, our proposed method achieves the most accurate results with the least runtime compared with the current state-of-the-art multimodal pedestrian detectors [26, 28,30].

The remainder of our chapter is organized as follows. Some existing research work as regards pedestrian detection using visible, thermal and multimodal images is summarized in Sect. 5.2. We present our proposed multimodal fusion architectures in detail in Sect. 5.3. The evaluation of a number of multimodal fusion architectures and the experimental comparison of multimodal pedestrian detectors are presented in Sect. 5.4. This chapter is concluded in Sect. 5.5.

5.2 Related Work

Pedestrian detection methods using visible, thermal and multimodal images are closely relevant to our work. We present a review of the recent researches on these topics below.

5.2.1 Visible Pedestrian Detection

A large variety of methods have been proposed to perform pedestrian detection using visible information in the last 20 years. Papageorgiou et al. [43] firstly abandoned motion cues to train a visible pedestrian detection system, utilizing Haar wavelets transform and support-vector machines (SVMs) [8]. The authors also collected a new visible pedestrian database (MIT) to evaluate their method. Viola and Jones [52] designed the integral images with a cascade structure for fast feature computation and efficient visible pedestrian detection, applying the AdaBoost classifier for automatic and accurate feature selection. Wu et al. [55] firstly proposed sliding window detectors for pedestrian detection, applying multiscale Haar wavelets and support-vector machines (SVMs) [8]. Dalal et al. [9] proposed the histograms of oriented gradient (HOG) descriptors along with a cascaded linear support-vector network for more accurate pedestrian detection. Dollar et al. [11] improved the HOG descriptors by designing the integrate channel features (ICFs) descriptors with an architecture of multichannel feature pyramids followed by the AdaBoost classifier. The feature representations of ICF have been further improved through multiple techniques including ACF [12], LDCF [39], SCF [1], and Checkerboards [62].

In recent years, DNN based approaches for object detection [19,18,45,25] have been widely adopted to improve the performance of visible pedestrian detection. Sermanet et al. [47] applied convolutional sparse coding to pretrain convolutional neural networks (CNNs) for feature extraction and fully-connected (FC) layers for classification. Tian et al. [50] developed extensive part detectors using weakly annotated humans to handle occlusion problems occurred in pedestrian detection. Li et al. [34] presented a scale-aware DNN which adaptively combines the outputs from a large-size sub-network and a small-size one to generate robust detection results at different scales. Cai et al. [5] designed the unified multiscale DNN to combine complementary scale-specific ones, applying various receptive fields to identify pedestrians in different scales. Zhang et al. [60] proposed a coarse-to-fine classification scheme for visible pedestrian detection by applying region proposal networks [45] to generate high-resolution convolutional feature maps, which are followed by the AdaBoost classifier for final classification. Mao et al. [38] designed a powerful deep convolutional neural networks architecture by utilizing the information of aggregating extra features to improve pedestrian detection performance without additional inputs in inference. Brazil et al. [4] developed a novel multitask learning scheme to improve visible pedestrian detection performance with the joint supervision on weakly box-based semantic segmentation and pedestrian detection, indicating that the box-based segmentation masks can provide sufficient supervision information to achieve additional performance gains.

We summarize the mentioned visible pedestrian detection methods in Table 5.1. However, pedestrian detectors trained on visible images are sensitive to changes of illumination, weather, and occlusions. These detectors are very likely to be stuck with images captured during nighttime.

Table 5.1: The summarization of visible pedestrian detection methods.

	Feature Extractor	Feature Classifier	Highlight	Year
Papageorgiou et al. [43]	Haar wavelets	SVM	No motion cues	1999
Viola and Jones [52]	Haar wavelets	AdaBoost	Cascade structure	2004
Wu et al. [55]	Haar wavelets	SVM	Sliding window	2005
Dalal et al. [9]	HOG	SVM	HOG descriptors	2005
Dollar et al. [11]	ICF	AdaBoost	ICF descriptors	2009
Dollar et al. [12]	ACF	AdaBoost	ACF descriptors	2012
Sermanet et al. [47]	CNN	FC	Convolutional sparse coding	2013
Nam et al. [39]	LDCF	AdaBoost	LDCF descriptors	2014
Benenson et al. [1]	SCF	AdaBoost	SCF descriptors	2014
Zhang et al. [62]	Checkerboards	AdaBoost	Checkerboards descriptors	2015
Tian et al. [50]	CNN	FC	Extensive part detectors	2015
Li et al. [34]	CNN	FC	Scale-aware mechanism	2015
Cai et al. [5]	CNN	FC	Multiscale CNN	2016
Mao et al. [38]	CNN	FC	Aggregating extra features	2017
Brazil et al. [4]	CNN	FC	Segmentation supervision	2017

5.2.2 Infrared Pedestrian Detection

Infrared imaging sensors, which provide excellent visible cues during nighttime, have found their importance for around-the-clock robotic applications, such as autonomous vehicle and surveillance system.

Nanda et al. [40] presented a real-time pedestrian detection system that works on low quality thermal videos. Probabilistic templates are utilized to capture the variations in human targets, working well especially for the case when object contrast is low and body parts are missing. Davis et al. [10] utilized a generalized template and an AdaBoosted ensemble classifier to detect people in widely varying thermal imagery. The authors also collected a challenging thermal pedestrian dataset (OSU-T) to test their method. Suard et al. [49] developed image descriptors, based on histograms of oriented gradients (HOG) with a support-vector machine (SVM) classifier, for pedestrian detection applied to stereo thermal images. This approach achieved good results for window classification in a video sequence. Zhang et al. [59] investigated the approaches derived from visible spectrum analysis for the task of thermal pedestrian detection. The author extended two feature classes (edgelets and HOG features) and two classification models (AdaBoost and SVM cascade) to the thermal images. Lee et al. [32] presented a nighttime part-based pedestrian detection method which divides a pedestrian into parts for a moving vehicle with one camera and one near-infrared lighting projector. The confidence of detected parts can be enhanced by analyzing the spatial relationship between every pair of parts, and the overall pedestrian detection result is refined by a block-based segmentation method. Zhao et al. [64] proposed a robust approach utilizing the shape distribution histogram (SDH) feature and the modified sparse representation classification (MSRC) for

Table 5.2: The summarization of infrared pedestrian detection methods.

	Feature Extractor	Feature Classifier	Highlight	Year
Nanda et al. [40]	Templates	Bayesian	Probabilistic templates	2003
Papageorgiou et al. [10]	Sobel	AdaBoost	Template-based method	2005
Suard et al. [49]	HOG	SVM	Stereo infrared application	2006
Zhang et al. [59]	Edgelets/HOG	AdaBoost/SVM	Experimental analysis	2007
Lee et al. [64]	HOG	SVM	Near-infrared application	2015
Zhao et al. [64]	SDH	MSRC	MSRC classifier	2015
Biswas et al. [3]	LSK	SVM	LSK descriptors	2017
Kim et al. [29]	TIR-ACF	SVM	TIR-ACF descriptors	2018

thermal pedestrian detection. Biswas et al. [3] proposed the multidimensional templates based on local steering kernel (LSK) descriptors to improve the pedestrian detection accuracies in low resolution and noisy thermal images. Kim et al. [29] presented a new thermal infrared radiometry aggregated channel feature (TIR-ACF) to detect pedestrians in the far thermal images at night.

We summarize the above-mentioned infrared pedestrian detection methods in Table 5.2. However, false detections are frequently caused due to strong solar radiation and background clutters in the daytime thermal images. With the development of multimodal sensing technology, it is possible to generate more stable detection results by simultaneously capturing multimodal information (e.g., visible, thermal and depth) of the same scene, which provide complementary information about objects of interest.

5.2.3 Multimodal Pedestrian Detection

It is experimentally demonstrated that multimodal images provide complementary information about objects of interest. As a result, pedestrian detectors trained using multimodal images can generate more robust detection results than using the visible or thermal images alone.

Grassi et al. [21] proposed a novel information fusion approach to detecting pedestrians, by determining the regions of interest in the video data through a lidar sensor and a thermal camera. Yuan et al. [58] proposed a multispectral based pedestrian detection approach which employs latent variable support-vector machines (L-SVMs) to train the multispectral (visible and near-thermal) pedestrian detection model. A large-size multispectral pedestrian dataset (KAIST) is presented by Hwang et al. [26]. The KAIST dataset contains well-aligned visible and thermal image pairs with dense pedestrian annotations. The authors further developed a new multimodal aggregated feature (ACF+T+THOG) followed by the AdaBoost classifier for target classification. The ACF+T+THOG concatenate the visible features generated by

Table 5.3: The summarization of multimodal pedestrian detection methods.

	Feature Extractor	Feature Classifier	Highlight	Year
Grassi et al. [21]	Invariant vectors	SVM	Invariant feature extraction	2011
Yuan et al. [58]	HOG	L-SVM	NIR-RGB application	2015
Hwang et al. [26]	ACF+T+THOG	AdaBoost	ACF+T+THOG descriptors	2015
Wagner et al. [53]	CNN	FC	Two-stream R-CNN	2016
Liu et al. [28]	CNN	FC	Two-stream Faster R-CNN	2016
Xu et al. [57]	CNN	FC	Cross-modal representations	2017
König et al. [30]	CNN	FC+BDT	Two-stream RPN	2017

ACF [12] and the thermal one generated by T+THOG, which contains the thermal image intensity (T) and the thermal HOG [9] features (THOG).

Recently, DNN based approaches for visible pedestrian detection have been widely adopted to design the multimodal pedestrian detectors. Wagner et al. [53] proposed the first application of DNN for multimodal pedestrian detection. The detections in [26] are considered as proposals, which are classified with a two-stream R-CNN [19] applying concatenation fusion in the late stage. The authors further evaluated the performance of architectures with different fusion stages, and the optimal architecture is in the late fusion stage. Liu et al. [28] investigated how to adopt the Faster R-CNN [45] model for the task of multimodal pedestrian detection and designed four different fusion architectures in which two-stream networks are integrated at different stages. Experimental results showed that the optimal architecture is the Halfway Fusion model which merges two-stream networks at a high-level convolutional stage. Xu et al. [57] designed a method to learn and transfer cross-modal deep representations in a supervised manner for robust pedestrian detection against bad illumination conditions. However, this method is based on information of visible channel only (during the testing stage) and its performance is not comparable with ones based multispectral data (e.g., Halfway Fusion model [28]) König et al. [30] modified the visible pedestrian detector RPN+BDT [60] to build Fusion RPN+BDT architecture for multimodal pedestrian detection. The Fusion RPN concatenates the two-stream RPN on the high-level convolutional stage and achieves the state-of-the-art performance on KAIST multimodal dataset.

We summarize the mentioned multimodal pedestrian detection methods in Table 5.3. It is worth it to mention in this chapter that our approach is definitely different from the above methods in two aspects. Firstly, a number of feature fusion strategies for multimodal pedestrian detection is firstly evaluated comparing with other basic fusion schemes. Secondly, we make use of semantic segmentation information as supervision for multimodal feature learning.

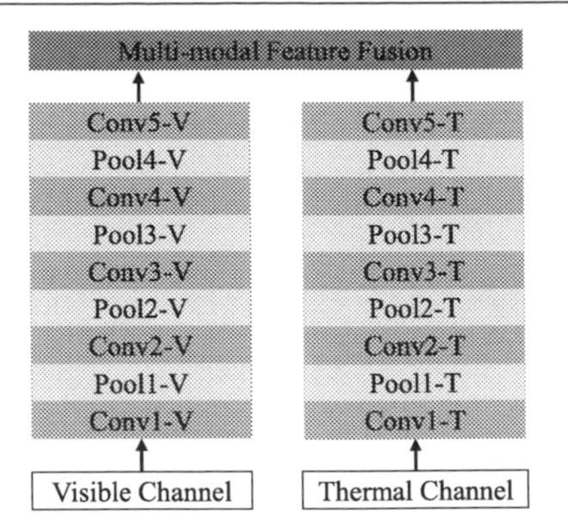

Figure 5.4 : The architecture of two-stream deep convolutional neural networks (TDCNNs) for multimodal feature learning and fusion. Please note that convolutional layers are represented by green boxes, pooling layers are represented by yellow boxes, and the feature fusion layer is represented by blue box. This figure is best seen in color.

5.3 Proposed Method

A multimodal fusion framework is presented for joint training of pedestrian detection and segmentation supervision. It consists of three major components including feature learning/fusion, pedestrian detection, and segmentation supervision. The details of each component are provided in the following subsections.

5.3.1 Multimodal Feature Learning/Fusion

The architecture of two-stream deep convolutional neural networks (TDCNNs) for multimodal feature learning and fusion is illustrated in Fig. 5.4. TDCNN learn semantic feature descriptions in the visible and thermal channels individually. The visible stream of TDCNN contains five convolutional layers (from Conv1-V to Conv5-V) and pooling ones (from Pool1-V to Pool4-V), which is the same as the thermal stream of TDCNN (from Conv1-T to Conv5-T and from Pool1-V to Pool4-V). Each stream of TDCNN takes the first five convolutional layers (from Conv1 to Conv5) and pooling ones (from Pool1 to Pool4) in VGG-16 as the backbone.

Previous researches revealed that feature fusion at a late stage can generate semantic feature maps which are appropriate for complex high-level detection tasks [53,28,30]. On the basis of this conclusion, the multimodal feature fusion layer is deployed after the Conv5-V and Conv5-T layers in TDCNN to combine the feature maps generated by the visible and thermal streams. The multimodal feature maps are generated as

$$\mathbf{m} = \mathcal{F}(\mathbf{v}, \mathbf{t}), \tag{5.1}$$

where $\mathcal{F}$ is the feature fusion function, $\mathbf{v} \in \mathbb{R}^{B_v \times C_v \times H_v \times W_v}$ and $\mathbf{t} \in \mathbb{R}^{B_t \times C_t \times H_t \times W_t}$ are the feature maps generated by the visible and thermal streams, respectively, and $\mathbf{m} \in \mathbb{R}^{B_m \times C_m \times H_m \times W_m}$ is the multimodal feature maps. It should be noted that B, C, H and W denote the number of batch size, channels, heights and widths of the respective feature maps. Considering that the feature maps generated by the visible and thermal streams are the same size, we set $B_t = B_c$, $C_v = C_t$, $W_v = W_t$, $H_v = H_t$. Three different feature fusion functions are considered: concatenation ($\mathcal{F}^{\text{concat}}$), maximization ($\mathcal{F}^{\text{max}}$) and summation ($\mathcal{F}^{\text{sum}}$).

Concatenation fusion. Concatenation function $\mathcal{F}^{\text{concat}}$ is the most widely used operation to integrate feature maps generated in visible and thermal channels [53,28,30]. $\mathbf{m}^{\text{concat}} = \mathcal{F}^{\text{concat}}(\mathbf{v}, \mathbf{t})$ stacks $\mathbf{v}$ and $\mathbf{t}$ across the individual channels c as

$$\begin{cases} \mathbf{m}^{\text{concat}}_{B,2c-1,H,W} = \mathbf{v}_{B,c,H,W}, \\ \mathbf{m}^{\text{concat}}_{B,2c,H,W} = \mathbf{t}_{B,c,H,W}, \end{cases} \tag{5.2}$$

where $c \in (1, 2, ..., C)$ and $\mathbf{m} \in \mathbb{R}^{B \times 2C \times H \times W}$. Considering that the feature maps generated in visible and thermal streams are directly stacked across feature channels using concatenation function, the correlation of features generated in visible and thermal streams will be further learned in subsequent convolutional layers.

Maximization fusion. $\mathbf{m}^{\text{max}} = \mathcal{F}^{\text{max}}(\mathbf{v}, \mathbf{t})$ takes the maximum of $\mathbf{v}$ and $\mathbf{t}$ at the same spatial locations h, w as

$$\mathbf{m}^{\text{max}}_{B,C,h,w} = \max\{\mathbf{v}_{B,C,h,w}, \mathbf{t}_{B,C,h,w}\}, \tag{5.3}$$

where $h \in (1, 2, ..., H)$, $w \in (1, 2, ..., W)$ and $\mathbf{m} \in \mathbb{R}^{B \times C \times H \times W}$. The maximization fusion function is utilized to select the features which are most distinct in either the visible or thermal streams.

Summation fusion. $\mathbf{m}^{\text{sum}} = \mathcal{F}^{\text{sum}}(\mathbf{p}, \mathbf{q})$ computes the summation of $\mathbf{v}$ and $\mathbf{t}$ at the same spatial locations h, w as

$$\mathbf{m}^{\text{max}}_{B,C,h,w} = \mathbf{v}_{B,C,h,w} + \mathbf{t}_{B,C,h,w}, \tag{5.4}$$

where $h \in (1, 2, ..., H)$, $w \in (1, 2, ..., W)$ and $\mathbf{m} \in \mathbb{R}^{B \times C \times H \times W}$. The summation fusion function is utilized to integrate the feature maps generated in visible and thermal streams using equal weighting scheme. Thus, multimodal feature maps will be stronger by combining the weak features generated in the visible and thermal streams. The performances of multimodal pedestrian detectors utilizing these three fusion functions ($\mathcal{F}^{\text{concat}}$, $\mathcal{F}^{\text{max}}$, $\mathcal{F}^{\text{sum}}$) are comparatively evaluated in Sect. 5.4.3.

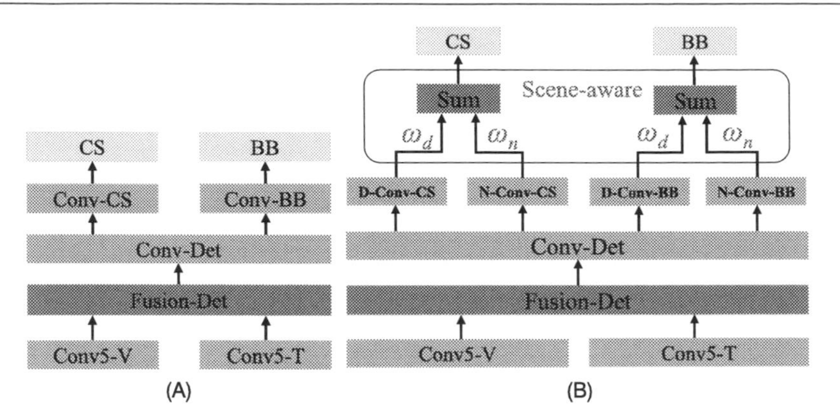

Figure 5.5 : Comparison of the baseline (A) and scene-aware (B) architectures. Please note that ω_d and ω_n are the computed scene-aware weighting parameters, convolutional layers are represented by green boxes, fusion layers are represented by blue boxes, and output layers are represented by orange boxes. This figure is best seen in color.

5.3.2 Multimodal Pedestrian Detection

In this subsection, we design two different DNN based models with and without a scene-aware mechanism for the task of multimodal pedestrian detection. Their overall architectures are shown in Fig. 5.5.

5.3.2.1 Baseline DNN model

The baseline DNN model is designed based on the TDCNN. The region proposal network (RPN) model [60] is adopted as two-stream region proposal network (TRPN) to generate multimodal pedestrian detection results due to its superior performance for large-scale target detection. Given a pair of well-aligned multimodal images, TRPN generate numbers of bounding boxes along with predicted scores following target classification and box regression. The architecture of TRPN is shown in Fig. 5.5A.

Given the multimodal feature maps generated utilizing the fusion layer (Fusion-Det), TRPN outputs classification scores (CSs) and bounding boxes (BBs) as multimodal pedestrian detections. A 3 × 3 convolutional layer (Conv-Det) is designed to encode pedestrian related features from the multimodal feature maps. Attached after the Conv-Det, two sibling 1 × 1 convolutional layers (Conv-CS and Conv-BB) are designed to generate multimodal pedestrian detections (CS and BB). In order to train the baseline DNN model, we utilize the loss term for detection $\mathcal{L}_{Det}$ as

$$\mathcal{L}_{Det} = \sum_{i \in S} \mathcal{L}_{Cls}(c_i, \hat{c}_i) + \lambda_r \sum_{i \in S} \hat{c}_i \cdot \mathcal{L}_{Reg}(b_i, \hat{b}_i), \tag{5.5}$$

where S denotes the set of training samples, c_i denotes the computed CS, b_i denotes the predicted BB, L_{cls} denotes the loss term for classification, L_{reg} denotes the loss term for box regression, and λ_r denotes the trade-off coefficient. The training label $\hat{c}_i$ is set to 1 for a positive sample. Otherwise, we set $\hat{c}_i = 0$ for a negative sample. A training sample is considered as a positive one in the case that the maximum intersection-over-union (IoU) ratio between every ground-truth label and the sample is larger than a set threshold. Otherwise, the training sample is considered a negative one. In Eq. (5.5), the loss term for classification L_{cls} is defined as

$$\mathcal{L}_{Cls}(c, \hat{c}) = -\hat{c} \cdot \log(c) - (1 - \hat{c})\log(1 - c), \tag{5.6}$$

and the loss term for box regression L_{reg} is defined as

$$\mathcal{L}_{Reg}(b, \hat{b}) = \sum_{j \in \{x,y,w,h\}} smooth_{L_1}(b_j, \hat{b}_j), \tag{5.7}$$

where $b = (b_x, b_y, b_w, b_h)$ represents the parameterized coordinates of the generated bounding box, $\hat{b} = (\hat{b}_x, \hat{b}_y, \hat{b}_w, \hat{b}_h)$ represents the coordinates of the bounding box label, and $smooth_{L_1}$ represents the robust $L1$ loss function defined in [18].

5.3.2.2 Scene-aware DNN model

Pedestrian samples have significantly different multimodal characteristics in daytime and nighttime scenes as shown in Fig. 5.6. Therefore, it is reasonable to deploy multiple scene-based sub-networks to handle intra-class object variances. Based on this observation, we further present a scene-aware DNN model to improve detection performance in daytime and nighttime scenes by considering the scene information encoded in multimodal images.

For this purpose, we firstly design a simple scene prediction model to estimate the probability of being daytime scene or nighttime one. As shown in Fig. 5.7, the scene prediction network (SPN) contains one pooling layer (SP-Pool), three continuous fully-connected layers (SP-FC1, SP-FC2, SP-FC3), and the classification layer (Soft-max). Each pair of multimodal images are entered into the first five convolutional layers and four pooling ones of TRPN to extract feature maps in individual visible and thermal channels. The two-stream feature maps are integrated utilizing the concatenate fusion layer (Concat). Inspired by the spatial pyramid pooling layer which can resize the feature maps to the same spatial resolution [24], the pooling layer SP-Pool resizes the multimodal feature maps to a fixed spatial size of 7×7 using symmetric bilinear interpolation from the nearest neighbors. Attached after the SP-FC3 is Soft-max, which is the classification layer of the scene prediction model. The outputs of Soft-max are ω_d and $\omega_n = (1 - \omega_d)$, which compute the probability of being day scene or night one. We define the error term of scene prediction $\mathcal{L}_{SP}$ as

$$\mathcal{L}_{SP} = -\hat{\omega}_d \cdot \log(\omega_d) - \hat{\omega}_n \cdot \log(\omega_n), \tag{5.8}$$

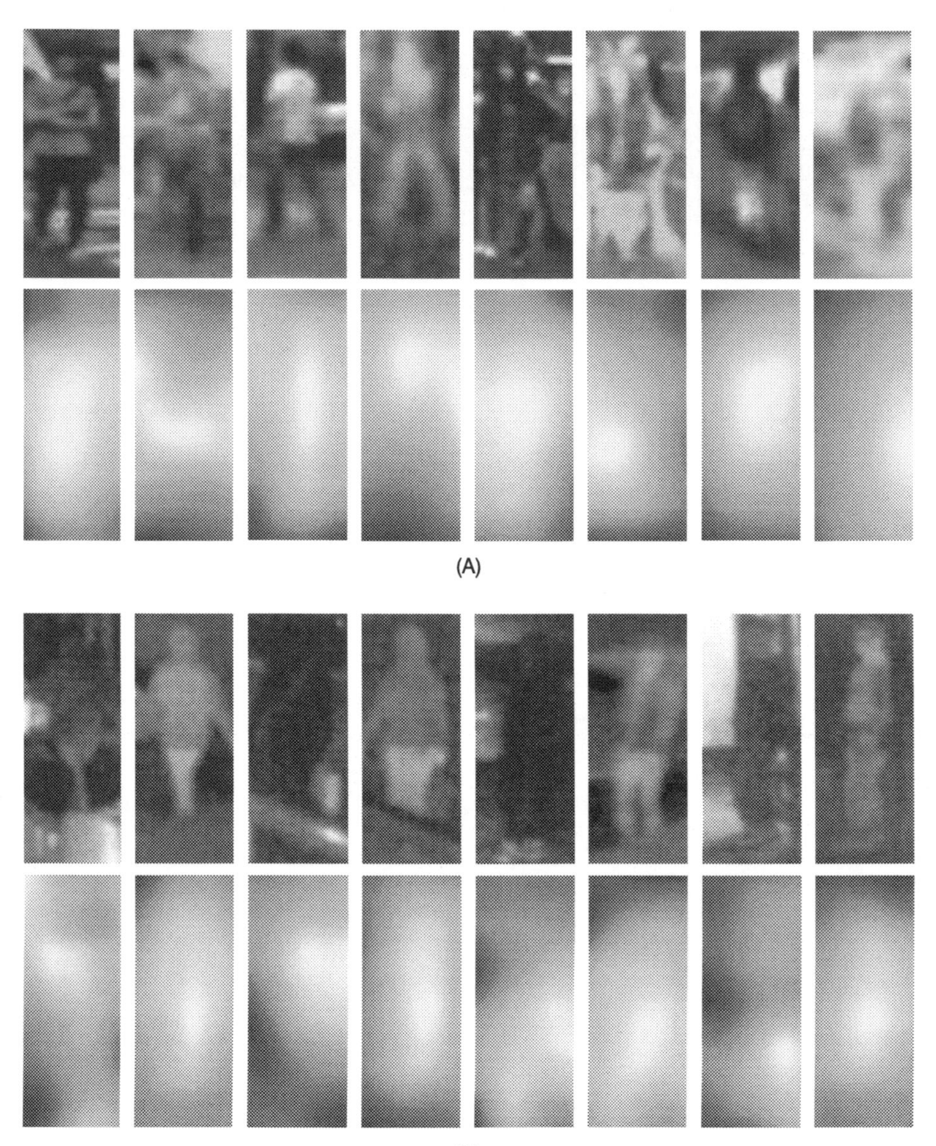

Figure 5.6 : Characteristics of multimodal pedestrian samples captured in (A) daytime and (B) nighttime scenes. The first rows display the multimodal images of pedestrian samples. The second rows illustrate the visualization of feature maps in the corresponding multimodal images. The feature maps are generated utilizing the RPN [60] well trained in their corresponding channels. It is observed that multimodal pedestrian samples have significantly different human-related characteristics under daytime and nighttime scenes.

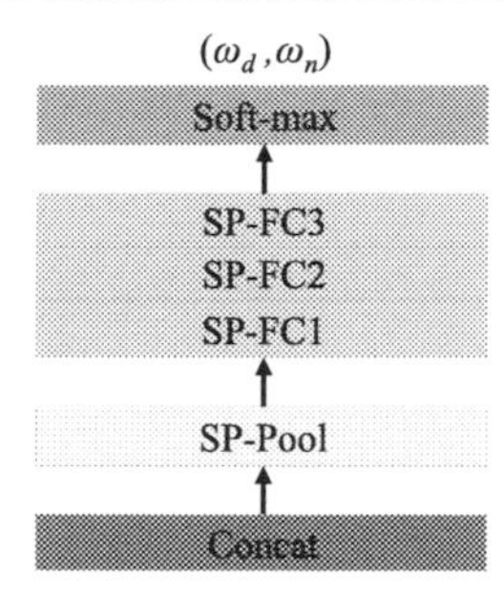

Figure 5.7 : Architecture of the scene prediction network (SPN). Please note that convolutional and fully-connected layers are represented by green boxes, pooling layers are represented by yellow boxes, fusion layers are represented by blue boxes, classification layer is represented by an orange box. This figure is best seen in color.

where ω_d and $\omega_n = (1 - \omega_d)$ are the predicted scene weights for daytime and nighttime, $\hat{\omega}_d$ and $\hat{\omega}_n = (1 - \hat{\omega}_d)$ are the scene labels. The $\hat{\omega}_d$ is set to 1 when the scene labels are annotated as daytime. Otherwise, we set $\hat{\omega}_d = 0$.

We incorporate scene information into the baseline DNN model in order to generate more accurate and robust results for various scene conditions. Specifically, the scene-aware two-stream region proposal networks (STRPNs) consist of four sub-networks (D-Conv-CS, N-Conv-CS, D-Conv-BB, and N-Conv-BB) to generate scene-aware detection results (CS and BB) as shown in Fig. 5.5B. D-Conv-CS and N-Conv-CS generate feature maps for classification under daytime and nighttime scenes, respectively. The outputs of D-Conv-CS and N-Conv-CS are combined utilizing the weights computed in the SPN model to produce the scene-ware classification scores (CSs). D-Conv-BB and N-Conv-BB generate feature maps for box regression in day and night scene conditions, respectively. The outputs of D-Conv-BB and N-Conv-BB are integrated using the weights computed in the SPN model to calculate the scene-ware bounding boxes (BBs). The loss term for detection L_{det} is also defined based on Eq. (5.5), while c_i^s is computed as the weighted sum of classification score in daytime scene c_i^d and one in nighttime scene c_i^n as

$$c_i^s = \omega_d \cdot c_i^d + \omega_n \cdot c_i^n, \tag{5.9}$$

and b_i^s is the scene-aware weighted combination of b_i^d and b_i^n, which are calculated by D-Conv-BB and N-Conv-BB sub-networks, respectively, as

$$b_i^s = \omega_d \cdot b_i^d + \omega_n \cdot b_i^n. \tag{5.10}$$

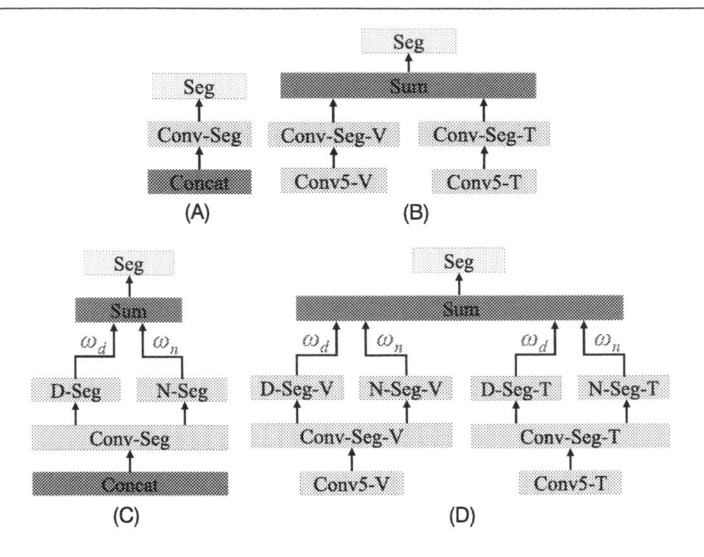

Figure 5.8 : The comparison of MS-F (A), MS (B), SMS-F (C) and SMS (D) architectures. Please note that ω_d and ω_n are the computed scene-aware weighting parameters, convolutional layers are represented by green boxes, fusion layers are represented by blue boxes, and output layers are represented by gray boxes. This figure is best seen in color.

5.3.3 Multimodal Segmentation Supervision

According to some recent research work [25,4], the performance of anchor-based object detectors can be improved using the information of semantic segmentation as a strong cue. The underlying reason is that box-based segmentation masks are able to supply additional effective supervision to make the feature maps in the shared layers more distinctive for the downstream object detectors. We integrate the segmentation supervision scheme with TRPN and STRPN to obtain more accurate multimodal pedestrian detection results.

Given feature maps generated from two-stream visible and thermal channels, different segmentation supervision results will be achieved by integrating the feature maps in different levels (feature-level and decision-level). In order to explore the optimal fusion architecture for the multimodal segmentation supervision task, we develop two different multimodal segmentation supervision architectures, denoted as feature-level multimodal segmentation supervision (MS-F) and decision-level multimodal segmentation supervision (MS). As illustrated in Fig. 5.8A, the MS-F contains a concatenate fusion layer (Concat), a 3×3 convolutional layer (Conv-Seg) and the output layer (Seg). Each pair of multimodal images are entered into the first five convolutional layers and four pooling ones of TRPN to extract feature maps in individual visible and thermal channels. The two-stream feature maps are integrated utilizing the Concat layer. The MS-F generate segmentation prediction (Seg) as a supervision to make the two-stream feature maps more distinctive for multimodal pedestrian detector. In comparison, as illustrated in Fig. 5.8B, the MS utilizes two 3×3 convolutional layers (Conv-Seg-V and Conv-Seg-T) to generate visible and thermal segmentation prediction as the supervision to make the visible and thermal feature maps more distinctive for multimodal pedestrian detector. The segmentation outputs of the MS is a combination of Conv-Seg-V and Conv-Seg-T.

It is reasonable to explore whether the performance of segmentation supervision able to be improved by integrating the scene information. Two different scene-aware multimodal segmentation supervision architectures (SMS-F and SMS) are developed based on the architectures of multimodal segmentation supervision models (MS-F and MS). As shown in Fig. 5.8C–D, multimodal segmentation outputs are generated using daytime and nighttime segmentation sub-networks (D-Seg and N-Seg) applying the scene-aware weighting mechanism. It should be noted that SMS-F consists of two sub-networks, while SMS consists of four sub-networks. The performance of these four different multimodal segmentation supervision architectures (SM-F, SM, SMS-F, and SMS) are evaluated in Sect. 5.4.

The loss term of segmentation supervision is defined as

$$L_{Seg} = \sum_{i \in C} \sum_{j \in B} [-\hat{s}_j \cdot \log(s_{ij}^s) - (1 - \hat{s}_j) \cdot \log(1 - s_{ij}^s)], \tag{5.11}$$

where s_{ij}^s is the segmentation prediction, C is the segmentation streams (MS-F and SMS-F consist of one multimodal stream while MS and SMS consist of one visible stream and one thermal stream), B are training samples for box-based segmentation supervision. We set $\hat{s}_j = 1$ if the sample is located in the region within any bounding box of ground truth. Otherwise, we set $\hat{s}_j = 0$. As for the scene-aware multimodal segmentation supervision architectures SMS-F and SMS, the scene-aware segmentation prediction s_{ij}^s is computed by the scene-aware weighted combination of daytime and nighttime segmentation prediction s_{ij}^d and s_{ij}^n, respectively, as

$$s_{ij}^s = \omega_d \cdot s_{ij}^d + \omega_n \cdot s_{ij}^n. \tag{5.12}$$

The loss terms defined in Eqs. (5.7), (5.2), and (5.11) are combined to conduct multitask learning of scene-aware pedestrian detection and segmentation supervision. The final multitask loss function is defined as

$$\mathcal{L} = \mathcal{L}_{Det} + \lambda_{sp} \cdot \mathcal{L}_{SP} + \lambda_{seg} \cdot \mathcal{L}_{Seg} \tag{5.13}$$

where λ_{sp} and λ_{seg} are the trade-off coefficient of loss term $\mathcal{L}_{SP}$ and $\mathcal{L}_{Seg}$, respectively.

5.4 Experimental Results and Discussion

5.4.1 Dataset and Evaluation Metric

The public KAIST multimodal pedestrian benchmark dataset [26] are utilized to evaluate our proposed methods. The KAIST dataset consists of 50,172 well-aligned visible-thermal image pairs with 13,853 pedestrian annotations in training set and 2252 image pairs with 1356 pedestrian annotations in the testing set. All the images in KAIST dataset are captured under daytime and nighttime lighting conditions. According to the related research work [28,30], we sample the training images every two frames. The original annotations under the "reasonable" evaluation setting (pedestrian instances are larger than 55 pixels height and over 50% visibility) are used for quantitative evaluation. Considering that many problematic annotations (e.g., missed pedestrian targets and inaccurate bounding boxes) are existed in the original KAIST testing dataset, we also utilize the improved annotations manually labeled by Liu et al. [37] for quantitative evaluation.

The log-average miss rate (MR) proposed by Dollar et al. [12] is used as the standard evaluation metric to evaluate the quantitative performance of multimodal pedestrian detectors. According to the related research work [26,28,30], a detection is considered as a true positive if the IoU ratio between the bounding boxes of the detection and any ground-truth label is greater than 50% [26,28,30]). Unmatched detections and unmatched ground-truth labels are considered as false positives and false negatives, respectively. The MR is computed by averaging miss rate at nine false positives per image (FPPI) rates which are evenly spaced in log-space from the range 10^{-2} to 10^{0} [26,28,30].

5.4.2 Implementation Details

According to the related research work [60,28,30], the image-centric training framework is applied to generate mini-batches and each mini-batch contains one pair of multimodal images and 120 randomly selected anchor boxes. In order to make the right balance between foreground and background training samples, we set the ratio of positive and negative anchor boxes to 1:5 in each mini-batch. The first five convolutional layers in each stream of TDCNN (from Conv1-V to Conv5-V in the visible stream and from Conv1-T to Conv5-T in the thermal one) are initialized using the weights and biases of VGG-16 [48] DNN pretrained on the ImageNet dataset [46] in parallel. Following the RPN designed by Zhang et al. [60], other convolutional layers are initialized with a standard Gaussian distribution. The number of channels in SP-FC1, SP-FC2, SP-FC3 are empirically set to 512, 64, 2, respectively. We set $\lambda_r = 5$ in Eq. (5.5) following [60] and $\lambda_s = 1$ in Eq. (5.13) according to the visible segmentation supervision method proposed by Brazil et al. [4]. All of multimodal pedestrian detectors are trained using the Caffe deep learning framework [27] with stochastic gradient descent [65]

Table 5.4: The quantitative comparison (MR [12]) of TRPN using different feature fusion functions on the original KAIST testing dataset [26].

Model	All-day (%)	Daytime (%)	Nighttime (%)
TRPN-Concat	32.60	33.80	30.53
TRPN-Max	31.54	32.66	29.43
TRPN-Sum	**30.49**	**31.27**	**28.29**

Table 5.5: The quantitative comparison (MR [12]) of TRPN using different feature fusion functions on the improved KAIST testing dataset [37].

Model	All-day (%)	Daytime (%)	Nighttime (%)
TRPN-Concat	21.12	20.66	22.81
TRPN-Max	19.90	18.45	23.29
TRPN-Sum	**19.45**	**17.94**	**22.37**

for the first two epochs in the learning rate (LR) of 0.001 and one more epoch in the LR of 0.0001. We clip gradients when the L2 norm of the back-propagation gradient is larger than 10 to avoid exploding gradient problems [44].

5.4.3 Evaluation of Multimodal Feature Fusion

In order to explore the optimal feature fusion scheme for multimodal pedestrian detection, we compare the TRPN with different feature fusion layers. As exposed in Sect. 5.3.1, three different feature fusion functions (concatenation, maximization, and summation) are utilized to build the three different TRPN models (TRPN-Concat, TRPN-Max, and TRPN-Sum). The TRPN models are trained using the loss term of detection $\mathcal{L}_{Det}$. The detection performances of TRPN-Concat, TRPN-Max, and TRPN-Sum are quantitatively compared in Table 5.4 and Table 5.5 using the log-average miss rate (MR) proposed by Dollar et al. [12]. In addition, qualitative comparisons of the detection performances of the three different TRPN models are conducted by showing some detection results in Fig. 5.9.

We can observe that the multimodal pedestrian detection performance is affected by the functions of feature fusion. Our experimental comparisons show that the performance of TRPN-Sum is better than the TRPN-Concat and TRPN-Max, resulting in lower MR on both original and improved KAIST testing datasets. Surprisingly, the widely used fusion function (concatenation) to integrate feature maps generated in visible and thermal channels [53,28,30] performs worst among the three different feature fusion functions. As described in Sect. 5.3.1, the feature maps generated in visible and thermal streams are directly stacked across feature

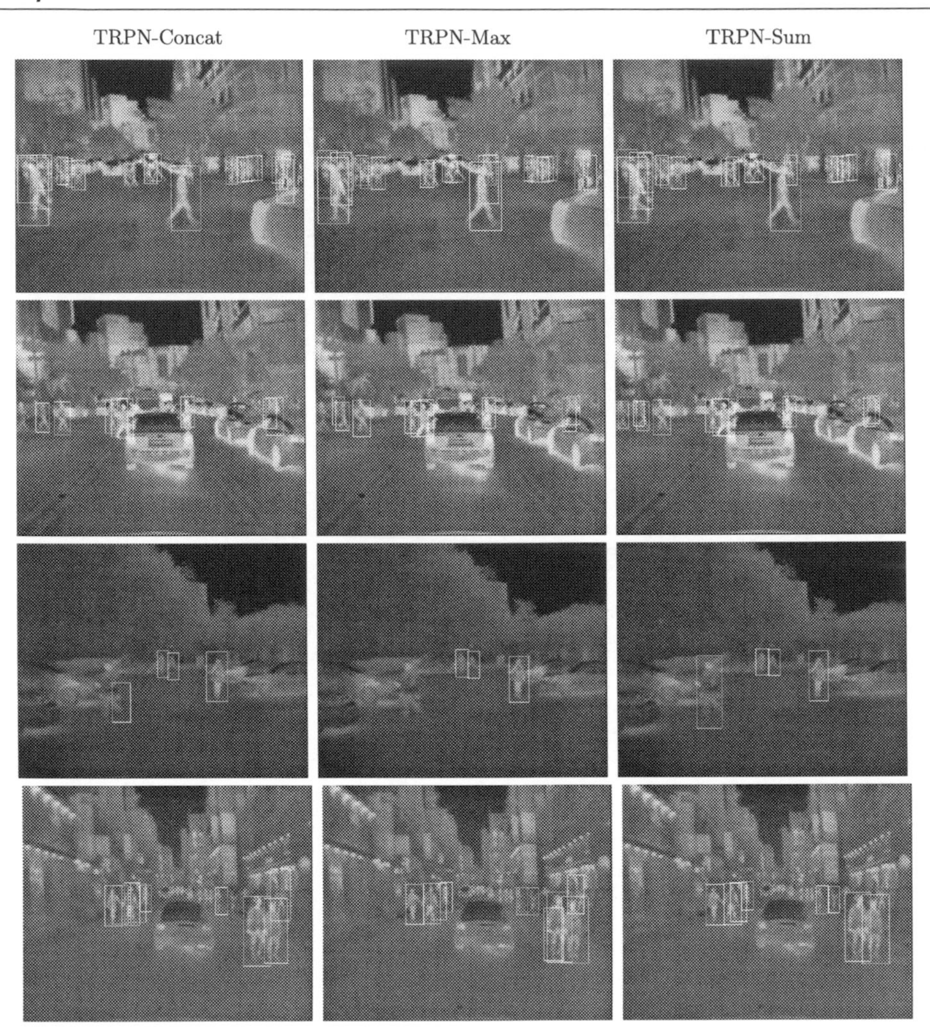

Figure 5.9 : Qualitative comparison of multimodal pedestrian detections of TRPN-Concat, TRPN-Max, and TRPN-Sum (only displayed in the thermal images). Note that yellow BB in solid line represents the pedestrian detections. This figure is best seen in color.

channels using concatenation function. Thus, the correlation of features generated in visible and thermal streams will be further learned using the Conv-Det layer in the TRPN-Concat model. As shown in Fig. 5.6, such correlation is different for daytime and nighttime multimodal pedestrian characteristics. It is difficult to build up the correlation between visible and thermal feature maps using a simple convolutional encoder (Conv-Det). On comparison, TRPN-Max and TRPN-Sum models achieve better detection results by using either maximum or summation function to define the correlation between visible and thermal feature maps. It

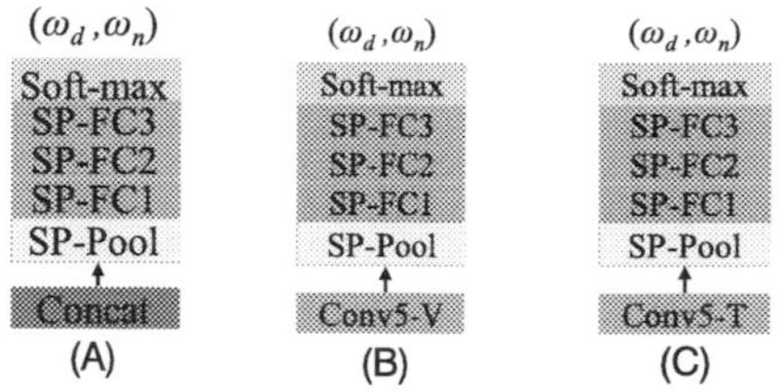

Figure 5.10 : The architecture of SPN (A), SPN-V (B) and SPN-T (C). Please note that convolutional and fully-connected layers are represented by green boxes, pooling layers are represented by yellow boxes, fusion layers are represented by blue boxes, classification layer is represented by an orange box. This figure is best seen in color.

should be noticed that the TRPN-Sum can successfully detect the pedestrian targets whose human-related characteristics are not distinct in both visible and thermal images. Different from the maximum function that selects the most distinct features in the visible and thermal streams, the summation function is able to integrate the weak features generated in the visible and thermal streams to generate a stronger multimodal one to facilitate more accurate pedestrian detection results.

5.4.4 Evaluation of Multimodal Pedestrian Detection Networks

The scene prediction network (SPN) is essential and fundamental in our proposed scene-aware multimodal pedestrian detection networks. We first evaluate whether the scene prediction networks (SPNs) can accurately compute the scene weights ω_d and $\omega_n = (1 - \omega_d)$, which supply vital information to integrate the scene-aware sub-networks. The KAIST testing dataset is utilized to evaluate thc performance of SPN. Please note that the KAIST testing dataset consists of 1455 image pairs captured in daytime scene conditions and 797 in nighttime. Given a pair of multimodal images, the SPN computes a daytime scene weight ω_d. The scene condition will be predicted correctly if $\omega_d > 0.5$ during daytime or $\omega_d < 0.5$ during nighttime. In order to investigate whether visible images or thermal ones can supply the most reliable information to predict the scene conditions, the performances of scene prediction utilizing the feature maps generated only in the visible stream (SPN-V) or thermal one (SPN-T) are evaluated individually. We illustrate the architectures of SPN-V, SPN-T, and SPN in Fig. 5.10. The prediction results of these three scene prediction networks are compared in Table 5.6.

We can observe that the SPN-V are able to generate accurate prediction of scene conditions for daytime scenes (97.94%) and nighttime ones (97.11%). The underlying reason is that the visible images display different brightness in daytime and nighttime scene conditions. The scene prediction performance using SPN-T are not comparable with SPN-T (daytime

Table 5.6: Accuracy of scene prediction utilizing SPN-V, SPN-T, and SPN.

	Daytime (%)	Nighttime (%)
SPN-V	97.94	97.11
SPN-T	93.13	94.48
SPN	**98.35**	**99.75**

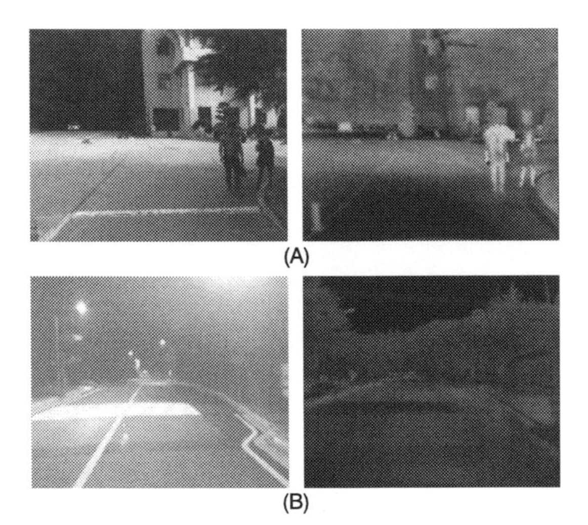

Figure 5.11 : Samples of false scene prediction results during (A) daytime and (B) nighttime. When the illumination condition is not good during daytime or street lights provide good illumination during nighttime, the SPN model will generate false prediction results.

93.13% vs. 97.94% and nighttime 94.48% vs. 97.11%). This is a reasonable result as the relative temperature between the pedestrian, and surrounding environment is not very different in daytime and nighttime scenes. Although thermal images cannot be utilized for scene prediction individually, it supplies supplementary information for the visible images to boost the performance of scene prediction. By integrating the complementary information of visible and thermal images, the SPN can generate more accurate prediction of scene conditions compared with SPN-V (daytime 98.35% vs. 97.94% and nighttime 99.75% vs. 97.11%). In addition, we show some fail cases using SPN in Fig. 5.11. The SPN may generate false scene predictions when the brightness is very low in daytime or the street lights provide high illumination in nighttime. Overall, the scene conditions can be accurately predicted utilizing the SPN by integrating visible and thermal features.

We further evaluate whether the performance of multimodal pedestrian detector can be boosted by applying our proposed scene-aware weighting mechanism by comparing the performance of the baseline TRPN and scene-aware TRPN (STRPN) models. The loss term for

Table 5.7: The calculated MRs of different TRPN and STRPN models on the original KAIST testing dataset [26].

	All-day (%)	Daytime (%)	Nighttime (%)
TRPN-Sum	30.49	31.27	28.29
STRPN-Sum	30.50	30.89	28.91
TRPN-Max	31.54	32.66	29.43
STRPN-Max	30.70	31.40	28.32
TRPN-Concat	32.60	33.80	30.53
STRPN-Concat	**29.62**	**30.30**	**26.88**

Table 5.8: The calculated MRs of Tdifferent TRPN and STRPN models on the improved KAIST testing dataset [37].

	All-day (%)	Daytime (%)	Nighttime (%)
TRPN-Sum	19.45	17.94	22.37
STRPN-Sum	17.67	17.76	18.16
TRPN-Maxn	19.90	18.45	23.29
STRPN-Max	18.40	18.15	18.69
TRPN-Concat	21.12	20.66	22.81
STRPN-Concat	**17.58**	**17.36**	**17.79**

scene prediction defined in Eq. (5.8) and loss term for detection defined in Eq. (5.2) are combined to jointly train SPN and STRPN. Although the TRPN model is an effective framework to integrate visible and thermal streams for robust pedestrian detection, it cannot differentiate human-related features under daytime and nighttime scenes when generating detection results. With comparison, STRPN utilize the scene-aware weighting mechanism to adaptively integrate multiple scene-aware sub-networks (D-Conv-CS, N-Conv-CS, and D-Conv-BB, N-Conv-BB) to generate detection results (CS and BB).

As shown in Tables 5.7 and 5.8, the quantitative comparative results of TRPN and STRPN are conducted using the log-average miss rate (MR) as the evaluation protocols. Applying the scene-aware weighting mechanism, the detection performances of the STRPN are significantly improved comparing with TRPN for all scenes on both original and improved KAIST testing datasets. We can observe that integrating the Concat fusion scheme with a global scene-aware mechanism, instead of a single Conv-Det layer, facilitates better learning of both human-related features and correlation between visible and thermal feature maps. It also worth mentioning that such performance gain (TRPN-Concat 32.60% MR v.s. STRPN-Concat 29.62% MR on the original KAIST testing dataset [26] and TRPN-Concat 21.12% MR v.s. STRPN-Concat 17.58% MR on the improved KAIST testing dataset [37]) is achieved at the cost of small computational overhead. Measured on a single Titan X GPU, the STRPN

Table 5.9: Detection results (MR) of STRPN, STRPN+MS-F, STRPN+MS, STRPN+SMS-F, and STRPN+SMS on the original KAIST testing dataset [26].

	All-day (%)	Daytime (%)	Nighttime (%)
STRPN	29.62	30.30	26.88
STRPN+MS-F	29.17	29.92	26.96
STRPN+MS	27.21	27.56	25.57
STRPN+SMS-F	28.51	28.98	27.52
STRPN+SMS	**26.37**	**27.29**	**24.41**

Table 5.10: Detection results (MR) of STRPN, STRPN+MS-F, STRPN+MS, STRPN+SMS-F, and STRPN+SMS on the improved KAIST testing dataset [26].

	All-day (%)	Daytime (%)	Nighttime (%)
STRPN	17.58	17.36	17.79
STRPN+MS-F	16.54	15.83	17.28
STRPN+MS	15.88	15.01	16.45
STRPN+SMS-F	16.41	15.17	16.91
STRPN+SMS	**14.95**	**14.67**	**15.72**

model takes 0.24 s to process a pair of multimodal images in KAIST dataset while TRPN needs 0.22 s. These experimental results show that the global scene information can be infused into scene-aware sub-networks to boost the performance of the multimodal pedestrian detector.

5.4.5 Evaluation of Multimodal Segmentation Supervision Networks

In this section, we investigate whether the performance of multimodal pedestrian detection can be improved by incorporating the segmentation supervision scheme with STRPN. As described in Sect. 5.3.3, four different multimodal segmentation supervision (MS) models including MS-F (feature-level MS), MS (decision-level MS), SMS-F (scene-aware feature-level MS) and SMS (scene-aware decision-level MS) are combined with STRPN to build STRPN, STRPN+MS-F, STRPN+MS and STRPN+SMS-F and STRPN+SMS respectively. Multimodal segmentation supervision models generate the box-based segmentation prediction and supply the supervision to make the multimodal feature maps more distinctive. The detection results (MR) of the STRPN, STRPN+MS-F, STRPN+MS and STRPN+SMS-F and STRPN+SMS are compared in Table 5.9 and Table 5.10.

Through the joint training of segmentation supervision and pedestrian detection, all the multimodal segmentation supervision models, except for STRPN+MS-F in nighttime scene, achieve performance gains. The reason is that semantic segmentation masks provide additional supervision to facilitate the training of more sophisticated features to enable more robust detection [4]. Meanwhile, we observe that the choice of fusion architectures (feature-level or decision-level) can significantly affect the performance of multimodal pedestrian detection results. As illustrated in Tables 5.9 and 5.10, the detection results of decision-level multimodal segmentation supervision models (MS and SMS) are much better than the feature-level models (MS-F and SMS-F). The reason is that decision-level models can generate more effective supervision information by infusing visible and thermal segmentation information directly for learning human-related features for multimodal pedestrian detection. It will be our future research work to explore the optimal segmentation fusion scheme for the effective supervision of multimodal pedestrian detection. More importantly, we can observe that the performance of segmentation supervision can be significantly improved by applying the scene-aware weighting mechanism. More accurate segmentation prediction can be generated by adaptively integrating the scene-aware segmentation sub-networks. Some comparative segmentation predictions utilizing MS-F, MS, SMS-F, and SMS are shown in Fig. 5.12. The STRPN+SMS can generate more accurate segmentation predictions which supply better supervision information to facilitate the training of human-related features for multimodal pedestrian detection task.

In order to show the improvements gains achieved by different scene-aware modules, we visualize the feature maps of TRPN, STRPN, and STRPN+SMS in Fig. 5.13. It can be observed that STRPN can generate more distinctive human-related features by incorporating scene-aware weighting mechanism into TRPN for better learning of multimodal human-related features. More importantly, further improvements can be achieved by STRPN+SMS through the scene-aware segmentation modules to supervise the learning of multimodal human-related features.

5.4.6 Comparison with State-of-the-Art Multimodal Pedestrian Detection Methods

We compare the designed STRPN and STRPN+SMS models with the current state-of-the-art multimodal pedestrian detection algorithms: ACF+T+THOG [26], Halfway Fusion [28] and Fusion RPN+BDT [30]. The log–log figure of MR vs. FPPI is plotted for performance comparison in Fig. 5.14. It can be observed that our proposed STRPN+SMS achieves the best detection accuracy (26.37% MR on the original KAIST testing dataset [26] and 14.95% MR on the improved KAIST testing dataset [37]) in all-day scenes (see Fig. 5.15). The performance gain on the improved KAIST testing dataset achieves a relative improvement rate of 18% compared with the best of current state-of-the-art algorithm Fusion RPN+BDT (18.23% MR on the improved KAIST testing dataset [37]). In addition, the detection performance of our proposed STRPN is comparable with the current state-of-art models. Furthermore, some

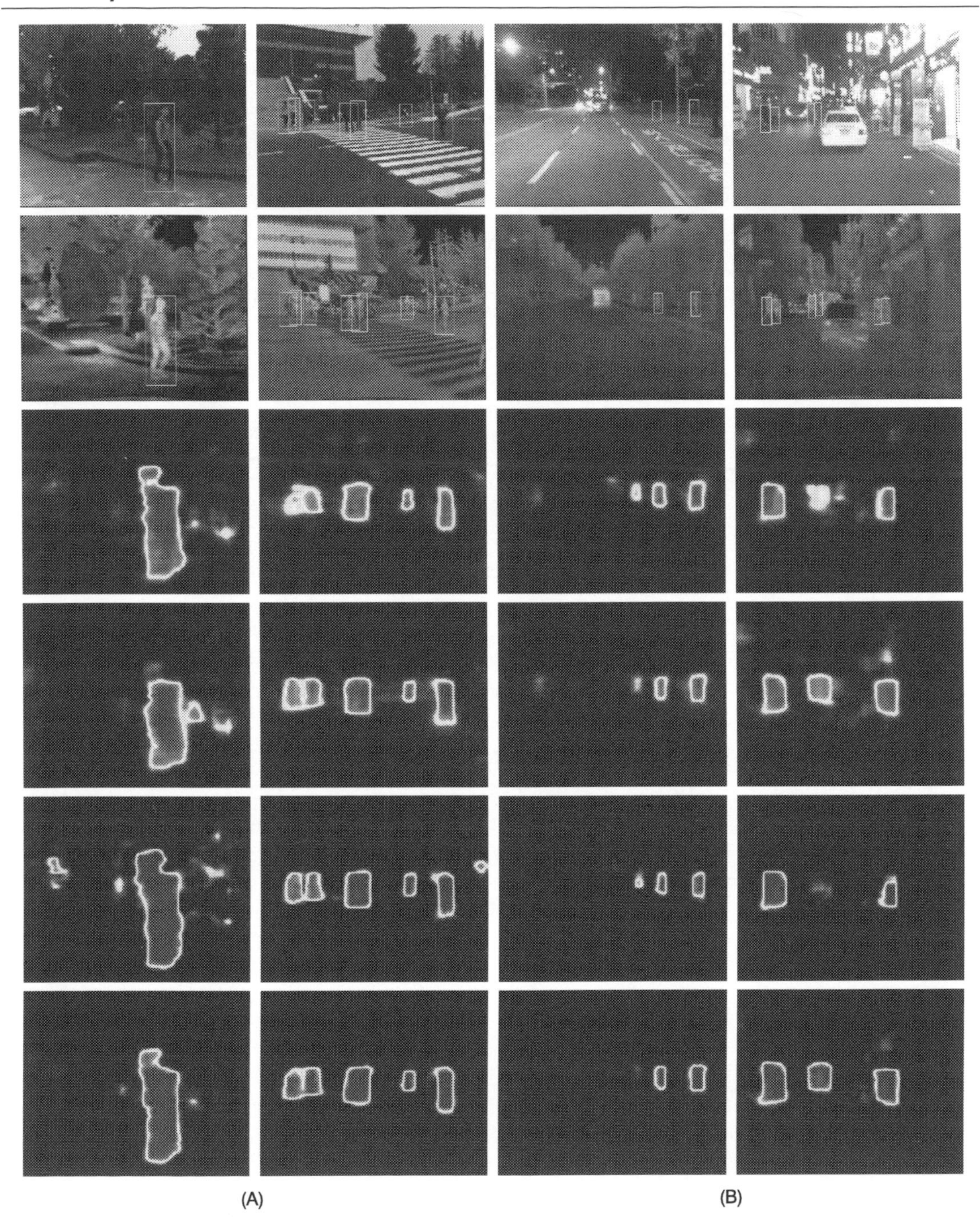

Figure 5.12 : Examples of pedestrian segmentation prediction utilizing four different multimodal segmentation supervision models. (A) Daytime. (B) Nighttime. The first two rows in (A) and (B) illustrate the visible and thermal images, respectively. Other rows in (A) and (B) illustrate the pedestrian segmentation prediction results of MS-F, MS, SMS-F, and SMS respectively. Please note that green BB presents pedestrian labels. This figure is best seen in color.

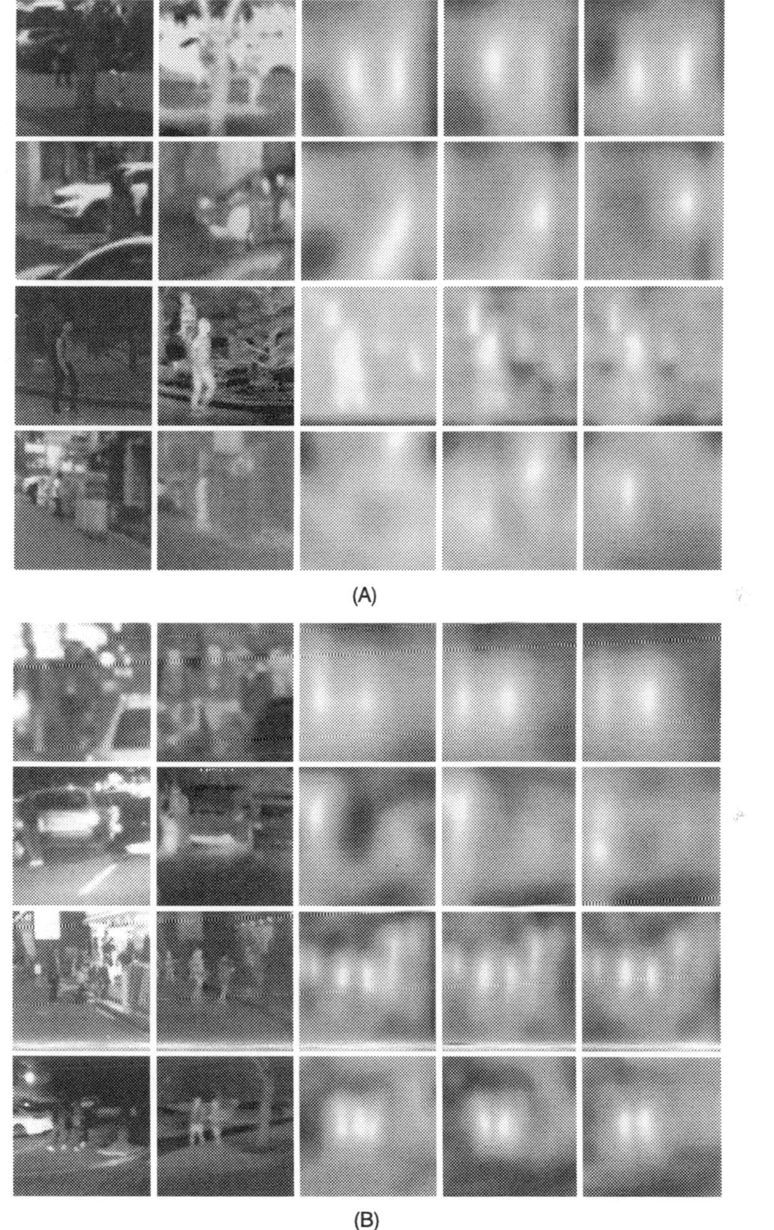

Figure 5.13 : Examples of multimodal pedestrian feature maps which are promoted by scene-aware mechanism captured in (A) daytime and (B) nighttime scenes. The first two columns in (A) and (B) show the pictures of visible and thermal pedestrian instances, respectively. The third to the fifth columns in (A) and (B) show the feature map visualizations generated from TRPN, STRPN, and STRPN+SMS respectively. It is noticed that the feature maps of a multimodal pedestrian become more distinct by using our proposed scene-aware modules (STRPN and SMS).

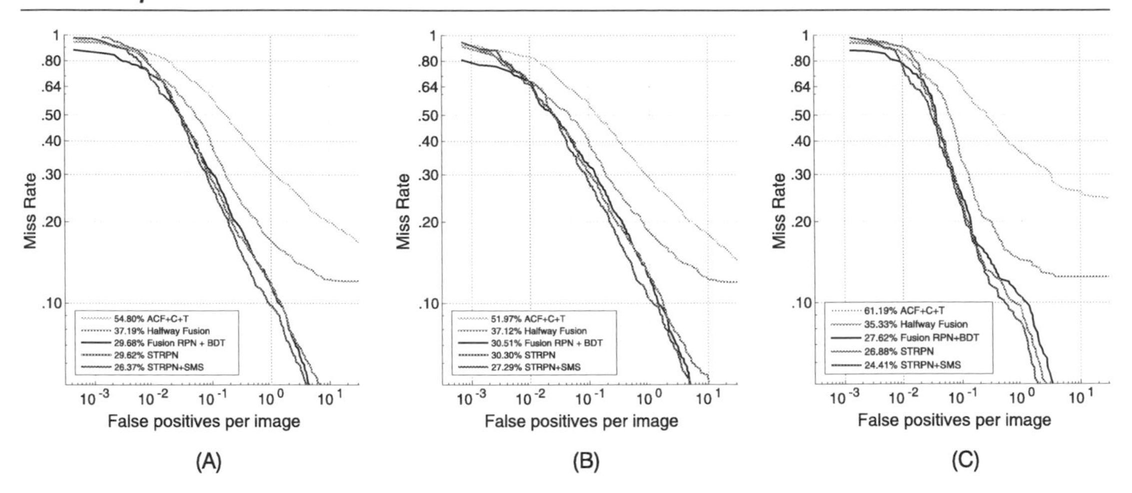

Figure 5.14 : Comparisons of the original KAIST test dataset under the reasonable evaluation setting during all-day (A), daytime (B), and nighttime (C) (legends indicate MR).

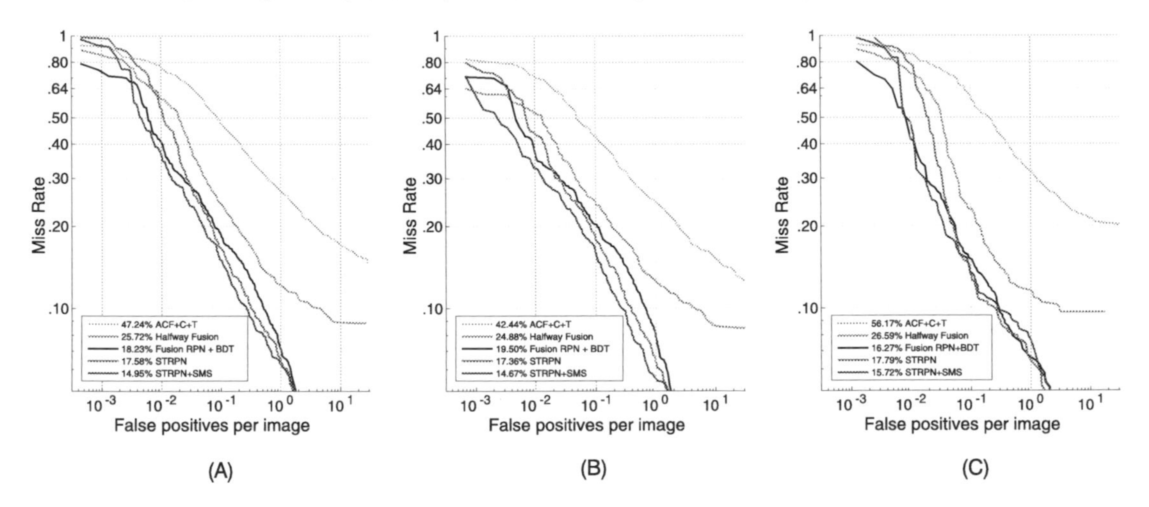

Figure 5.15 : Comparisons of the improved KAIST test dataset under the reasonable evaluation setting during all-day (A), daytime (B), and nighttime (C) (legends indicate MR).

detection results of the Fusion RPN+BDT and STRPN+SMS are visualized for qualitative comparison as shown in Fig. 5.16. We can observe that our proposed STRPN+SMS model can generate more accurate detection results in both daytime and nighttime scene conditions.

The computation efficiency of STRPN, STRPN+SMS and the current state-of-the-art methods [28,30] are illustrated in Table 5.11. Every method is executed 1000 times to compute the averaged runtime. We can observe that our proposed STRPN+SMS results in least runtime comparing with the current state-of-the-art methods [28,30] (STRPN+SMS 0.25 s vs.

Figure 5.16 : Comparison with the current state-of-the-art multimodal pedestrian detector (Fusion RPN+BDT). The first column shows the multimodal images along with the ground truth (displayed in the visible image) and the other columns show the detection results of Fusion RPN+BDT, STRPN, and STRPN+SMS (displayed in the thermal image). Please note that green BBs in solid line show positive labels, green BBs in dashed line show ignore ones, yellow BB in solid line show true positives, and red BB show false positives. This figure is best seen in color.

Halfway Fusion 0.40 s vs. Fusion RPN+BDT 0.80 s). The reason is that both the extra Fast R-CNN module [18] utilized in the Halfway Fusion model and the boosted decision trees [14] algorithm utilized in Fusion RPN+BDT model significantly decrease the computation efficiency. It should be noticed that our proposed scene-aware architectures can significantly

Table 5.11: Comprehensive comparison of STRPN and STRPN+SMS with state-of-the-art multimodal pedestrian detectors [28,30] on the improved KAIST testing dataset. A single Titan X GPU is used to evaluate the computation efficiency. Note that DL represents deep learning and BDT represents boosted decision trees [14].

	MR (%)	Runtime (s)	Method
Halfway Fusion [28]	25.72	0.40	DL
Fusion RPN+BDT [30]	18.23	0.80	DL+BDT
TRPN	21.12	0.22	DL
STRPN	17.58	0.24	DL
STRPN+SMS	14.95	0.25	DL

boost the multimodal pedestrian detection results while only cause a small computational overhead (TRPN 0.24 s vs. STRPN 0.24 s vs. STRPN+SMS 0.25 s).

5.5 Conclusion

In this chapter, we present a new multimodal pedestrian detection method based on multi-task learning of scene-aware target detection and segmentation supervision. To achieve better learning of both human-related features and correlation between visible and thermal feature maps, we propose a feature fusion scheme utilizing the concatenation with a global weighting scene-aware mechanism. Experimental results illustrate that multimodal pedestrian detections can be improved by applying our proposed scene-aware weighting mechanism. Moreover, we design four different multimodal segmentation supervision architectures and conclude that scene-aware decision-level multimodal segmentation (SMS) module can generate the most accurate prediction as supervision for visual feature learning. Experimental evaluation on public KAIST benchmark shows that our method achieves the most accurate results with least runtime compared with the current state-of-the-art multimodal pedestrian detectors.

Acknowledgment

This research was supported by the National Natural Science Foundation of China (No. 51605428, No. 51575486, U1664264) and DFG (German Research Foundation) YA 351/2-1.

References

[1] Rodrigo Benenson, Mohamed Omran, Jan Hosang, Bernt Schiele, Ten years of pedestrian detection, what have we learned?, in: European Conference on Computer Vision, 2014, pp. 613–627.

[2] M. Bilal, A. Khan, M.U. Karim Khan, C.M. Kyung, A low-complexity pedestrian detection framework for smart video surveillance systems, IEEE Transactions on Circuits and Systems for Video Technology 27 (10) (2017) 2260–2273.
[3] Sujoy Kumar Biswas, Peyman Milanfar, Linear support tensor machine with lsk channels: pedestrian detection in thermal infrared images, IEEE Transactions on Image Processing 26 (9) (2017) 4229–4242.
[4] Garrick Brazil, Xi Yin, Xiaoming Liu, Illuminating pedestrians via simultaneous detection & segmentation, in: IEEE International Conference on Computer Vision, IEEE, 2017.
[5] Zhaowei Cai, Quanfu Fan, Rogerio S. Feris, Nuno Vasconcelos, A unified multi-scale deep convolutional neural network for fast object detection, in: European Conference on Computer Vision, 2016, pp. 354–370.
[6] Yanpeng Cao, Dayan Guan, Weilin Huang, Jiangxin Yang, Yanlong Cao, Yu Qiao, Pedestrian detection with unsupervised multispectral feature learning using deep neural networks, Information Fusion 46 (2019) 206–217.
[7] Marius Cordts, Mohamed Omran, Sebastian Ramos, Timo Rehfeld, Markus Enzweiler, Rodrigo Benenson, Uwe Franke, Stefan Roth, Bernt Schiele, The cityscapes dataset for semantic urban scene understanding, in: IEEE Conference on Computer Vision and Pattern Recognition, IEEE, 2016, pp. 3213–3223.
[8] Corinna Cortes, Vladimir Vapnik, Support-vector networks, Machine Learning 20 (3) (1995) 273–297.
[9] Navneet Dalal, Bill Triggs, Histograms of oriented gradients for human detection, in: IEEE Conference on Computer Vision and Pattern Recognition, vol. 1, 2005, pp. 886–893.
[10] James W. Davis, Mark A. Keck, A two-stage template approach to person detection in thermal imagery, in: Application of Computer Vision, 2005, vol. 1, Seventh IEEE Workshops on, WACV/MOTIONS'05, IEEE, 2005, pp. 364–369.
[11] Piotr Dollár, Zhuowen Tu, Pietro Perona, Serge Belongie, Integral channel features, in: British Machine Vision Conference, 2009.
[12] Piotr Dollár, Christian Wojek, Bernt Schiele, Pietro Perona, Pedestrian detection: an evaluation of the state of the art, IEEE Transactions on Pattern Analysis and Machine Intelligence 34 (4) (2012) 743–761.
[13] Andreas Ess, Bastian Leibe, Luc Van Gool, Depth and appearance for mobile scene analysis, in: IEEE International Conference on Computer Vision, IEEE, 2007, pp. 1–8.
[14] Yoav Freund, Robert E. Schapire, A decision-theoretic generalization of on-line learning and an application to boosting, in: European Conference on Computational Learning Theory, 1995, pp. 23–37.
[15] Andreas Geiger, Philip Lenz, Raquel Urtasun, Are we ready for autonomous driving? The kitti vision benchmark suite, in: IEEE Conference on Computer Vision and Pattern Recognition, IEEE, 2012, pp. 3354–3361.
[16] David Geronimo, Antonio M. Lopez, Angel D. Sappa, Thorsten Graf, Survey of pedestrian detection for advanced driver assistance systems, IEEE Transactions on Pattern Analysis and Machine Intelligence 32 (7) (2010) 1239–1258.
[17] Samuel Gidel, Paul Checchin, Christophe Blanc, Thierry Chateau, Laurent Trassoudaine, Pedestrian detection and tracking in an urban environment using a multilayer laser scanner, IEEE Transactions on Intelligent Transportation Systems 11 (3) (2010) 579–588.
[18] Ross Girshick, Fast r-cnn, in: IEEE International Conference on Computer Vision, 2015, pp. 1440–1448.
[19] Ross Girshick, Jeff Donahue, Trevor Darrell, Jitendra Malik, Rich feature hierarchies for accurate object detection and semantic segmentation, in: IEEE Conference on Computer Vision and Pattern Recognition, 2014, pp. 580–587.
[20] Alejandro González, Zhijie Fang, Yainuvis Socarras, Joan Serrat, David Vázquez, Jiaolong Xu, Antonio M. López, Pedestrian detection at day/night time with visible and fir cameras: a comparison, Sensors 16 (6) (2016) 820.
[21] F.P. León, A.P. Grassi, V. Frolov, Information fusion to detect and classify pedestrians using invariant features, Information Fusion 12 (2011) 284–292.
[22] Dayan Guan, Yanpeng Cao, Jiangxin Yang, Yanlong Cao, Christel-Loic Tisse, Exploiting fusion architectures for multispectral pedestrian detection and segmentation, Applied Optics 57 (2018) 108–116.

[23] Dayan Guan, Yanpeng Cao, Jiangxin Yang, Yanlong Cao, Michael Ying Yang, Fusion of multispectral data through illumination-aware deep neural networks for pedestrian detection, Information Fusion 50 (2019) 148–157.

[24] Kaiming He, Xiangyu Zhang, Shaoqing Ren, Jian Sun, Spatial pyramid pooling in deep convolutional networks for visual recognition, in: European Conference on Computer Vision, Springer, 2014, pp. 346–361.

[25] Kaiming He, Georgia Gkioxari, Piotr Dollar, Ross Girshick, Mask r-cnn, in: IEEE International Conference on Computer Vision, 2017.

[26] Soonmin Hwang, Jaesik Park, Namil Kim, Yukyung Choi, In So Kweon, Multispectral pedestrian detection: benchmark dataset and baseline, in: IEEE Conference on Computer Vision and Pattern Recognition, 2015, pp. 1037–1045.

[27] Yangqing Jia, Evan Shelhamer, Jeff Donahue, Sergey Karayev, Jonathan Long, Ross Girshick, Sergio Guadarrama, Trevor Darrell, Caffe: convolutional architecture for fast feature embedding, in: International Conference on Multimedia, 2014, pp. 675–678.

[28] Liu Jingjing, Zhang Shaoting, Wang Shu, Metaxas Dimitris, Multispectral deep neural networks for pedestrian detection, in: British Machine Vision Conference, 2016, pp. 73.1–73.13.

[29] Taehwan Kim, Sungho Kim, Pedestrian detection at night time in fir domain: comprehensive study about temperature and brightness and new benchmark, Pattern Recognition 79 (2018) 44–54.

[30] Daniel König, Michael Adam, Christian Jarvers, Georg Layher, Heiko Neumann, Michael Teutsch, Fully-convolutional region proposal networks for multispectral person detection, in: IEEE Conference on Computer Vision and Pattern Recognition Workshops, 2017, pp. 243–250.

[31] Stephen J. Krotosky, Mohan Manubhai Trivedi, Person surveillance using visual and infrared imagery, IEEE Transactions on Circuits and Systems for Video Technology 18 (8) (2008) 1096–1105.

[32] Yi-Shu Lee, Yi-Ming Chan, Li-Chen Fu, Pei-Yung Hsiao, Near-infrared-based nighttime pedestrian detection using grouped part models, IEEE Transactions on Intelligent Transportation Systems 16 (4) (2015) 1929–1940.

[33] Alex Leykin, Yang Ran, Riad Hammoud, Thermal-visible video fusion for moving target tracking and pedestrian classification, in: IEEE Conference on Computer Vision and Pattern Recognition, IEEE, 2007, pp. 1–8.

[34] J. Li, X. Liang, S. Shen, T. Xu, J. Feng, S. Yan, Scale-aware fast r-cnn for pedestrian detection, IEEE Transactions on Multimedia 20 (4) (2018) 985–996.

[35] Xiaofei Li, Lingxi Li, Fabian Flohr, Jianqiang Wang, Hui Xiong, Morys Bernhard, Shuyue Pan, Dariu M. Gavrila, Keqiang Li, A unified framework for concurrent pedestrian and cyclist detection, IEEE Transactions on Intelligent Transportation Systems 18 (2) (2017) 269–281.

[36] Xudong Li, Mao Ye, Yiguang Liu, Feng Zhang, Dan Liu, Song Tang, Accurate object detection using memory-based models in surveillance scenes, Pattern Recognition 67 (2017) 73–84.

[37] Jingjing Liu, Shaoting Zhang, Shu Wang, Dimitris Metaxas, Improved annotations of test set of kaist, http://paul.rutgers.edu/~jl1322/multispectral.htm/, 2018.

[38] Jiayuan Mao, Tete Xiao, Yuning Jiang, Zhimin Cao, What can help pedestrian detection?, in: IEEE Conference on Computer Vision and Pattern Recognition, July 2017.

[39] Woonhyun Nam, Piotr Dollár, Joon Hee Han, Local decorrelation for improved pedestrian detection, in: Advances in Neural Information Processing Systems, 2014, pp. 424–432.

[40] H. Nanda, L. Davis, Probabilistic template based pedestrian detection in infrared videos, in: Intelligent Vehicle Symposium, vol. 1, 2003, pp. 15–20.

[41] Miguel Oliveira, Vitor Santos, Angel D. Sappa, Multimodal inverse perspective mapping, Information Fusion 24 (2015) 108–121.

[42] Michael Oren, Constantine Papageorgiou, Pawan Sinha, Edgar Osuna, Tomaso Poggio, Pedestrian detection using wavelet templates, in: IEEE Conference on Computer Vision and Pattern Recognition, IEEE, 1997, pp. 193–199.

[43] Constantine Papageorgiou, Tomaso Poggio, Trainable pedestrian detection, in: Image Processing, 1999, Proceedings. 1999 International Conference on, vol. 4, ICIP 99, IEEE, 1999, pp. 35–39.

[44] Razvan Pascanu, Tomas Mikolov, Yoshua Bengio, On the difficulty of training recurrent neural networks, in: International Conference on Machine Learning, 2013, pp. 1310–1318.
[45] Shaoqing Ren, Kaiming He, Ross Girshick, Jian Sun, Faster r-cnn: towards real-time object detection with region proposal networks, in: Advances in Neural Information Processing Systems, 2015, pp. 91–99.
[46] Olga Russakovsky, Jia Deng, Hao Su, Jonathan Krause, Sanjeev Satheesh, Sean Ma, Zhiheng Huang, Andrej Karpathy, Aditya Khosla, Michael Bernstein, et al., Imagenet large scale visual recognition challenge, International Journal of Computer Vision 115 (3) (2015) 211–252.
[47] Pierre Sermanet, Koray Kavukcuoglu, Soumith Chintala, Yann Lecun, Pedestrian detection with unsupervised multi-stage feature learning, in: Computer Vision and Pattern Recognition, 2013, pp. 3626–3633.
[48] Karen Simonyan, Andrew Zisserman, Very deep convolutional networks for large-scale image recognition, in: International Conference on Learning Representations, 2015.
[49] Frédéric Suard, Alain Rakotomamonjy, Abdelaziz Bensrhair, Alberto Broggi, Pedestrian detection using infrared images and histograms of oriented gradients, in: Intelligent Vehicles Symposium, 2006 IEEE, IEEE, 2006, pp. 206–212.
[50] Yonglong Tian, Ping Luo, Xiaogang Wang, Xiaoou Tang, Deep learning strong parts for pedestrian detection, in: IEEE International Conference on Computer Vision, 2015, pp. 1904–1912.
[51] Atousa Torabi, Guillaume Massé, Guillaume-Alexandre Bilodeau, An iterative integrated framework for thermal–visible image registration, sensor fusion, and people tracking for video surveillance applications, Computer Vision and Image Understanding 116 (2) (2012) 210–221.
[52] P. Viola, M. Jones, Robust real-time face detection, International Journal of Computer Vision 57 (2) (2004) 137–154.
[53] Jörg Wagner, Volker Fischer, Michael Herman, Sven Behnke, Multispectral pedestrian detection using deep fusion convolutional neural networks, in: European Symposium on Artificial Neural Networks, 2016, pp. 509–514.
[54] Xiaogang Wang, Meng Wang, Wei Li, Scene-specific pedestrian detection for static video surveillance, IEEE Transactions on Pattern Analysis and Machine Intelligence 36 (2) (2014) 361–374.
[55] B. Wu, R. Nevatia, Detection of multiple, partially occluded humans in a single image by bayesian combination of edgelet part det, in: IEEE Intl. Conf. Computer Vision, 2005.
[56] Bichen Wu, Forrest Iandola, Peter H. Jin, Kurt Keutzer, Squeezedet: unified, small, low power fully convolutional neural networks for real-time object detection for autonomous driving, arXiv preprint, arXiv:1612.01051, 2016.
[57] Dan Xu, Wanli Ouyang, Elisa Ricci, Xiaogang Wang, Nicu Sebe, Learning cross-modal deep representations for robust pedestrian detection, in: Proceedings of the IEEE Conference on Computer Vision and Pattern Recognition, 2017, pp. 5363–5371.
[58] Yuan Yuan, Xiaoqiang Lu, Xiao Chen, Multi-spectral pedestrian detection, Signal Processing 110 (2015) 94–100.
[59] Li Zhang, Bo Wu, R. Nevatia, Pedestrian detection in infrared images based on local shape features, in: Computer Vision and Pattern Recognition, 2007, IEEE Conference on, CVPR '07, 2007, pp. 1–8.
[60] Liliang Zhang, Liang Lin, Xiaodan Liang, Kaiming He, Is faster r-cnn doing well for pedestrian detection?, in: European Conference on Computer Vision, 2016, pp. 443–457.
[61] S. Zhang, R. Benenson, M. Omran, J. Hosang, B. Schiele, Towards reaching human performance in pedestrian detection, IEEE Transactions on Pattern Analysis and Machine Intelligence 40 (4) (2018) 973–986.
[62] Shanshan Zhang, Rodrigo Benenson, Bernt Schiele, Filtered channel features for pedestrian detection, in: IEEE Conference on Computer Vision and Pattern Recognition, 2015.
[63] Shanshan Zhang, Rodrigo Benenson, Bernt Schiele, Citypersons: a diverse dataset for pedestrian detection, in: The IEEE Conference on Computer Vision and Pattern Recognition, vol. 1, CVPR, 2017, p. 3.
[64] Xinyue Zhao, Zaixing He, Shuyou Zhang, Dong Liang, Robust pedestrian detection in thermal infrared imagery using a shape distribution histogram feature and modified sparse representation classification, Pattern Recognition 48 (6) (2015) 1947–1960.
[65] Martin Zinkevich, Markus Weimer, Lihong Li, Alex J. Smola, Parallelized stochastic gradient descent, in: Advances in Neural Information Processing Systems, 2010, pp. 2595–2603.

CHAPTER 6

Multispectral Person Re-Identification Using GAN for Color-to-Thermal Image Translation

Vladimir V. Kniaz*,†, Vladimir A. Knyaz*,†

*State Res. Institute of Aviation Systems (GosNIIAS), Moscow, Russia †Moscow Institute of Physics and Technology (MIPT), Moscow, Russia

Contents

Multimodal Scene Understanding
https://doi.org/10.1016/B978-0-12-817358-9.00012-3

6.1 Introduction

Person re-identification (ReID) is of primary importance for tasks such as video surveillance and photo-tagging. It has been the focus of intense research in recent years. While modern methods provide excellent results in a well-lit environment, their performance is not robust without a suitable light source.

Infrared cameras perceive thermal emission that is invariant to the lighting conditions. However, challenges of cross-modality color-infrared matching reduce benefits of night mode infrared cameras. Recently, cross-modality color-to-thermal matching received a lot of scholar attention [1–4]. Multiple datasets with infrared images [5,1,2,6] were developed for cross-modality infrared-to-color person ReID. Thermal cameras operate in longwave infrared (LWIR, 8–14 μ) and provide real temperatures of a person body which are more stable to viewpoint changes than near-infrared images [7–9].

This chapter is focused on the development of a `ThermalGAN` framework for color-thermal cross-modality person ReID. We use assumptions of Bhuiyan [10] and Zhu [11] as a starting point for our research to develop a color-to-thermal transfer framework for cross-modality person ReID. We perform matching using calibrated thermal images to benefit from the stability of surface temperatures to changes in light intensity and viewpoint. Matching is performed in two steps. Firstly, we model a person's appearance in a thermal image conditioned by a color image. We generate a multimodal thermal probe set from a single color probe image using a generative adversarial network (GAN). Secondly, we perform matching in thermal images using the synthesized thermal probe set and a real thermal gallery set (Fig. 6.1).

We collected a new ThermalWorld dataset to train our GAN framework and to test the ReID performance. The dataset contains two parts: ReID and Visual Objects in Context (VOC). The ReID split includes 15,118 pairs of color and thermal images and 516 person ID. The VOC part is designed for training color-to-thermal translation using a GAN framework. It consists of 5098 pairs of color and thermal images and pixel-level annotations for ten classes: person, car, truck, van, bus, building, cat, dog, tram, boat.

We perform an evaluation of baselines and our framework on the ThermalWorld dataset. The results of the evaluation are encouraging and show that our `ThermalGAN` framework achieves and surpasses the state of the art in the cross-modality color-thermal ReID. The new `ThermalGAN` framework will be able to perform matching of color probe image with thermal gallery set in video surveillance applications.

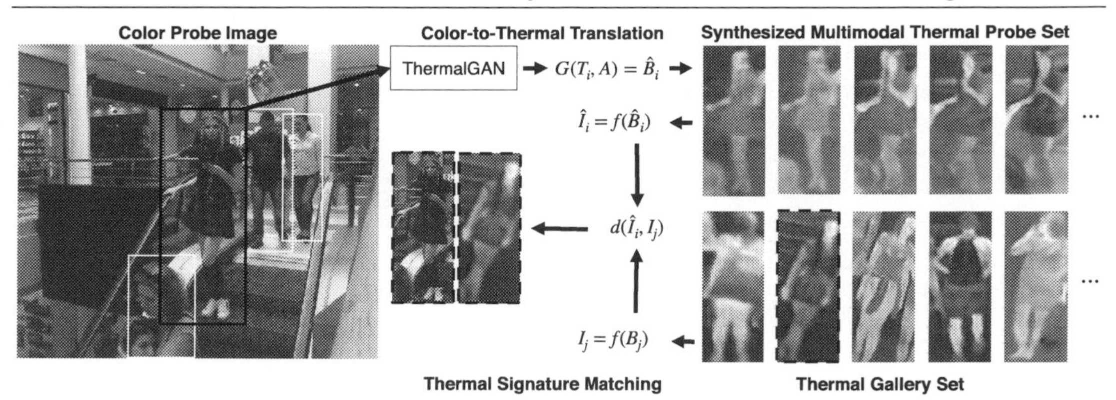

Figure 6.1 : Overview of color-thermal ReID using our ThermalGAN framework. We transform a single color probe image A to multimodal thermal probe set $\{B_1, \ldots, B_i\}$. We use thermal signatures I to perform matching with real thermal gallery set.

We present three main contributions: (1) the ThermalGAN framework for color-to-thermal image translation and ReID, (2) a large-scale multispectral ThermalWorld dataset with two splits: ReID with 15,118 color-thermal image pairs and 516 person ID, and VOC with 5098 pairs color-thermal image pairs with ground truth pixel-level object annotations of ten object classes, (3) an evaluation of the modern cross-modality ReID methods on ThermalWorld ReID dataset.

Sect. 6.2 presents the structure of the developed ThermalWorld dataset. In Sect. 6.3 we describe the ThermalGAN framework and thermal signature-based matching. In Sect. 6.4 we give an evaluation of ReID baselines and the ThermalGAN framework on the ThermalWorld dataset.

6.2 Related Work

6.2.1 Person Re-Identification

Person re-identification has been intensively studied by computer vision society recently [10, 12–15,6]. While new methods improve the matching performance steadily, video surveillance applications still pose challenges for ReID systems. Recent methods regarding person ReID can be divided into three kinds of approaches [10]: direct methods, metric learning methods and transform learning methods.

In [10] an overview of modern ReID methods was performed, and a new transform learning-based method was proposed. The method models the appearance of a person in a new camera using a cumulative weight brightness transfer function (CWBTF). The method leverages a robust segmentation technique to segment the human image into meaningful parts. Matching of features extracted only from the body area provides an increased ReID performance. The method also exploits multiple pedestrian detections to improve the matching rate.

While the method provides the state-of-the-art performance on color images, night-time application requires additional modalities to perform robust matching in low-light conditions. Cross-modality color-infrared matching is gaining increasing attention. Multiple multispectral datasets were collected in recent years [5,1,2,6,4]. The SYSU-MM01 dataset [6] includes unpaired color and near-infrared images. The RegDB dataset [4] presents color and infrared images for evaluation of cross-modality ReID methods. Evaluation of modern methods on these datasets has shown that color-infrared matching is challenging. While modern ReID methods achieve ReID rate ($r = 20$) of 90 in the visible range [10], the same rate for cross-modality ReID methods reach only 70 [4]. Nevertheless, cross-modality approach provides an increase in ReID robustness during the night-time.

Thermal camera has received a lot of scholar attention in the field of video surveillance [16, 17]. While thermal cameras provide a significant boost in pedestrian detection [18,19] and ReID with paired color and thermal images [5], cross-modality person ReID is challenging [1, 2,20,5,21] due to severe changes in a person's appearance in color and thermal images.

Recently, generative adversarial networks (GANs) [22] have demonstrated encouraging results in arbitrary image-to-image translation applications. We hypothesize that color-to-thermal image translation using a dedicated GAN framework can increase color-thermal ReID performance.

6.2.2 Color-to-Thermal Translation

Transformation of the spectral range of an image has been intensively studied in such fields as transfer learning, domain adaptation [23–26] and cross-domain recognition [6,8,27–32]. In [33] a deep convolutional neural network (CNN) was proposed for translation of a near-infrared image to a color image. The approach was similar to image colorization [34,35] and style transfer [36,37] problems that were actively studied in recent years. The proposed architecture succeeded in a translation of near-infrared images to color images. Transformation of LWIR images is more challenging due to the low correlation between the red channel of a color image and a thermal image.

6.2.3 Generative Adversarial Networks

GANs increased the quality of image-to-image translation significantly [38,8,9] using an antagonistic game approach [22]. Isola et al. [38] proposed a `pix2pix` GAN framework for arbitrary image transformations using geometrically aligned image pairs sampled from source and target domains. The framework succeeded in performing arbitrary image-to-image translations such as season change and object transfiguration. Zhang et al. [8,9] trained the `pix2pix` network to transform the thermal image of a human face to the color image. The translation improved the quality of a face recognition performance in a cross-modality thermal to visible range setting. While human face has a relatively stable temperature, color-thermal image translation for the whole human body with an arbitrary background is more ambiguous and conditioned by the sequence of events that have occurred with a person.

We hypothesize that multimodal image generation methods can model multiple possible thermal outputs for a single color probe image. Such modeling can improve the ReID performance. Zhu et al. proposed a `BicycleGAN` framework [39] for modeling a distribution of possible outputs in a conditional generative modeling setting. To resolve the ambiguity of the mapping Zhu et al. used a randomly sampled low-dimension latent vector. The latent vector is produced by an encoder network from the generated image and compared with the original latent vector to provide an additional consistency loss. We propose a conditional color-to-thermal translation framework for modeling of a set of possible appearances of a person in a thermal image conditioned by a single color image.

6.3 ThermalWorld Dataset

We collected a new ThermalWorld dataset to train and evaluate our cross-modality ReID framework. The dataset was collected with multiple FLIR ONE PRO cameras and divided into two splits: ReID and visual objects in context (VOC). The ReID split includes 15,118 aligned color and thermal image pairs of 516 IDs. The VOC split was created for effective color-to-thermal translation GAN training. It contains 5098 color and thermal image pairs and pixel-level annotations of ten object classes: person, car, truck, van, bus, building, cat, dog, tram, boat.

Initially, we have tried to train a color-to-thermal translation GAN model using only the ReID split. However, the trained network has poor generalization ability due to a limited number of object classes and backgrounds. This stimulated us to collect a large-scale dataset with aligned pairs of color and thermal images. The rest of this section presents a brief dataset overview.

6.3.1 ThermalWorld ReID Split

The ReID split includes pairs of color and thermal images captured by 16 FLIR ONE PRO cameras. Sample images from the dataset are presented in Fig. 6.2. All cameras were located

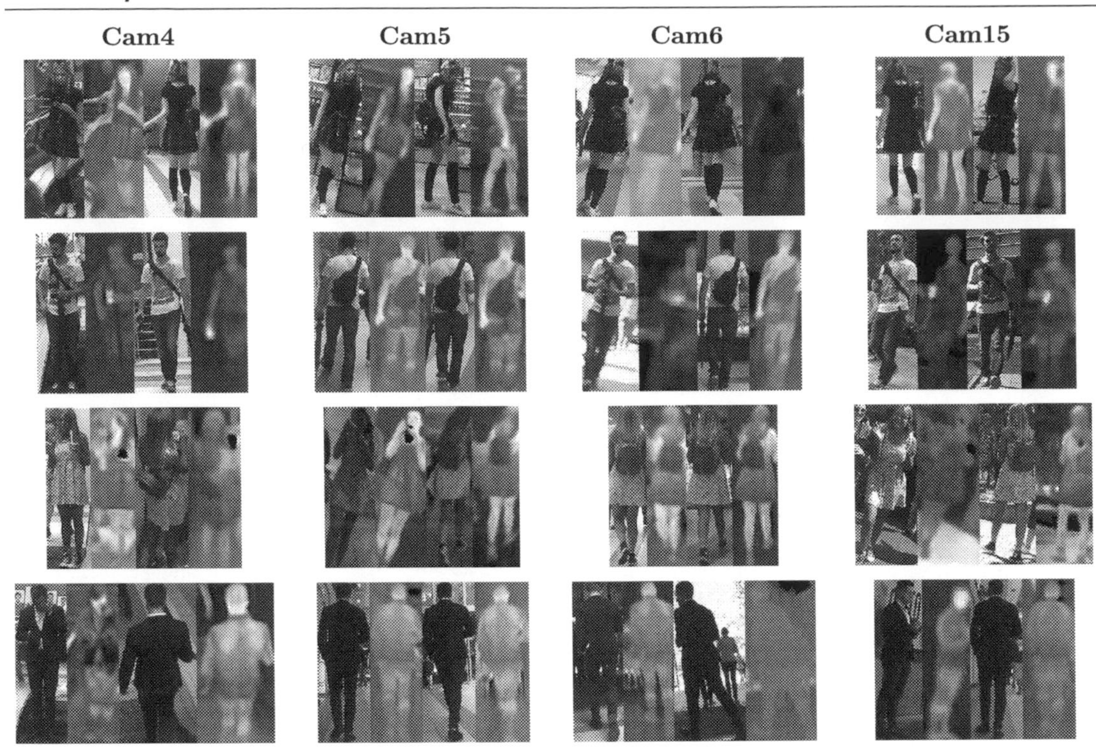

Figure 6.2 : Examples of person images from ThermalWorld ReID dataset.

in a shopping mall area. Cameras #2, 9, 13 are located in underground passages with low-light conditions. Cameras #1, 3, 7, 8, 10, 12, 14 are located at the entrances and present both day-time and night-time images. Cameras #15, 16 are placed in the garden. The rest of the cameras are located inside the mall.

We have developed a dedicated application for a smartphone to record sequences of thermal images using FLIR ONE PRO. A comparison with previous ReID datasets is presented in Table 6.1.

6.3.2 ThermalWorld VOC Split

The VOC split of the dataset was collected using two FLIR ONE PRO cameras. We use insights of developers of previous multispectral datasets [32,30,49,50,17,51] and provide new object classes with pixel-level annotations. The images were collected in different cities (Paris, Strasbourg, Riva del Garda, Venice) during all seasons and in different weather conditions (sunny, rain, snow). Captured object temperatures range from −20°C to +40°C.

Table 6.1: Comparison with previous ReID datasets. #/# represents the number of color/infrared images or cameras.

Dataset	#ID	#images	#cam.	Color	NIR	Thermal
iLDS [40]	119	476	2	✓	×	×
CAVIAR [41]	72	610	2	✓	×	×
PRID2011 [42]	200	971	2	✓	×	×
VIPER [43]	632	1264	2	✓	×	×
CUHK01 [44]	972	1942	2	✓	×	×
CUHK03 [45]	1467	13,164	6	✓	×	×
SYSU [46]	502	24,448	2	✓	×	×
Market [47]	1501	32,668	6	✓	×	×
MARS [48]	1261	1,191,003	6	✓	×	×
RegDB [1]	412	4120/4120	1/1	✓	×	✓
SYSU-MM01 [6]	491	287,628/15,792	4/2	✓	✓	×
ThermalWorld	516	15,818/15,818	16/16	✓	×	✓

Figure 6.3 : Examples of annotated images in ThermalWorld VOC dataset.

We hypothesized that training a GAN to predict the relative temperature contrasts of an object (e.g., clothes/skin) instead of its absolute temperature can improve the translation quality. We were inspired by the previous work on the explicit encoding of multiple modes in the output [39], and we assumed that the thermal segmentation that provides average temperatures of

Figure 6.4 : Examples of annotated images in ThermalWorld VOC dataset.

the emitting objects in the scene could resolve the ambiguity of the generated thermal images. Examples from ThermalWorld VOC dataset are presented in Figs. 6.3, 6.4 and 6.5.

We manually annotated the dataset, to automatically extract an object's temperature from the thermal images. Comparison with previous multispectral datasets and examples of all classes are presented in Table 6.1.

6.3.3 Dataset Annotation

Both datasets were manually annotated with pixel-level segmentations of the visible region of all contained objects. The annotation guidelines [52] were used. Annotators were pro-

Figure 6.5 : Examples of annotated images in ThermalWorld VOC dataset.

vided with a semi-automated tool. The tool performs a rough segmentation using the graph cut method [53]. The rough segmentation is refined manually. To provide the optimal balance between the annotation time and its accuracy we add a pixel border similar to [54].

Annotation of objects covered with tree branches is time-consuming during the winter time. To keep the annotation time short, we exclude from the annotated object only the trunks. Small branches without leaves are annotated as the object behind them.

6.3.4 Comparison of the ThermalWorld VOC Split with Previous Datasets

During the dataset design, we have studied the structure and class definitions of the related datasets. The primary motivation for the collection of the new dataset was the absence of the

Table 6.2: Comparison with previous thermal datasets. The type of labels provided is listed: object bounding boxes (B), dense pixel-level semantic labels (D), coarse labels (C) that do not aim to label the whole image. The type of thermal data: (tp) thermal preview, (°C) absolute temperature.

Dataset	Labels	Color	Thermal	Camera	Scene	#images	#classes
VAIS [32]	×	✓	8-bit tp	Mixed	Sea	1242	15
KAIST [30]	B	✓	8-bit tp	Car	Urban	95,000	1
INO [49]	B, C	✓	8-bit tp	Mixed	Urban	12,695	1
STTM [50]	B	×	16-bit tp	Pedestrian	Mixed	11,375	1
OSU [17]	B	×	8-bit tp	Pedestrian	Urban	284	1
TSEOD [51]	×	✓	8-bit tp	Facades	Urban	17,089	2
ThermalWorld	D	✓	16-bit°C	Mixed	Mixed	5098	10

Table 6.3: Comparison with previous thermal datasets in terms of sensor resolution.

Dataset	Resolution color	Resolution thermal	Thermal range
VAIS [32]	5056×5056	1024×768	7.5–13 μm
KAIST [30]	640×480	640×480	8–14 μm
INO [49]	328×254	328×254	3–11 μm
STTM [50]	×	640×512	17 μm
OSU [17]	×	360×240	7–14 μm
TSEOD [51]	320×240	360×240	7–14 μm
Ours	1185×889	640×480	8–14 μm

datasets with thermal images representing absolute temperature and dense semantic labels. We provide the comparison with related thermal datasets in terms of the spectral range, the types of annotation available and the number of images in Table 6.2. The sensor resolutions are given in Table 6.3.

6.3.5 Dataset Structure

The dataset is provided as folders with images representing four types of data: (1) color images, (2) absolute temperature images, (3) class annotations, (4) object instance annotations.

The FLIR ONE PRO camera has different resolutions for the color sensor and the thermal sensor. All annotation was performed on the high-resolution color images cropped to the field of view of the thermal sensor.

The actual resolution of the FLIR ONE PRO camera is 160×120 pixels. The camera uses the internal super-resolution filter to upscale the image to 640×480 pixels. We provide the thermal images as 32-bit float arrays in the NumPy library format [55]. All other images are provided in PNG format either color or grayscale.

We provide a Python toolkit that includes the dataset viewer and scripts for data access and export in various formats (e.g. `pix2pix`, `CycleGAN`).

6.3.6 Data Processing

The FLIR ONE camera produces as a standard output thermal preview images that present temperature of captured objects as monochrome (or pseudo-colors) images with a reference temperature scale bar. Also the FLIR ONE camera provides raw 16-bit data and the EXIF information for acquired images. Values of the raw data represent the object emission in the wavelengths 8–14 μ. The EXIF information contains necessary parameters for calculating the absolute temperature for objects in the scene. To convert raw images to calibrated temperatures, we use the EXIF data and the inversion of the Planck function given by [56]

$$T = \frac{c_2}{\lambda \ln(\frac{c_1 \cdot \lambda^{-5}}{M(\lambda, T)} + 1)}, \tag{6.1}$$

in which $M(\lambda, T)$ is the radiant exitance from a blackbody with temperature T and wavelength λ, c_1, c_2 are constants given by

$$c_1 = 2\pi h c^2 = 3.741 \cdot 10^{-16}\ \mathrm{W\,m^2}, \tag{6.2}$$

$$c_2 = \frac{hc}{k} = 1.4393 \cdot 10^{-2}\ \mathrm{mK}, \tag{6.3}$$

in which h is Planck's constant ($6.6256 \cdot 10^{-34}$ J s), c is the speed of light ($2.9979 \cdot 10^8$ m s^{-1}), and k is the Boltzmann gas constant ($1.38 \cdot 10^{-23}$ J K^{-1}).

6.4 Method

Color-to-thermal image translation is challenging because there are multiple possible thermal outputs for a single color input. For example, a person in a cold autumn day and a hot summer afternoon may have a similar appearance in the visible range, but the skin temperature will be different.

We have experimented with multiple state-of-the-art GAN frameworks [57,58,38,39] for multimodal image translation on the color-to-thermal task. We have found that GANs can predict object temperature with accuracy of approximately 5°C.

However, thermal images must have accuracy of 1°C to make local body temperature contrasts (e.g., skin/cloth) distinguishable. To improve the translation quality we developed two-step approach inspired by [59]. We have observed that relative thermal contrasts (e.g.,

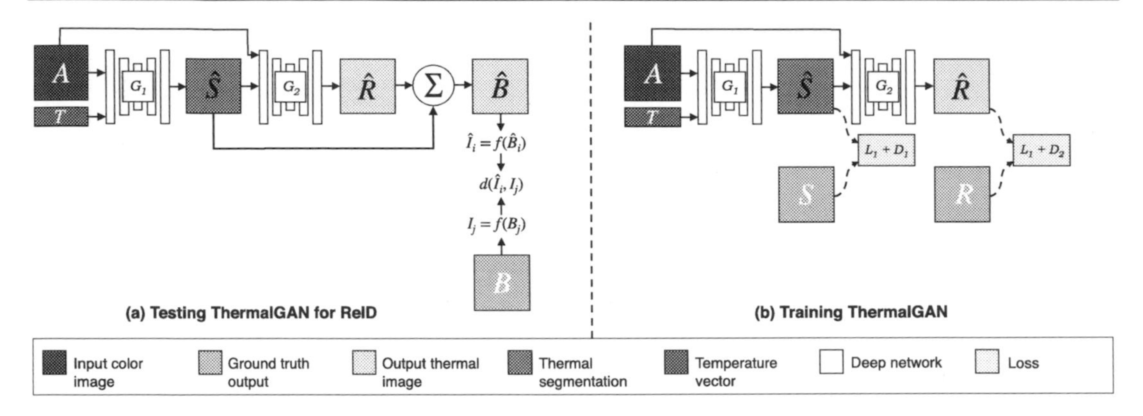

Figure 6.6 : Overview of our `ThermalGAN` **framework.**

eyes/brow) are nearly invariant to changes in the average body temperature due to different weather conditions.

We hypothesize that a prediction of relative thermal contrasts can be conditioned by an input color image and average temperatures for each object. Thus, we predict the absolute object temperature in two steps (Fig. 6.6). Firstly, we predict average object temperatures from an input color image. We term the resulting image a "thermal segmentation" image $\hat{S}$.

Secondly, we predict the relative local temperature contrasts $\hat{R}$, conditioned by a color image A and a thermal segmentation $\hat{S}$. The sum of a thermal segmentation and temperature contrasts provides the thermal image: $\hat{B} = \hat{S} + \hat{R}$.

The sequential thermal image synthesis has two advantages. Firstly, the problem remains multimodal only in the first step (generation of thermal segmentation). Secondly, the quality of thermal contrasts prediction is increased due to lower standard deviation and reduced range of temperatures.

To address the multimodality in color-to-thermal translation, we use a modified `BicyleGAN` framework [39] to synthesize multiple color segmentation images for a single color image. Instead of a random noise sample, we use a temperature vector T_i, which contains the desired background and object temperatures.

The rest of this section presents a brief introduction to conditional adversarial networks, details on the developed `ThermalGAN` framework and features used for thermal signature matching.

6.4.1 Conditional Adversarial Networks

Generative adversarial networks produce an image $\hat{B}$ for a given random noise vector z, $G : z \to \hat{B}$ [38,22]. Conditional GAN receives extra bits of information A in addition to the vector z, $G : \{A, z\} \to \hat{B}$. Usually, A is an image that is transformed by the generator network G. The discriminator network D is trained to distinguish "real" images from target domain B from the "fakes" $\hat{B}$ produced by the generator. The two networks are trained simultaneously. Discriminator provides the adversarial loss that enforces the generator to produce "fakes" $\hat{B}$ that cannot be distinguished from "real" images B.

We train two GAN models. The first generator $G_1 : \{T_i, A\} \to \hat{S}_i$ synthesizes multiple thermal segmentation images $\hat{S}_i \in \mathbb{R}^{W \times H}$ conditioned by a temperature vector T_i and a color image $A \in \mathbb{R}^{W \times H \times 3}$. The second generator $G_2 : \{\hat{S}_i, A\} \to \hat{R}_i$ synthesizes thermal contrasts $\hat{R}_i \in \mathbb{R}^{W \times H}$ conditioned by the thermal segmentation $\hat{S}_i$ and the input color image A. We can produce multiple realistic thermal outputs for a single color image by changing only the temperature vector T_i.

6.4.2 Thermal Segmentation Generator

We use the modified `BicycleGAN` framework for thermal segmentation generator G_1. Our contribution to the original U-Net generator [60] is an addition of one convolutional layer and one deconvolutional layer to increase the output resolution. We use average background temperatures instead of the random noise sample to be able to control the appearance of the thermal segmentation. The loss function for the generator G_1 is given by [39]

$$G_1^*(G_1, D_1) = \arg\min_{G_1} \max_{D_1} \mathcal{L}_{GAN}^{VAE}(G_1, D_1, E) + \lambda \mathcal{L}_1(G_1, E) \\ + \mathcal{L}_{GAN}(G_1, D_1) + \lambda_{\text{thermal}} \mathcal{L}_1^{\text{thermal}}(G_1, E) + \lambda_{KL} \mathcal{L}_{KL}(E), \quad (6.4)$$

where $\mathcal{L}_{GAN}^{VAE}$ is the variational autoencoder-based loss [39], which stimulates the output to be multimodal, $\mathcal{L}_1$ is an L^1 loss, $\mathcal{L}_{GAN}$ the loss provided by the discriminator D_1, $\mathcal{L}_1^{\text{thermal}}$ is a loss of the latent temperature domain, $\mathcal{L}_{KL}$ the Kullback–Leibler-divergence loss, E the encoder network, and λ weight parameters. We train both generators independently.

6.4.3 Relative Thermal Contrast Generator

We hypothesize that the distribution of relative thermal contrasts conditioned by the thermal segmentation and a color image is unimodal (compare images B and R for various background temperatures in Fig. 6.7). Hence, we use a unimodal `pix2pix` framework [38] as a

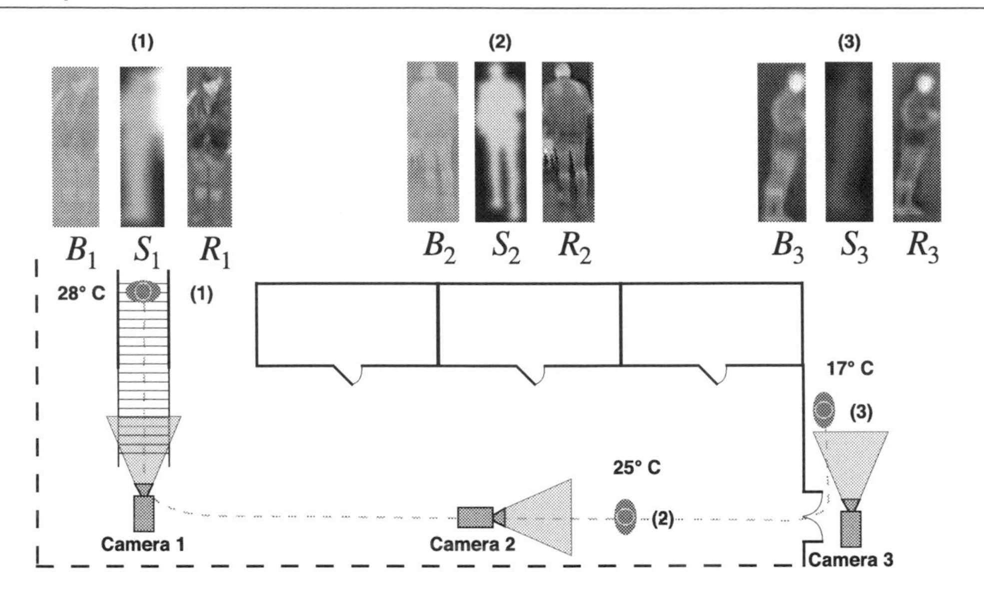

Figure 6.7 : Comparison of relative contrast R and absolute temperature B images for various camera locations. The relative contrast image R is equal to the difference of the absolute temperature B and the thermal segmentation S. Note that the person's appearance is invariant to background temperature in relative contrast images.

starting point for our relative contrast generator G_2. Our contribution to the original framework is two-fold. Firstly, we made the same modifications to the generator G_2 as for the generator G_1. Secondly, we use a four channel input tensor. The first three channels are RGB channels of an image A, the fourth channel is thermal segmentation $\hat{S}_i$ produced by generator G_1. We sum the outputs from the generators to obtain an absolute temperature image.

6.4.4 Thermal Signature Matching

We use an approach similar to [10] to extract discriminative features from thermal images. The original feature signature is extracted in three steps [10]: (1) a person's appearance is transferred from camera C_k to camera C_l, (2) the body region is separated from the background using structure element component analysis (SCA) [61], (3) feature signature is extracted using color histograms [14] and maximally stable color regions (MSCRs) [62].

However, thermal images contain only a single channel. We modify color features for absolute temperature domain. We use the monochrome ancestor of MSCR – maximally stable extremal regions (MSERs) [63]. We use temperature histograms instead of color histograms. The resulting matching method includes four steps. Firstly, we transform the person's appearance from a single color probe image A to multiple thermal images $\hat{B}_i$ using the `ThermalGAN` framework. Various images model possible variations of the temperature from camera to camera. Unlike the original approach [10], we do not train the method to transfer a person as regards appearance from camera to camera. Secondly, we extract body regions using SCA from real thermal images B_j in the gallery set and synthesized images $\hat{B}_i$. After that, we extract thermal signatures I from the body regions,

$$I = f(B) = \left[H_t(B), f_{\text{MSER}}(B) \right], \tag{6.5}$$

where H_t is a histogram of body temperatures, f_{MSER} is for MSER blobs for an image B.

Finally, we compute a distance between two signatures using the Bhattacharyya distance for temperature histograms and MSER distance [63,41]

$$d(\hat{I}_i, I_j) = \beta_H \cdot d_H(H_t(\hat{B}_i), H_t(B_j)) + (1 - \beta_H) \cdot d_{\text{MSER}}(f_{\text{MSER}}(\hat{B}_i), f_{\text{MSER}}(B_j)), \tag{6.6}$$

where d_H is the Bhattacharyya distance, d_{MSER} is the MSER distance [63], and β_H is a calibration weight parameter.

6.5 Evaluation

6.5.1 Network Training

The ThermalGAN framework was trained on the VOC split of the ThermalWorld dataset using the PyTorch library [64]. The VOC split includes indoor and outdoor scenes to avoid domain shift. The training was performed using the NVIDIA 1080 Ti GPU and took 76 hours for G_1, D_1 and 68 hours for G_2, D_2. For network optimization, we use minibatch SGD with an Adam solver. We set the learning rate to 0.0002 with momentum parameters $\beta_1 = 0.5$, $\beta_2 = 0.999$ similar to [38].

6.5.2 Color-to-Thermal Translation

6.5.2.1 Qualitative comparison

For a qualitative comparison of the ThermalGAN model on the color-to-thermal translation, we generate multiple thermal images from the independent ReID split of ThermalWorld dataset. Our goal is to keep the resulting images both realistic and diverse in terms of person relative thermal contrasts. We compare our framework with five baselines: pix2pix+noise [38], cLR-GAN [57,39], cVAE-GAN [58,39], cVAE-GAN++ [58,39], BicycleGAN [39]. All baselines were trained to convert color image to grayscale image representing perceptual thermal contrasts (8-bit, grayscale). Our ThermalGAN framework was trained to produce thermal images in

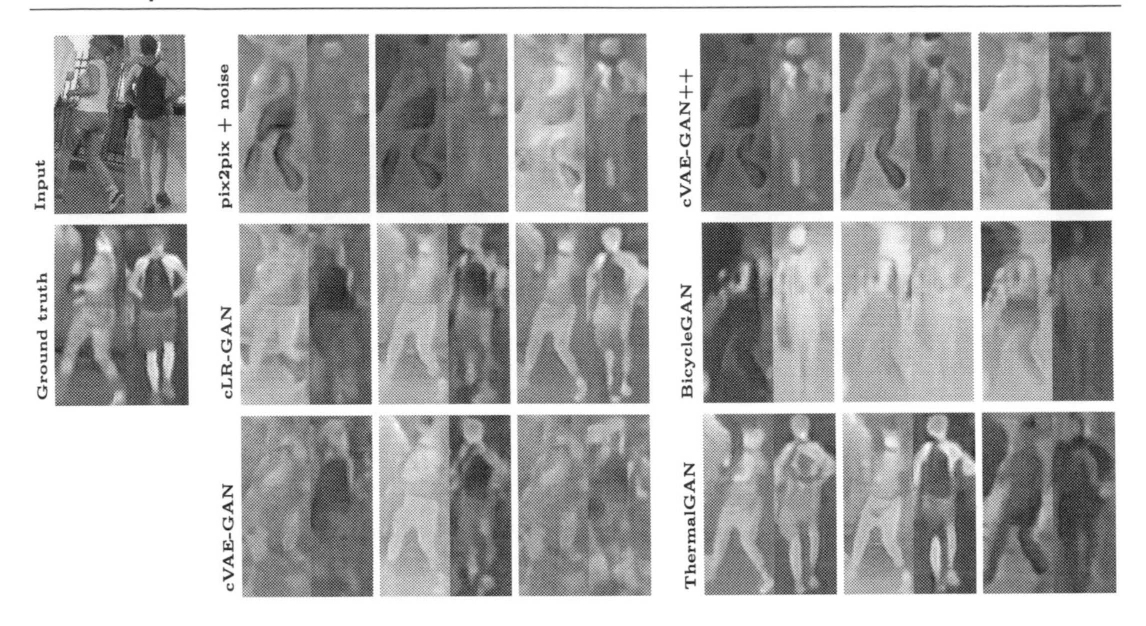

Figure 6.8 : Qualitative method comparison. We compare performance of various multimodal image translation frameworks on the ThermalWorld ReID dataset. For each model, we present three random output. The output of ThermalGAN framework is realistic, diverse, and shows the small temperatures contrasts that are important for robust ReID. Please note that only ThermalGAN framework produces output as calibrated temperatures that can be used for thermal signature matching.

degree Celsius. For comparison, they were converted to perceptual thermal intensities. The results of multimodal thermal image generation are presented in Fig. 6.8.

The results of pix2pix+noise are unrealistic and do not provide changes of thermal contrast. cLR-GAN and cVAE-GAN produce a slight variation of thermal contrasts but do not translate meaningful features for ReID. cVAE-GAN++ and BicycleGAN produce a diverse output, which fails to model thermal features present in real images. Our ThermalGAN framework combines the power of BicycleGAN method with two-step sequential translation to produce the output that is both realistic and diverse.

6.5.2.2 Quantitative evaluation

We use the generated images to perform a quantitative analysis of our ThermalGAN framework and the baselines. We measure two characteristics: diversity and perceptual realism. To measure multimodal reconstruction diversity, we use the averaged learned perceptual image patch similarity (LPIPS [65]) distance as proposed in [65,39]. For each baseline and our method, we calculate the average distance between 1600 pairs of random output thermal

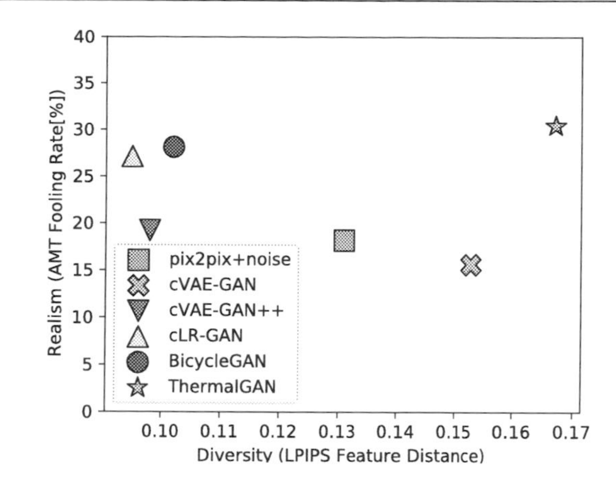

Figure 6.9 : Realism vs. diversity for synthesized thermal images.

Table 6.4: Comparison with state-of-the-art multimodal image-to-image translation methods.

	Realism	Diversity
Method	**AMT Fooling Rate [%]**	**LPIPS Distance**
Random real images	50.0%	
pix2pix+noise [38]	18.17	0.131
cVAE-GAN [58,39]	15.61	0.153
cVAE-GAN++ [58,39]	19.21	0.098
cLR-GAN [57,39]	27.10	0.095
BicycleGAN [39]	28.12	0.102
ThermalGAN	30.41	0.167

images, conditioned by 100 input color images. We measure perceptual realism of the synthesized thermal images using the human experts utilizing an approach similar to [35]. Real and synthesized thermal images are presented to human experts in a random order for one second. Each expert must indicate if the image is real or not. We perform the test on Amazon Mechanical Turk (AMT). We summarize the results of the quantitative evaluation in Fig. 6.9 and Table 6.4. Results of cLR-GAN, BicycleGAN and our ThermalGAN framework were most realistic. Our ThermalGAN model outperforms baselines in terms of both diversity and perceptual realism.

6.5.3 ReID Evaluation Protocol

We use 516 ID from the ReID split for testing the ReID performance. Please note that we use independent VOC split for training color-to-thermal translation. We use cumulative matching characteristic (CMC) curves and normalized area-under-curve (nAUC) as a performance

Table 6.5: Experiments on ThermalWorld ReID dataset in single-shot setting.

Methods	ThermalWorld ReID single-shot					
	r = 1	r = 5	r = 10	r = 15	r = 20	nAUC
TONE_2 [4]	15.10	29.26	38.95	42.40	44.48	37.98
TONE_1 [3]	8.87	13.71	21.27	27.48	31.86	23.64
HOG [66]	14.29	23.56	33.45	40.21	43.92	34.86
TSN [6]	3.59	5.13	8.85	13.97	18.56	12.25
OSN [6]	13.29	23.11	33.05	40.06	42.76	34.27
DZP [6]	15.37	22.53	30.81	36.80	39.99	32.28
ThermalGAN	**19.48**	**33.76**	**42.69**	**46.29**	**48.19**	**41.84**

measure. The CMC curve presents a recognition performance versus re-identification ranking score. nAUC is an integral score of a ReID performance of a given method. To keep our evaluation protocol consistent with related work [10], we use 5 pedestrians in the validation set. We also keep independent the gallery set and the probe set according to common practice.

We use images from color cameras for a probe set and images from thermal cameras for a gallery set. We exclude images from cameras #2, 9, 13 from the probe set, because they do not provide meaningful data in the visible range. We use a single color image in the single-shot setting. `ThermalGAN` ReID framework uses this single color image to generate 16 various thermal images. Baseline methods use the single input color image according to the common practice. For the multishot setting, we use ten color images for the probe set. Therefore `ThermalGAN` framework generates 16 thermal images for each color probe image and generates 160 thermal images for the probe set.

6.5.4 Cross-modality ReID Baselines

We compare our framework with six baseline models including hand-crafted features HOG [66] and modern deep-learning-based cross-modality matching methods: one-stream network (OSN) [6], two-stream network (TSN) [6], deep zero-padding (DZP) [6], two-stream CNN network (TONE_1) [3], and modified two-stream CNN network (TONE_2) [4].

6.5.5 Comparison and Analysis

We show the results of a comparison of our framework and baselines on ThermalWorld ReID datasets in Table 6.5 for a single-shot setting and in Table 6.6 for the multishot setting. Results are given in terms of the top-ranked matching rate and nAUC. We present the results in terms of CMC curves in Fig. 6.10.

Table 6.6: Experiments on ThermalWorld ReID dataset in multishot setting.

Methods	ThermalWorld ReID multishot					
	r = 1	r = 5	r = 10	r = 15	r = 20	nAUC
TONE_2 [4]	20.11	38.19	51.62	56.73	59.38	50.30
TONE_1 [3]	11.10	17.79	24.18	31.58	36.66	27.46
HOG [66]	16.08	27.10	40.10	48.64	51.41	40.82
TSN [6]	8.71	14.97	21.10	26.30	29.87	23.21
OSN [6]	15.36	25.17	39.14	47.65	50.04	39.85
DZP [6]	14.62	24.14	33.09	39.57	44.08	34.78
ThermalGAN	**22.59**	**48.24**	**59.40**	**62.85**	**66.12**	**57.35**

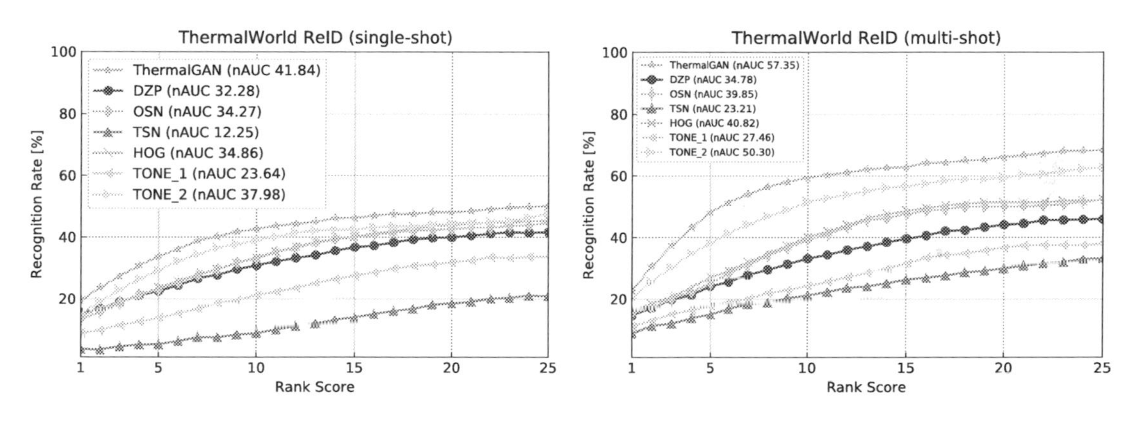

Figure 6.10 : CMC plot and nAUC for evaluation of baselines and `ThermalGAN` method in single-shot setting (left) and multishot setting (right).

We make the following observations from the single-shot evaluation. Firstly, the two-stream network [6] performs the worst among other baselines. We assume that the reason is that fine-tuning of the network from near-infrared data to thermal range is not sufficient for effective matching. Secondly, hand-crafted HOG [66] descriptor provided discriminative features that present both in color and thermal images and can compete with some of modern methods. Finally, our `ThermalGAN` ReID framework succeeds in the realistic translation of meaningful features from color to thermal images and provides discriminative features for effective color-to-thermal matching.

Results in the multishot setting are encouraging and prove that multiple person detection improves the matching rate with a cross-modality setup. We conclude to the following observations from the results presented in Table 6.6 and Fig. 6.10. Firstly, the performance of deep-learning-based baselines is improved in average in 5%. Secondly, multishot setting improves the rank-5 and rank-10 recognition rates. Finally, our `ThermalGAN` method benefits from the multishot setting and can be used effectively with multiple person images provided by robust pedestrian detectors for thermal images [19].

Figure 6.11 : Examples of the dataset augmentation for the Cityscapes dataset.

6.5.6 Applications

The ThermalWorld VOC dataset can be also used for evaluation of methods for conversion of the spectral range and multispectral semantic segmentation. This chapter explores only color photo to thermal image problem. The evaluation of reverse operation (*relative temperature→color*) and performance of semantic segmentation methods on multispectral imagery is a subject to future research. Below we demonstrate the performance of the `pix2pix` model with relative temperature target domain.

Dataset augmentation. Models trained on the ThermalGAN dataset can be used to augment large existing datasets with thermal imagery. The images produced with relative temperature target domain are mostly realistic enough to use them as the source data for training new models. Examples of dataset augmentation for the Cityscapes dataset are presented in Fig. 6.11.

6.6 Conclusion

We showed that conditional generative adversarial networks are effective for cross-modality prediction of a person's appearance in thermal image conditioned by a probe color image. Furthermore, discriminative features can be extracted from real and synthesized thermal images for effective matching of thermal signatures. Our main observation is that thermal cameras coupled with a GAN ReID framework can significantly improve the ReID performance in low-light conditions.

We developed a `ThermalGAN` framework for cross-modality person ReID in the visible range and LWIR images. We have collected a large-scale multispectral ThermalWorld dataset to train our framework and compare it with baselines. We made the dataset publicly available. Evaluation of modern cross-modality ReID methods and our framework proved that our `ThermalGAN` method achieves the state of the art and outperforms it in the cross-modality color-thermal ReID.

Acknowledgments

The reported study was funded by Russian Foundation for Basic Research (RFBR) according to the research project No. 17-29-03185 and by the Russian Science Foundation (RSF) according to the research project No. 16-11-00082.

References

[1] S. Bagheri, J.Y. Zheng, S. Sinha, Temporal mapping of surveillance video for indexing and summarization, Computer Vision and Image Understanding 144 (2016) 237–257.

[2] D. Nguyen, K. Park, Body-based gender recognition using images from visible and thermal cameras, Sensors 16 (2) (2016) 156.

[3] M. Ye, Z. Wang, X. Lan, P.C. Yuen, Visible thermal person re-identification via dual-constrained top-ranking, in: Proceedings of the Twenty-Seventh International Joint Conference on Artificial Intelligence, International Joint Conferences on Artificial Intelligence Organization, California, 2018, pp. 1092–1099.

[4] M. Ye, X. Lan, J. Li, P.C. Yuen, Hierarchical discriminative learning for visible thermal person re-identification, in: AAAI, 2018.

[5] D. Nguyen, H. Hong, K. Kim, K. Park, Person recognition system based on a combination of body images from visible light and thermal cameras, Sensors 17 (3) (2017) 605–629.

[6] A. Wu, W.-S. Zheng, H.-X. Yu, S. Gong, J. Lai, RGB-infrared cross-modality person re-identification, in: The IEEE International Conference on Computer Vision, ICCV, 2017.

[7] H.M. Vazquez, C.S. Martín, J. Kittler, Y. Plasencia, E.B.G. Reyes, Face recognition with LWIR imagery using local binary patterns, in: ICB, 5558, 2009, pp. 327–336 (Chapter 34)

[8] H. Zhang, V.M. Patel, B.S. Riggan, S. Hu, Generative adversarial network-based synthesis of visible faces from polarimetrie thermal faces, in: 2017 IEEE International Joint Conference on Biometrics, IJCB, IEEE, 2017, pp. 100–107.

[9] T. Zhang, A. Wiliem, S. Yang, B.C. Lovell, TV-GAN: generative adversarial network based thermal to visible face recognition, arXiv:1712.02514.

[10] A. Bhuiyan, A. Perina, V. Murino, Exploiting multiple detections for person re-identification, Journal of Imaging 4 (2) (2018) 28.

[11] N. Gheissari, T.B. Sebastian, R. Hartley, Person reidentification using spatiotemporal appearance, in: 2006 IEEE Computer Society Conference on Computer Vision and Pattern Recognition, vol. 2, CVPR'06, 2006, pp. 1528–1535.

[12] B. Prosser, S. Gong, T. Xiang, Multi-camera matching using bi-directional cumulative brightness transfer functions, in: Proceedings of the British Machine Vision Conference 2008, BMVC 2008, Queen Mary, University of London, London, United Kingdom, British Machine Vision Association, 2008, pp. 64.1–64.10.

[13] A. Bhuiyan, A. Perina, V. Murino, Person re-identification by discriminatively selecting parts and features, in: European Conference on Computer Vision 2014, ECCV, Istituto Italiano di Tecnologia, Genoa, Italy, in: Lecture Notes in Computer Science (including subseries Lecture Notes in Artificial Intelligence and Lecture Notes in Bioinformatics), Springer International Publishing, Cham, 2015, pp. 147–161.

[14] M. Farenzena, L. Bazzani, A. Perina, V. Murino, M. Cristani, Person re-identification by symmetry-driven accumulation of local features, in: 2010 IEEE Conference on Computer Vision and Pattern Recognition, CVPR, IEEE, 2010, pp. 2360–2367.

[15] S. Gong, M. Cristani, S. Yan, Person re-identification, in: Advances in Computer Vision and Pattern Recognition, Springer London, London, 2014.

[16] A. Yilmaz, K. Shafique, M. Shah, Target tracking in airborne forward looking infrared imagery, Image and Vision Computing 21 (7) (2003) 623–635.

[17] J.W. Davis, M.A. Keck, A two-stage template approach to person detection in thermal imagery, in: Application of Computer Vision, 2005, vol. 1, Seventh IEEE Workshops on, WACV/MOTIONS'05, IEEE, 2005, pp. 364–369.

[18] M. San-Biagio, A. Ulas, M. Crocco, M. Cristani, U. Castellani, V. Murino, A multiple kernel learning approach to multi-modal pedestrian classification, in: Pattern Recognition, 2012 21st International Conference on, ICPR, IEEE, 2012, pp. 2412–2415.
[19] D. Xu, W. Ouyang, E. Ricci, X. Wang, N. Sebe, Learning cross-modal deep representations for robust pedestrian detection, in: 2017 IEEE Conference on Computer Vision and Pattern Recognition, CVPR, IEEE, 2017, pp. 4236–4244.
[20] D. Nguyen, K. Park, Enhanced gender recognition system using an improved histogram of oriented gradient (HOG) feature from quality assessment of visible light and thermal images of the human body, Sensors 16 (7) (2016) 1134.
[21] D. Nguyen, K. Kim, H. Hong, J. Koo, M. Kim, K. Park, Gender recognition from human-body images using visible-light and thermal camera videos based on a convolutional neural network for image feature extraction, Sensors 17 (3) (2017) 637.
[22] I. Goodfellow, J. Pouget-Abadie, M. Mirza, B. Xu, D. Warde-Farley, S. Ozair, A. Courville, Y. Bengio, Generative adversarial nets, in: Advances in Neural Information Processing Systems, 2014, pp. 2672–2680.
[23] P. Morerio, J. Cavazza, V. Murino, Minimal-entropy correlation alignment for unsupervised deep domain adaptation, arXiv preprint, arXiv:1711.10288.
[24] V.V. Kniaz, V.S. Gorbatsevich, V.A. Mizginov, Thermalnet: a deep convolutional network for synthetic thermal image generation, The International Archives of the Photogrammetry, Remote Sensing and Spatial Information Sciences XLII-2/W4 (2017) 41–45, https://doi.org/10.5194/isprs-archives-XLII-2-W4-41-2017, https://www.int-arch-photogramm-remote-sens-spatial-inf-sci.net/XLII-2-W4/41/2017/.
[25] V.V. Kniaz, V.A. Mizginov, Thermal texture generation and 3d model reconstruction using sfm and gan, The International Archives of the Photogrammetry, Remote Sensing and Spatial Information Sciences XLII-2 (2018) 519–524, https://doi.org/10.5194/isprs-archives-XLII-2-519-2018, https://www.int-arch-photogramm-remote-sens-spatial-inf-sci.net/XLII-2/519/2018/.
[26] V.A. Knyaz, O. Vygolov, V.V. Kniaz, Y. Vizilter, V. Gorbatsevich, T. Luhmann, N. Conen, Deep learning of convolutional auto-encoder for image matching and 3d object reconstruction in the infrared range, in: The IEEE International Conference on Computer Vision, Workshops, ICCV, 2017.
[27] Z. Xie, P. Jiang, S. Zhang, Fusion of LBP and HOG using multiple kernel learning for infrared face recognition, in: ICIS, 2017.
[28] C. Herrmann, T. Müller, D. Willersinn, J. Beyerer, Real-time person detection in low-resolution thermal infrared imagery with MSER and CNNs, in: D.A. Huckridge, R. Ebert, S.T. Lee (Eds.), SPIE Security + Defence, SPIE, 2016, 99870I.
[29] V. John, S. Tsuchizawa, Z. Liu, S. Mita, Fusion of thermal and visible cameras for the application of pedestrian detection, Signal, Image and Video Processing 11 (3) (2016) 517–524.
[30] S. Hwang, J. Park, N. Kim, Y. Choi, I. So Kweon, Multispectral pedestrian detection: benchmark dataset and baseline, in: The IEEE Conference on Computer Vision and Pattern Recognition, CVPR, 2015.
[31] A. Berg, J. Ahlberg, M. Felsberg, A thermal infrared dataset for evaluation of short-term tracking methods, in: Swedish Symposium on Image Analysis, 2015.
[32] M.M. Zhang, J. Choi, K. Daniilidis, M.T. Wolf, C. Kanan, VAIS – a dataset for recognizing maritime imagery in the visible and infrared spectrums, in: CVPR Workshops, 2015, pp. 10–16.
[33] M. Limmer, H.P. Lensch, Infrared colorization using deep convolutional neural networks, in: Machine Learning and Applications, 2016 15th IEEE International Conference on, ICMLA, IEEE, 2016, pp. 61–68.
[34] S. Guadarrama, R. Dahl, D. Bieber, M. Norouzi, J. Shlens, K. Murphy, Pixcolor: pixel recursive colorization, arXiv preprint, arXiv:1705.07208.
[35] R. Zhang, P. Isola, A.A. Efros, Colorful image colorization, in: ECCV 9907, 2016, pp. 649–666 (Chapter 40).
[36] C. Li, M. Wand, Precomputed real-time texture synthesis with Markovian generative adversarial networks, arXiv.org, arXiv:1604.04382v1.
[37] D. Ulyanov, V. Lebedev, A. Vedaldi, V.S. Lempitsky, Texture networks – feed-forward synthesis of textures and stylized images, CoRR, arXiv:1603.03417, 2016, arXiv:1603.03417.

[38] P. Isola, J.-Y. Zhu, T. Zhou, A.A. Efros, Image-to-image translation with conditional adversarial networks, in: 2017 IEEE Conference on Computer Vision and Pattern Recognition, CVPR, IEEE, 2017, pp. 5967–5976.

[39] J.-Y. Zhu, R. Zhang, D. Pathak, T. Darrell, A.A. Efros, O. Wang, E. Shechtman, Toward multimodal image-to-image translation, in: I. Guyon, U.V. Luxburg, S. Bengio, H. Wallach, R. Fergus, S. Vishwanathan, R. Garnett (Eds.), Advances in Neural Information Processing Systems, vol. 30, Curran Associates, Inc., 2017, pp. 465–476, http://papers.nips.cc/paper/6650-toward-multimodal-image-to-image-translation.pdf.

[40] W.S. Zheng, S. Gong, T. Xiang, Associating groups of people, in: British Machine Vision Conference, 2009.

[41] D.S. Cheng, M. Cristani, M. Stoppa, L. Bazzani, V. Murino, Custom pictorial structures for re-identification, in: Proceedings of the British Machine Vision Conference 2011, BMVC 2011, Universita degli Studi di Verona, Verona, Italy, 2011.

[42] M. Hirzer, C. Beleznai, P.M. Roth, H. Bischof, Person re-identification by descriptive and discriminative classification, in: Scandinavian Conference on Image Analysis 2011, SCIA, Technische Universitat Graz, Graz, Austria, in: Lecture Notes in Computer Science (including subseries Lecture Notes in Artificial Intelligence and Lecture Notes in Bioinformatics), Springer Berlin Heidelberg, Berlin, Heidelberg, 2011, pp. 91–102.

[43] D. Gray, S. Brennan, H. Tao, Evaluating appearance models for recognition, reacquisition, and tracking, in: IEEE International Workshop on Performance Evaluation for Tracking and Surveillance, Rio de Janeiro, 2007.

[44] W. Li, R. Zhao, X. Wang, Human Reidentification with Transferred Metric Learning, in: Asian Conference on Computer Vision 2012, ACCV, Chinese University of Hong, Kong, Hong Kong, China, in: Lecture Notes in Computer Science (including subseries Lecture Notes in Artificial Intelligence and Lecture Notes in Bioinformatics), Springer Berlin Heidelberg, Berlin, Heidelberg, 2013, pp. 31–44.

[45] W. Li, R. Zhao, T. Xiao, X. Wang, DeepReID: Deep filter pairing neural network for person re-identification, in: 2013 IEEE Conference on Computer Vision and Pattern Recognition, Chinese University of Hong, Kong, Hong Kong, China, IEEE, 2014, pp. 152–159.

[46] C.-C. Guo, S.-Z. Chen, J.-H. Lai, X.-J. Hu, S.-C. Shi, Multi-shot person re-identification with automatic ambiguity inference and removal, in: 2014 22nd International Conference on Pattern Recognition, 2014, pp. 3540–3545.

[47] L. Zheng, L. Shen, L. Tian, S. Wang, J. Wang, Q. Tian, Scalable person re-identification: a benchmark, in: Proceedings of the IEEE International Conference on Computer Vision, Tsinghua University, Beijing, China, IEEE, 2015, pp. 1116–1124.

[48] L. Zheng, Z. Bie, Y. Sun, J. Wang, C. Su, S. Wang, Q. Tian, Mars: a video benchmark for large-scale person re-identification, in: B. Leibe, J. Matas, N. Sebe, M. Welling (Eds.), Computer Vision, ECCV 2016, Springer International Publishing, Cham, 2016, pp. 868–884.

[49] F. Généreux, B. Tremblay, M. Girard, J.-E. Paultre, F. Provençal, Y. Desroches, H. Oulachgar, S. Ilias, C. Alain, On the figure of merit of uncooled bolometers fabricated at ino, in: Infrared Technology and Applications XLII, vol. 9819, International Society for Optics and Photonics, 2016, p. 98191U.

[50] A. Berg, J. Ahlberg, M. Felsberg, A thermal object tracking benchmark, in: 2015 12th IEEE International Conference on Advanced Video and Signal Based Surveillance, AVSS, 2015, pp. 1–6, http://ieeexplore.ieee.org/document/7301772/.

[51] L. St-Laurent, X. Maldague, D. Prévost, Combination of colour and thermal sensors for enhanced object detection, in: Information Fusion, 2007 10th International Conference on, IEEE, 2007, pp. 1–8.

[52] M. Everingham, The pascal visual object classes challenge 2007 (voc2007) annotation guidelines, http://pascallin.ecs.soton.ac.uk/challenges/VOC/voc2007/guidelines.html.

[53] Y. Boykov, V. Kolmogorov, An experimental comparison of min-cut/max-flow algorithms for energy minimization in vision, IEEE Transactions on Pattern Analysis and Machine Intelligence 26 (9) (2004) 1124–1137.

[54] M. Everingham, L. Van Gool, C.K.I. Williams, J. Winn, A. Zisserman, The pascal visual object classes (voc) challenge, International Journal of Computer Vision 88 (2) (2010) 303–338, https://doi.org/10.1007/s11263-009-0275-4.

[55] T.E. Oliphant, A Guide to NumPy, vol. 1, Trelgol Publishing, USA, 2006.
[56] A. Harris, Thermal Remote Sensing of Active Volcanoes, Cambridge University Press, London, 2013.
[57] X. Chen, Y. Duan, R. Houthooft, J. Schulman, I. Sutskever, P. Abbeel, Infogan: interpretable representation learning by information maximizing generative adversarial nets, in: Advances in Neural Information Processing Systems, 2016, pp. 2172–2180.
[58] A.B.L. Larsen, S.K. Sønderby, H. Larochelle, O. Winther, Autoencoding beyond pixels using a learned similarity metric, in: M.F. Balcan, K.Q. Weinberger (Eds.), Proceedings of the 33rd International Conference on Machine Learning, in: Proceedings of Machine Learning Research, vol. 48, PMLR, New York, New York, USA, 2016, pp. 1558–1566, http://proceedings.mlr.press/v48/larsen16.html.
[59] J. Wu, Y. Wang, T. Xue, X. Sun, W.T. Freeman, J.B. Tenenbaum, MarrNet: 3D shape reconstruction via 2.5D sketches, arXiv.org, arXiv:1711.03129v1.
[60] O. Ronneberger, P. Fischer, T. Brox, U-net: convolutional networks for biomedical image segmentation, in: International Conference on Medical Image Computing and Computer-Assisted Intervention, Springer, 2015, pp. 234–241.
[61] N. Jojic, A. Perina, M. Cristani, V. Murino, B. Frey, Stel component analysis: modeling spatial correlations in image class structure, in: Computer Vision and Pattern Recognition, IEEE Conference on, CVPR 2009, IEEE, 2009, pp. 2044–2051.
[62] P.-E. Forssén, Maximally stable colour regions for recognition and matching, in: Computer Vision and Pattern Recognition, 2007, IEEE Conference on, CVPR'07, IEEE, 2007, pp. 1–8.
[63] J. Matas, O. Chum, M. Urban, T. Pajdla, Robust wide baseline stereo from maximally stable extremal regions, in: Proceedings of the British Machine Vision Conference 2002, British Machine Vision Association, 2002, pp. 36.1–36.10.
[64] A. Paszke, S. Gross, S. Chintala, G. Chanan, E. Yang, Z. DeVito, Z. Lin, A. Desmaison, L. Antiga, A. Lerer, Automatic Differentiation in Pytorch, 2017.
[65] R. Zhang, P. Isola, A.A. Efros, E. Shechtman, O. Wang, The unreasonable effectiveness of deep features as a perceptual metric, in: The IEEE Conference on Computer Vision and Pattern Recognition, CVPR, 2018.
[66] N. Dalal, B. Triggs, Histograms of oriented gradients for human detection, in: Computer Vision and Pattern Recognition, 2005, IEEE Computer Society Conference on, vol. 1, CVPR 2005, IEEE, 2005, pp. 886–893.

CHAPTER 7

A Review and Quantitative Evaluation of Direct Visual–Inertial Odometry

Lukas von Stumberg[*,†], **Vladyslav Usenko**[*], **Daniel Cremers**[*,†]

[*]Department of Informatics, Technical University of Munich, Munich, Germany [†]*Artisense GmbH, Garching, Germany*

Contents

Multimodal Scene Understanding
https://doi.org/10.1016/B978-0-12-817358-9.00013-5

7.1 Introduction

One of the most important tasks in modern robotics is to determine the motion of the robot and a reconstruction of the environment. Applications can be found in robot navigation [1] and autonomous driving [2] [3]. There are several different sensors available for this task, but cameras offer several advantages. They are cheap, lightweight, small and passive, which explains why they have been used very frequently for this task during the last years. While current visual odometry methods are already capable of providing highly accurate pose-estimates, they are not optimal in terms of robustness yet. Especially fast motions and low-textured areas present challenges for current solutions. This is where a combination with an inertial measurement unit (IMU) can help. An IMU contains a gyroscope and an accelerometer, which measure rotational velocities and linear acceleration at a very high rate (typically more than 100 Hz). This provides accurate short-term motion constraints and, unlike vision, is not prone to outliers.

In this chapter we will describe a state-of-the-art direct and sparse approach to visual–inertial odometry. The system is based on Direct Sparse Odometry (DSO) which was developed by Engel et al. [4]. DSO represents the geometry with a sparse set of points, each associated with an estimated inverse depth. They are jointly optimized together with 3D poses and camera parameters in a combined energy functional. We will describe how to complement this function with an additional IMU error term. This greatly improves the robustness of the system, as the inertial error term is convex and not prone to outliers.

A key drawback of monocular visual odometry is that it is not possible to obtain the metric scale of the environment. Adding an IMU enables us to observe the scale. Yet, depending on the performed motions this can take infinitely long, making the monocular visual–inertial initialization a challenging task. Rather than relying on a separate IMU initialization we show how to include the scale as a variable into the model of our system and jointly optimize it together with the other parameters.

Quantitative evaluation on the EuRoC dataset [5] demonstrates that we can reliably determine camera motion and sparse 3D structure (in metric units) from a visual–inertial system on a rapidly moving micro aerial vehicle (MAV) despite challenging illumination conditions (Fig. 7.1).

Note that this book chapter builds up on and significantly extends the paper [6], which was presented at the International Conference on Robotics and Automation (ICRA) 2018.

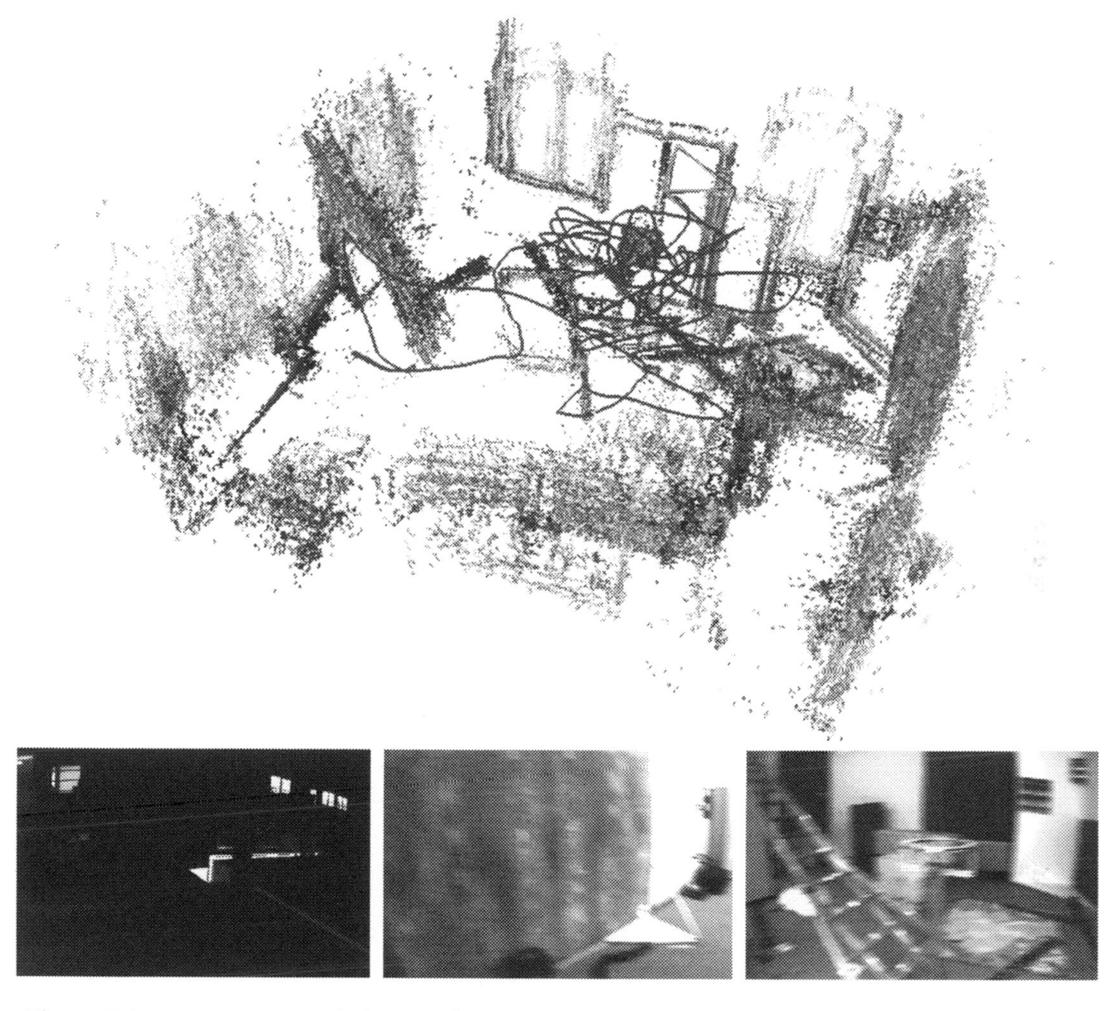

Figure 7.1 : Bottom: example images from the EuRoC dataset: low illumination, strong motion blur and little texture impose significant challenges for odometry estimation. Still our method is able to process all sequences with a RMSE of less than 0.23 m. Top: reconstruction, estimated pose (red camera) and ground-truth pose (green camera) at the end of V1_03_difficult.

7.2 Related Work

Motion estimation in 3D is a very important topic in robotics, and an IMU can be very useful when combined with vision data properly. Consequently there have been many works revolving around this topic in the past. In the following subsections we will present a short summary of visual and visual–inertial odometry methods.

When combining a monocular camera with IMU measurements a challenging task is the visual–inertial initialization. Therefore we will also discuss existing approaches, where the scale of the scene, initial orientation and velocity are not known from the beginning.

7.2.1 Visual Odometry

The task of incrementally tracking the camera motion from video is commonly called visual odometry. It was established by Nister et al. [7]. They proposed to match a sparse set of features from frame to frame, yielding multiple observations of the same point in different images. The 3D camera motion is then computed by minimizing the reprojection error. Afterwards there were many other works based on this feature-based approach. The general idea is to extract and track features over the images and then minimize the reprojection error of them. Later works like MonoSLAM [8] and PTAM [9] added the ability to create a consistent map in real-time. A very recent example is ORB-SLAM [10]. It is capable of tracking the camera motion and creating a consistent map with the ability to close trajectory loops and re-localization, outperforming previous methods in accuracy and robustness.

Another approach to visual odometry are the *direct* methods. Instead of the tracking features and minimizing the reprojection error, they *directly* use the image data to obtain the camera motion by optimizing a photometric error. In the work of [11] this error was minimized in real-time in a stereo setting by formulating quadrifocal constraints on the pixels. Especially for tracking with RGB-D cameras direct solutions were proposed by Newcombe et al. [12] and Kerl et al. [13]. Recently, several direct approaches were proposed, which use only a monocular camera. They differ in the density of the image information used. Newcombe et al. [14] proposed a dense approach which uses the complete image for tracking. However, for monocular visual odometry mainly image regions with a sufficient gradient contribute information. Therefore Engel et al. [15] have presented a semi-dense SLAM system exploiting this fact. Although sparse methods have traditionally been keypoint-based, direct sparse methods have advantages discussed in two recent publications. SVO proposed by Forster et al. [16] uses a keypoint-based approach for mapping keyframes and performs sparse direct image alignment to track non-keyframes. In contrast DSO proposed by Engel et al. [4] is a purely direct and sparse system. By using a sparse set of points it can efficiently perform bundle adjustment on a window of keyframes yielding superior accuracy and robustness compared to the semi-dense approach.

7.2.2 Visual–Inertial Odometry

IMU sensors are in many ways complementary to vision. The accelerometer and gyroscope measurements provide good short-term motion information. Furthermore they make scale as well as the roll and pitch angles observable, which is relevant for some control tasks, like the navigation of drones. Therefore there were several methods which combine them with vision. They can be classified into two general approaches.

The first one is the so-called *loosely coupled* fusion. Here the vision system is used as a blackbox, providing pose measurements, which are then combined with the inertial data, usually using an (extended or unscented) Kalman filter. Examples of this are shown in [1], [17] and [18].

Recently, *tightly coupled* works became more popular. Here the motion of the camera is computed by jointly optimizing the parameters in a combined energy function consisting of a visual and an inertial error term. The advantage is that the two measurement sources are combined on a very deep level of the system yielding a very high accuracy and especially robustness. Especially for vision systems that are based on nonlinear optimization the inertial error term can help overcome nonlinearities in the energy functional. The disadvantage is that the tightly coupled fusion has to be done differently for different vision systems making it much harder to implement. Still a tightly-coupled fusion has been done for filtering-based approaches [19] [20] [21] as well as for energy-minimization based systems [22] [23] [24] [25] [26].

Especially for monocular energy-minimization based approaches it is important to start with a good visual–inertial initialization. In practical use-cases the system does not have any prior knowledge about initial velocities, poses and geometry. Therefore apart from the usual visual initialization procedure most visual–inertial systems need a specific visual–inertial initialization. However, the problem is even more difficult, because several types of motion do not allow one to uniquely determine all variables. In [27] a closed-form solution for visual–inertial initialization was presented, as well as an analysis of the exceptional cases. It was later extended to work with IMU biases by [28]. In [25] a visual–inertial initialization procedure was proposed that works on the challenging EuRoC dataset. They use 15 seconds of flight data for initialization to make sure that all variables are observable on this specific dataset. Also in [26] and [29] IMU initialization methods are proposed which work on the EuRoC dataset, the latter even works without prior initialization of IMU extrinsics.

7.3 Background: Nonlinear Optimization and Lie Groups

7.3.1 Gauss–Newton Algorithm

The goal of the Gauss–Newton algorithm is to solve optimization problems of the form

$$\min_{\boldsymbol{x}} E(\boldsymbol{x}) = \min_{\boldsymbol{x}} \sum_i w_i \cdot r_i(\boldsymbol{x})^2 \tag{7.1}$$

where each r_i is a residual and w_i is the associated weight.

We will start with an explanation of the general Newton methods. The idea is based on a Taylor approximation of the energy function around $\boldsymbol{x}_0$,

$$E(\boldsymbol{x}) \approx E(\boldsymbol{x}_0) + \boldsymbol{b}^T(\boldsymbol{x} - \boldsymbol{x}_0) + \frac{1}{2}(\boldsymbol{x} - \boldsymbol{x}_0)^T \mathbf{H}(\boldsymbol{x} - \boldsymbol{x}_0) \tag{7.2}$$

where $\boldsymbol{b} = \frac{dE}{dx}$ is the gradient and $\mathbf{H} = \frac{d^2E}{dx^2}$ is the Hessian.

Using this approximation the optimal solution is where

$$0 \stackrel{!}{=} \frac{dE}{d\boldsymbol{x}} = \boldsymbol{b} + \mathbf{H}(\boldsymbol{x} - \boldsymbol{x}_0). \tag{7.3}$$

The optimal solution is therefore at

$$\boldsymbol{x} = \boldsymbol{x}_0 - \mathbf{H}^{-1} \cdot \boldsymbol{b}. \tag{7.4}$$

Of course, this is not the exact solution because we used an approximation to derive it. Therefore the optimization is performed iteratively by starting with an initial solution $\boldsymbol{x}_0$ and successively calculating

$$\boldsymbol{x}_{t+1} = \boldsymbol{x}_t - \lambda \cdot \mathbf{H}^{-1} \cdot \boldsymbol{b} \tag{7.5}$$

where λ is the step size.

Calculating the Hessian is very time-intensive. Therefore the idea of the Gauss–Newton method is to use an approximation for it instead. In the case where the problem has the form of Eq. (7.1) we can compute

$$b_j = 2\sum_i w_i \cdot r_i(\boldsymbol{x}) \frac{\partial r_i}{\partial x_j}, \tag{7.6}$$

$$\mathbf{H}_{jk} = 2\sum_i w_i \left(\frac{\partial r_i}{\partial x_j} \frac{\partial r_i}{\partial x_k} + r_i \frac{\partial^2 r_i}{\partial x_j \partial x_k} \right) \approx 2\sum_i w_i \left(\frac{\partial r_i}{\partial x_j} \frac{\partial r_i}{\partial x_k} \right). \tag{7.7}$$

We define the stacked residual vector $\boldsymbol{r}$ by stacking all residuals r_i. The Jacobian $\mathbf{J}$ of $\boldsymbol{r}$ therefore contains at each element

$$\mathbf{J}_{ij} = \frac{\partial r_i}{x_j}. \tag{7.8}$$

We also stack the weights into a diagonal matrix $\mathbf{W}$.

Then the approximated Hessian can be calculated with

$$\mathbf{H} \approx 2\sum_i w_i \mathbf{J}_{ij} \mathbf{J}_{ik} = 2\mathbf{J}^T \mathbf{W} \mathbf{J}. \tag{7.9}$$

Similarly the gradient of the energy function is

$$b = 2\mathbf{J}^T \mathbf{W} \boldsymbol{r}. \tag{7.10}$$

In summary the Gauss–Newton algorithm iteratively computes updates using Eq. (7.5), where the Hessian is approximated using Eq. (7.9).

7.3.2 Levenberg–Marquandt Algorithm

When initialized to far off from the optimum it can happen that the Gauss–Newton algorithm does not converge. The idea of the Levenberg–Marquandt algorithm is to dampen the Gauss–Newton updates. It still uses the same definition of **H** and $\boldsymbol{b}$ as the Gauss–Newton algorithm.

When computing the iterative update, however, it uses

$$\boldsymbol{x}_{t+1} = \boldsymbol{x}_t - (\mathbf{H} + \lambda \cdot \mathrm{diag}(\mathbf{H}))^{-1} \cdot \boldsymbol{b}. \tag{7.11}$$

The rough idea is to create a mixture between the Gauss–Newton algorithm and a simple gradient descent. Intuitively for a value of $\lambda = 0$ this will degenerate to a normal Gauss–Newton step and for a high value it will be dampened to something similar to gradient descent.

Clearly there is no optimal value for λ so it is usually adapted on the fly. After each update it is calculated whether or not the update reduces the energy (and the update is discarded if not). When an update is discarded λ is increased and when it is accepted λ gets reduced.

Compared to the Gauss–Newton algorithm the Levenberg–Marquandt method is in general more robust to a bad initialization.

7.4 Background: Direct Sparse Odometry

Direct sparse odometry (DSO) is a method that was very recently proposed by Engel et al. [4] (see Fig. 7.2). The novelty is that it optimizes a direct error term but still uses only a sparse set of points making it very accurate and at the same time real-time capable. The main advantage over previous direct methods like LSD-SLAM [15] is that it performs a bundle adjustment like optimization, which is computationally feasible because the set of points is sparse. In this section we will explain the theory behind DSO as well as the general structure.

7.4.1 Notation

Throughout the chapter we will use the following notation: bold upper case letters **H** represent matrices, bold lower case $\boldsymbol{x}$ vectors and light lower case λ represent scalars. Transformations

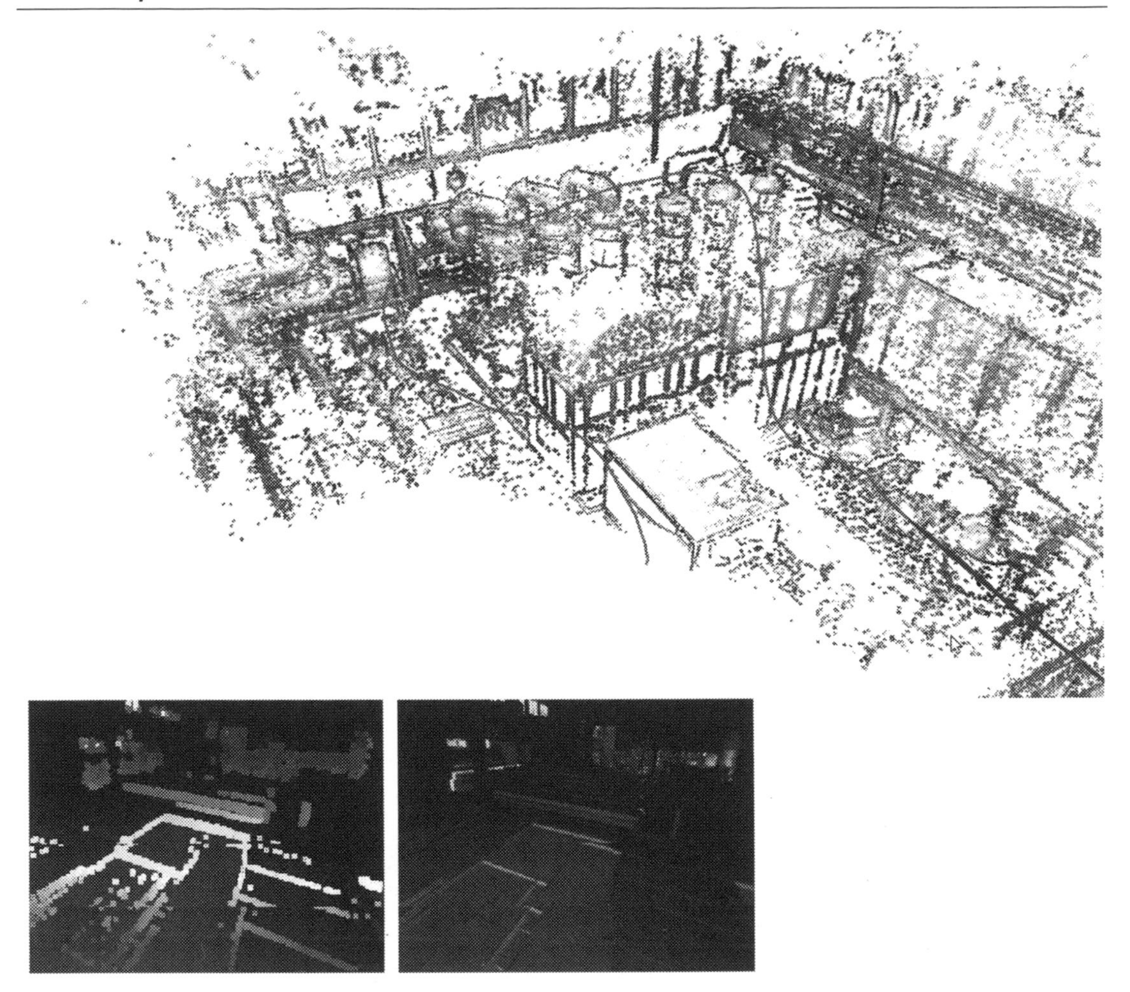

Figure 7.2 : Direct sparse odometry (DSO) running on the EuRoC dataset.

between coordinate frames are denoted as $\mathbf{T}_{i\_j} \in \mathbf{SE}(3)$ where point in coordinate frame i can be transformed to the coordinate frame j using the following equation: $\mathbf{p}_i = \mathbf{T}_{i\_j}\mathbf{p}_j$. We denote Lie algebra elements as $\hat{\boldsymbol{\xi}} \in \mathfrak{se}(3)$, where $\boldsymbol{\xi} \in \mathbb{R}^6$, and use them to apply small increments to the 6D pose $\boldsymbol{\xi}'_{i\_j} = \boldsymbol{\xi}_{i\_j} \boxplus \boldsymbol{\xi} := \log\left(e^{\hat{\boldsymbol{\xi}}_{i\_j}} \cdot e^{\hat{\boldsymbol{\xi}}}\right)^{\vee}$. We define the *world* as a fixed inertial coordinate frame with gravity acting along the negative Z axis. We also assume that the transformation from camera to IMU frame $T_{\text{imu\_cam}}$ is fixed and calibrated in advance. Factor graphs are expressed as a set G of factors and we use $G_1 \cup G_2$ to denote a factor graph containing all factors that are either in G_1 or in G_2.

7.4.2 Photometric Error

The main idea behind direct methods is that the intensity of a point should be constant over the images it is observed in. However, changes in illumination and photometric camera parameters such as vignetting, auto-exposure and response function can produce a change of the pixel value even when the real-world intensity is constant. To account for this, DSO can perform a photometric calibration and account for illumination changes in the cost function.

To do this it uses the following error definition. The photometric error of a point $p \in \Omega_i$ in reference frame i observed in another frame j is defined as follows:

$$E_{pj} = \sum_{\mathbf{p} \in \mathcal{N}_p} \omega_{\mathbf{p}} \left\| (I_j[\mathbf{p}'] - b_j) - \frac{t_j e^{a_j}}{t_i e^{a_i}} (I_i[\mathbf{p}] - b_i) \right\|_\gamma , \tag{7.12}$$

where $\mathcal{N}_p$ is a set of eight pixels around the point $\boldsymbol{p}$, I_i and I_j are images of respective frames, t_i, t_j are the exposure times, a_i, b_i, a_j, b_j are the coefficients to correct for affine illumination changes, γ is the Huber norm, ω_p is a gradient-dependent weighting and $\boldsymbol{p}'$ is the point projected into I_j.

The key idea here is that brightness constancy of a pixel is assumed when observed in different frames. However, the method compensates for illumination changes in the image, the vignette of the camera and the response function and the exposure time.

With that we can formulate the photometric error as

$$E_{\text{photo}} = \sum_{i \in \mathcal{F}} \sum_{p \in \mathcal{P}_i} \sum_{j \in \text{obs}(\boldsymbol{p})} E_{pj}, \tag{7.13}$$

where $\mathcal{F}$ is a set of keyframes that we are optimizing, $\mathcal{P}_i$ is a sparse set of points in keyframe i, and $obs(\mathbf{p})$ is a set of observations of the same point in other keyframes.

7.4.3 Interaction Between Coarse Tracking and Joint Optimization

For the accuracy of visual odometry it is important to optimize geometry and camera poses jointly. In DSO this can be done efficiently because the Hessian matrix of the system is sparse (see Sect. 7.4.5). However, the joint optimization still takes relatively much time which is why it cannot be done for each frame in real-time. Therefore the joint optimization is only performed for keyframes.

As we want to obtain poses for all frames however, we also perform a coarse tracking. Contrary to the joint estimation, we fix all point depths in the coarse tracking and only optimize the pose of the newest frame, as well as its affine illumination parameters (a_i and b_i). This is done using normal direct image alignment in a pyramid scheme where we track against the latest keyframe (see Sect. 7.4.4).

In summary the system has the structure depicted in Fig. 7.3. For each frame we perform the coarse tracking yielding a relative pose of the newest frame with respect to the last keyframe. For each frame we also track candidate points (during the "trace points" procedure). This serves as an initialization when some of the candidate points are later activated during the joint optimization.

For some frames we create a new keyframe, depending on the distance and visual change since the last keyframe. Then the joint estimation is performed, optimizing the poses of all active keyframes as well as the geometry. In the real-time version of the system coarse tracking and joint optimization are performed in parallel in two separate threads. Active keyframes are all keyframes which have not been marginalized yet.

7.4.4 Coarse Tracking Using Direct Image Alignment

The coarse tracking is based on normal direct image alignment.

In order to speed up computation all active points are projected into the frame of the latest keyframe and tracking is performed only with reference to this keyframe.

As in the rest of the system the coarse tracking minimizes the photometric error in order to get a pose estimate. The optimized parameters include the 6 DOF pose which is represented as the transformation from the new frame to the reference frame.

This error is minimized using the Levenberg–Marquandt algorithm. In order to increase the convergence radius the well known pyramid scheme is used. This means that the image is scaled down multiple times (using Gaussian blurring). When tracking, at first the lowest resolution is used. After convergence the optimization is proceeded with the next level of the image pyramid (using the result from the previous level as an initialization). This is done until the original image is reached and the parameters are optimized using it.

This procedure has two advantages. First, tracking on a low resolution first is very fast, as less pixels have to be handled. The main advantage, however, is that it increases the convergence radius. The Taylor approximation of the energy, which in practice uses the image gradient, is only valid in the vicinity of the optimum. When using a smaller-scaled image the gradient contains information from a larger region (in terms of the original image), and therefore increases the convergence basin.

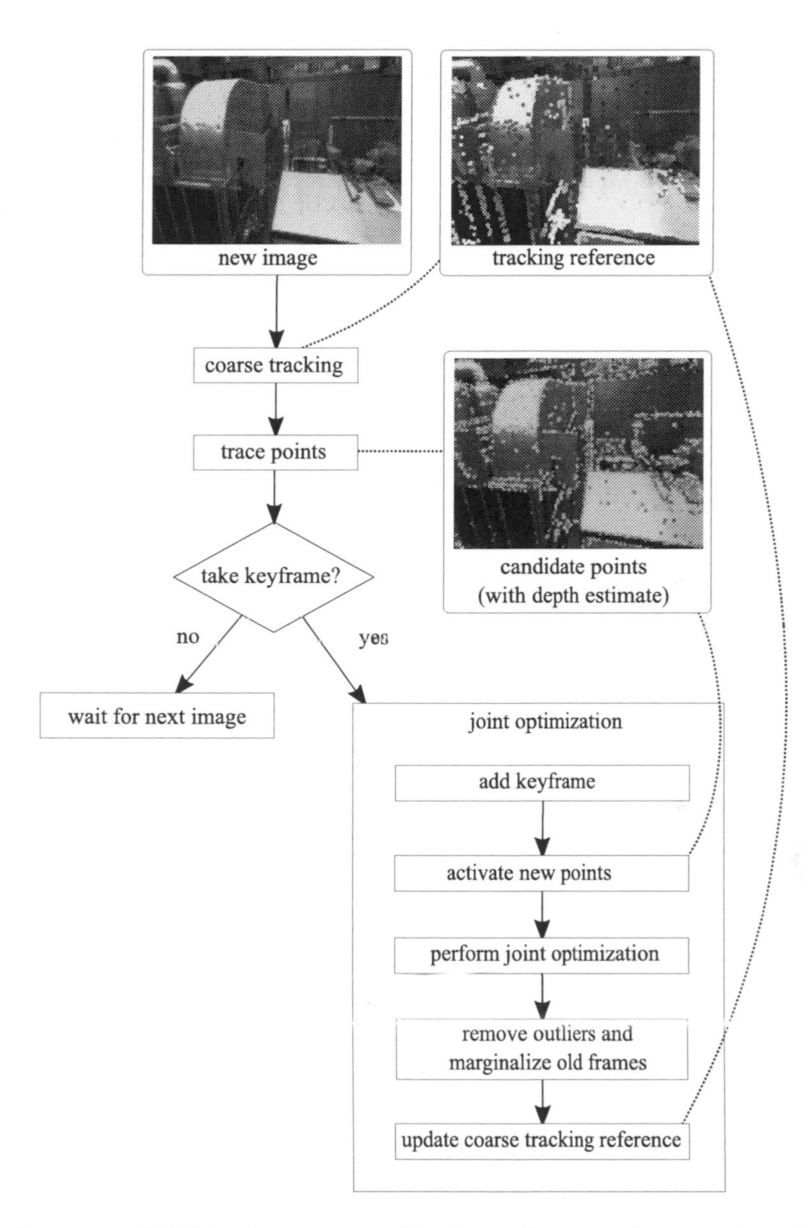

Figure 7.3 : Structure of DSO. The coarse tracking is run for every frame and tracks the current image against the reference frame. Meanwhile the joint optimization is performed only when a new keyframe is created. Note that for the real-time-version the coarse tracking runs in a separate thread.

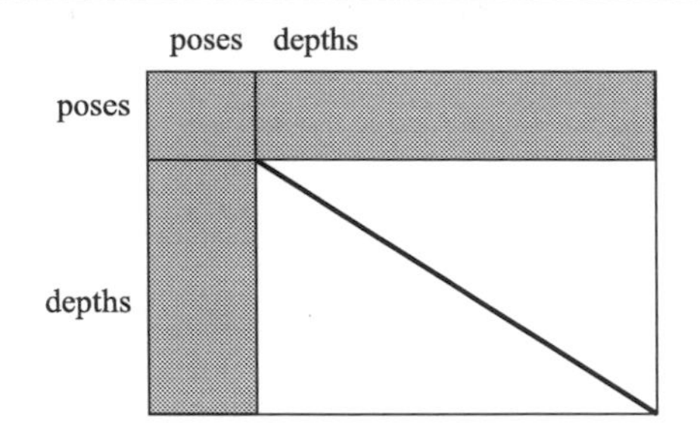

Figure 7.4 : Structure of the Hessian in the joint optimization. While the pose-pose and pose-depths correlations can be dense, the depth-depth-part of the Hessian is purely diagonal.

7.4.5 Joint Optimization

In contrast to the coarse tracking the joint optimization not only optimizes poses, but also the geometry, making the task significantly more difficult.

However, here all variables are initialized close to the optimum: The approximate pose of the newest frame is obtained from the previously executed coarse tracking. Newly activated points have a depth estimated by the tracing-procedure and all other variables have already been optimized in the previous execution of the joint optimization. Therefore DSO uses the Gauss–Newton algorithm for this task (instead of Levenberg–Marquandt), and no image pyramid.

When looking at the update step

$$\delta \boldsymbol{x} = \mathbf{H}^{-1} \cdot \boldsymbol{b} \tag{7.14}$$

performed in each iteration it is clear that the most calculation-intensive part is inverting the Hessian matrix.

However, we can use the sparsity of the structure (see Fig. 7.4). The reason for this specific structure is that each residual only depends on the pose of two frames, the affine lighting parameters and one point depth (see also Eq. (7.12)). Therefore there is no residual depending on two different points, meaning that the off-diagonal part of $\mathbf{J}^T\mathbf{J}$ is 0 for all depths.

This information can be used to efficiently solve the problem. We can partition the linear system into

$$\begin{bmatrix} \mathbf{H}_{\alpha\alpha} & \mathbf{H}_{\alpha\beta} \\ \mathbf{H}_{\beta\alpha} & \mathbf{H}_{\beta\beta} \end{bmatrix} \begin{bmatrix} \delta\boldsymbol{x}_\alpha \\ \delta\boldsymbol{x}_\beta \end{bmatrix} = \begin{bmatrix} \boldsymbol{b}_\alpha \\ \boldsymbol{b}_\beta \end{bmatrix} \tag{7.15}$$

where $\boldsymbol{x}_\beta$ are the depth variables and $\boldsymbol{x}_\alpha$ are all other variables. Note that this means that $\mathbf{H}_{\beta\beta}$ is a diagonal matrix.

We can now apply the Schur complements, which yields the new linear system $\widehat{\mathbf{H}_{\alpha\alpha}}\boldsymbol{x}_\alpha = \widehat{\boldsymbol{b}_\alpha}$, with

$$\widehat{\mathbf{H}_{\alpha\alpha}} = \mathbf{H}_{\alpha\alpha} - \mathbf{H}_{\alpha\beta}\mathbf{H}_{\beta\beta}^{-1}\mathbf{H}_{\beta\alpha}, \tag{7.16}$$

$$\widehat{\boldsymbol{b}_\alpha} = \boldsymbol{b}_\alpha - \mathbf{H}_{\alpha\beta}\mathbf{H}_{\beta\beta}^{-1}\boldsymbol{b}_\beta. \tag{7.17}$$

In this procedure the only inversion needed is $\mathbf{H}_{\beta\beta}^{-1}$, which is trivial because $\mathbf{H}_{\beta\beta}$ is diagonal.

The resulting $\widehat{\mathbf{H}_{\alpha\alpha}}$ is much smaller than the original $\mathbf{H}$ because there are typically much more points, than other variables. With it we can efficiently compute the update for the non-depth variables with

$$\delta\boldsymbol{x}_\alpha = \widehat{\mathbf{H}_{\alpha\alpha}}^{-1} \cdot \widehat{\boldsymbol{b}_\alpha}. \tag{7.18}$$

The update step of for the depth variables is done by reinserting

$$\mathbf{H}_{\beta\alpha}\delta\boldsymbol{x}_\alpha + \mathbf{H}_{\beta\beta}\delta\boldsymbol{x}_\beta = \boldsymbol{b}_\beta \Leftrightarrow \delta\boldsymbol{x}_\beta = \mathbf{H}_{\beta\beta}^{-1}\left(\boldsymbol{b}_\beta - \mathbf{H}_{\beta\alpha}\delta\boldsymbol{x}_\alpha\right). \tag{7.19}$$

7.5 Direct Sparse Visual–Inertial Odometry

The visual–inertial odometry system we describe in this chapter was proposed in [6]. It is based on DSO [4] and could also be viewed as a direct formulation of [22]. The approach differs from [24] and [23] by jointly estimating poses and geometry in a bundle adjustment like optimization.

As mentioned in the introduction it is important to integrate IMU data on a very tight level to the system. In the odometry system poses and geometry are estimated by minimizing a photometric error function E_{photo}. We integrate IMU data by instead optimizing the following energy function:

$$E_{\text{total}} = \lambda \cdot E_{\text{photo}} + E_{\text{inertial}}. \tag{7.20}$$

The summands E_{photo} and E_{inertial} are explained in Sects. 7.4.2 and 7.5.1, respectively.

The system contains two main parts running in parallel:

- The coarse tracking is executed for every frame and uses direct image alignment combined with an inertial error term to estimate the transformation between the new frame and the latest keyframe.
- When a new keyframe is created we perform a visual–inertial bundle adjustment like optimization that estimates the poses of all active keyframes as well as the geometry.

In contrast to [25] this method does not wait for a fixed amount of time before initializing the visual–inertial system but instead jointly optimizes all parameters including the scale. This yields a higher robustness as inertial measurements are used right from the beginning.

7.5.1 Inertial Error

In this section we derive the nonlinear dynamic model that we use to construct the error term, which depends on rotational velocities measured by gyroscope and linear acceleration measured by accelerometer. We decompose the 6D pose from the state $\boldsymbol{s}$ into rotation and translational part:

$$\mathbf{T}_{w\_imu} = \begin{pmatrix} \mathbf{R} & \mathbf{p} \\ 0 & 1 \end{pmatrix}, \tag{7.21}$$

with that we can write the following dynamic model:

$$\dot{\boldsymbol{p}} = \boldsymbol{v}, \tag{7.22}$$

$$\dot{\boldsymbol{v}} = \mathbf{R}\left(\boldsymbol{a}_z + \boldsymbol{\epsilon}_a - \boldsymbol{b}_a\right) + \boldsymbol{g}, \tag{7.23}$$

$$\dot{\mathbf{R}} = \mathbf{R}\left[\boldsymbol{\omega}_z + \boldsymbol{\epsilon}_\omega - \boldsymbol{b}_\omega\right]_\times, \tag{7.24}$$

$$\dot{\boldsymbol{b}}_a = \boldsymbol{\epsilon}_{b,a}, \tag{7.25}$$

$$\dot{\boldsymbol{b}}_\omega = \boldsymbol{\epsilon}_{b,\omega}, \tag{7.26}$$

where $\boldsymbol{\epsilon}_a$, $\boldsymbol{\epsilon}_\omega$, $\boldsymbol{\epsilon}_{b,a}$, and $\boldsymbol{\epsilon}_{b,\omega}$ denote the Gaussian white noise that affects the measurements and $\boldsymbol{b}_a$ and $\boldsymbol{b}_\omega$ denote slowly evolving biases. $[\cdot]_\times$ is the skew-symmetric matrix such that, for vectors $\boldsymbol{a}$, $\boldsymbol{b}$, $[\boldsymbol{a}]_\times \boldsymbol{b} = \boldsymbol{a} \times \boldsymbol{b}$.

As IMU data is obtained with a much higher frequency than images, we follow the preintegration approach proposed in [30] and improved in [31] and [23]. We integrate the IMU measurements between timestamps i and j in the IMU coordinate frame and obtain pseudo-measurements $\Delta \boldsymbol{p}_{i\to j}$, $\Delta \boldsymbol{v}_{i\to j}$, and $\mathbf{R}_{i\to j}$.

We initialize pseudo-measurements with $\Delta \boldsymbol{p}_{i\to i} = 0$, $\Delta \boldsymbol{v}_{i\to i} = 0$, $\mathbf{R}_{i\to i} = \mathbf{I}$, and assuming the time between IMU measurements is Δt we integrate the raw measurements:

$$\Delta \boldsymbol{p}_{i\to k+1} = \Delta \boldsymbol{p}_{i\to k} + \Delta \boldsymbol{v}_{i\to k}\Delta t, \tag{7.27}$$

$$\Delta \boldsymbol{v}_{i\to k+1} = \Delta \boldsymbol{v}_{i\to k} + \mathbf{R}_{i\to k}\left(\boldsymbol{a}_z - \boldsymbol{b}_a\right)\Delta t, \tag{7.28}$$

$$\mathbf{R}_{i\to k+1} = \mathbf{R}_{i\to k}\exp(\left[\boldsymbol{\omega}_z - \boldsymbol{b}_\omega\right]_\times \Delta t). \tag{7.29}$$

Given the initial state and integrated measurements the state at the next time-step can be predicted:

$$\boldsymbol{p}_j = \boldsymbol{p}_i + (t_j - t_i)\boldsymbol{v}_i + \frac{1}{2}(t_j - t_i)^2 \boldsymbol{g} + \mathbf{R}_i \Delta \boldsymbol{p}_{i\to j}, \tag{7.30}$$

$$\boldsymbol{v}_j = \boldsymbol{v}_i + (t_j - t_i)\boldsymbol{g} + \mathbf{R}_i \Delta \boldsymbol{v}_{i\to j}, \tag{7.31}$$

$$\mathbf{R}_j = \mathbf{R}_i \mathbf{R}_{i\to j}. \tag{7.32}$$

For the previous state s_{i-1} (based on the state definition in Eq. (7.39)) and IMU measurements $\boldsymbol{a}_{i-1}, \boldsymbol{\omega}_{i-1}$ between frames i and $i-1$, the method yields a prediction

$$\widehat{s}_i := h\left(\boldsymbol{\xi}_{i-1}, \boldsymbol{v}_{i-1}, \boldsymbol{b}_{i-1}, \boldsymbol{a}_{i-1}, \boldsymbol{\omega}_{i-1}\right) \tag{7.33}$$

of the pose, velocity, and biases in frame i with associated covariance estimate $\widehat{\Sigma}_{s,i}$. Hence, the IMU error function terms are

$$E_{\text{inertial}}(\boldsymbol{s}_i, \boldsymbol{s}_j) := \left(\boldsymbol{s}_j \boxminus \widehat{\boldsymbol{s}}_j\right)^T \widehat{\boldsymbol{\Sigma}}_{s,j}^{-1} \left(\boldsymbol{s}_j \boxminus \widehat{\boldsymbol{s}}_j\right). \tag{7.34}$$

7.5.2 IMU Initialization and the Problem of Observability

In contrast to a purely monocular system the usage of inertial data enables us to observe more parameters such as scale and gravity direction. While this on the one hand is an advantage, on the other hand it also means that these values have to be correctly estimated as otherwise the IMU-integration will not work correctly. This is why other methods such as [25] need to estimate the scale and gravity direction in a separate initialization procedure before being able to use inertial measurements for tracking.

Especially the metric scale, however, imposes a problem as it is only observable after the camera has performed certain motions. An example of a degenerate case is when the system is initialized while the camera moves with a fixed speed that does not change for the first minute. Then the IMU will measure no acceleration (except gravity and the bias) for the first minute rendering it useless for scale estimation. This shows that in theory it can take an arbitrary amount of time until all parameters are observable. Reference [27, Tables I and II] contains an excellent summary, showing the different cases where a unique solution for the problem cannot be found. Although these completely degenerate scenarios usually do not occur, also in practice it can often take multiple seconds for the scale to be observable. In [25] the authors say that they wait for 15 seconds of camera motion in order to be sure that everything is observable when processing the popular EuRoC dataset. Obviously it is not optimal to wait for such a long time before using any inertial measurements for tracking especially as on other datasets it might take even longer. However, as explained previously, using a wrong scale will render the inertial data useless. Therefore in [6] a novel strategy for handling this issue has been proposed. We explicitly include scale (and gravity direction) as a parameter in our visual–inertial system and jointly optimize them together with the other values such as poses and geometry. This means that we can initialize with an arbitrary scale instead of waiting until it is observable. We initialize the various parameters as follows:

- We use the same visual initializer as [4] which computes a rough pose estimate between two frames as well as approximate depths for several points. They are normalized so that the average depth is 1.

- The initial gravity direction is computed by averaging up to 40 accelerometer measurements, yielding a sufficiently good estimate even in cases of high acceleration.
- We initialize the velocity and IMU-biases with zero and the scale with 1.0.

All these parameters are then jointly optimized during a bundle adjustment like optimization.

7.5.3 SIM(3)-based Model

In order to be able to start tracking and mapping with a preliminary scale and gravity direction we need to include them into our model. Therefore in addition to the metric coordinate frame we define the DSO coordinate frame to be a scaled and rotated version of it. The transformation from the DSO frame to the metric frame is defined as $\mathbf{T}_{m\_d} \in \{\mathbf{T} \in \mathbf{SIM}(3) \mid \text{translation}(\mathbf{T}) = 0\}$, together with the corresponding $\xi_{m\_d} = \log(\mathbf{T}_{m\_d}) \in \mathfrak{sim}(3)$. We add a superscript D or M to all poses denoting in which coordinate frame they are expressed. In the optimization the photometric error is always evaluated in the DSO frame, making it independent of the scale and gravity direction. The inertial error on the other hand has to be evaluated in the metric frame.

7.5.4 Scale-Aware Visual–Inertial Optimization

We optimize the poses, IMU-biases and velocities of a fixed number of keyframes. Fig. 7.5A shows a factor graph of the problem. Note that there are in fact many separate visual factors connecting two keyframes each, which we have combined to one big factor connecting all the keyframes in this visualization. Each IMU-factor connects two subsequent keyframes using the preintegration scheme described in Sect. 7.5.1. As the error of the preintegration increases with the time between the keyframes we ensure that the time between two consecutive keyframes is not bigger than 0.5 seconds, which is similar to what [25] have done. Note that in contrast to their method, however, we allow the marginalization procedure described in Sect. 7.5.4.2 to violate this constraint which ensures that long-term relationships between keyframes can be properly observed.

An important property of our algorithm is that the optimized poses are not represented in the metric frame but in the DSO frame. This means that they do not depend on the scale of the environment. If we were optimizing them in the metric frame instead a change of the scale would mean that the translational part of all poses also changes which does not happen during nonlinear optimization.

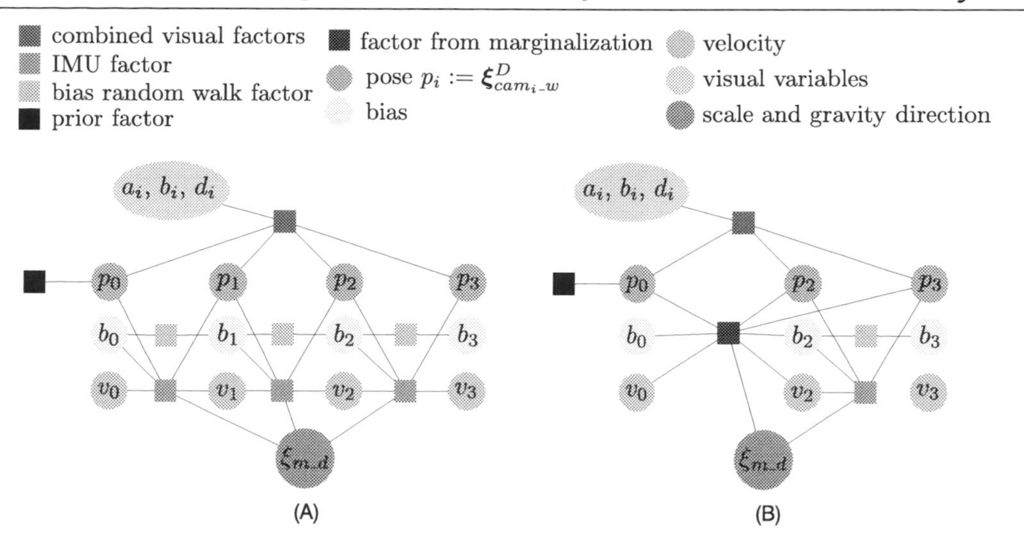

Figure 7.5 : Factor graphs for the visual–inertial joint optimization before and after the marginalization of a keyframe. (A) Factor graph for the visual–inertial joint optimization. (B) Factor graph after keyframe 1 has been marginalized.

7.5.4.1 Nonlinear optimization

We perform nonlinear optimization using the Gauss–Newton algorithm. For each active keyframe we define a state vector

$$s_i := [\left(\boldsymbol{\xi}^D_{cam_i\_w}\right)^T, \boldsymbol{v}_i^T, \boldsymbol{b}_i^T, a_i, b_i, d_i^1, d_i^2, ..., d_i^m]^T \tag{7.35}$$

where $\boldsymbol{v}_i \in \mathbb{R}^3$ is the velocity, $\boldsymbol{b}_i \in \mathbb{R}^6$ is the current IMU bias, a_i and b_i are the affine illumination parameters used in Eq. (7.12) and d_i^j are the inverse depths of the points hosted in this keyframe.

The full state vector is then defined as

$$s = [\boldsymbol{c}^T, \boldsymbol{\xi}_{m\_d}^T, s_1^T, s_2^T, ..., s_n^T]^T \tag{7.36}$$

where $\boldsymbol{c}$ contains the geometric camera parameters and $\boldsymbol{\xi}_{m\_d}$ denotes the translation-free transformation between the DSO frame and the metric frame as defined in Sect. 7.5.3. In order to simplify the notation we define the operator $s \boxplus s'$ to work on state vectors by applying the concatenation operation $\boldsymbol{\xi} \boxplus \boldsymbol{\xi}'$ for Lie algebra components and a plain addition for other components.

Using the stacked residual vector $\boldsymbol{r}$ we define the Jacobian

$$\mathbf{J} = \left.\frac{d\boldsymbol{r}\,(s \boxplus \boldsymbol{\epsilon})}{d\boldsymbol{\epsilon}}\right|_{\boldsymbol{\epsilon}=0} \tag{7.37}$$

where $\mathbf{W}$ is a diagonal weight matrix. Now we define $\mathbf{H} = \mathbf{J}^T\mathbf{W}\mathbf{J}$ and $\boldsymbol{b} = -\mathbf{J}^T\mathbf{W}\boldsymbol{r}$ where $\mathbf{W}$ is a diagonal weight matrix. Then the update that we compute is $\boldsymbol{\delta} = \mathbf{H}^{-1}\boldsymbol{b}$.

Note that the visual energy term E_{photo} and the inertial error term E_{imu} do not have common residuals. Therefore we can divide $\mathbf{H}$ and $\boldsymbol{b}$ each into two independent parts

$$\mathbf{H} = \mathbf{H}_{\text{photo}} + \mathbf{H}_{\text{imu}} \text{ and } \boldsymbol{b} = \boldsymbol{b}_{\text{photo}} + \boldsymbol{b}_{\text{imu}}. \tag{7.38}$$

As the inertial residuals compare the current relative pose to the estimate from the inertial data they need to use poses in the metric frame relative to the IMU. Therefore we define additional state vectors (containing only IMU-relevant data) for the inertial residuals.

$$\boldsymbol{s}'_i := [\boldsymbol{\xi}^M_{w\_imu_i}, \boldsymbol{v}_i, \boldsymbol{b}_i]^T, \tag{7.39}$$

$$\boldsymbol{s}' = \left[\boldsymbol{s}'^T_1, \boldsymbol{s}'^T_2, \ldots, \boldsymbol{s}'^T_n\right]^T. \tag{7.40}$$

The inertial residuals lead to

$$\mathbf{H}'_{\text{imu}} = \mathbf{J}'^T_{\text{imu}}\mathbf{W}_{\text{imu}}\mathbf{J}'_{\text{imu}} \tag{7.41}$$

$$\boldsymbol{b}'_{\text{imu}} = -\mathbf{J}'^T_{\text{imu}}\mathbf{W}_{\text{imu}}\boldsymbol{r}_{\text{imu}}. \tag{7.42}$$

For the joint optimization, however, we need to obtain $\mathbf{H}_{\text{imu}}$ and $\boldsymbol{b}_{\text{imu}}$ based on the state definition in Eq. (7.36). As the two definitions mainly differ in their representation of the poses we can compute $\mathbf{J}_{\text{rel}}$ such that

$$\mathbf{H}_{\text{imu}} = \mathbf{J}^T_{\text{rel}} \cdot \mathbf{H}'_{\text{imu}} \cdot \mathbf{J}_{\text{rel}} \text{ and } \boldsymbol{b}_{\text{imu}} = \mathbf{J}^T_{\text{rel}} \cdot \boldsymbol{b}'_{\text{imu}}. \tag{7.43}$$

The computation of $\mathbf{J}_{\text{rel}}$ is detailed in Sect. 7.6. Note that we represent all transformations as elements of $\mathfrak{sim}(3)$ and fix the scale to 1 for all of them except $\boldsymbol{\xi}_{m\_d}$.

7.5.4.2 Marginalization using the Schur complement

In order to compute Gauss–Newton updates in a reasonable time-frame we perform partial marginalization for older keyframes. This means that all variables corresponding to this keyframe (pose, bias, velocity and affine illumination parameters) are marginalized out using the Schur complements. Fig. 7.5B shows an example how the factor graph changes after the marginalization of a keyframe.

The marginalization of the visual factors is handled as in [4] by dropping residual terms that affect the sparsity of the system and by first marginalizing all points in the keyframe before marginalizing the keyframe itself.

We denote the set of variables that we want to marginalize as s_β and the ones that are connected to them in the factor graph as s_α. Using this the linear system becomes

$$\begin{bmatrix} \mathbf{H}_{\alpha\alpha} & \mathbf{H}_{\alpha\beta} \\ \mathbf{H}_{\beta\alpha} & \mathbf{H}_{\beta\beta} \end{bmatrix} \begin{bmatrix} s_\alpha \\ s_\beta \end{bmatrix} = \begin{bmatrix} b_\alpha \\ b_\beta \end{bmatrix}. \tag{7.44}$$

The application of the Schur complements yields the new linear system $\widehat{\mathbf{H}_{\alpha\alpha}} s_\alpha = \widehat{b_\alpha}$, with

$$\widehat{\mathbf{H}_{\alpha\alpha}} = \mathbf{H}_{\alpha\alpha} - \mathbf{H}_{\alpha\beta}\mathbf{H}_{\beta\beta}^{-1}\mathbf{H}_{\beta\alpha}, \tag{7.45}$$

$$\widehat{b_\alpha} = b_\alpha - \mathbf{H}_{\alpha\beta}\mathbf{H}_{\beta\beta}^{-1} b_\beta. \tag{7.46}$$

Note that this factor resulting from the marginalization requires the linearization point of all connected variables in s_α to remain fixed. However, we can still approximate the energy around further linearization points $s' = s \boxplus \Delta s$ using

$$\begin{aligned} \mathbf{E}(s' \boxplus \epsilon) &= \mathbf{E}(s \boxplus \Delta s \boxplus \epsilon) \\ &\approx \mathbf{E}(s \boxplus (\Delta s + \epsilon)) \\ &\approx (\Delta s + \epsilon)^T \widehat{\mathbf{H}_{\alpha\alpha}} (\Delta s + \epsilon) + 2(\Delta s + \epsilon)^T \widehat{b_\alpha} + \text{const} \\ &= \Delta s^T \widehat{\mathbf{H}_{\alpha\alpha}} \Delta s + \Delta s^T \widehat{\mathbf{H}_{\alpha\alpha}} \epsilon + \epsilon^T \widehat{\mathbf{H}_{\alpha\alpha}} \Delta s + \epsilon^T \widehat{\mathbf{H}_{\alpha\alpha}} \epsilon + 2\Delta s^T \widehat{b_\alpha} + 2\epsilon^T \widehat{b_\alpha} + \text{const} \\ &= \epsilon^T \widehat{\mathbf{H}_{\alpha\alpha}} \epsilon + 2\epsilon^T \underbrace{\left(\widehat{b_\alpha} + \widehat{\mathbf{H}_{\alpha\alpha}} \Delta s\right)}_{=:\widehat{b'_\alpha}} + \text{const}. \end{aligned} \tag{7.47}$$

In order to maintain consistency of the system it is important that Jacobians are all evaluated at the same value for variables that are connected to a marginalization factor as otherwise the nullspaces get eliminated. Therefore we apply "first estimates Jacobians". For the visual factors we follow [4] and evaluate $\mathbf{J}_{\text{photo}}$ and $\mathbf{J}_{\text{geo}}$ at the linearization point. When computing the inertial factors we fix the evaluation point of $\mathbf{J}_{\text{rel}}$ for all variables which are connected to a marginalization factor. Note that this always includes ξ_{m_d}.

7.5.4.3 Dynamic marginalization for delayed scale convergence

The marginalization procedure described in Sect. 7.5.4.2 has two purposes: reduce the computation complexity of the optimization by removing old states and maintain the information about the previous states of the system. This procedure fixes the linearization points of the states connected to the old states, so they should already have a good estimate. In our scenario this is the case for all variables except of scale.

The main idea of "dynamic marginalization" is to maintain several marginalization priors at the same time and reset the one we currently use when the scale estimate moves too far from

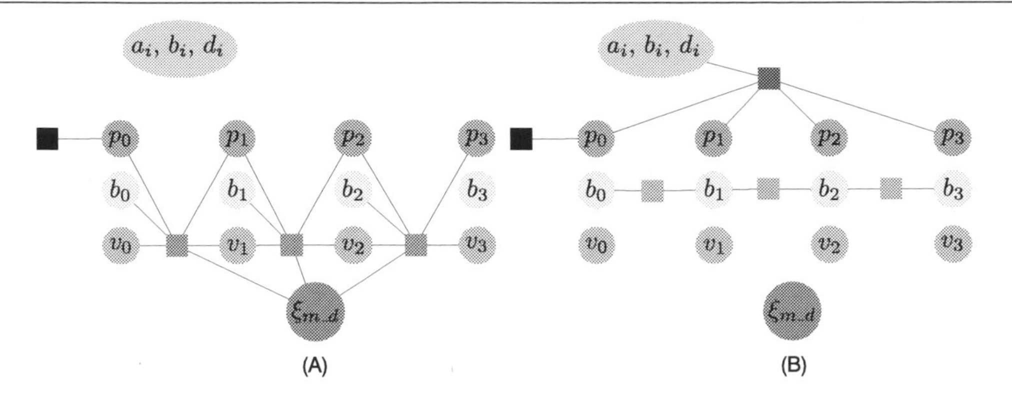

Figure 7.6 : Partitioning of the factor graph from Fig. 7.5A into G_{metric} (A) and G_{visual} (B). G_{metric} contains all IMU-factors while G_{visual} contains the factors that do not depend on ξ_{m_d}. Note that both of them do not contain any marginalization factors.

the linearization point in the marginalization prior. To get an idea of the method we strongly recommend watching the animation at https://youtu.be/GoqnXDS7jbA?t=2m6s.

In our implementation we use three marginalization priors: M_{visual}, M_{curr} and M_{half}. M_{visual} contains only scale independent information from previous states of the vision and cannot be used to infer the global scale. M_{curr} contains all information since the time we set the linearization point for the scale and M_{half} contains only the recent states that have a scale close to the current estimate.

When the scale estimate deviates too much from the linearization point of M_{curr}, the value of M_{curr} is set to M_{half} and M_{half} is set to M_{visual} with corresponding changes in the linearization points. This ensures that the optimization always has some information about the previous states with consistent scale estimates. In the remaining part of the section we provide the details of our implementation.

We define G_{metric} to contain only the visual–inertial factors (which depend on ξ_{m_d}) and G_{visual} to contain all other factors, except the marginalization priors. Then

$$G_{\text{full}} = G_{\text{metric}} \cup G_{\text{visual}}. \tag{7.48}$$

Fig. 7.6 depicts the partitioning of the factor graph.

We define three different marginalization factors M_{curr}, M_{visual} and M_{half}. For the optimization we always compute updates using the graph

$$G_{ba} = G_{\text{metric}} \cup G_{\text{visual}} \cup M_{\text{curr}}. \tag{7.49}$$

When keyframe i is marginalized we update M_{visual} with the factor arising from marginalizing frame i in $G_{\text{visual}} \cup M_{\text{visual}}$. This means that M_{visual} contains all marginalized visual factors and no marginalized inertial factors making it independent of the scale.

For each marginalized keyframe i we define

$$s_i := \text{scale estimate at the time, } i \text{ was marginalized.} \tag{7.50}$$

We define $i \in M$ if and only if M contains an *inertial* factor that was marginalized at time i. Using this we enforce the following constraints for inertial factors contained in M_{curr} and M_{half}:

$$\forall i \in M_{\text{curr}} : s_i \in \left[s_{\text{middle}}/d_i, s_{\text{middle}} \cdot d_i\right], \tag{7.51}$$

$$\forall i \in M_{\text{half}} : s_i \in \begin{cases} \left[s_{\text{middle}}, s_{\text{middle}} \cdot d_i\right], & \text{if } s_{\text{curr}} > s_{\text{middle}}, \\ \left[s_{\text{middle}}/d_i, s_{\text{middle}}\right], & \text{otherwise,} \end{cases} \tag{7.52}$$

where s_{middle} is the current middle of the allowed scale interval (initialized with s_0), d_i is the size of the scale interval at time i, and s_{curr} is the current scale estimate.

We update M_{curr} by marginalizing frame i in G_{ba} and we update M_{half} by marginalizing i in $G_{\text{metric}} \cup G_{\text{visual}} \cup M_{\text{half}}$.

Algorithm 1 Constrain Marginalization

```
upper ← s_curr > s_middle
if upper ≠ lastUpper then                         ▷ Side changes.
    M_half ← M_visual
end if
if s_curr > s_middle · d_i then                   ▷ Upper boundary exceeded.
    M_curr ← M_half
    M_half ← M_visual
    s_middle ← s_middle · d_i
end if
if s_curr < s_middle / d_i then                   ▷ Lower boundary exceeded.
    M_curr ← M_half
    M_half ← M_visual
    s_middle ← s_middle / d_i
end if
lastUpper ← upper
```

In order to preserve the constraints in Eqs. (7.51) and (7.52) we apply Algorithm 1 every time a marginalization happens. By following these steps on the one hand we make sure that the constraints are satisfied which ensures that the scale difference in the currently used marginalization factor stays smaller than d_i^2. On the other hand the factor always contains some inertial

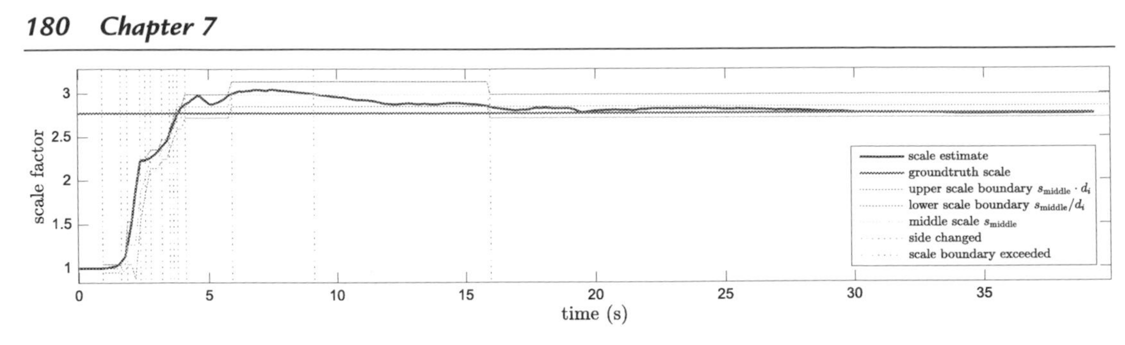

Figure 7.7 : The scale estimation running on the V1_03_difficult sequence from the EuRoC dataset. We show the current scale estimate (bold blue), the groundtruth scale (bold red) and the current scale interval (light lines). The vertical dotted lines denote when the side changes (blue) and when the boundary of the scale interval is exceeded (red). In practice this means that M_{curr} contains the inertial factors since the last blue or red dotted line that is before the last red dotted line. For example at 16 s it contains all inertial data since the blue line at 9 seconds.

factors so that the scale estimation works at all times. Note also that M_{curr} and M_{half} have separate first estimate Jacobians that are employed when the respective marginalization factor is used. Fig. 7.7 shows how the system works in practice.

An important part of this strategy is the choice of d_i. It should be small, in order to keep the system consistent, but not too small so that M_{curr} always contains enough inertial factors. Therefore we chose to dynamically adjust the parameter as follows. At all time steps i we calculate

$$d_i = \min\{d_{\text{min}}^j \mid j \in \mathbb{N} \setminus \{0\}, \frac{s_i}{s_{i-1}} < d_i\}. \tag{7.53}$$

This ensures that it cannot happen that the M_{half} gets reset to M_{visual} at the same time that M_{curr} is exchanged with M_{half}. Therefore it prevents situations where M_{curr} contains no inertial factors at all, making the scale estimation more reliable. On the one hand d_{min} should not be too small as otherwise not enough inertial information is accumulated. On the other hand it should not be too big, because else the marginalization can get inconsistent. In our experiments we chose $d_{\text{min}} = \sqrt{1.1}$ which is a good tradeoff between the two.

7.5.4.4 Measuring scale convergence

For many applications it is very useful to know when the scale has converged, especially as this can take an arbitrary amount of time in theory. The lack of this measure is also one of the main drawbacks of the initialization method described in [25]. Here we propose a simple but effective strategy. With s_i being the scale at time i and n being the maximum queue size (60

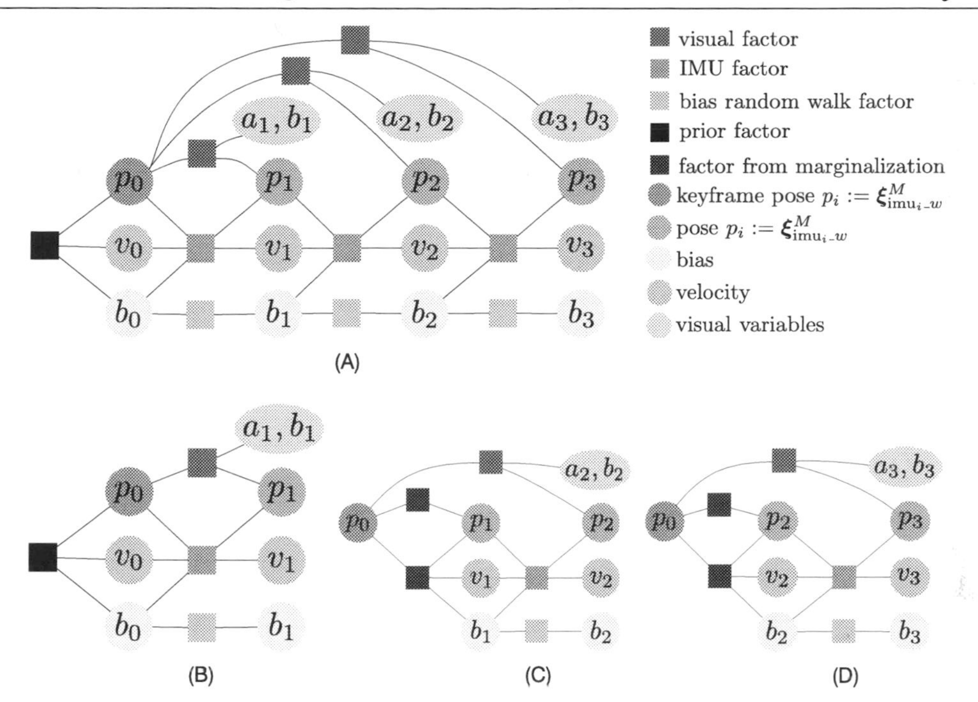

Figure 7.8 : Factor graphs for the coarse tracking. (A) General factor graph for the problem. (B) First pose estimate after new keyframe. (C) Second pose estimate. (D) Third pose estimate.

in all our experiments) we compute

$$c = \frac{\max^n_{j:=i-n+1} s_j}{\min^n_{j:=i-n+1} s_j} - 1. \tag{7.54}$$

Fig. 7.13B shows a plot of this measure. When c is below a certain threshold c_{min} (0.005 in our experiments) we consider the scale to have converged and fix ξ_{m_d}. While this does not influence the accuracy of the system, fixing the scale might be useful for some applications.

7.5.5 Coarse Visual–Inertial Tracking

The coarse tracking is responsible for computing a fast pose estimate for each frame that also serves as an initialization for the joint estimation detailed in Sect. 7.5.4. We perform conventional direct image alignment between the current frame and the latest keyframe, while keeping the geometry and the scale fixed. The general factor graph for the coarse tracking is shown in Fig. 7.8A. Inertial residuals using the previously described IMU preintegration scheme are placed between subsequent frames. Note that in contrast to the factor graph for the joint optimization, the visual factor for the coarse tracking does not connect all keyframes and depths, but instead just connects each frame with the current keyframe.

Every time the joint estimation is finished for a new frame, the coarse tracking is reinitialized with the new estimates for scale, gravity direction, bias, and velocity as well as the new keyframe as a reference for the visual factors. Similar to the joint estimation we perform partial marginalization to keep the update time constrained. After estimating the variables for a new frame we marginalize out all variables except the keyframe pose and the variables of the newest frame. Also we immediately marginalize the photometric parameters a_i and b_i as soon as the pose estimation for frame i is finished. Fig. 7.8B–D shows how the coarse factor graph changes over time. In contrast to the joint optimization we do not need to use dynamic marginalization because the scale is not included in the optimization.

7.6 Calculating the Relative Jacobians

As mentioned in Sect. 7.5.4.1 we want to compute $\mathbf{J}_{\text{rel}}$, such that (7.43) holds.

We convert between the poses using

$$\boldsymbol{\xi}_{w\_imu}^{M} = \boldsymbol{\xi}_{m\_d} \boxplus \left(\boldsymbol{\xi}_{cam\_w}^{D}\right)^{-1} \boxplus \boldsymbol{\xi}_{m\_d}^{-1} \boxplus \boldsymbol{\xi}_{cam\_imu}^{M} =: \Psi(\boldsymbol{\xi}_{cam\_w}^{D}, \boldsymbol{\xi}_{m\_d}). \tag{7.55}$$

For any function $\Omega(\boldsymbol{\xi}) : \mathfrak{sim}(3) \to \mathfrak{sim}(3)$ we define the derivative $\frac{d\Omega(\boldsymbol{\xi} \boxplus \epsilon)}{d\epsilon}$ implicitly using

$$\Omega(\boldsymbol{\xi})^{-1} \boxplus \Omega(\boldsymbol{\xi} \boxplus \epsilon) = \frac{d\Omega(\boldsymbol{\xi} \boxplus \epsilon)}{d\epsilon} \cdot \epsilon + \eta(\epsilon) \cdot \epsilon \tag{7.56}$$

where the error function $\eta(\epsilon)$ goes to 0 as $||\epsilon||$ goes to 0.

Note that in principle there are three other ways to define this derivative (as you can place the increment with ϵ as well as the multiplication with the inverse on either side). However, it can be shown (see Sect. 7.6.1) that only with this version the following chain rule holds for $f(\boldsymbol{\xi}) : \mathfrak{sim}(3) \to \mathbb{R}$:

$$\frac{df\left(\Omega(\boldsymbol{\xi} \boxplus \epsilon)\right)}{d\epsilon} = \frac{df\left(\Omega(\boldsymbol{\xi}) \boxplus \delta\right)}{\delta} \cdot \frac{d\Omega(\boldsymbol{\xi} \boxplus \epsilon)}{\epsilon}. \tag{7.57}$$

Using these definitions the relevant derivatives for Ψ can be computed (see Sects. 7.6.2 and 7.6.3)

$$\frac{\partial \Psi(\boldsymbol{\xi}_{cam\_w}^{D} \boxplus \epsilon, \boldsymbol{\xi}_{m\_d})}{\partial \epsilon} = -\text{Adj}\left(\mathbf{T}_{cam\_imu}^{-1} \cdot \mathbf{T}_{m\_d} \cdot \mathbf{T}_{cam\_w}\right), \tag{7.58}$$

$$\frac{\partial \Psi(\boldsymbol{\xi}^D_{cam\_w}, \boldsymbol{\xi}_{m\_d} \boxplus \boldsymbol{\epsilon})}{\partial \boldsymbol{\epsilon}} = \mathrm{Adj}(\mathbf{T}^{-1}_{cam\_imu} \cdot \mathbf{T}_{m\_d} \cdot \mathbf{T}_{cam\_w}) - \mathrm{Adj}(\mathbf{T}^{-1}_{cam\_imu} \cdot \mathbf{T}_{m\_d}). \quad (7.59)$$

Stacking these derivatives correctly we can compute a Jacobian $\mathbf{J}_{\mathrm{rel}}$ such that

$$\mathbf{J}_{\mathrm{imu}} = \mathbf{J}'_{\mathrm{imu}} \cdot \mathbf{J}_{\mathrm{rel}}. \quad (7.60)$$

Using this we can finally compute

$$\begin{aligned} \mathbf{H}_{\mathrm{imu}} &= \mathbf{J}^T_{\mathrm{imu}} \mathbf{W}_{\mathrm{imu}} \mathbf{J}_{\mathrm{imu}} \\ &= \mathbf{J}^T_{\mathrm{rel}} \cdot \left(\mathbf{J}'_{\mathrm{imu}}\right)^T \cdot \mathbf{W}_{\mathrm{imu}} \cdot \mathbf{J}'_{\mathrm{imu}} \cdot \mathbf{J}_{\mathrm{rel}} \\ &= \mathbf{J}^T_{\mathrm{rel}} \cdot \mathbf{H}'_{\mathrm{imu}} \cdot \mathbf{J}_{\mathrm{rel}}, \end{aligned} \quad (7.61)$$

$$\begin{aligned} \boldsymbol{b}_{\mathrm{imu}} &= -\mathbf{J}^T_{\mathrm{imu}} \mathbf{W}_{\mathrm{imu}} \boldsymbol{r}_{\mathrm{imu}} \\ &= -\mathbf{J}^T_{\mathrm{rel}} \left(\mathbf{J}'_{\mathrm{imu}}\right)^T \mathbf{W}_{\mathrm{imu}} \boldsymbol{r}_{\mathrm{imu}} \\ &= \mathbf{J}^T_{\mathrm{rel}} \cdot \boldsymbol{b}'_{\mathrm{imu}}. \end{aligned} \quad (7.62)$$

7.6.1 Proof of the Chain Rule

In this section we will prove the chain rule (7.57).

We define the multivariate derivatives implicitly:

$$f(\Omega(\boldsymbol{\xi}) \boxplus \boldsymbol{\delta}) - f(\Omega(\boldsymbol{\xi})) = \frac{df\left(\Omega(\boldsymbol{\xi}) \boxplus \boldsymbol{\delta}\right)}{\boldsymbol{\delta}} \cdot \boldsymbol{\delta} + \mu(\boldsymbol{\delta}) \cdot \boldsymbol{\delta}, \quad (7.63)$$

where the error function $\mu(\boldsymbol{\delta})$ goes to 0 as $\boldsymbol{\delta}$ goes to zero. We can rewrite Eq. (7.56) by multiplying $\Omega(\boldsymbol{\xi})$ from the left,

$$\Omega(\boldsymbol{\xi} \boxplus \boldsymbol{\epsilon}) = \Omega(\boldsymbol{\xi}) \boxplus \left(\frac{d_l \Omega(\boldsymbol{\xi} \boxplus \boldsymbol{\epsilon})}{d\boldsymbol{\epsilon}} \cdot \boldsymbol{\epsilon} + \eta(\boldsymbol{\epsilon}) \cdot \boldsymbol{\epsilon}\right). \quad (7.64)$$

Using this we can compute

$$\begin{aligned} f(\Omega(\boldsymbol{\xi} \boxplus \boldsymbol{\epsilon})) - f(\Omega(\boldsymbol{\xi})) &\overset{(7.64)}{=} f\left(\Omega(\boldsymbol{\xi}) \boxplus \underbrace{\left(\frac{d_l \Omega(\boldsymbol{\xi} \boxplus \boldsymbol{\epsilon})}{d\boldsymbol{\epsilon}} \cdot \boldsymbol{\epsilon} + \eta(\boldsymbol{\epsilon}) \cdot \boldsymbol{\epsilon}\right)}_{=:\boldsymbol{\delta}_\epsilon}\right) - f(\Omega(\boldsymbol{\xi})) \\ &= f\left(\Omega(\boldsymbol{\xi}) \boxplus \boldsymbol{\delta}_\epsilon\right) - f(\Omega(\boldsymbol{\xi})) \overset{(7.63)}{=} \frac{df\left(\Omega(\boldsymbol{\xi}) \boxplus \boldsymbol{\delta}\right)}{\boldsymbol{\delta}} \cdot \boldsymbol{\delta}_\epsilon + \mu(\boldsymbol{\delta}_\epsilon) \cdot \boldsymbol{\delta}_\epsilon \\ &= \frac{df\left(\Omega(\boldsymbol{\xi}) \boxplus \boldsymbol{\delta}\right)}{\boldsymbol{\delta}} \cdot \left(\frac{d_l \Omega(\boldsymbol{\xi} \boxplus \boldsymbol{\epsilon})}{d\boldsymbol{\epsilon}} \cdot \boldsymbol{\epsilon} + \eta(\boldsymbol{\epsilon}) \cdot \boldsymbol{\epsilon}\right) + \mu(\boldsymbol{\delta}_\epsilon) \cdot \left(\frac{d_l \Omega(\boldsymbol{\xi} \boxplus \boldsymbol{\epsilon})}{d\boldsymbol{\epsilon}} \cdot \boldsymbol{\epsilon} + \eta(\boldsymbol{\epsilon}) \cdot \boldsymbol{\epsilon}\right) \end{aligned}$$

$$
\begin{aligned}
&= \frac{df\left(\Omega(\xi)\boxplus\delta\right)}{\delta}\cdot\frac{d_l\Omega(\xi\boxplus\epsilon)}{d\epsilon}\cdot\epsilon+\frac{df\left(\Omega(\xi)\boxplus\delta\right)}{\delta}\cdot\eta(\epsilon)\cdot\epsilon\\
&+\mu(\delta_\epsilon)\cdot\frac{d_l\Omega(\xi\boxplus\epsilon)}{d\epsilon}\cdot\epsilon+\mu(\delta_\epsilon)\cdot\eta(\epsilon)\cdot\epsilon\\
&= \frac{df\left(\Omega(\xi)\boxplus\delta\right)}{\delta}\cdot\frac{d_l\Omega(\xi\boxplus\epsilon)}{d\epsilon}\cdot\epsilon\\
&+\underbrace{\left(\frac{df\left(\Omega(\xi)\boxplus\delta\right)}{\delta}\cdot\eta(\epsilon)+\mu(\delta_\epsilon)\cdot\frac{d_l\Omega(\xi\boxplus\epsilon)}{d\epsilon}+\mu(\delta_\epsilon)\cdot\eta(\epsilon)\right)}_{=:\gamma(\epsilon)}\cdot\epsilon.
\end{aligned}
\tag{7.65}
$$

When ϵ goes to 0, then δ_ϵ also goes to 0 (as can be seen from its definition). Using this it follows by the definition of the derivative that $\eta(\epsilon)$ and $\mu(\delta_\epsilon)$ go to 0 as well. This shows that $\gamma(\epsilon)$ goes to 0 when ϵ goes to 0. Therefore the last line of Eq. (7.65) is in line with our definition of the derivative and

$$
\frac{df\left(\Omega(\xi\boxplus\epsilon)\right)}{d\epsilon}=\frac{df\left(\Omega(\xi)\boxplus\delta\right)}{\delta}\cdot\frac{d_l\Omega(\xi\boxplus\epsilon)}{\epsilon}. \quad \square
\tag{7.66}
$$

7.6.2 Derivation of the Jacobian with Respect to Pose in Eq. (7.58)

In this section we will show how to derive the Jacobians $\frac{\partial\Psi(\xi^D_{cam\_w}\boxplus\epsilon,\xi_{m\_d})}{\partial\epsilon}$ using the implicit definition of the derivative shown in Eq. (7.56).

In order to do this we need the definition of the adjoint.

$$
\mathbf{T}\cdot\exp(\epsilon)=\exp(\mathrm{Adj}_{\mathbf{T}}\cdot\epsilon)\cdot\mathbf{T}
\tag{7.67}
$$

with $\mathbf{T}\in\mathbf{SIM}(3)$. It follows that

$$
\log\left(\mathbf{T}\cdot\exp(\epsilon)\cdot\mathbf{T}^{-1}\right)=\mathrm{Adj}_{\mathbf{T}}\cdot\epsilon.
\tag{7.68}
$$

Using this we can compute

$$
\begin{aligned}
&\Psi(\xi^D_{cam\_w},\xi_{m\_d})^{-1}\boxplus\Psi(\xi^D_{cam\_w}\boxplus\epsilon,\xi_{m\_d})\\
&\overset{(7.55)}{=}\left(\xi_{m\_d}\boxplus\left(\xi^D_{cam\_w}\right)^{-1}\boxplus\xi^{-1}_{m\_d}\boxplus\xi^M_{cam\_imu}\right)^{-1}\boxplus\\
&\left(\xi_{m\_d}\boxplus\left(\xi^D_{cam\_w}\boxplus\epsilon\right)^{-1}\boxplus\xi^{-1}_{m\_d}\boxplus\xi^M_{cam\_imu}\right)\\
&=\left(\xi^M_{cam\_imu}\right)^{-1}\boxplus\xi_{m\_d}\boxplus\xi^D_{cam\_w}\boxplus\underbrace{\xi^{-1}_{m\_d}\boxplus\xi_{m\_d}}_{=0}\boxplus\epsilon^{-1}\boxplus\left(\xi^D_{cam\_w}\right)^{-1}\boxplus\\
&\xi^{-1}_{m\_d}\boxplus\xi^M_{cam\_imu}
\end{aligned}
$$

$$= \log\left(\left(\mathbf{T}^M_{cam\_imu}\right)^{-1} \cdot \mathbf{T}_{m\_d} \cdot \mathbf{T}^D_{cam\_w} \cdot \exp(\epsilon)^{-1} \cdot \left(\mathbf{T}^D_{cam\_w}\right)^{-1} \cdot \mathbf{T}^{-1}_{m\_d} \cdot \mathbf{T}^M_{cam\_imu}\right)$$

$$\stackrel{(7.68)}{=} \mathrm{Adj}\left(\left(\mathbf{T}^M_{cam\_imu}\right)^{-1} \cdot \mathbf{T}_{m\_d} \cdot \mathbf{T}^D_{cam\_w}\right) \cdot (-\epsilon). \tag{7.69}$$

After moving the minus sign to the left this is in line with our definition of the derivative in Eq. (7.56) and it follows that

$$\frac{\partial \Psi(\boldsymbol{\xi}^D_{cam\_w} \boxplus \epsilon, \boldsymbol{\xi}_{m\_d})}{\partial \epsilon} = -\mathrm{Adj}\left(\left(\mathbf{T}^M_{cam\_imu}\right)^{-1} \cdot \mathbf{T}_{m\_d} \cdot \mathbf{T}^D_{cam\_w}\right). \tag{7.70}$$

7.6.3 Derivation of the Jacobian with Respect to Scale and Gravity Direction in Eq. (7.59)

In order to derive the Jacobian $\frac{\partial \Psi(\boldsymbol{\xi}^D_{cam\_w}, \boldsymbol{\xi}_{m\_d} \boxplus \epsilon)}{\partial \epsilon}$ we need the Baker–Campbell–Hausdorff formula: Let $\boldsymbol{a}, \boldsymbol{b} \in \mathfrak{sim}(3)$, then

$$\log(\exp(\boldsymbol{a}) \cdot \exp(\boldsymbol{b})) = \boldsymbol{a} + \boldsymbol{b} + \frac{1}{2}[\boldsymbol{a}, \boldsymbol{b}] + \frac{1}{12}([\boldsymbol{a}, [\boldsymbol{a}, \boldsymbol{b}]] + [\boldsymbol{b}, [\boldsymbol{b}, \boldsymbol{a}]])$$
$$+ \frac{1}{48}([\boldsymbol{b}, [\boldsymbol{a}, [\boldsymbol{b}, \boldsymbol{a}]]] + [\boldsymbol{a}, [\boldsymbol{b}, [\boldsymbol{b}, \boldsymbol{a}]]]) + \ldots. \tag{7.71}$$

Here $[\boldsymbol{a}, \boldsymbol{b}] := \boldsymbol{ab} - \boldsymbol{ba}$ denotes the Lie bracket.

In this section we will omit the superscripts to simplify the notation. We have

$$\Psi(\boldsymbol{\xi}_{cam\_w}, \boldsymbol{\xi}_{m\_d})^{-1} \boxplus \Psi(\boldsymbol{\xi}_{cam\_w}, \boldsymbol{\xi}_{m\_d} \boxplus \epsilon)$$
$$\stackrel{(7.55)}{=} \left(\boldsymbol{\xi}_{m\_d} \boxplus \boldsymbol{\xi}^{-1}_{cam\_w} \boxplus \boldsymbol{\xi}^{-1}_{m\_d} \boxplus \boldsymbol{\xi}_{cam\_imu}\right)^{-1} \boxplus$$
$$\left(\boldsymbol{\xi}_{m\_d} \boxplus \epsilon \boxplus \boldsymbol{\xi}^{-1}_{cam\_w} \boxplus (\boldsymbol{\xi}_{m\_d} \boxplus \epsilon)^{-1} \boxplus \boldsymbol{\xi}_{cam\_imu}\right)$$
$$= \boldsymbol{\xi}^{-1}_{cam\_imu} \boxplus \boldsymbol{\xi}_{m\_d} \boxplus \boldsymbol{\xi}_{cam\_w} \boxplus \boldsymbol{\xi}^{-1}_{m\_d} \boxplus \boldsymbol{\xi}_{m\_d} \boxplus \epsilon \boxplus \boldsymbol{\xi}^{-1}_{cam\_w} \boxplus \epsilon^{-1} \boxplus \boldsymbol{\xi}^{-1}_{m\_d} \boxplus \boldsymbol{\xi}_{cam\_imu}$$
$$= \boldsymbol{\xi}^{-1}_{cam\_imu} \boxplus \boldsymbol{\xi}_{m\_d} \boxplus \boldsymbol{\xi}_{cam\_w} \boxplus \epsilon \boxplus \boldsymbol{\xi}^{-1}_{cam\_w} \boxplus$$
$$\underbrace{\left(\boldsymbol{\xi}^{-1}_{cam\_imu} \boxplus \boldsymbol{\xi}_{m\_d}\right)^{-1} \boxplus \left(\boldsymbol{\xi}^{-1}_{cam\_imu} \boxplus \boldsymbol{\xi}_{m\_d}\right)}_{=0} \boxplus \epsilon^{-1} \boxplus \boldsymbol{\xi}^{-1}_{m\_d} \boxplus \boldsymbol{\xi}_{cam\_imu}$$
$$= \underbrace{\boldsymbol{\xi}^{-1}_{cam\_imu} \boxplus \boldsymbol{\xi}_{m\_d} \boxplus \boldsymbol{\xi}_{cam\_w} \boxplus \epsilon \boxplus \boldsymbol{\xi}^{-1}_{cam\_w} \boxplus \boldsymbol{\xi}^{-1}_{m\_d} \boxplus \boldsymbol{\xi}_{cam\_imu}}_{:=\boldsymbol{a}} \boxplus$$
$$\underbrace{\boldsymbol{\xi}^{-1}_{cam\_imu} \boxplus \boldsymbol{\xi}_{m\_d} \boxplus \epsilon^{-1} \boxplus \boldsymbol{\xi}^{-1}_{m\_d} \boxplus \boldsymbol{\xi}_{cam\_imu}}_{:=\boldsymbol{b}}. \tag{7.72}$$

Using Eq. (7.68) we can compute

$$a = \mathrm{Adj}\left(\mathbf{T}_{cam\_imu}^{-1} \cdot \mathbf{T}_{m\_d} \cdot \mathbf{T}_{cam\_w}\right) \cdot \boldsymbol{\epsilon}, \tag{7.73}$$

and

$$b = -\mathrm{Adj}\left(\mathbf{T}_{cam\_imu}^{-1} \cdot \mathbf{T}_{m\_d}\right) \cdot \boldsymbol{\epsilon}. \tag{7.74}$$

Now we have to prove that all terms of the Baker–Campbell–Hausdorff formula which contain a Lie bracket can be written as $\mu(\boldsymbol{\epsilon}) \cdot \boldsymbol{\epsilon}$, where $\mu(\boldsymbol{\epsilon})$ goes to zero, when $\boldsymbol{\epsilon}$ goes to zero.

We can compute

$$a \cdot b = \underbrace{\mathrm{Adj}\left(\mathbf{T}_{cam\_imu}^{-1} \cdot \mathbf{T}_{m\_d} \cdot \mathbf{T}_{cam\_w}\right) \cdot \boldsymbol{\epsilon} \cdot \left(-\mathrm{Adj}\left(\mathbf{T}_{cam\_imu}^{-1} \cdot \mathbf{T}_{m\_d}\right)\right)}_{\mu_1(\boldsymbol{\epsilon})} \cdot \boldsymbol{\epsilon}, \tag{7.75}$$

$$b \cdot a = \underbrace{-\mathrm{Adj}\left(\mathbf{T}_{cam\_imu}^{-1} \cdot \mathbf{T}_{m\_d}\right) \cdot \boldsymbol{\epsilon} \cdot \mathrm{Adj}\left(\mathbf{T}_{cam\_imu}^{-1} \cdot \mathbf{T}_{m\_d} \cdot \mathbf{T}_{cam\_w}\right)}_{\mu_2(\boldsymbol{\epsilon})} \cdot \boldsymbol{\epsilon}, \tag{7.76}$$

$$[a, b] = ab - ba = (\mu_1(\boldsymbol{\epsilon}) + \mu_2(\boldsymbol{\epsilon})) \cdot \boldsymbol{\epsilon}. \tag{7.77}$$

Obviously $\mu_1(\boldsymbol{\epsilon})$ and $\mu_2(\boldsymbol{\epsilon})$ go to zero when $\boldsymbol{\epsilon}$ goes to zero. For the remaining summands of Eq. (7.71) the same argumentation can be used. It follows that

$$\begin{aligned}(7.72) &= a + b + \mu(\boldsymbol{\epsilon}) \cdot \boldsymbol{\epsilon} \\ &= \left(\mathrm{Adj}\left(\mathbf{T}_{cam\_imu}^{-1} \cdot \mathbf{T}_{m\_d} \cdot \mathbf{T}_{cam\_w}\right) - \mathrm{Adj}\left(\mathbf{T}_{cam\_imu}^{-1} \cdot \mathbf{T}_{m\_d}\right) \cdot \boldsymbol{\epsilon}\right) \cdot \boldsymbol{\epsilon} + \mu(\boldsymbol{\epsilon}) \cdot \boldsymbol{\epsilon}\end{aligned} \tag{7.78}$$

where $\mu(\boldsymbol{\epsilon})$ goes to zero when $\boldsymbol{\epsilon}$ goes to zero. According to (7.56) this means that

$$\frac{\partial \Psi(\boldsymbol{\xi}_{cam\_w}^{D}, \boldsymbol{\xi}_{m\_d} \boxplus \boldsymbol{\epsilon})}{\partial \boldsymbol{\epsilon}} = \mathrm{Adj}(\mathbf{T}_{cam\_imu}^{-1} \cdot \mathbf{T}_{m\_d} \cdot \mathbf{T}_{cam\_w}) - \mathrm{Adj}(\mathbf{T}_{cam\_imu}^{-1} \cdot \mathbf{T}_{m\_d}). \tag{7.79}$$

7.7 Results

We evaluate the described VI-DSO on the publicly available EuRoC dataset [5]. The performance is compared to [4], [20], [10], [24], [22], [32] and [33]. We also recommend watching the video at vision.in.tum.de/vi-dso.

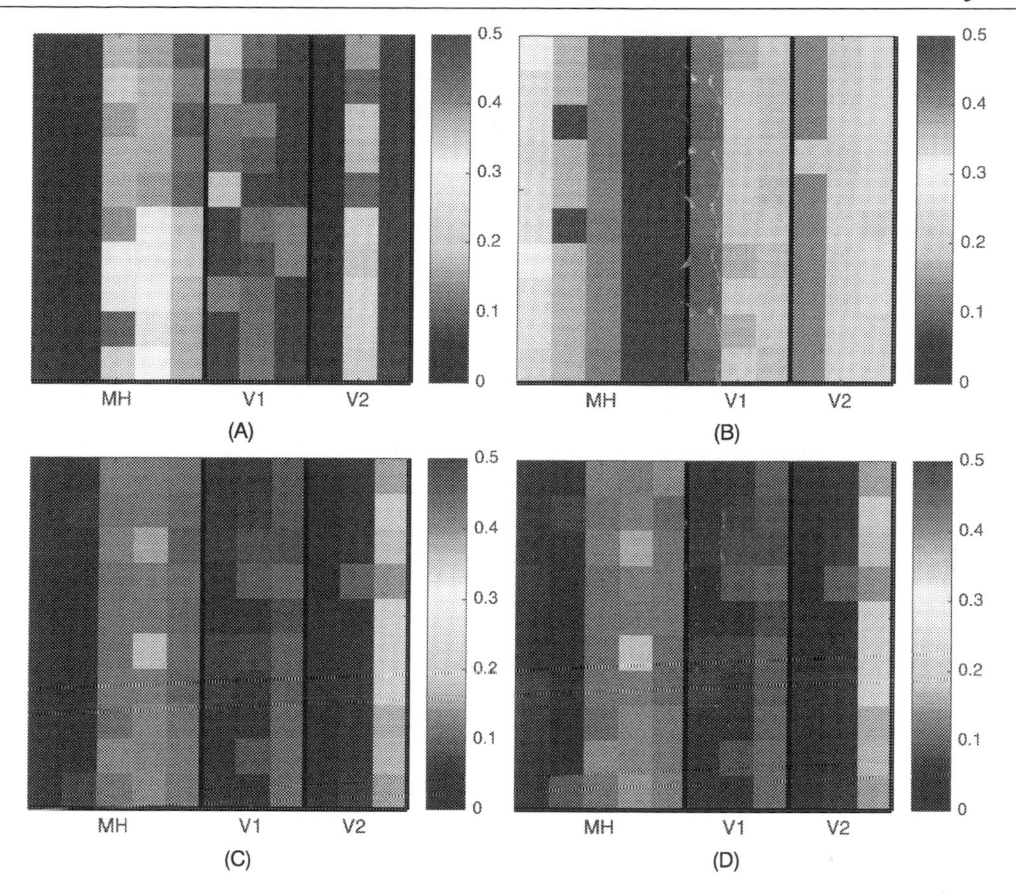

Figure 7.9 : RMSE for different methods run 10 times (lines) on each sequence (columns) of the EuRoC dataset. (A) DSO [4] realtime gt-scaled. (B) ROVIO real-time [20]. (C) VI-DSO real-time gt-scaled. (D) VI-DSO real-time.

7.7.1 Robust Quantitative Evaluation

In order to obtain an accurate evaluation we run our method 10 times for each sequence of the dataset (using the left camera). We directly compare the results to visual-only DSO [4] and ROVIO [20]. As DSO cannot observe the scale we evaluate using the optimal ground truth scale in some plots (with the description "gt-scaled") to enable a fair comparison. For all other results we scale the trajectory with the final scale estimate (our method) or with 1 (other methods). For DSO we use the results published together with their paper. We use the same start and end times for each sequence to run our method and ROVIO. Note that the drone has a high initial velocity in some sequences when using these start times making it especially challenging for our IMU initialization. Fig. 7.9 shows the root mean square error (RMSE) for every run and Fig. 7.10 displays the cumulative error plot. Clearly our method significantly

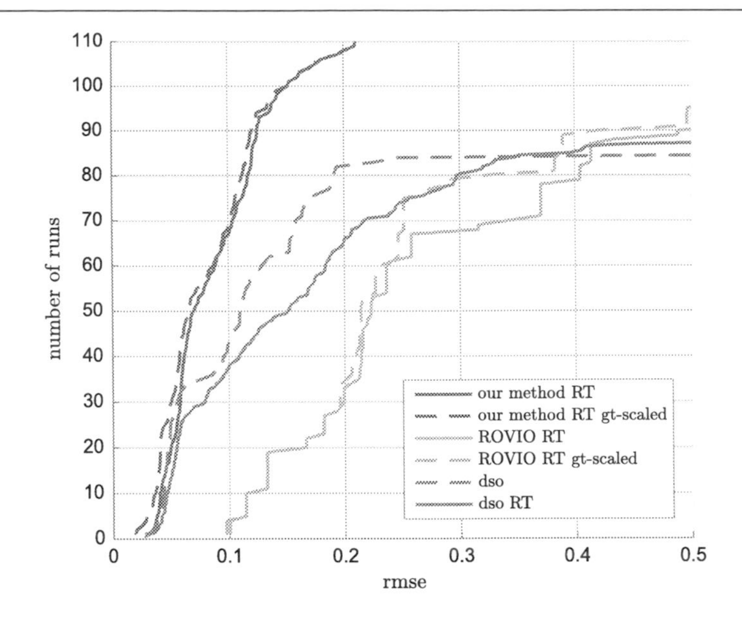

Figure 7.10 : Cumulative error plot on the EuRoC dataset (RT means real-time). This experiment demonstrates that the additional IMU not only provides a reliable scale estimate, but that it also significantly increases accuracy and robustness.

outperforms DSO and ROVIO. Without inertial data DSO is not able to work on all sequences especially on V1_03_difficult and V2_03_difficult and it is also not able to scale the results correctly. ROVIO on the other hand is very robust but as a filtering-based method it cannot provide sufficient accuracy.

Table 7.1 shows a comparison to several other methods. For our results we have displayed the median error for each sequence from the 10 runs plotted in Fig. 7.9C. This makes the results very meaningful. For the other methods unfortunately only one result was reported so we have to assume that they are representative as well. The results for [22] and [32] were taken from [32] and the results for visual–inertial ORB-SLAM were taken from [10]. For VINS-Mono [26] the results were taken from their paper [33]. The results for [10] differ slightly from the other methods as they show the error of the keyframe trajectory instead of the full trajectory. This is a slight advantage as keyframes are bundle adjusted in their method which does not happen for the other frames.

In comparison with VI ORB-SLAM our method outperforms it in terms of RMSE on several sequences. As ORB-SLAM is a SLAM system while ours is a pure odometry method this is a remarkable achievement especially considering the differences in the evaluation. Note that the Vicon room sequences (V*) are executed in a small room and contain a lot of loopy motions where the loop closures done by a SLAM system significantly improve the performance. Also

Table 7.1: Accuracy of the estimated trajectory on the EuRoC dataset for several methods. Note that ORB-SLAM does a convincing job showing leading performance on some of the sequences. Nevertheless, since VI-DSO directly works on the sensor data (colors and IMU measurements), we observe similar precision and a better robustness—even without loop closuring. Moreover, the proposed method is the only one not to fail on any of the sequences (except ROVIO).

Sequence		MH1	MH2	MH3	MH4	MH5	V11	V12	V13	V21	V22	V23
VI-DSO (RT) (median of 10 runs each)	RMSE	**0.062**	**0.044**	0.117	**0.132**	0.121	0.059	0.067	**0.096**	0.040	0.062	0.174
	RMSE gt-scaled	0.041	0.041	0.116	0.129	0.106	0.057	0.066	0.095	0.031	0.060	0.173
	Scale Error (%)	1.1	**0.5**	**0.4**	**0.2**	0.8	1.1	1.1	**0.8**	1.2	**0.3**	**0.4**
VI ORB-SLAM (keyframe trajectory)	RMSE	0.075	0.084	**0.087**	0.217	**0.082**	**0.027**	**0.028**	X	**0.032**	**0.041**	**0.074**
	RMSE gt-scaled	0.072	0.078	0.067	0.081	0.077	0.019	0.024	X	0.031	0.026	0.073
	Scale Error (%)	**0.5**	0.8	1.5	3.4	**0.5**	**0.9**	**0.8**	X	**0.2**	1.4	0.7
VINS-MONO [33]	RMSE	0.12	0.12	0.13	0.18	0.21	0.068	0.084	0.19	0.081	0.16	X
VI odometry [22], mono	RMSE	0.34	0.36	0.30	0.48	0.47	0.12	0.16	0.24	0.12	0.22	X
VI odometry [22], stereo	RMSE	0.23	0.15	0.23	0.32	0.36	0.04	0.08	0.13	0.10	0.17	X
VI SLAM [32], mono	RMSE	0.25	0.18	0.21	0.30	0.35	0.11	0.13	0.20	0.12	0.20	X
VI SLAM [32], stereo	RMSE	0.11	0.09	0.19	0.27	0.23	0.04	0.05	0.11	0.10	0.18	X

our method is more robust as ORB-SLAM fails to track one sequence. Even considering only sequences where ORB-SLAM works our approach has a lower maximum RMSE.

Compared to [22] and [32] our method obviously outperforms them. It is better than the monocular versions on every single sequence and it beats even the stereo and SLAM-versions on 9 out of 11 sequences. Also VINS-Mono [33] is outperformed by VI-DSO on every sequence of the EuRoC dataset.

In summary our method is the only one which is able to track all the sequences successfully except ROVIO.

We also compare the relative pose error with [10] and [24] on the V1_0*-sequences of EuRoC (Figs. 7.11 and 7.12). While our method cannot beat the SLAM system and the stereo method on the easy sequence we outperform [24] and we are as good as [10] on the medium sequence. On the hard sequence we outperform both of the contenders even though we neither use stereo nor loop closures.

7.7.2 Evaluation of the Initialization

Especially in the initialization there is a difference in how methods are evaluated on the EuRoC dataset. The machine hall sequences (MH*) all start with a part where the drone is shaken manually and then put onto the floor again. A few seconds later the drone is started

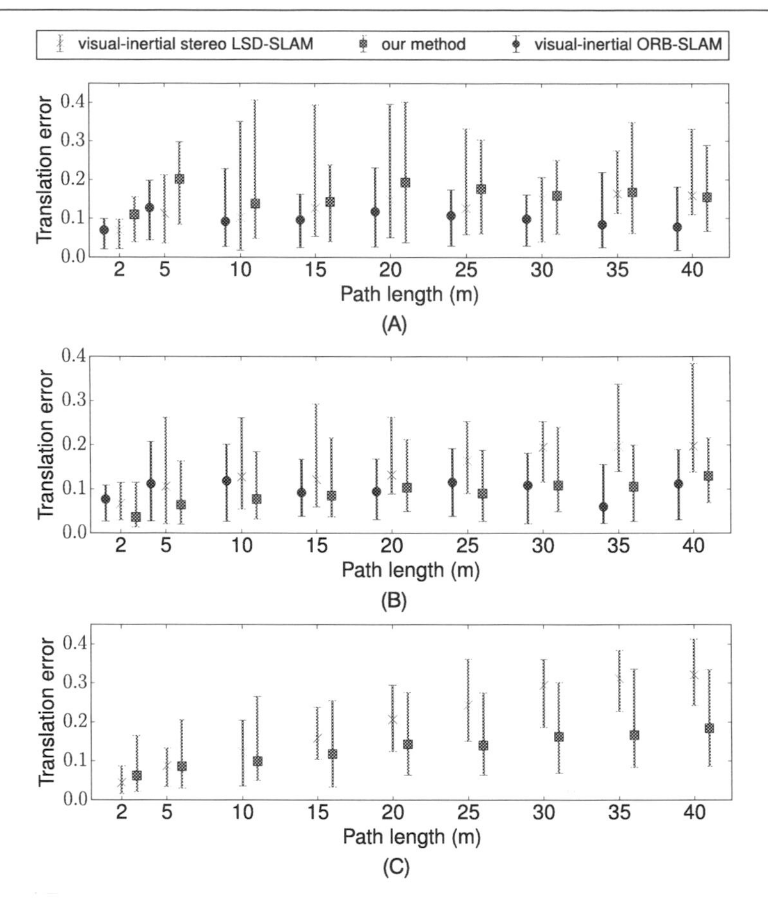

Figure 7.11 : Relative pose error (translation) evaluated on three sequences of the EuRoC dataset for visual-inertial ORB-SLAM [10], visual-inertial stereo LSD-SLAM [24] and our method. (A) Translation error V1_01_easy. (B) Translation error V1_02_medium. (C) Translation error V1_03_difficult. Note that [10] does loop closures and [24] uses stereo.

and the main part of the sequence begins. In our evaluation we follow visual–inertial ORB-SLAM [10] and monocular DSO [4] and crop the first part of the sequence and only use the part where the drone is flying. On the other hand VINS-Mono [33] uses the whole sequence including the part where the drone is shaken. Note that both starting times involve different challenges for visual–inertial initialization: Using the first part of the sequence is a challenge especially for the visual part, as the drone is accelerated very quickly and there is a lot of motion blur involved. On the other hand the motions in this first part are almost purely translational and are quite fast, which is perfect for a fast initialization of scale and IMU biases.

Using only the cropped sequence makes the task for the visual initialization relatively easy, because the motions are relatively slow. However, the task of visual–inertial initialization be-

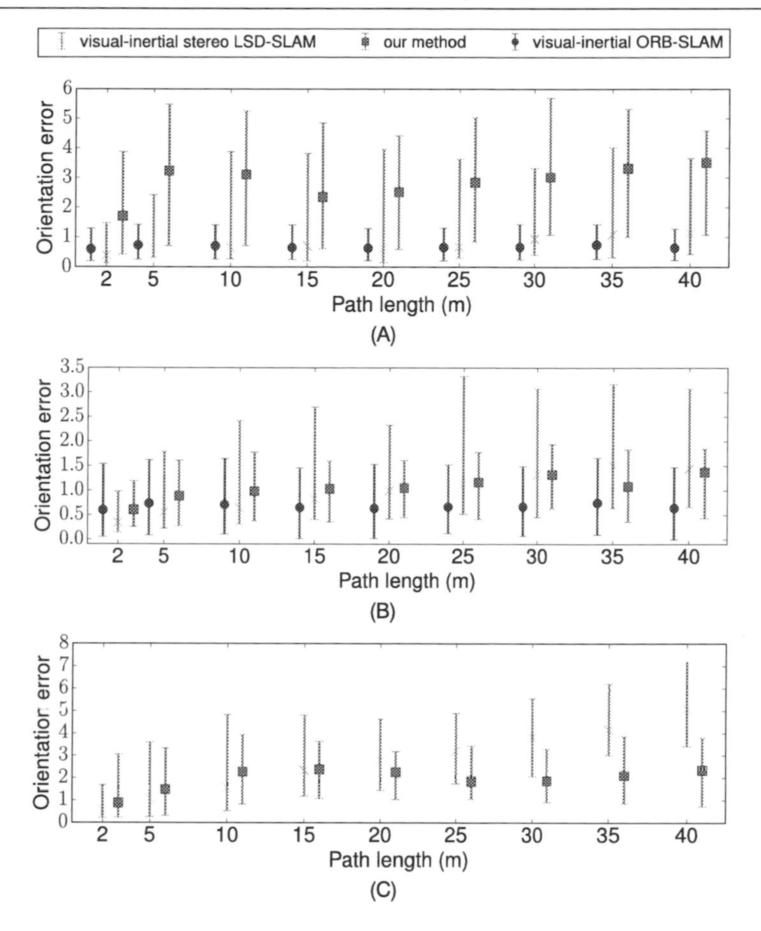

Figure 7.12 : Relative pose error (orientation) evaluated on three sequences of the EuRoC dataset. (A) Orientation error V1_01_easy. (B) Orientation error V1_02_medium. (C) Orientation error V1_03_difficult.

comes very hard, because it takes quite a lot of time for the parameters (especially the scale) to become observable.

This difference can also explain why the initializer of visual–inertial ORB-SLAM [10] needs more than 10 seconds on some sequences of the EuRoC dataset, whereas the initializer of VINS-Mono [33] is much faster. The reason is most likely not so much the algorithm, but more how the dataset is used.

In this comparison we will focus on initialization methods that use only the cropped part of the machine hall sequences—namely VI ORB-SLAM, as only then the problem of late scale observability is present. We do note, however, that an extensive evaluation of different initialization methods has not been done yet and would be a great contribution at this point.

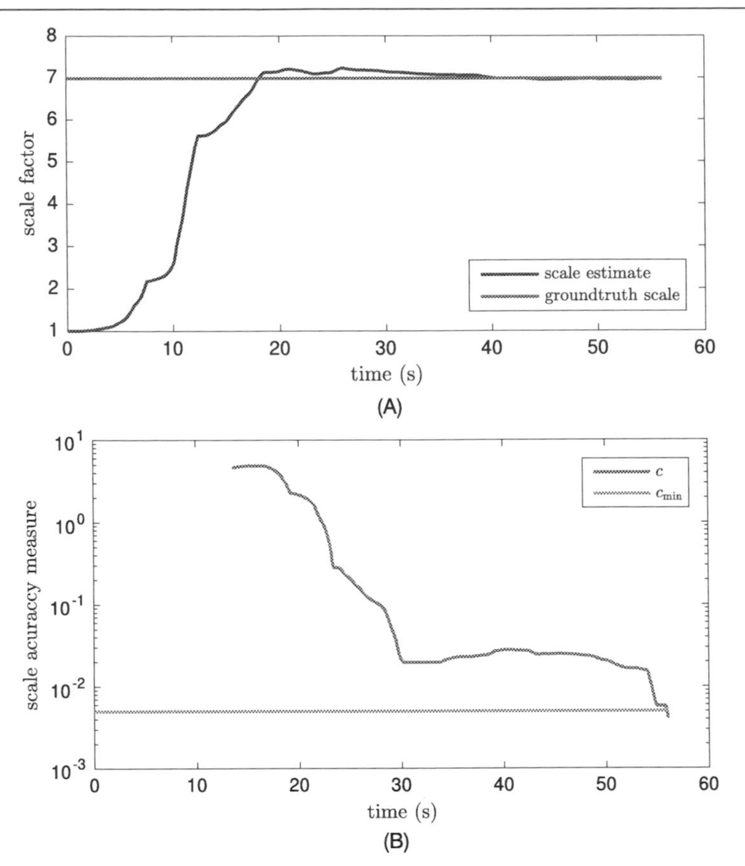

Figure 7.13 : Scale estimate (A) and scale accuracy measure c (B) for MH_04_difficult (median result of 10 runs in terms of tracking accuracy). Note how the estimated scale converges to the correct value despite being initialized far from the optimum.

In comparison to [10] our estimated scale is better overall (Table 7.1). On most sequences our method provides a better scale, and our average scale error (0.7% compared to 1.0%) as well as our maximum scale error (1.2% compared to 3.4%) is lower. In addition our method is more robust as the initialization procedure of [10] fails on V1_03_difficult.

Apart from the numbers we argue that our approach is superior in terms of the general structure. While [10] have to wait for 15 seconds until the initialization is performed, our method provides an approximate scale and gravity direction almost instantly that gets enhanced over time. Whereas in [10] the pose estimation has to work for 15 seconds without any IMU data, in our method the inertial data is used to improve the pose estimation from the beginning. This is probably one of the reasons why our method is able to process V1_03_difficult. Finally, our method is better suited for robotics applications. For example an autonomous drone

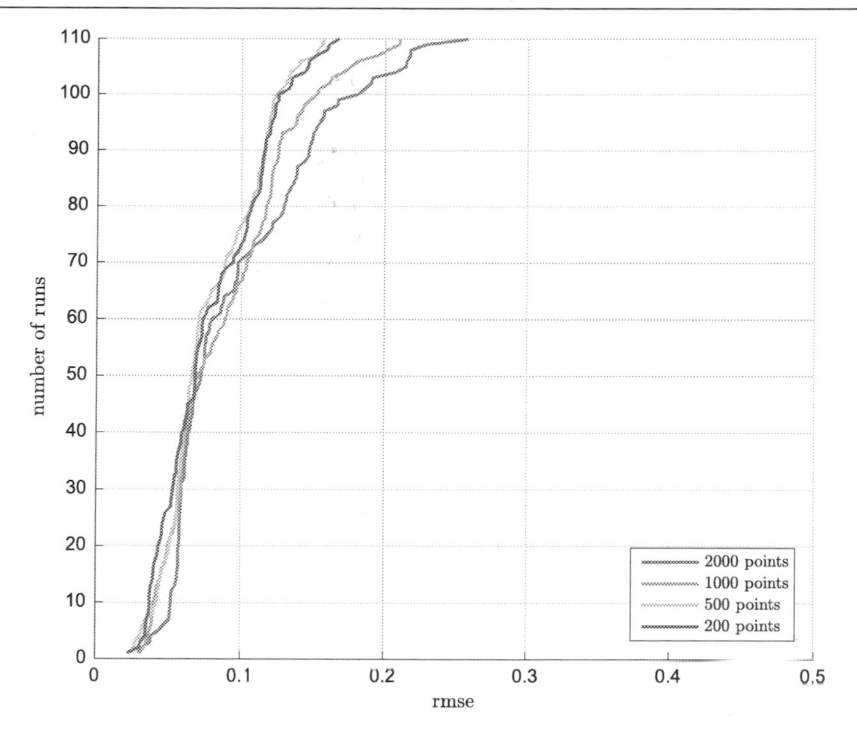

Figure 7.14 : RMSE when running the method (not in real-time) with different numbers of active points. Interestingly focusing on fewer (but more reliable points) actually improves the performance, both in precision and in runtime.

is not able to fly without gravity direction and scale for 15 seconds and hope that afterwards the scale was observable. In contrast our method offers both of them right from the start. The continuous rescaling is also not a big problem as an application could use the unscaled measurements for building a consistent map and for providing flight goals, whereas the scaled measurements can be used for the controller. Finally, the measure c defined in Eq. (7.54) allows one to detect how accurate the current scale estimate is which can be particularly useful for robotics applications. Fig. 7.13 shows the estimated scale and c for the sequence MH_04 of the EuRoC dataset.

7.7.3 Parameter Studies

In order to further evaluate the method we provide several parameter studies in this section. As in the previous sections, we have run all variants 10 times for each sequence of the EuRoC dataset and present either the accumulation of all results or the median of the 10 runs. First of all we show the dependence on the number of points in Fig. 7.14. The main takeaway of this

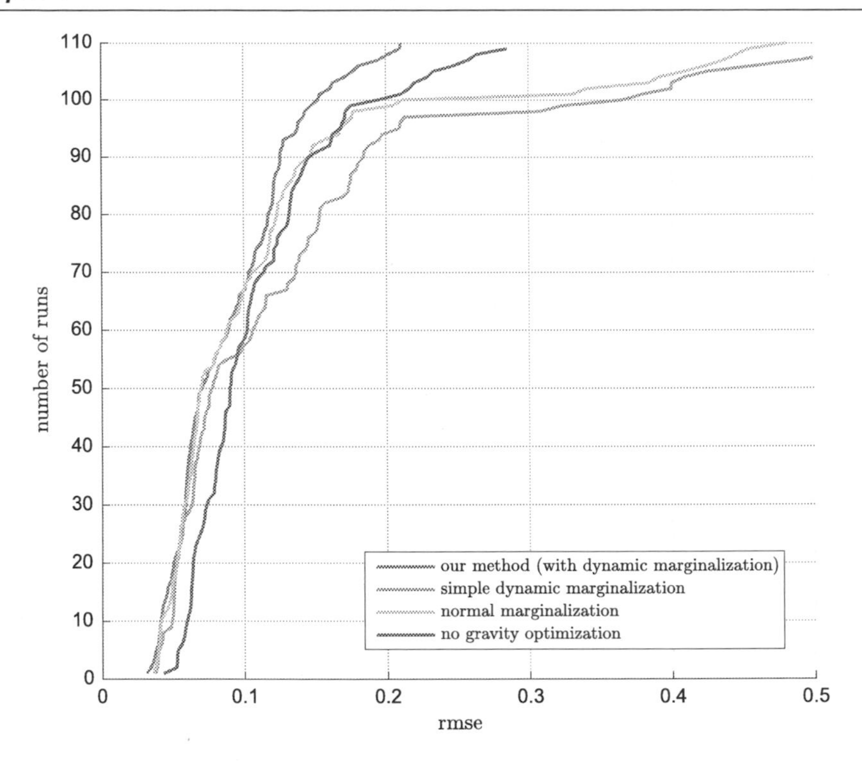

Figure 7.15 : The method run (in real-time) with different changes. "Simple dynamic marginalization" means that we have changed the marginalization to not use M_{half}, but rather replace M_{curr} directly with M_{visual}, when the scale interval is exceeded. For "normal marginalization" we have used the normal marginalization procedure with just one marginalization prior. For the pink line we have turned off the gravity direction optimization in the system (and only optimize for scale). This plot shows that the individual parts presented in this chapter, in particular the joint optimization with scale and gravity, and the dynamic marginalization procedure are important for the accuracy and robustness of the system.

should be that because of the addition of inertial data, we can significantly reduce the number of points without losing tracking performance.

In order to evaluate the importance of the dynamic marginalization strategy we have replaced it with two alternatives (Fig. 7.15). For one example we have used normal marginalization instead where only one marginalization prior is used. Furthermore we have tried a simpler dynamic marginalization strategy, where we do not use M_{half}, but instead directly reset the marginalization prior with M_{visual}, as soon as the scale interval is exceeded. Clearly dynamic marginalization yields the most robust result. Especially on sequences with a large initial scale error, the other strategies do not work well. Fig. 7.16 shows the difference in the scale convergence. When using the simple marginalization strategy the marginalization

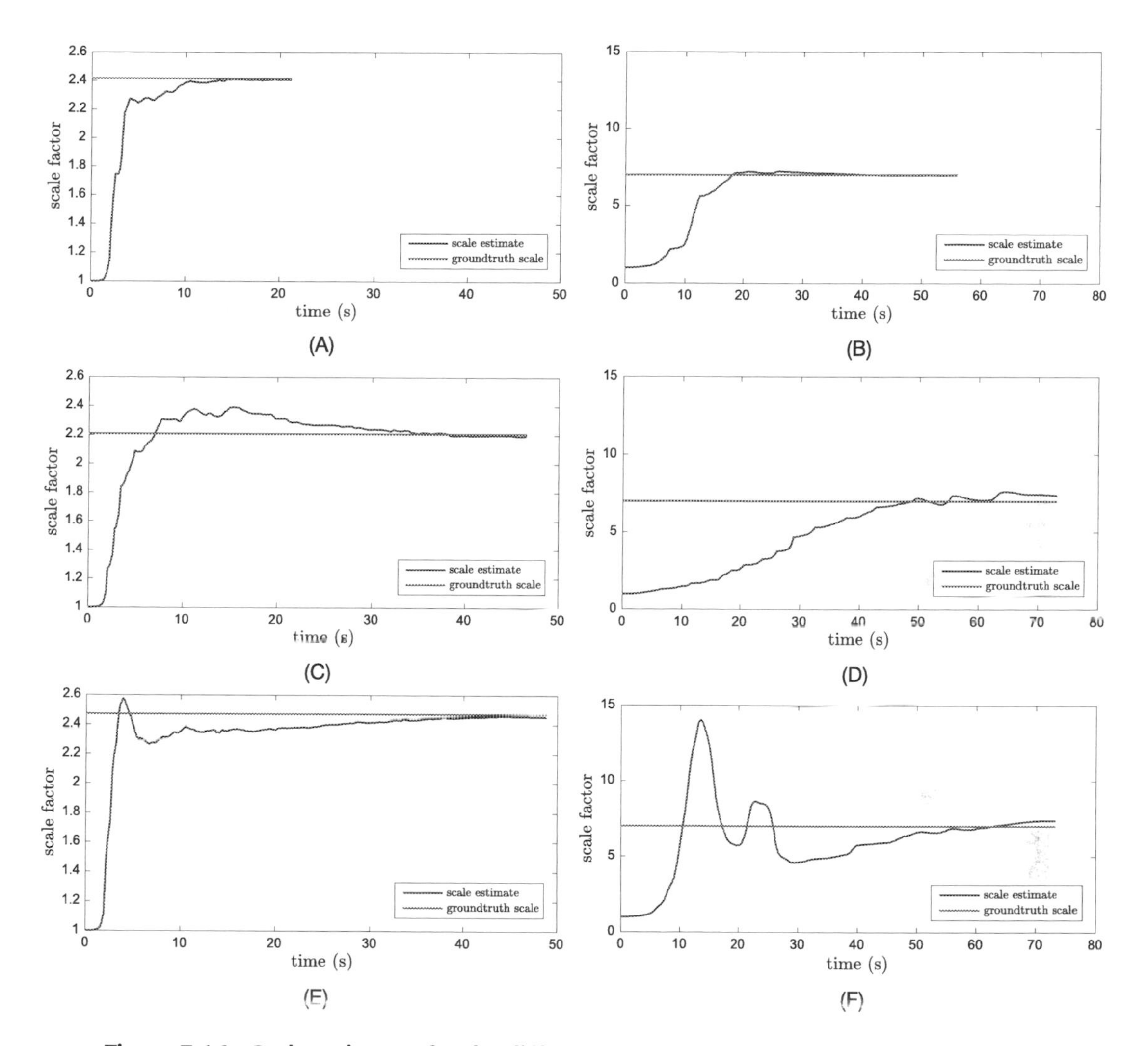

Figure 7.16 : Scale estimates for the different marginalization strategies evaluated on V2_03_difficult (left) and MH_04_difficult (right). (A) Scale estimate for V2_03_difficult with our method. After 21 seconds our system considered the scale converged (using the measure defined in Sect. 7.5.4.4) and fixed it for the rest of the sequence. (B) Scale estimate for MH_04_difficult with our method. (C) Scale estimate for V2_03_difficult with the simple dynamic marginalization strategy (see also Fig. 7.15). (D) Scale estimate for MH_04_difficult with the simple dynamic marginalization strategy. (E) Scale estimate for V2_03_difficult with normal marginalization. (F) Scale estimate for MH_04_difficult with normal marginalization. All figures show the median result of the 10 runs that were accumulated in Fig. 7.15.

prior gets reset directly to M_{visual} resulting in a slower scale convergence and oscillations (Figs. 7.16C and 7.16D). With a normal marginalization the scale estimate overshoots and it takes a long time for the system to compensate for the factors with a wrong scaled that were initially marginalized (Figs. 7.16E and 7.16F). With our dynamic marginalization implementation, however, the scale converges much faster and with almost no overshoot or oscillations (Figs. 7.16A and 7.16B).

We have also disabled the joint optimization of gravity direction in the main system (Fig. 7.15). Clearly the simple initialization of gravity direction (described in Sect. 7.5.2) is not sufficient to work well, without adding the direction of gravity to the model.

7.8 Conclusion

We have described a state-of-the-art approach to direct sparse visual–inertial odometry. The combination of direct image residuals with IMU data enables our system to outperform state of the art methods especially in terms of robustness. We explicitly include scale and gravity direction in our model in order to deal with cases where the scale is not immediately observable. As the initial scale can be very far from the optimum we have described a technique called dynamic marginalization where we maintain multiple marginalization priors and constrain the maximum scale difference. Extensive quantitative evaluation on the EuRoC dataset demonstrates that the described visual–inertial odometry method outperforms other state-of-the-art methods, both the complete system and the IMU initialization procedure. In particular, experiments confirm that inertial information not only provides a reliable scale estimate, but it also drastically increases precision and robustness.

References

[1] S. Weiss, M. Achtelik, S. Lynen, M. Chli, R. Siegwart, Real-time onboard visual–inertial state estimation and self-calibration of MAVs in unknown environments, in: IEEE Int. Conf. on Robot. and Autom., ICRA, 2012.

[2] A. Stelzer, H. Hirschmüller, M. Görner, Stereo-vision-based navigation of a six-legged walking robot in unknown rough terrain, The International Journal of Robotics Research 31 (4) (2012) 381–402.

[3] A. Geiger, P. Lenz, R. Urtasun, Are we ready for autonomous driving? The KITTI vision benchmark suite, in: IEEE Int. Conf. Comput. Vision and Pattern Recognition, CVPR, 2012.

[4] J. Engel, V. Koltun, D. Cremers, Direct sparse odometry, IEEE Transactions on Pattern Analysis and Machine Intelligence 40 (3) (2018) 611–625, https://doi.org/10.1109/TPAMI.2017.2658577.

[5] M. Burri, J. Nikolic, P. Gohl, T. Schneider, J. Rehder, S. Omari, M.W. Achtelik, R. Siegwart, The euroc micro aerial vehicle datasets, The International Journal of Robotics Research 35 (10) (2016) 1157–1163, https://doi.org/10.1177/0278364915620033, http://ijr.sagepub.com/content/early/2016/01/21/0278364915620033.full.pdf+html, http://ijr.sagepub.com/content/early/2016/01/21/0278364915620033.abstract.

[6] L. von Stumberg, V. Usenko, D. Cremers, Direct sparse visual–inertial odometry using dynamic marginalization, in: IEEE Int. Conf. on Robot. and Autom., ICRA, 2018.

[7] D. Nister, O. Naroditsky, J. Bergen, Visual odometry, in: IEEE Int. Conf. Comput. Vision and Pattern Recognition, vol. 1, CVPR, 2004, pp. I-652–I-659.
[8] A. Davison, I. Reid, N. Molton, O. Stasse, MonoSLAM: real-time single camera SLAM, IEEE Transactions on Pattern Analysis and Machine Intelligence 29 (6) (2007) 1052–1067.
[9] G. Klein, D. Murray, Parallel tracking and mapping for small AR workspaces, in: Proc. of the Int. Symp. on Mixed and Augmented Reality, ISMAR, 2007.
[10] R. Mur-Artal, J.M.M. Montiel, J.D. Tardós, Orb-slam: a versatile and accurate monocular slam system, IEEE Transactions on Robotics 31 (5) (2015) 1147–1163, https://doi.org/10.1109/TRO.2015.2463671.
[11] A. Comport, E. Malis, P. Rives, Accurate quadri-focal tracking for robust 3D visual odometry, in: IEEE Int. Conf. on Robot. and Autom., ICRA, 2007.
[12] R.A. Newcombe, S. Izadi, O. Hilliges, D. Molyneaux, D. Kim, A.J. Davison, P. Kohli, J. Shotton, S. Hodges, A. Fitzgibbon, KinectFusion: real-time dense surface mapping and tracking, in: Proc. of the Int. Symp. on Mixed and Augmented Reality, ISMAR, 2011.
[13] C. Kerl, J. Sturm, D. Cremers, Robust odometry estimation for RGB-D cameras, in: IEEE Int. Conf. on Robot. and Autom., ICRA, 2013.
[14] R. Newcombe, S. Lovegrove, A. Davison, DTAM: dense tracking and mapping in real-time, in: IEEE Int. Conf. Comp. Vision, ICCV, 2011.
[15] J. Engel, T. Schöps, D. Cremers, LSD-SLAM: large-scale direct monocular SLAM, in: Proc. of European Conf. on Comput. Vision, ECCV, 2014.
[16] C. Forster, M. Pizzoli, D. Scaramuzza, SVO: fast semi-direct monocular visual odometry, in: IEEE Int. Conf. on Robot. and Autom., ICRA, 2014.
[17] L. Meier, P. Tanskanen, F. Fraundorfer, M. Pollefeys, Pixhawk: a system for autonomous flight using onboard computer vision, in: IEEE Int. Conf. on Robot. and Autom., ICRA, 2011.
[18] J. Engel, J. Sturm, D. Cremers, Camera-based navigation of a low-cost quadrocopter, in: IEEE/RSJ Int. Conf. on Intell. Robots and Syst., IROS, 2012.
[19] M. Li, A. Mourikis, High-precision, consistent EKF-based visual–inertial odometry, The International Journal of Robotics Research 32 (6) (2013) 690–711.
[20] M. Bloesch, S. Omari, M. Hutter, R. Siegwart, Robust visual inertial odometry using a direct EKF-based approach, in: IEEE/RSJ Int. Conf. on Intell. Robots and Syst., IROS, 2015.
[21] P. Tanskanen, T. Naegeli, M. Pollefeys, O. Hilliges, Semi-direct EKF-based monocular visual–inertial odometry, in: IEEE/RSJ Int. Conf. on Intell. Robots and Syst., IROS, 2015, pp. 6073–6078.
[22] S. Leutenegger, S. Lynen, M. Bosse, R. Siegwart, P. Furgale, Keyframe-based visual–inertial odometry using nonlinear optimization, The International Journal of Robotics Research 34 (3) (2015) 314–334.
[23] C. Forster, L. Carlone, F. Dellaert, D. Scaramuzza, IMU preintegration on manifold for efficient visual–inertial maximum-a-posteriori estimation, in: Robot.: Sci. and Syst., RSS, 2015.
[24] V. Usenko, J. Engel, J. Stückler, D. Cremers, Direct visual–inertial odometry with stereo cameras, in: IEEE Int. Conf. on Robot. and Autom., ICRA, 2016.
[25] R. Mur-Artal, J.D. Tardós, Visual–inertial monocular slam with map reuse, IEEE Robotics and Automation Letters 2 (2) (2017) 796–803, https://doi.org/10.1109/LRA.2017.2653359.
[26] T. Qin, P. Li, S. Shen, Vins-mono: a robust and versatile monocular visual–inertial state estimator, arXiv preprint, arXiv:1708.03852.
[27] A. Martinelli, Closed-form solution of visual–inertial structure from motion, International Journal of Computer Vision 106 (2) (2014) 138–152, https://doi.org/10.1007/s11263-013-0647-7.
[28] J. Kaiser, A. Martinelli, F. Fontana, D. Scaramuzza, Simultaneous state initialization and gyroscope bias calibration in visual inertial aided navigation, IEEE Robotics and Automation Letters 2 (1) (2017) 18–25, https://doi.org/10.1109/LRA.2016.2521413.
[29] W. Huang, H. Liu, Online initialization and automatic camera-imu extrinsic calibration for monocular visual–inertial slam, in: IEEE Int. Conf. on Robot. and Autom., ICRA, 2018.

[30] T. Lupton, S. Sukkarieh, Visual–inertial-aided navigation for high-dynamic motion in built environments without initial conditions, IEEE Transactions on Robotics 28 (1) (2012) 61–76, https://doi.org/10.1109/TRO.2011.2170332.

[31] L. Carlone, Z. Kira, C. Beall, V. Indelman, F. Dellaert, Eliminating conditionally independent sets in factor graphs: a unifying perspective based on smart factors, in: IEEE Int. Conf. on Robot. and Autom., ICRA, 2014.

[32] A. Kasyanov, F. Engelmann, J. Stückler, B. Leibe, Keyframe-based visual–inertial online SLAM with relocalization, arXiv e-prints, arXiv:1702.02175.

[33] T. Qin, P. Li, S. Shen, Relocalization, global optimization and map merging for monocular visual–inertial slam, in: IEEE Int. Conf. on Robot. and Autom., ICRA, 2018.

CHAPTER 8

Multimodal Localization for Embedded Systems: A Survey

Imane Salhi[*,†], Martyna Poreba[†], Erwan Piriou[*], Valerie Gouet-Brunet[†], Maroun Ojail[*]

[*]CEA, LIST, Gif-sur-Yvette Cedex, France [†]Univ. Paris-Est, LASTIG MATIS, IGN, ENSG, Saint-Mande, France

Contents

Multimodal Scene Understanding
https://doi.org/10.1016/B978-0-12-817358-9.00014-7

8.1 Introduction

The localization of people, objects, and vehicles (robot, drone, car, *etc.*) in their environment represents a key challenge in several topical and emerging applications requiring the analysis and understanding of the surrounding scene, such as autonomous navigation, augmented reality for industry or people assistance, mapping, entertainment, *etc*. The surrounding conditions, coupled with the moving system in order to localize, often make it difficult to determine its localization with the desired precision from the sensors provided. In addition, the notion of autonomy for sensor-based mobile systems makes navigation and/or action decisions possible, while being aware of their environment and the time of availability of resources allowing for this comprehension/perception and allowing for spatial and temporal mobility. These two aspects make the task of localization, and more generally of scene understanding, in an embedded context even more complex.

Computer vision (CV) methods have long been confined to the world of the personal computer (PC) (fixed or farm computing machines) and often for offline computing. The dataset was then acquired by mobile systems that aggregated the raw data from the perception and/or possibly positioning sensors. Heavy processing such as in photogrammetry has then been carried out in post-acquisition. Nowadays, the trends show that some parts of the processing chain more and more migrate to embedded targets, which seems to become compatible with the emergence of mobile embedded systems that are increasingly powerful in terms of computing and relatively energy efficiency. This massive migration is a benefit for the aforementioned applications, where a real-time and *in situ* comprehension of the scene is helpful to act or interact in this environment more efficiently, while providing new capabilities and consequently opening the way to new services. At the same time, such systems can nowadays benefit from a panel in constant evolution of heterogeneous sensors (*e.g.* global navigation satellite system (GNSS), inertial navigation system (INS), camera, laser imaging detection and ranging (LiDAR), *etc.*), each of them capable of providing information of variable quality about localization.

This chapter presents a survey on the scientific and technological advances related to three essential aspects of the problem of localization dedicated to an embedded system, essentially coming from the domains of robotics, computer vision, photogrammetry, and embedded architecture:

The sensors. The first analysis concerns the choice of sensors, within a panel in constant evolution, according to the function to be performed and the desired accuracy. Sect. 8.2 presents the main sensors available nowadays for scene understanding and, in particular, for localization. Several of their characteristics are discussed in order to highlight their heterogeneity and to take into consideration their adequacy with embedded systems: synchronization of the sensors on the embedded system, raw data processing, topology of the sensor network, and coupling of different sensors.

Localization with heterogeneous sensors. Then, the second study, addressed in Sect. 8.3, concerns the different localization solutions existing in the literature. They have become very numerous for several decades, with various regular improvements and new proposals depending on the available sensors, datasets, and their multimodal combination. Because of the embedded systems context, we deal with *online* localization solutions, *i.e.* solutions where the localization of the system is obtained from local measurements (*e.g.* as part of a trajectory) or from a single acquisition (*e.g.* photography), by opposition to *offline* methods (often called *full* methods in the simultaneous localization and mapping (SLAM) context) where the different localizations of a moving system are globally optimized during post-processing according to all the measurements previously acquired. Because some applications, for instance augmented reality, require a very precise information on localization, we define localization as a *pose* with six degrees of freedom (6-DoF), *i.e.* the 3-DoF spatial position and 3-DoF orientation of the system, in either an absolute environment (absolute pose) or a relative reference (relative pose).

Algorithm-architecture co-design for embedded localization. In the case of embedded localization systems, we study in Sect. 8.4 the adequacy of the correspondence between the aforementioned multimodal algorithms of localization and the main embedded computing architectures available on the market. We review several characteristics and constraints inherent to the sensors, the computation architectures, and the field of application, such as the form factor, the criticality of the system, and the compromise between performance and consumption, otherwise known as energy efficiency. Other criteria, such as safety and determinism, are integral parts and govern the entire design process, particularly for the transport sector (*e.g.* autonomous car).

In view of the continuity of these three aspects, Sect. 8.5 is dedicated to the presentation of selected topical or emerging domains of application of scene understanding through the prism of embedded systems, developed for professionals or the general public in different sectors. In particular, we illustrate that the combination of some of the sensors, algorithms, and computing architectures previously presented may provide powerful systems dedicated to scene mapping, pedestrian localization, autonomous navigation, and mixed reality. Finally, Sect. 8.6 summarizes the main trends encountered and gives perspectives on the topic of scene understanding for embedded systems.

8.2 Positioning Systems and Perception Sensors

Very heterogeneous sensors can provide information about the environment surrounding the device and can be exploited for localization purposes, such as INS, GNSS, cameras (*i.e.* red–green–blue (RGB), RGB-Depth (RGB-D), infrared (IR), or event-based camera), and LiDAR sensors. These are generally small, low-energy, and low-cost sensors that communicate with each others in *ad hoc* fashion and form a wireless sensor network.

This section establishes a state of the art of the different types of input for multimodal localization. First, the most relevant aspects of GNSS and INS positioning systems are presented in Sect. 8.2.1. Then, Sect. 8.2.2 focuses on characteristics of perception sensors, widely used in vision positioning systems for drones or unmanned aerial vehicles (UAVs), robots, autonomous driving (AD) vehicles, smart glasses, *etc.* The discussion is followed by an overview of the sensor configuration types in Sect. 8.2.3.1, sensor coupling approaches in Sect. 8.2.3.2, and sensor fusion architectures in Sect. 8.2.3.3 for multimodal localization systems.

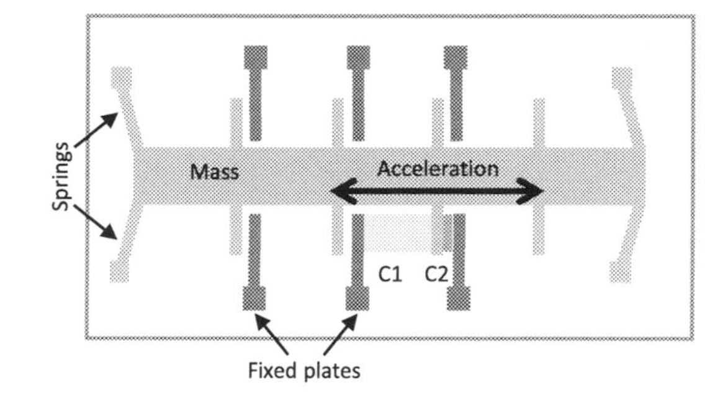

Figure 8.1 : Micro structure of an accelerometer.

8.2.1 Positioning Systems

This section revisits the two main existing families of positioning systems: INS in Sect. 8.2.1.1 and GNSS in Sect. 8.2.1.2.

8.2.1.1 Inertial navigation systems

An INS is a self contained navigation technique consisting of a processor and of an inertial measurement unit (IMU). Primarily, INSs were designed for inertial navigation systems of guided missiles, aircraft, spacecraft, and watercraft. A typical operation cycle of an INS includes three phases: inertial measurements by IMU, the calibration and alignment phase, and the navigation phase, where a trajectory is calculated using a fusion algorithm called the navigation algorithm. This algorithm is based on a data fusion filter integrated in the IMU which merges its various, filtered and processed, measurements. In fact, in order to carry out these measurements, an IMU is equipped with accelerometers, gyroscopes and magnetometers, as well as other sensors, if required, such as barometers or temperature sensors:

Accelerometer. This device determines the object proper acceleration independently in three axes by measuring the capacity variation. The accelerometer consists of a mass fixed to a mobile part which moves along one direction and of fixed outer plates, as illustrated in Fig. 8.1. These two elements form a variable capacity. During an acceleration in a particular direction the mass moves and the capacity between the plates and the mass changes. This capacitance change is measured, processed, and results in a particular acceleration value.

Gyroscope. This device measures the angular position (according to one, two, or three axes) with respect to an inertial reference frame, as illustrated in Fig. 8.2. It measures the angular velocity. According to the same principle as the accelerometer, the capacity varies due to a perpendicular mass displacement which is caused by the application of an external angular velocity. In the microscopic structure of the gyroscope, there is a mass that is constantly in

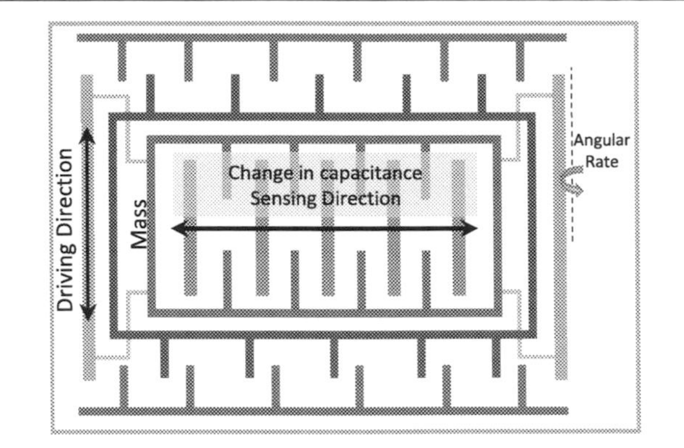

Figure 8.2 : Micro structure of gyroscope.

motion, and at the application of an external angular velocity, a flexible part of the mass displaces itself making the displacement perpendicular. The capacitance change is measured, processed, and it results in a particular angular velocity.

Magnetometer. Sometimes called a magnetic compass, this device measures the direction and intensity of the Earth's magnetic field using the Hall effect or magneto-resistive effect. In fact, nearly 90% of sensors on the market use the Hall effect.

Signals from accelerometers, gyroscopes, and magnetometers are processed at high rates (200–500 Hz). The measurements are summed, giving a total acceleration and rotation for the IMU sample period. These measurements are then transmitted to the INS and used as input to the INS filter. Nevertheless, the high degree of position accuracy is only achieved on a short time scale because the INS suffers from a cumulative, infinitely increasing, error over time. This is due to the errors relevant for IMU, *e.g.* random errors (noise) and systematic errors (sensor non-orthogonality (misalignment), scale factor), which can be minimized by careful manufacturing and calibration. A key to having a high performance INS is to understand what the errors are in the system. Once the nature of the errors is known, one can mitigate them by proper modeling and invoking an error compensation technique as proposed by [1,2]. Despite these efforts, it is still difficult for a single inertial sensor to provide good quality continuous positioning. However, with a view on the main advantages of this system *i.e.* it being free from infrastructure deployment, of relatively low cost, very autonomous, of small size, and of low power consumption, the INS finds many applications.

The INS market shows a large diversity of sensors specifications, mainly measurement range, sensitivity, noise density, resolution, temperature range, bandwidth, and bias. These characteristics are the guide for the choice of the relevant device for a specific application. The price is also a criterion to be taken into consideration, for instance, in Table 8.1 two IMU ranges

Table 8.1: Examples of different IMUs ranges.

Low cost IMU				
Range	**Sensors**	**Measurement range**	**Unit**	**Application examples**
Bosch BNO55	Accelerometer Gyroscope Magnetometer	(+/-){2, 4, 8, 16} (+/-){125, 250, 500, 1000} B_{xy}(+/-){1200, 1300} B_z(+/-){2000, 2500}	g °/s μT	Augmented Reality (e.g. Smart gloves) Navigation
Bosch BMI088	Accelerometer Gyroscope	(+/-){3, 6, 12, 24} (+/-){125, 250, 500, 1000, 2000}	g °/s	Navigation (e.g. flight controller for drone)
High-cost IMU				
Range	**Sensors**	**Measurement Range**	**Unit**	**Application examples**
ADIS (16400/ 16405)	Accelerometer Gyroscope Magnetometer	(+/-){18} (+/-){300, 350} (+/-){300, 400}	g °/s μT	Unmanned aerial vehicles Platform control Navigation
ADIS 16488A	Accelerometer Gyroscope Magnetometer Barometer	(+/-){18} (+/-){450, 480} (+/-) 300 [300, 1100]	g °/s μT mbar	Navigation Platform stabilization and control

are depicted: Bosch, which is a low-cost IMU range, and ADIS, the high-cost IMU range. The difference between a high-cost and a low-cost IMU is the nature of the motion to be measured, so it is mainly noticeable in terms of measurement and sensitivity ranges. For systems such as smart phones and smart glasses, the most used IMUs are low-cost ones; they are often micro electro-mechanical systems (MEMS) characterized by their small size and the ability to measure small motions, and one uses their sensitivity to detect small vibrations and shaking. However, for systems that are more influenced by large movements but require high accuracy, such as aircraft, high-cost IMU are used, since they offer high accuracy with precise measurement ranges.

8.2.1.2 Global navigation satellite systems

The use of global satellite-based navigation systems (SNSs) for in-flight determination of position is a fully functional solution for several years now. A number of countries continue to make their own contribution in order to build the systems that meet users' needs for outdoor positioning. Together with United States' global positioning system (GPS) and its future evolutions, China's Beidou navigation satellite system (BDS), and Russia's GLONASS and European Galileo, SNSs are generally referred to as GNSS. Further augmentations of such systems are still in the process of development for regional navigation purposes, in particular

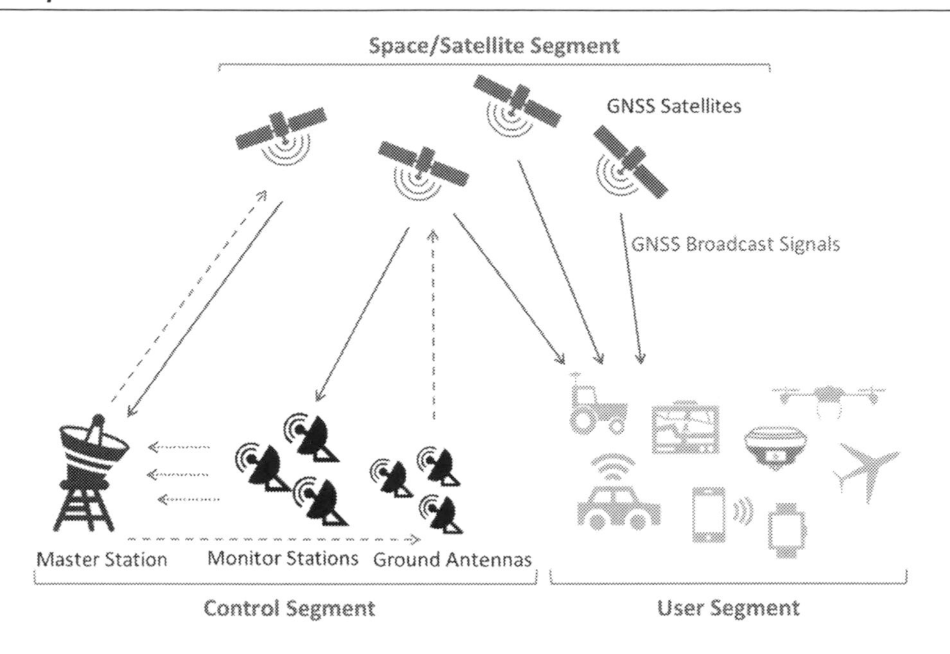

Figure 8.3 : GNSS segments.

Indian regional navigation satellite system (IRNSS) or Japan's quasi-zenith satellite system (QZSS). Together, they will more than double the number of navigation satellites in orbit around Earth. The basic principle behind all these systems is based on techniques common to conventional surveying trilateration and works in the presence of pseudo-range measurement for an arbitrary number of satellites in view. The goal is therefore to allow for real-time navigation (user position and velocity) anywhere on Earth.

GNSS includes three major segments: a space segment, a control segment, and a user segment, as illustrated in Fig. 8.3. The space segment consists of GNSS satellites distributed uniformly in orbit around the earth at the altitude of approximately 20,000 km. Each satellite in the constellation broadcasts a pseudo random code (PRC) with encoded and encrypted data (*i.e.* its orbit, time, and status) at multiple frequencies in the L-band of the radio spectrum, namely in the lower L-band (1151–1310 MHz) where the GPS L2 and L5, Galileo E5, GLONASS G2 are located, and upper L-Band (1559–1610 MHz) where the GPS L1, Galileo E1 and GLONASS G1 are located. The role of the control segment is to maximize the accuracy, reliability, and continuity of GNSS services. The user segment consists of GNSS antennas and receivers. They are used to receive a GNSS signal, measure the time that signal takes to reach its position from the satellite, then determine pseudo-ranges (distance to the satellite) and solve the navigation equations. Hence, the system needs at least three satellites to determine the user position (given in latitude, longitude and ellipsoidal height) within

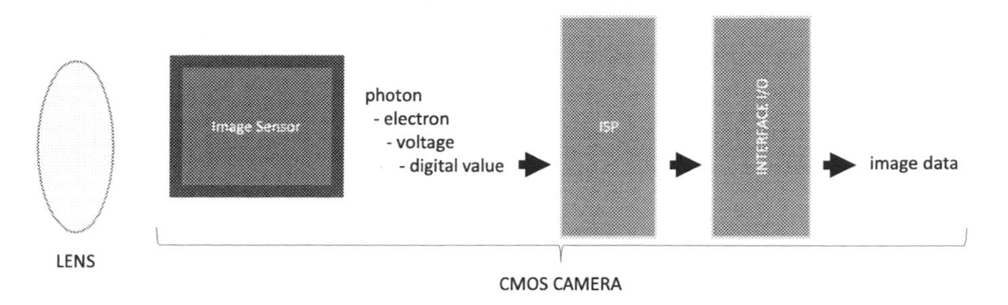

Figure 8.4 : An image sensor in a camera system.

the general global reference frame World Geodetic System 84 (WGS84). A fourth satellite is needed to determine the clock error.

GNSS position suffers from several systematic errors, such as delay caused by the crossing of the atmosphere (tropospheric and ionospheric errors), errors of the satellite clocks, orbital solution, and ephemeris prediction [3]. Most of these errors are either corrected or mitigated using algorithms within the receiver or using some differential correction techniques, *i.e.* differential GNSS (dGNSS) or real time kinematic GNSS (RTK-GNSS) techniques, as explained in Sect. 8.3.1.2.

8.2.2 Perception Sensors

This section focuses on the characteristics of perception sensors widely used in vision positioning systems for UAVs, robots, autonomous driving vehicles, smart glasses, etc. We revisit different categories of cameras: visible light cameras (Sect. 8.2.2.1), IR cameras (Sect. 8.2.2.2), event-based cameras (Sect. 8.2.2.3), and RGB-D cameras (Sect. 8.2.2.4).

8.2.2.1 Visible light cameras

A visible light camera is an electronic device mainly based on an image sensor plus other components as illustrated in Fig. 8.4. It converts an optical image (photons) into an electronic signal; this signal is then processed by the camera or the imaging device processor to be transformed into a digital image.

Image sensors are distinguished from each others by technological, commercial, or practical intrinsic parameters. Here some of the important parameters for selecting an image sensor are listed [4].

Technology. When the image sensor converts the optical image into electrons, it reads each cell's value in the image. There are two types of sensor depending on their built-in technology (Fig. 8.5 represents a comparison between these two types):

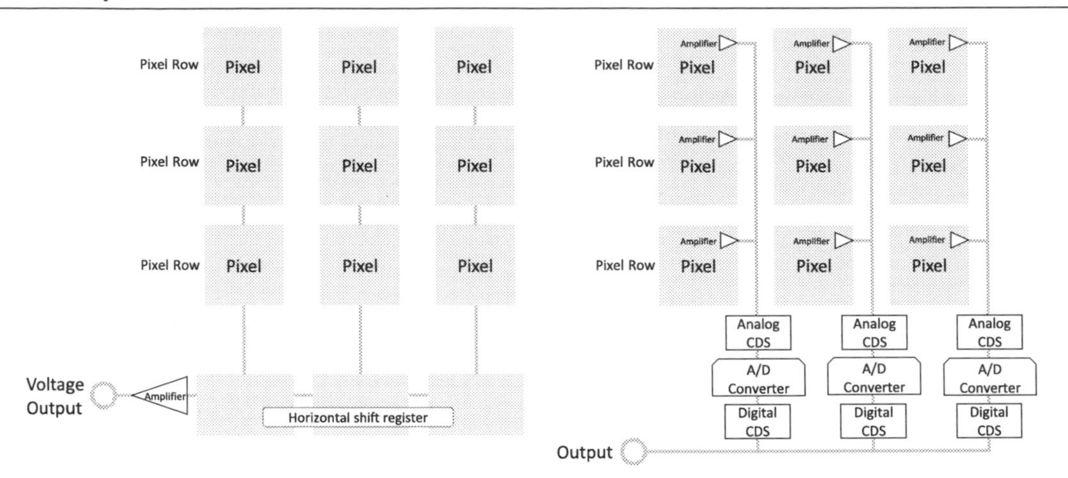

Figure 8.5 : CCDs transport charge across sensor pixels and convert it to voltage at output node. Therefore the CMOS image sensors the conversion that is done inside each pixel.

- Charge-coupled device (CCD): it transports the charge across the image pixels and reads it at the horizontal shift register. Then, it sends it to the floating diffusion amplifier to be converted to a voltage at the output node.
- Complementary metal-oxide-semiconductor (CMOS): in order to transport and amplify the charge with more conventional wires, this kind of devices uses transistors at each pixel and a smaller analog-to-digital converter (ADC) for each column allowing for a higher frame rates than CCD.

Resolution. It is defined by the number of pixels per unit length, which is actually the image pixel density. It is expressed in pixel per inch (*ppi*). Resolution is a crucial criterion when choosing a camera, it depends on its use and its purpose. The aspect ratio also must be taken into consideration; it represents the ratio between the number of horizontal pixels and the number of vertical pixels and it also affects the sensor price.

Frame rate. It represents the number of frames taken by the camera per second. It can be decisive to have a maximum number of frames per second and therefore have an up-to-date image information.

Shutter technology. The shutter mode describes how the image is read from the sensor. Mainly there exist two shutter technologies which are described thus:

- Global shutter: all pixels of a sensor are exposed at the same time and over the same duration, but readout still happens line by line. Global shutter technology allows one to produce non-distorted images and is especially used for capturing high-speed moving objects.

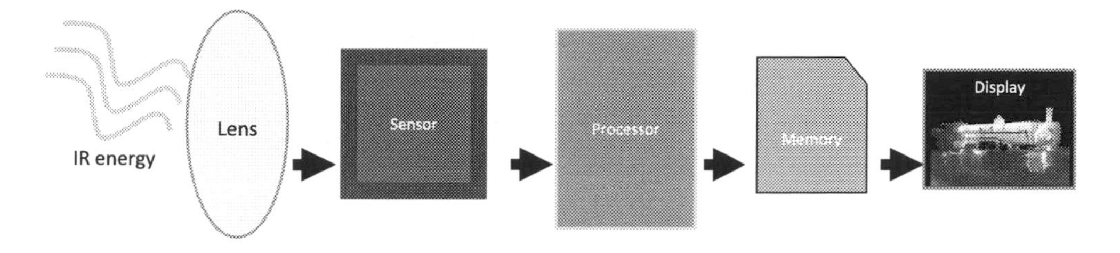

Figure 8.6 : Infrared imaging sensor in an IR camera.

- Rolling shutter: this shutter type exposes an image sequentially, line by line, at different moments in time. This technique causes image distortion in the case that either the camera or the target moves. Consequently, rolling shutter is used for imaging static objects or objects in slow motion and it offers, in this case, an excellent sensitivity.

Field of view (FoV). FoV is the observable area seen by an optical device. In other words, it is the angle through which the device can pick up electromagnetic radiation. The FoV can be determined horizontally, vertically, or diagonally and depends on the sensor size and the focal length.

Dynamic range. It refers to the light intensity range that the sensor can capture in a single exposure. This parameter is well known to be particularly important when the image sensor is used under uncontrolled conditions.

Sensitivity. It indicates the signal reaction when the sensor absorbs the light. A greater sensor sensitivity involves a higher signal level than a lower sensor sensitivity under the same lighting conditions.

A cutting-edge smart camera achieving a framerate of 5500 fps has been presented in the Very Large Scale Integration (VLSI) 2018 conference. The design exhibits a compact and flexible parallel single instruction, multiple data (SIMD) architecture, increasing image signal processing performances in an embedded environment. It consists of a 2-layer, 3D stacked, back side illuminated vision chip performing high-speed programmable parallel computing. It exploits in-focal-plane pixel readout circuits. This integrated circuit (IC) exhibits a frame rate 5 times higher than previous works without reducing ADC resolution [5].

8.2.2.2 IR cameras

The IR type of imaging sensors is characterized by its ability to operate independently of light conditions. For this reason, it is mainly used for applications in the marine, land army, and air forces, as well as other dedicated applications. It allows for obtaining thermal images by measuring the objects temperature with no contact. As depicted in Fig. 8.6, an IR camera is equipped with a germanium optic (lens) that enables infrared radiation to pass through, unlike a visible light camera. The optic concentrates this radiation on the infrared imaging sensor that sends this information to the electronic component (processor). This part converts the data into a visible image.

In most embedded localization systems using visual data [6], the most commonly used cameras are the standard cameras based on visible light imaging sensors. But this does not prevent one to use IR cameras or study their interest in a night vision context, for example, in spite of their dimensions and their weight. The META glasses [7] for example, in addition to a light-based camera, are also equipped with an IR camera.

8.2.2.3 Event-based cameras

Unlike a standard camera, an event-based camera, such as a dynamic vision sensor (DVS), is a silicon retina which generates an asynchronous peak flow, each with pixel position, sign, and precise timing, indicating when each pixel records a change in the log threshold intensity [8]. In addition, an event-based camera offers significant advantages such as a very high dynamic range, no motion blur, low latency, and a compressed visual information that requires lower transmission bandwidth, storage capacity, processing time, and power consumption. However, it requires new computer-vision algorithms, adapted to the asynchronous mode of the sensor, and it is characterized by its high price. Recently, the Prophesee company introduced its ONBOARD camera [9], which is able to run at 10,000 fps while producing up to 1000 time fewer data than equivalent frame-based cameras. With its video graphics array (VGA) resolution (640×480 pixels), the camera has a dynamic range of 120 dB with a low-light cutoff <1 lux.

8.2.2.4 RGB-D cameras

RGB-D camera systems are well suited for 3D mapping and localization, path planning, autonomous navigation, object recognition, and people tracking. A stereo camera setup can be used as an RGB-D sensor but it requires visual texture in the scene. It is a passive 3D depth estimation setup that typically involves the use of a stereo calibration procedure in order to compute a projection matrix that will transform a 2D point into a 3D point in the left camera coordinate system. These stereo cameras are used for outdoor applications. For example, the Arcure company proposes a smart camera "Blaxtair" [10], mounted on a construction vehicle, able to distinguish a person from another obstacle in real time and alert the operator in case of danger.

Active 3D depth sensors such as Microsoft Kinect or Asus Xtion Pro use the time of flight (ToF) technique. They both draw on CMOS camera and IR lighting to provide both color and dense depth images. However, these devices suffer from several drawbacks: a high sensitivity to the natural lighting, being only usable for short range, and not being easily wearable. They are thus considered only for indoor scene understanding. In 2014, [11] proposes an IR stereo Kinect that allows one to lower the minimal distance for acquisition (from 0.8 to 0.5 m) and reduce drastically the number of invalid pixels in the depth map (*e.g.* remove problems with materials such as glass).

8.2.2.5 LiDAR sensors

LiDAR allows one to collect 3D information about the shape of the Earth and its features by means of a laser scanner. This is an electronic distance measuring device that uses laser light to measure the distance from the sensor to the object (the emitter/receiver principle). A LiDAR generally uses light in the near-infrared, visible, or ultraviolet spectrum (frequency>10 THz), unlike a radar, which operates in the microwave domain (1 GHz<frequency<100 GHz), and a sonar, which uses sound waves.

Different principles can be applied to measure the sensor–target distance by means of a laser scanner. They differ in precision but all have their justification for a certain range envelope. It is either done by comparing the ToF (time that the emitted light takes to reach an object and reflect back to the sensor), phase shift (difference of phase between emitted and received laser beams), or triangulation (it records a 3D image of the laser light reflection in the object being scanned, so the principle is close to the concept of stereo vision). ToF-based scanners are much slower but suitable for long distances, while phase- or triangulation-based scanners are very fast and preferred for short-range measurements.

As a result, a laser scanner measures a collection of points (3D points cloud) by determining range and the direction (orientation of the mirror) as regards reflecting surfaces. Each point has its own set of X, Y, and Z coordinates in the reference system of the sensor itself. Optionally, some additional attributes like intensity, color, or angle of incidence are also recorded.

The high-resolution map from LiDAR finds its applications in different fields such as geodesy, forestry, seismology (seismic risk studies), archeology, meteorology, control (traffic regulation, automatic guidance), as well as navigation for autonomous vehicles, *etc*.

Among the benefits of LiDAR-based solutions are their high accuracy in distance measurement (centimeter-accuracy level for the largest ranges, and millimeter-accuracy level for the shortest ones), possible use in different states of brightness (laser scanners are not affected by external light sources), and rapid acquisition that allows 3D mapping in real-time.

However, LiDAR sensors have disadvantages such as sensitivity not only to atmospheric conditions (adverse weather conditions like heavy rain, snow, fog, high humidity) but also to the object properties, *i.e.* the material (metal plates or high reflectivity surfaces) and the shape, which might disturb reflection of the laser beam. In addition, they are still high priced and consume much energy.

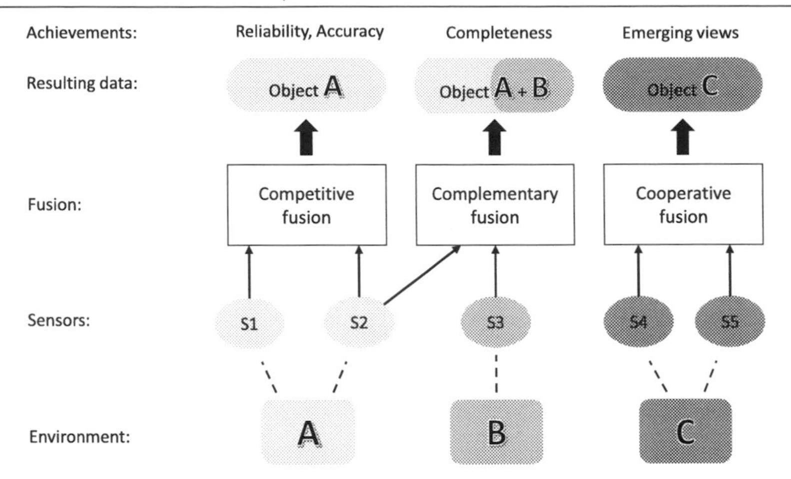

Figure 8.7 : Competitive, complementary and cooperative fusion [13].

Having said that, LiDAR sensors for embedded systems have been appearing in increasing numbers recently. Examples of these sensors include LiDAR-Lite v3—a compact LiDAR by Garmin, which is well suited for use on drones, robots, or smart glasses—and Velodyne LiDAR—designed to be used in autonomous vehicles, vehicle safety systems, 3D mobile mapping, 3D aerial mapping, and security applications.

8.2.3 Heterogeneous Sensor Data Fusion Methods

The different categories of sensors described in the previous sections are less and less used alone, and there exist many possibilities to exploit them jointly. Sensor data fusion can be performed using different cross-sectional methods. Theoretically, it is possible to categorize sensor data fusion according to where the fusion is made, what type of data is fused and how it is fused. Subsequently, the methods presented in this section are discussed according to the sensor configuration type, the sensor coupling, and the sensor architecture.

8.2.3.1 Sensor configuration types

Three categories have been proposed by Durrant-Whyte [12]: complementary fusion, competitive fusion, and cooperative fusion [13] [14]. Such a categorization is based on how the sensors are configured:

Complementary fusion. Sensors in this category are distinct from each other. Their data fusion can be useful to have a global and complete view of the environment, the situation, or the observed phenomenon. This combination is simple to implement as the sensors are independent and so data can be mixed together [13] [14]. In Fig. 8.7, sensors S2 and S3 represent the complementary fusion configuration. For example, let us consider a set of two cameras on the front and rear of a vehicle. The fusion of each view brings independent information of the non-recovering part of the environment.

Competitive fusion. In this category, sensors provide independent measurements of the same or similar properties. These measurements can be taken at different moments for a single sensor. This method is used to increase the reliability of the measurements [13] [14]. In Fig. 8.7, sensors S1 and S2 represent the competitive fusion configuration. One can site the used of consecutive frames output by one or several high-speed camera to reduce noise effects for the observation.

Cooperative fusion. Using sensors in a cooperative manner involves using information provided by two or more sensors to derive information that would not be available from individual sensors. This category is considered the most difficult to implement due to the sensitivity of the resulting data to inaccuracies related to each individual participating sensor [15]. Thus, unlike competitive fusion, cooperative fusion generally reduces accuracy and reliability. In Fig. 8.7, sensors S4 and S5 represent the cooperative fusion configuration. A system consisting of a camera sensor and depth sensor (such as LIDAR or IR) outputs this emerging view of the environment. Nowadays, such devices are available in mirror base of a vehicle.

8.2.3.2 Sensor coupling approaches

Sensor coupling in multisensor systems include basically two forms: tight coupling and loose coupling. These two types of coupling can be used in an orthogonal way, thus it is possible to have systems integrating sensors coupled in a tightly way forming a single block which is in turn coupled in a loose way with one or more sensors. These two types of couplings are explained below.

Loose approach. In a loose coupling approach, the sensors data processing blocks are independent. This coupling allows individual states to be estimated from each sensor before the information is combined [16] [17].

Tight approach. Contrary to loose coupling, the data from different sensors are processed in the same block using raw sensor measurements and a single high order estimation filter. The status vector in this case will be large because it will include different information and sensor-related parameters. This causes computational and implementation constraints as the sensor frequencies are different. The extended Kalman filter (EKF) and the unscented Kalman filter (UKF) are examples of solutions to this problem. This type of coupling is more precise than the first one but it is more complicated and requires more effort for its execution and implementation [18–21].

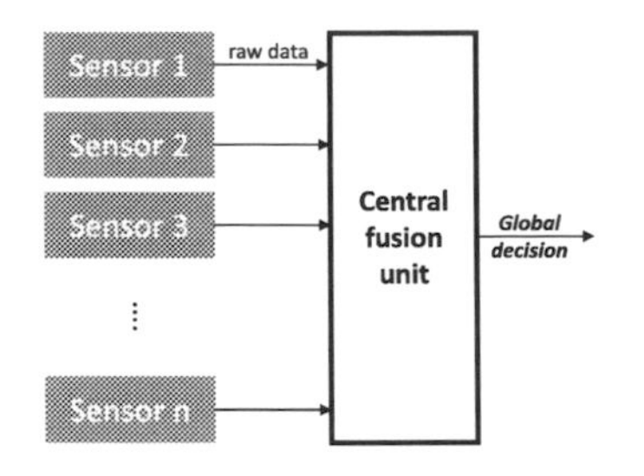

Figure 8.8 : Centralized architecture.

8.2.3.3 Sensors fusion architectures

Parallel to the sensor configuration and coupling modes explained above, there are different sensor fusion architectures. So to ensure a sensor fusion providing good results, it is necessary to choose the best fusion architecture appropriate to the handled application. Referring to [22], four architectures can be defined:

Centralized architecture. This structure uses all the sensors' information by grouping them together and processing them in a central unit (Fig. 8.8) which makes the system more bandwidth consuming. This central unit is used for the representation and fusion of all sensor measurements. Although it has the benefit of taking advantage of global system information and fusion at different levels of information, this type of architecture is not robust against failures that can easily influence the central processor. It is also characterized by a large data flow to be processed. Consequently, the technological and economical development that data fusion and embedded computers have undergone has made it possible to develop new alternative architectures that remedy these problems of centralized fusion.

Hierarchical architecture. There are two types of hierarchical architectures. The first one, on only one level, is characterized by a single fusion processor. The second, at multiple levels, includes multiple fusion processors (local fusion and group fusion) (Fig. 8.9). This architecture provides a significant decrease in central fusion unit load as well as better system robustness than the centralized architecture. This architecture presents, however, many inconveniences: the central fusing unit still suffers from the calculation load which is imposed and depends on the size of the system. Moreover, the proper functioning of the entire system is linked to the proper functioning of the central fusion unit.

Distributed architecture. This architecture is motivated by the intention to make the system more modular and flexible, as well as cheaper, in terms of communication and calculations, than centralized or hierarchical architectures. Each sensor within this architecture has a local processor allowing it to process its information and extract the most useful, while using

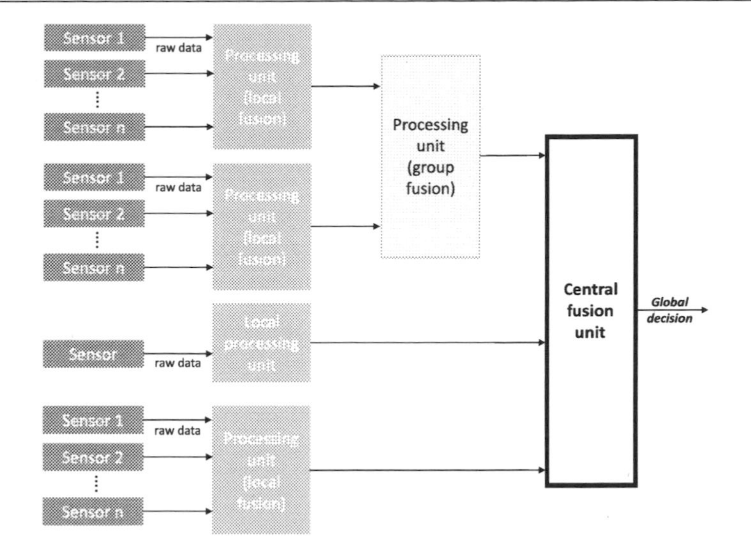

Figure 8.9 : Hierarchical architecture (fusion on multiple levels).

Figure 8.10 : Distributed *Blackboard* architecture.

the blackboard technique (Fig. 8.10). This technique first appeared in the field of Hearsay's speech comprehension, and is now used in various fields mainly artificial intelligence and data fusion. This is a blackboard used as a common communication facility or shared memory resource. Each sensor can read and/or write, without restriction, from this resource, any information that is interesting for the performed task. This architecture is modular and allows for the suppression and the addition of a sensor if needed, without modification of the system's architecture or operation. The use of a common memory or communication resource implies the need for central control where a single decision maker is used to organize information reading/writing from the shared resource. Consequently, this central control mechanism imposes problems and limitations that make this architecture similar to a hierarchical one-level fusion architecture with limitations in flexibility, system evolution, and other ones related to central resource utilization.

Decentralized architecture. Durrant-Whyte proposed an architecture in [22] which relies on the principle that each sensor in the set has its own local processor that fuses the local observations and the information communicated by its neighbors. A system based on this architecture has three constraints:

- No central fusion unit can exist in the system.
- No common communication exists between all the components of the system, the information communication is only done between two nodes.
- No node knows the overall structure of the network, so each node can only communicate with its neighbors.

This solution eliminates any central fusing unit. As a result, this architecture is well known for its flexibility, evolution and robustness. However, it requires a delicate implementation especially in terms of communications between the various system nodes. This kind of architectures is used for very popular applications such as automated industrial installations, control of spacecraft operations, *etc.* [14].

Depending on the application, different system implementations are proposed for the different fusion classifications. These implementations also differ in the detection, description, fusion, or other used algorithms.

8.2.4 Discussion

In this section, a concise presentation of the different positioning systems and a wide range of sensors available for perception has been provided. The INS, GNSS, a set of camera-based systems, and the LiDAR sensor principles have been described. Then, the different techniques and methods of data fusion of heterogeneous sensors were classified. All these systems have their own characteristics, especially with regards to accuracy, data rate, and depth range measurement for perception solutions.

The GNSS, based on a constellation of active satellites, is the most widely used technology for geo-spatial positioning and outdoor navigation applications. When necessary, other satellite systems, especially assisted-GPS (A-GPS), repeaters or pseudolites (pseudo-satellites) [23,24] offer increased accuracy and GNSS signal availability, even for indoor localization. Meanwhile, in indoor environments, several other families of positioning techniques have been proposed so far, such as:

- Networks of sensors: passive radio-frequency identification (RFID), ultrasound [25,26], or infrared [27,28].
- Wireless communication networks: wireless fidelity (Wi-Fi) [29,30], bluetooth [31,32], ultra-wide band (UWB) [33], ZigBee [34], or active RFID.
- Mobile phone networks: global system for mobile (GSM), Cell-ID, enhanced observed time difference (E-OTD). The size and complexity of cellular networks continue to grow with upcoming 5G networks (around 2020).

Table 8.2: Specification by technique, in terms of localization precision [35].

Techniques	Indoor	Outdoor
Network of sensors	1 cm to 5 m	Not suitable
RFID	<1 m	<1 m
WLAN	Few m	Not suitable
UWB	~10 cm	Not suitable
Cell-ID	500 m to 10 km	100 m to 10 km
E-OTD 2G/TDOA (3G)	≫200 m	<100 m
GNSS	Not available	Few m
A-GPS	10 m to not available	Few m
Pseudolites	~10 cm	Few m
Repeaters	~1 to 2 m	Few m
INS	<1 m (time dependent)	<1 m (time dependent)

Table 8.2 presents precisions of localization that can be achieved by some of these systems, in indoor and outdoor environments [35]. This table can also be completed by expected performances for network-based localization, which would be supported in 5G. As reported in the next generation mobile network (NGMN) 5G White Paper [36], an accuracy from 10 m to less than 1 m on 80% of occasions, and less than 1 m for indoor cases, could be achieved.

This multitude of sensors makes the localization step efficient, but also complex. Thus, the design and adjustment of the sensor fusion step for localization is a tedious task. The different methods of localization as well as the implementation of these fusion algorithms are discussed in the following section.

8.3 State of the Art on Localization Methods

In this section, we present the main algorithmic solutions existing for the problem of the localization of a system equipped with one of several sensors. We do not consider specific architectures here, this aspect being considered in Sect. 8.4. Because of the embedded localization context, we address here *online* localization solutions, *i.e.* solutions where the localization of the system is obtained from local measurements (*e.g.* as part of a trajectory) or from a single acquisition (*e.g.* a photography), by opposition to *offline* methods (often called *full* methods in the SLAM context) where the different localizations of a moving system are globally optimized according to all the measurements previously acquired.

The literature on this topic is particularly large, because of the variety of sensors, because of the various disciplines that have been interested in this issue for years or even decades, and because of the numerous applications where localization is a central step. When existing, we refer to surveys dedicated to some families of approaches.

The section is organized in two sections, respectively, one focusing on monomodal localization (Sect. 8.3.1) and one on multimodal localization (Sect. 8.3.2), depending on whether the approach considered is based on one or several sensors of different natures. The approaches are categorized according to the nature of the sensors considered, namely INS, GNSS, cameras and LiDAR. Note that the different types of data involved refer to sensors located on the system to localize as well as to the geo-referenced datasets employed in some localization approaches.

8.3.1 Monomodal Localization

The most commonly used outdoor localization system for many applications is the GNSS (see Sect. 8.2.1.2). Hoverer, in order to deal with more complex use, such as precise localization, indoor navigation or urban canyon challenges, there exist also a lot of other techniques based on light, radio waves, magnetic fields, and acoustic signals, as presented in Sect. 8.2, which may be exploited for localization. For some of these sensors, we revisit here the best-known approaches of localization.

8.3.1.1 INS-based localization

INS-based localization methods do not provide the absolute position, but rather the position relative to the starting point. However, as reported in Sect. 8.2.1.1, the problem with this technology is that localization errors grow with time. Hence, it is difficult for a single IMU to provide accurate prolonged positioning information.

Since decades ago, INS has been successfully applied to navigation purposes, *i.e.* attitude estimation of space vehicles [37,38]. With the development of MEMS, which leads to a reduction of the size and cost of IMUs, it is a new surge to INS-based localization. Numerous are the studies reporting that the solely INS-based system has been successfully applied to positioning, for example the pedestrian dead-reckoning (PDR) technique, which is based on the kinematics of human walking to track location and orientation of a pedestrian in indoor environments. This can be done either by using waist-mounted and shoe-mounted IMUs [39–41] or smart phone's IMU [42–44]. For a more detailed description of PDR technology, the reader may refer to Sect. 8.5.2.1. In recent years, IMUs were also used for unmanned vehicles tracking [45,46] or human motion tracking [47]. The latter has several applications, such as for example rehabilitation, localization, and human–robot interaction.

8.3.1.2 GNSS-based localization

GNSS offers a cheap and easily accessible solution to estimate absolute position of moving or static devices. It is successfully used only in outdoor and open-sky applications. In contrast

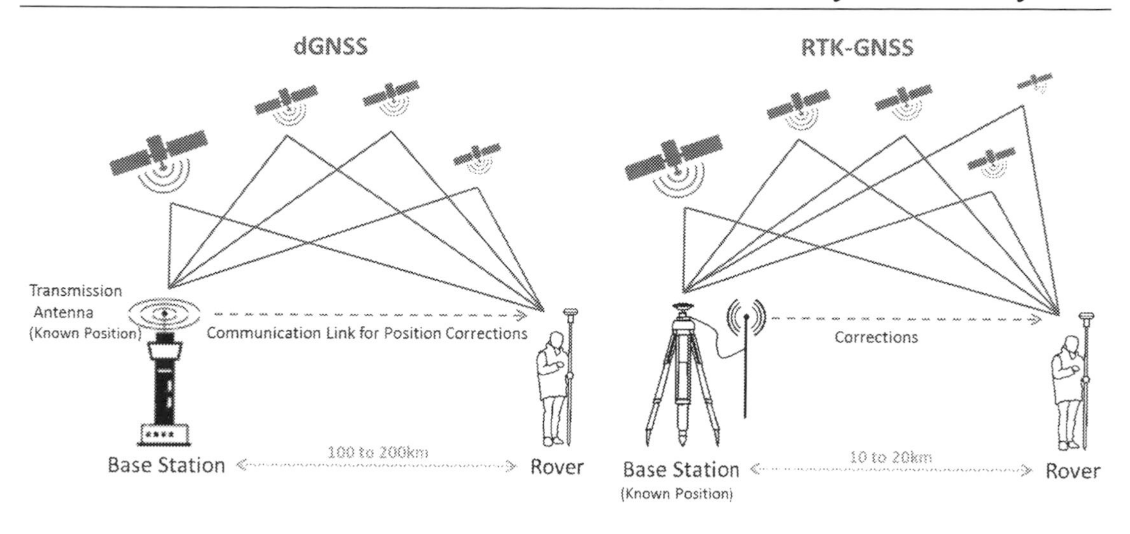

Figure 8.11 : RTK-GNSS vs. dGNSS.

to INS-based solutions, GNSS continuous survey mode allows for ongoing data collection that does not present accumulative errors. The nominal position accuracy for a single-band receiver ranges from a few meters to above 20 m. So, it can be useful for applications that only need an approximate position, for example in car navigation systems.

Differential positioning. The GNSS accuracy can be improved upon by using dGNSS (sub-meter-accuracy level) or RTK-GNSS (sub-centimeter-accuracy level) techniques. The difference in the accuracy is due to how each of them calculates the position using the same broadcast GNSS signal (see Fig. 8.11). dGNSS, which stands for differential GNSS, is referred to as code-based positioning. In other words, it aligns the received code to its own and calculates the propagation delay. The dGNSS is composed of three elements: one user receiver (rover) at an unknown location, at least one base station (reference GNSS receiver/antenna) at a known location, and a communication link between them. Both, reference and rover receivers must receive the same GNSS signals from at least four common satellites in view. Since the position of the base station is surveyed very accurately, it compares this position with the position calculated by the GNSS signal. The differences, which reflect the amount of error in the signal from each satellite, are broadcast to the user. The transmitted pseudo-range corrections are then used to improve the accuracy of position fix.

RTK-GNSS follows the same general concept, but uses the satellite signal's carrier wave. Thus, the multifrequency rover units compare their own phase measurements with those received from the base station. RTK-GNSS provides accuracy enhancements up to about 20 km from the base station [48], while dGNSS works up to distances of 100 km [49]. Short baseline lengths limit the usefulness of the RTK-GNSS method for general localization. This technology is also more expensive than dGNSS, because it requires at least two RTK-capable receivers (one base station and one or more rovers) whose cost per receiver is higher than that of dGNSS-capable only. Nevertheless, such technology, unlike dGNSS, helps to filter out multipath signals (signals not coming directly from satellites but reflected or diffracted from local objects). All these characteristics make the method perfectly suited to surveying applications for example.

Augmentation systems: pseudolites and assisted-GNSS (A-GNSS) concepts. The main disadvantage of satellite-based localization is that the GNSS signal is not available everywhere and it will be degraded or not fully available in hostile areas (dense urban streets, a.k.a. urban canyon), building structures and underwater environments. In such GNSS-denied areas, it is generally difficult to provide reliable positioning information from satellite systems. However, in order to make positioning signals ubiquitous, integration between GNSS and indoor positioning can be made. Existing approaches can be divided into two classes: with and without additional architecture. For the former architecture, pseudolites or repeaters that must be deployed locally, exist. This requirement is no longer valid for the high sensitivity GNSS/GPS receivers that are able to extract and calculate a position from a weak signal in most indoor situations. In turn, the A-GPS, which is extensively used with GPS-capable smart phones and other mobile devices, improves the positioning performance when GPS signals are weak or not available. The system draws its information from local cell towers instead of radio connection from the satellites. In this way, the almanac and other pieces of information are transferred through a cellular telephone network. The main benefit of such a system is the significant improvement of the start-up performance *i.e.* time-to-first-fix (TTFF), which refers to the time needed by the receiver to perform the first position fix. This initial TTFF, often called a cold start, can take from 30 seconds to a couple of minutes to acquire satellite signals. According to [50], the A-GPS receivers can provide position solutions in less than 30 seconds, especially the newer receivers which have TTFF typically less than 10 seconds. However, A-GPS provides less position accuracy, compared to dGNSS for example.

8.3.1.3 Image-based localization

Image-based localization consists in retrieving an information about the location of a visual query material within a known space representation, which can be absolute or relative to an initial location. It is an increasingly dynamic research subject in the last decade, because of the availability of large geo-localized image databases (see Sect. 8.3.2.2), and of the multiplication of embedded visual acquisition systems (*e.g.* cameras on smart phones), facing the limitation of usual localization systems in urban environment (*e.g.* the GNSS signal failure in a cluttered environment). It is involved in several topical practical applications, such as GPS-like localization system, indoor or outdoor navigation, 3D reconstruction, mapping, robotics, consumer photography and augmented reality. This family of approaches is studied in several research domains of computer science and artificial intelligence: computer vision, robotics, photogrammetry, machine learning and content-based image retrieval. In these domains, the approaches are various, they may involve different kinds of information extracted from one or several images (sequences of images or stereoscopic images), including optical, geometrical and semantic information, or a combination of them (studied in Sect. 8.3.2 on multimodal localization).

As a very well-studied topic, many recent surveys propose overviews of the image-based localization domain, according to different viewpoints: the article of Brejcha and Cadik [51] classifies visual-based localization solutions depending on the environment for which the particular method was developed, while Zamir et al. [52] gather recent articles to draw a panorama of visual-based localization at large scale. Conversely, in [53], the authors focus on the heterogeneity of the available data which can be exploited for the problem of localization in challenging scenarios, similar to [54], where several recent approaches are benchmarked. The importance of this topic is shown by regarding the many tutorials and workshops on the subject present in high impact international conferences, such as the recent editions of CVPR from 2014 (the IEEE International Conference on Computer Vision and Pattern Recognition).

Before presenting the main approaches devoted to 6-DoF pose estimation, we revisit the different representations of the image content that are usually employed. It is a central step in the pose estimation chain.

Classical hand-crafted representations of the image content. In the literature, the representations encountered are clearly dominated by **local features**, which usually take the form of interest points, patches or blobs. These low-level features have interesting properties, such as their robustness to many geometric and photometric image transformations, that make them suitable for visual localization purposes. They are automatically extracted in the image, then described before being compared to other features in order to determine the similarity between images (in the indirect approaches) or the geometry of cameras (in the direct approaches). In the context of topological mapping for robotics, a comprehensive overview of local feature descriptors can be found in the survey [55]. The most used interest point feature remains the Hessian-affine detector [56] combined with a scale-invariant feature transform (SIFT) [57] descriptor. The discriminative histogram of oriented gradients (HOG) [58] descriptor has been used in image-based localization for capturing architectural cues of building and landmarks [59,60]. In the work of [61], a maximally stable extremal region (MSER) blob detector by [62] is used to extract visual information. All these traditional techniques of description are usually labeled as *hand-crafted*, in contrast with some more recent approaches that involve *learned* local features.

Learned and dedicated features. Some recent approaches involve *learned* local features, such as in [63] where features are described through convolutional neural networks (CNNs) trained for the task of association of similar features. The reader may refer to the recent comparison of hand-crafted and learned feature proposed in [64], which highlights some advantages of each category according to the scenario of use. Other types of representation exist to represent specific contents, or to address specific scenarios of localization. For example, some of them involve **geometrical features**, *e.g.* vertical lines are a convenient descriptor in urban environment to represent buildings [65,60]; contour extraction has also been employed in [66] to recover the pose of an image in sites of archaeological excavations. By considering the image content as a whole, **global features** are also exploited, because they are computationally less intensive to extract and because they capture a simple description of the visual data that can be used as a filter to eliminate more complex features (*e.g.* local features) efficiently: the most popular hand-crafted global description exploited for localization is the generating interaction between schemata and text (GIST) descriptor introduced by Oliva and Torralba [67], which describes the shape of the scene compactly, and which is used for localization purposes in [68,69] for instance. Since the 2000 s, the democratization of CNN in CV has given the birth to efficient learned global features, which are obtained by grouping weights of a given layer into a single vector [70]. Some of them are exploited in content-based image retrieval dedicated to urban scenes [71–74].

Exploitation of semantic information. To localize a visual content, the semantic information can also be a key element. The extraction of **semantic features** from images, a task involved in object detection, object recognition or image categorization, has been a intensely studied subject of research [75], and it has recently received growing interest in visual localization and navigation [76] research, supported by the explosion of deep learning approaches [77]. Indeed, the semantic objects extracted can be exploited as visual landmarks that are able to be robust to many kinds of challenging appearance changes (night/day, season, temporal gap, *etc.*). For example, in [78], the extraction of traffic signs is injected into a local bundle adjustment processing in order to provide a very precise online 6-DoF pose estimation during navigation in urban environments. In [79], skyline features are extracted to describe mountain panoramas and exploited for localization in such environments. Differently, image categorization can be seen as a global description associated to semantic labels. It provides classes of semantically identical images, which can be exploited for localization purposes, as for example in [80,81], at a very large scale (world-wide level) for [81]. Benefiting from the recent progress in machine learning with deep learning, such approaches are numerous; the reader may consult the surveys in [52,51,53].

Based on the features extracted from images, we revisit here the main approaches of the literature that provide the 6-DoF image location information, *i.e.* a 3-DoF spatial position and a

3-DoF rotation in the world coordinate (absolute localization) or in a relative reference (generally related to an initial pose), for a given image or a sequence of images belonging to a moving system:

Odometry. This name gathers a family of low-level techniques used to estimate the relative position and orientation of a moving vehicle. It comes from robotics and is based on the use of data from motion sensors, or camera images when considering visual odometry, to estimate change in pose over time. Traditional visual odometry approaches rely on the extraction and tracking of visual features such as interest points across the sequence, according to a motion model. For example, one can cite its application to Martian exploration by the rovers Spirit and Opportunity [82]. The latest advances in this field include, for example, the fast and accurate extraction of features for embedded devices, such as in [83] with application applied to micro-aerial vehicle in GPS-denied environments, or the combination of different sources for improving tracking, such as in [16] with IMU and camera data.

SLAM. Formalized in the 1980 s in robotics, SLAM (simultaneous localization and mapping) represents a family of widely used approaches which provide a robot with the possibility to localize its movements in an unknown environment, by simultaneously and incrementally building a map of its surroundings. To perform these combined actions, the robot has to have at least one sensor able to collect data about its surroundings, *i.e.* a camera, a laser scanner, a sonar or any other sensor. The approaches dedicated to SLAM are numerous [84]. The most representative real-time SLAM system is probably PTAM, for parallel tracking and mapping [85]. It was the first work to introduce the idea of splitting camera tracking and mapping in parallel threads, and it was demonstrated to be very accurate for real-time augmented reality applications in small environments. As for more recent cases, one may mention oriented FAST and rotated BRIEF-SLAM (ORB-SLAM) [86], a monocular SLAM approach having some interesting properties: it operates in real-time, in small and large indoor and outdoor environments. The system generates a compact and trackable long-term map and is robust to severe motion clutter, allows for wide baseline loop closing and relocalization based on indexing approaches, and includes full automatic initialization. It has been recently optimized to deal with stereoscopic and RGB-D contents [87].

Structure from motion. In the community of photogrammetry, the family of techniques gathered under the term of structure from motion (SfM) [88] also provides a relative pose of a sequence of data. It consists in reconstructing a scene from a set of images taken from an unknown position. Here, it is also necessary to determine simultaneously the positions of the consecutive shots and the environment in the form of a dense or sparse reconstruction, such as points clouds. With the point cloud model, the localization problem can be formulated as a 2D-to-3D registration process between 2D image features and the 3D point cloud [89]. To deal with large points sets, it has to be optimized, such as in [90], which exploits scalable indexing tools inspired by a bag of features. The SfM techniques also strongly makes use of geometric relationships and bundle adjustment to achieve correct localization and reconstruction, as explained below.

Refinement of the pose. To refine the pose of the image input, more complex processes—generally heavier—can be applied. In particular, *bundle adjustment* (BA) is the widely used technique when dealing with 3D structures or point cloud obtained from images. It consists in simultaneously refining the 3D points describing the scene and the calibration of the camera(s), according to an optimality criterion based on the image projection of these points; see for example [91,92] or [86,93], where *local* BA is employed when real-time performances are targeted (contrary to *full* BA, which considers the cloud globally). Another classical refinement method is the *iterative closest points algorithm* (ICP), which consists in iteratively transforming a set of features to match better with another one, by minimizing an error metric usually based on a distance (most commonly Euclidean distance). It is largely employed for image-based localization with various types of data, including alignment between images and point cloud, between point cloud and 3D model [66,94,60].

Content-based image retrieval. Recently, to perform absolute localization at large scale of one image, some approaches have relied on content-based image retrieval [95,96], by retrieving the nearest neighbors of the input image in a geo-referenced image dataset. The final pose is obtained by propagation to the query image of the location information associated to the similar images retrieved. According to the previous approaches, this family of methods are scalable, benefiting from scalable image retrieval methods [97], the area covered can be very large (a city, the world), but generally they suffer from the precision of the pose estimated by fusion of pose. Nevertheless, the solution obtained can serve as initial input location in a more classical approach providing a more precise pose. Recently, the proposal presented in [98] provided precise localization by fusing candidate relative poses together by minimizing a well-defined geometry error. The reader may consult the survey in [53] for an overview on such approaches.

Direct pose regression. Finally, with the development of researches in deep learning, approaches estimating the pose by pose regression directly from input visual data to its corresponding pose become more widespread, often providing more accurate and large-scale results: the best-known CNN architecture is probably PoseNet [99] and its improvements [100]. Most other standard previous regression techniques used to rely on regression forests; see for example [101,102].

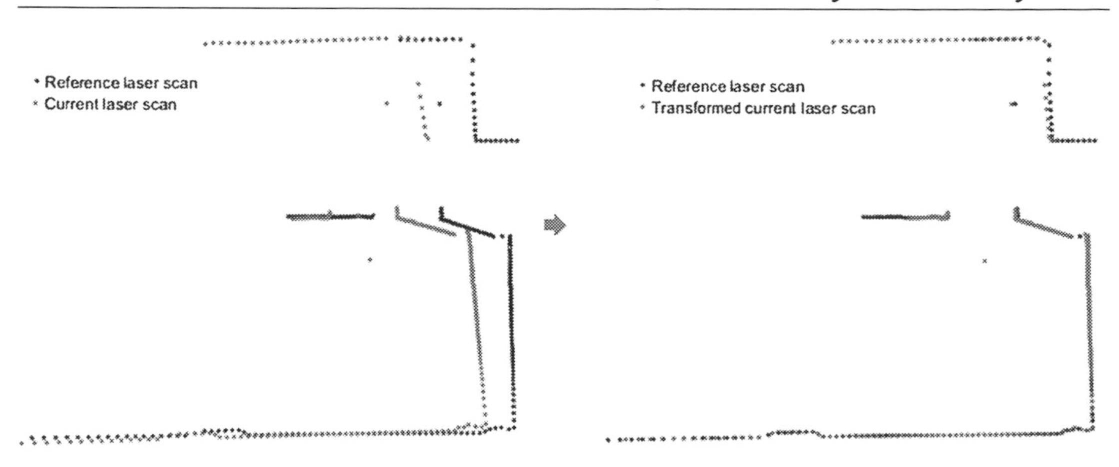

Figure 8.12 : Scan matching.

8.3.1.4 LiDAR-map based localization

As the camera-based techniques are sensitive to changes in illumination conditions (which may cause the system fails), LiDAR has the advantage of being independent of external lighting and making use of full 3D. The use of a 3D map, collected by laser scanner, for estimating pose starts attracting more and more attention. Especially the Velodyne LiDAR, getting increasingly smaller and cheaper (*e.g.* VLP-16, the 600-gram LiDAR sensor at about 4000 dollars), is popular in many robocar teams. For instance, in the recent state-of-the-art literature, laser scanners are rarely considered as the only sensor for localization. This may also result from an emerging challenge of real-time sensing and pose estimation from collected point clouds. It is well known that LiDAR sensors provide three dimensional scans consisting of over one million points each second, which requires powerful hardware.

The most common technique to estimate pose change from LiDAR measurements is the **3D scan matching** (3D point cloud matching). Its goal is to find the relative pose between the two device positions where the scans were taken, as illustrated in Fig. 8.12.

Generally, existing methods can be divided into two types: local matching (*e.g.* ICP [103], and iterative dual correspondence (IDC) [104]), which requires the initial position and is free of these constrained global matching techniques. Examples of global matching are signature-based scan matching [105], the cross correlation function (CCF) approach [106], which reduces the searching problem of 3D to three problems of 1D, or anchor point relation (APR) matching [107]. Regardless of this classification, algorithms can be grouped into three categories according to the data type being matched [108]: point-based matching, feature-based matching and mathematical property-based matching. The last of the three can be based on histograms [106,109] effectively summarizing a map's salient features, cross-correlation

[110,111], Hough transforms [112], normal distribution transform [113,114], or polar coordinates [115].

The point-based approach deals with the raw LiDAR point clouds by searching and matching corresponding points from two overlapping 3D scans. The solution often employed to solve this task and estimate the best transformation is the ICP algorithm or its variants. Being a local minimization method, a drawback of ICP algorithm is its dependence to a sufficient mutual overlap and initial alignment of two scans to be matched. In order to tackle the limitations, several algorithms that provide a satisfactory coarse alignment are proposed. This can be achieved by establishing global correspondences based on local feature descriptors, which describe the local geometry around a point of 3D point cloud, as suggested in [116,117].

In turn, feature-based matching first transforms original point clouds into a more simplified form by extracting the higher-level features like for example line segments [118,119], planes [120], corners and jump edges [121]. By tracking the feature parameters among two point clouds, the pose change can be derived more reliably.

As an example of research work in a domain related to the LiDAR-based localization, [121] propose a parallel feature-based indoor localization algorithm to derive the position of a robot—a moving SLAM-type platform. The acquired 3D point cloud is first processed to extract corners and line features. Then, the point registration procedure, *i.e.* a weighted parallel iterative closed point (WPICP), is applied. A similar problem is handled by [122] for vehicle localization of self-driving cars that have no a priori knowledge of the environment. In their approach, the world is modeled as a mixture of several Gaussians that characterize the z-height and reflectivity distribution of the environment. It allows one also to reduce large point clouds into a compact, parametric representation, which can then be integrated into an EKF to estimate vehicle state. Finally, a landmark-based SLAM known as FastSLAM is proposed by [123]. First, the algorithm extracts uniquely identifiable features from the LiDAR scans *i.e.* vertical objects, which serve as landmarks. Then, they are associated robustly from frame to frame in FastSLAM using a RaoBlackwellized particle filter to estimate the positions of landmarks.

8.3.2 Multimodal Localization

Nowadays, localization is more and more becoming based on multimodal techniques. This is a way to take advantage of different sources and overcome their limitations. As presented in Sects. 8.2 and 8.3.1, GNSS, INS, LiDAR and camera methods (monocular, stereoscopic or RGB-D images) provide useful sources of information for localization, where each one has its own drawbacks. For instance, GNSS-based localization system must be used in a region with satellite coverage, otherwise it is expected to fail such as in tunnels, caves or complex urban environments. The same for INS-based localization systems, which are efficient in the case of short distance, but in other case these devices can be noisy and biased, skewing the final result. Camera-based localization systems may fail because of challenging conditions such as extreme lighting, long-term localization, season change or brutal and fast motion. In this way, a large variety of multimodal localization systems and algorithms have been proposed in the state of the art.

Therefore, multimodal systems require a discussion on how data acquisition, sensor coupling techniques and data fusion algorithms should be carried out. In the following sections, we revisit the most widespread techniques exploiting different natures of data jointly: Sect. 8.3.2.1 is dedicated to the most classical data fusion algorithms that can be applied to multimodal localization, while Sect. 8.3.2.2 revisits the main benchmarks existing with application to multimodal localization. Sect. 8.3.2.3 presents the main families of solutions encountered in the literature for multimodal localization, by finishing in Sect. 8.3.2.4 on more high-level techniques based on graphs.

8.3.2.1 Classical data fusion algorithms

Several classical algorithms dedicated to the fusion of heterogeneous data with application to localization exist. They first appeared in military applications. The fusion gives access to more reliable and precise information thanks to the complementarity of the information provided by the different sources. Ongoing research has a great interest in data fusion given the diversity of its applications including localization. Data fusion algorithms are diverse, but generally proceed through the same steps:

- **Modeling and aligning the data** to be fused.
- **Combining and estimating data** to generate assumptions, evaluate them, and select them to find better information.
- **Making a decision** by evaluating the outcome of the fusion carried out through the preceding steps.

For the same purpose, a data fusion system uses a **control function** to manage and control resources. This function depends on the steps mentioned above.

Various fusion filters exist and are implemented for different applications, including localization. Choosing the appropriate filter depends on the application requirements and constraints.

Kalman filter (KF). In 1960, Rudolf Kalman proposed a fusion filter [124]. Its objective is to estimate the system state $x(k)$, while knowing the observations $z(k)$. The Kalman filter is applied under certain assumptions in order to provide the optimal problem solution:

- The studied system must be dynamic, linear, and noisy so that its evolution can be expressed by Eq. (8.1).

- Noise w_k and v_k are white, centered, and independent, and covariance matrices are finite. We have

$$x_{k+1} = F_k x_k + B_k u_k + w_k \qquad k \geq 0, \tag{8.1}$$

$$z_k = H_k x_k + v_k, \tag{8.2}$$

with x_k the state vector, F_k the state matrix, B_k the command matrix, u_k the command vector, w_k the considered noise model (a zero mean Gaussian white noise, its covariance matrix is Q_k), z_k the observation vector, H_k the observation matrix, and v_k the noise related to the observation z_k (a zero mean Gaussian white noise; its covariance matrix is R_k).

The KF generally consists of two main steps, *the prediction step* and *the correction step*:

Prediction step: this is a prior estimate of the state labeled $x_{k/k-1}$ and the covariance matrix $P_{k/k-1}$ of the system, without taking into account the measure z_k, as the following formulas show:

$$x_{k/k-1} = F_{k-1} x_{k-1/k-1} + B_{k-1} u_{k-1}, \tag{8.3}$$

$$P_{k/k-1} = F_{k-1} P_{k-1/k-1} F_{k-1}^T + Q_{k-1}. \tag{8.4}$$

Correction step: the first step estimate is updated in this correction step using the measure z_k. Using the following calibration equations, we obtain the posterior estimate, $x_{k/k}$:

$$x_{k/k} = x_{k/k-1} + k_k (z_k - H_k x_{k/k-1}), \tag{8.5}$$

$$P_{k/k} = P_{k/k-1} - k_k S_k k_k^T = (I - k_k H_k) P_{k/k-1}, \tag{8.6}$$

$$k_k = P_{k/k-1} H_k^T S_k^{-1}, \tag{8.7}$$

$$S_k = H_k P_{k/k-1} H_k^T + R_k, \tag{8.8}$$

with $z_k - H_k x_{k/k-1}$ the innovation, S_k the innovation covariance matrix, and K_k Kalman's gain.

Therefore, if the models of Eqs. (8.3) and (8.4) mentioned above are no longer linear, but are linearizable in the range of the real state, the Kalman filter could not be applied. So alternative filters are needed for these cases.

Extended Kalman filter (EKF). This filter is a suitable solution for non-linear systems that allow one to give an estimated state close to the real state after their linearization. The solution proposed by EKF is based on the linearization of the equation representing the dynamic model $f(.)$ and the equation representing the observation $h(.)$.

First, the equations for $f(.)$ and $h(.)$ must be linearized. They are assumed to be derivable with respect to the variable x_k. This linearization is then made according to their Taylor development to the first order. As with a Kalman filter, the EKF is also carried out in two steps: a *prediction step* and a *correction step*.

EKF is a good alternative to the Kalman filter in the case of non-linear systems, especially when the estimated states considered by the filter are close to the actual state, which is not always valid. Another limitation of this algorithm is the computational cost imposed by the calculation of Jacobian matrices.

Unscented Kalman filter (UKF). Proposed in 1997 by Julier and Uhlmann [125], UKF is another fusion alternative for non-linear systems. This algorithm was introduced in order to alleviate the robustness of EKF. This filter is based on the same logic as the KF for linear systems. However, in non-linear systems, in order to avoid the use of Jacobians, this filter proceeds by creating the unscented transform (UT). UKF applies the UT approach, based on deterministic sampling, without a linearization approximation of the non-linear function. This approach is characterized by a set of $(2n + 1)$ points, called sigma-points $S = \{\mathcal{X}_i, W_i\}$. This set is composed of $\{\mathcal{X}_i\}$ points, chosen in a deterministic way, and their weights W_i, where n is the state space dimension. These points precisely evaluate the mean and covariance matrix of the predicted state (with Taylor precision order three) [126]. The approximations are based on the system equation interpolations with sigma-points.

Actually, UKF represents the recursive form of the UT [125] which consists in three steps: *initialization step*, *prediction step* and *correction step*. It provides a good robustness and the same or slightly better accuracy in estimating system state in various applications than EKF. In terms of its implementation, it is simpler than EKF, since the linearization of the system model is not used.

Sequential Monte Carlo methods for non-linear filtering (particular filter). Sequential Monte Carlo methods represent an important element used in tracking, localization, vision, *etc*. They consist of a set of tools that solves inference problems related to probabilistic and statistical models. The process of these methods consists in constructing a set of samples distributed according to a probability density (the investigated probability density). These samples are propagated over time using a large distribution and are weighted with the measurement available at time t.

- **Standard Monte Carlo sampling** and **weighted Monte Carlo sampling** are two techniques of the Monte Carlo method where the latter represents a generalization of the first one, which is a standard technique.
- **The particular filter (PF)** [127] [128] technique is a Monte Carlo-based data fusion algorithm dedicated to non-linear systems. It approximates the Bayesian filter using numerical methods. PF performs this in two steps: a *prediction step* and a *correction step*.

Table 8.3: Characteristics table of the different data fusion algorithms mentioned before [129] [130] (n: the number of state vector parameters, n_a: the size of the vector state x_k^a in the UT, M: the sample number).

Algorithm	Computational complexity	Non-Linearity
KF	$O(n^{2.376})$	Unused
EKF	$\max(O(n^{2.376}), O(g(s_{t-1})), O(h(s_t)))$	Linearization
UKF	$O(n_a^3)$	UT
PF	$O(M^n)$	Sampling

To conclude, multisensor system design is based on data fusion. Indeed, to conceive this type of system, it is necessary to choose the way in which the sensor measurements will be processed, so the most adapted data fusion algorithm is used depending on the applicative conditions.

Data fusion algorithms are many and vary according to their computational complexity, precision, and use case. The methods presented above are the most commonly used algorithms for localization based on visual/inertial data fusion.

As shown in Table 8.3, KF is the least complex in calculation. The total complexity of a single application of the Kalman filter is $O(9n^{2.376} + 10n^2 + 5n) \in O(n^{2.376})$, given that the best-known algorithm for multiplying two $n \times n$ matrices runs as $O(n^{2.376})$ (see [129] for more details as regards the KF time complexity calculation). Nevertheless, KF cannot be applied to non-linear systems in contrast to EKF, UKF, and PF. According to the literature, EKF and UKF depend on the targeted system, they have results that are not too far from each other in precision and computational complexity. It is worth noting that EKF is a bit less complex than UKF while UKF is relatively more accurate than EKF. A PF is the most computing complex filter due to the use of the sampling method used to solve the non-linearity of the systems.

8.3.2.2 Reference multimodal benchmarks

With the democratization of mobile systems capable of mapping the territory, several international benchmarks have appeared for multipurpose applications, among them localization, tracking and mapping. They can be employed to evaluate algorithms of localization or as a geo-reference for localization. Recent proposals thoroughly consider the major change in appearance that a scene can undergo across days, seasons and years, and then try to gather several modalities made available by the development of efficient sensors. Most of them focus on images (sequences, monocular or stereo, high definition (HD) panoramas) which are geo-referenced; more recent ones include laser information, and the very recent ones may include additional information such as semantic, or long-term data including changing conditions such as illumination (day/night), weather (sunny/rain/snow), and seasons (summer/winter) [54]. There exist about 30 datasets in the literature, and we briefly describe the main ones, which are most often exploited for localization purposes:

- *KITTI* [131] is composed of geo-localized stereo images and LiDAR data collected from a moving terrestrial vehicle. The dataset was enriched by considering aerial information in [132].
- The *Oxford Robotcar* public dataset [133] is a common multimodal dataset used for image-based localization. This dataset contains LiDAR point cloud and geo-localized stereo images captured from a terrestrial vehicle. A recent version, *RobotCar Seasons*, was proposed to deal with long-term localization [54].
- The *TorontoCity* benchmark [134] gathers both airborne and ground HD images and LiDAR, GPS and semantic information, vector maps, 3D models, aimed at the coverage of a large city.
- The dataset *EuRoC* [135] has been recorded in the context of the European Robotics Challenge (EuRoC), to assess the contestant's visual–inertial SLAM and 3D reconstruction capabilities on micro aerial vehicles. It consists of a stereo camera data (wide video graphics array (WVGA), global shutter, 20 fps), IMU data (MEMS at 200 Hz) and the ground-truth instruments data (laser+Vicon motion capture system).
- The dataset *7-Scenes* is a collection of tracked RGB-D camera frames [101], such as the TUM RGB-D benchmark [136], which contains RGB-D data and ground-truth data for the evaluation of visual odometry and visual SLAM systems.

A great part of the approaches of localization presented in Sects. 8.3 and 8.4 exploit these datasets, as benchmarks for their evaluation, or as reference dataset for approaches requiring such a kind of reference.

8.3.2.3 A panorama of multimodal localization approaches

Many of the aforementioned vision-based and IMU-based methods are founded on relative measurements, where the random errors may be progressively accumulated through the whole trajectory of the moving system. Also GNSS masks are corrected by an INS that can induce drifts, *i.e.* a gap between the trajectory supplied by the system and the real trajectory. In addition, some of the monomodal solutions that provide absolute pose (*e.g.* content-based image retrieval or pose regression) may suffer from the precision of the pose estimated. Consequently, with the development or improvement of new sensors and the construction of many databases describing the territory spatially, the approaches mixing different modalities—at the level of the input sensors and of the reference databases queried—are very numerous. We revisit in this section the main trends encountered in the recent years.

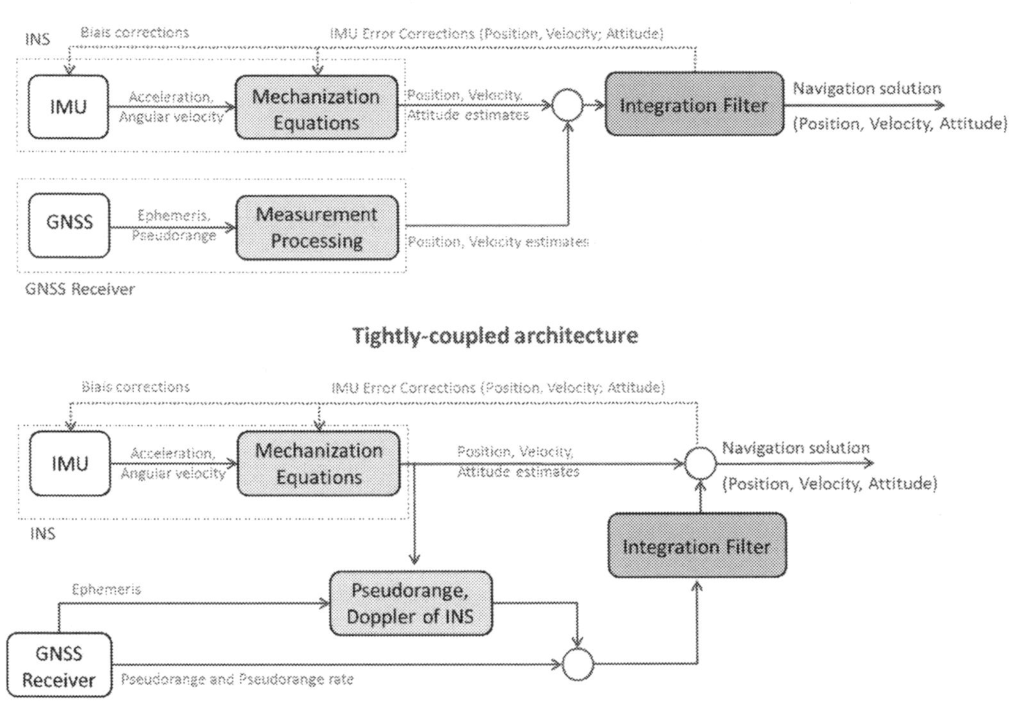

Figure 8.13 : Block diagram of GNSS/INS loosely-coupled and tightly-coupled architecture.

Localization without visual references. The methods able to decrease regularly the drift effect by associating constraints from other positioning systems are numerous. Therefore, it is preferable to employ multiple sensors instead of a single sensor. In order to reinforce indoor pedestrian navigation (see Sect. 8.3.1.1) accuracy and robustness, several hybrid INS/WiFi localization approaches are proposed. For example, [137] use a multidimensional dynamic time warping (MDTW)-based weighted least squares to fuse INS and WiFi positioning informations. In [138], the same problem is solved by tightly coupled integration of WiFi and IMU measurements. Thus, the gyro noise is removed by using PDR and INS integration into an EKF to compute position.

For outdoor localization, GNSS and dead reckoning fusion are often suggested in order to get reliable position data. The state estimation by means of KF [139,140], and by EKF [141], UKF [142,143] and PF [144] have been developed and tested for vehicle positioning. Simultaneously, different integrated schemes including a loosely-coupled and a tightly-coupled INS/GNSS integrated strategy (see Fig. 8.13) have been researched and developed [145,146]. Also, a comparison of performances assessed with loose and tight GNSS/INS integration in real urban scenarios has been performed recently [146]. In addition to this, some alternative approaches for GNSS/INS data fusion have proven to be effective such as for example a method based on the state-dependent Riccati equation (SDRE), non-linear filtering for unmanned aerial vehicles (UAV) localization [147], or a method based on a set of predictive models and occupancy grid constraints for a driver-less car [148].

Nonetheless, the quality of these methods is directly related to the precision of GNSS measurements and still need to be improved for autonomous vehicle under urban environments. To meet the needs and get more stable position, the UKF-based [149,150] or KF-based [151] GNSS/IMU/idstance-measuring instrument (DMI) are also popular sensor fusion methods. In the same vein, [150] propose to detect additionally GNSS spoofing attack. As the spoofed GNSS signal leads to fake navigation solution, it is critical topic for emerging applications such as autonomous vehicle navigation. The approach consists on a robust KF applied in a loosely-coupled reduced IMU and odometer (RIO) integration system that navigation solution is used to cross check with the GNSS measurements.

Localization coupling visual and inertial sensors. To improve localization, a very classical combination of sensors is to exploit image camera and INS data jointly; some of the methods are usually categorized under the name VIO, for visual inertial odometry. This configuration is quite common in professional and consumer systems, where each modality can benefit from the other. For example, the SLAM approach ORB-SLAM [86], presented in Sect. 8.3.1.3, has been optimized by exploiting INS information during tracking in its new version Visual–Inertial ORB-SLAM (VIORBSLAM) [152], illustrated in Fig. 8.14. Similarly, in [153], visual-based localization of a vehicle is performed with visual landmarks (mainly road signs), and then the raw vehicle pose is adjusted with the help of the inertial sensors; the system PIRVS [19] provides a visual–inertial SLAM system characterized by a flexible sensor fusion and hardware co-design. Reference [154] performs an evaluation of publicly available monocular VIO algorithms on hardware configurations found in flying robot systems (*i.e.* a light and low-power platform). Thus, they compared VIO pipelines such as MSCKF [155], OKVIS [156], ROVIO [157], VINS-Mono [158], semi direct visual odometry+MSF (SVO+MSF) [159], and SVO+GTSAM [159] [160].

A similar problem is handled by [161] to online estimate the position of underwater vehicles. The proposed method is based on a monocular video camera, aided by a low-cost MEMS-IMU and a depth sensor. To iteratively estimate ego-motion, they integrate optical flow computation based on pyramidal implementation of the Lucas–Kanade feature tracker [162] and corrected by inertial and pressure measurements. The latter is used to compute the scale factor of the observed scene. The main advantage of this method is that the position error does not grow with time, but only with the covered distance. Still, for underwater localization purposes, the vehicle position can also be estimated using approaches that fuse measurements from IMU and camera into an EKF [163,164].

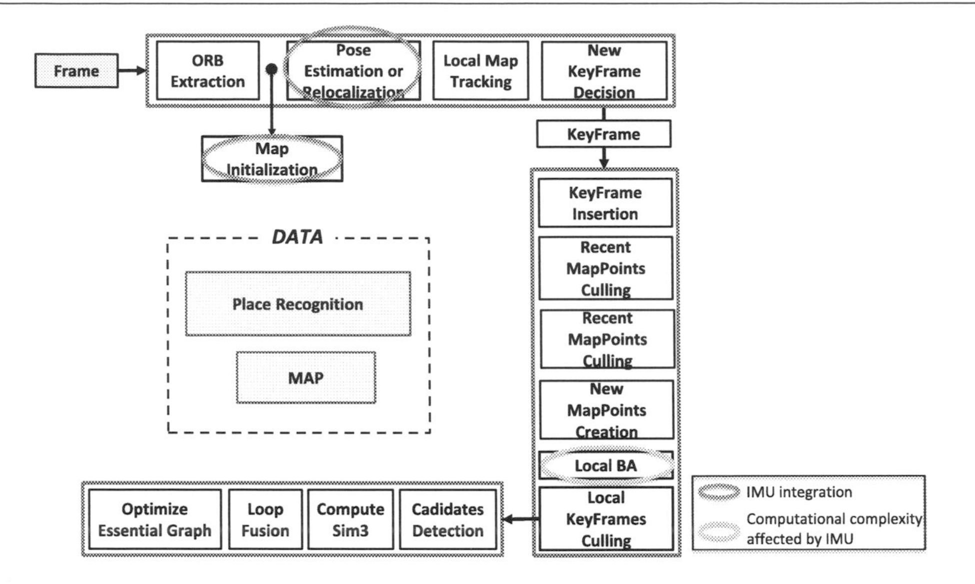

Figure 8.14 : Functional diagram of VIORB-SLAM [152], focusing on the exploitation of inertial information in the global visual chain.

Reference [165] develops a UAV-based wind-turbine blade inspection system for on-site outdoor localization and mapping. It is composed of an inexpensive RPLiDAR laser range scanner and 9-DOF IMU. Instead of using a conventional SLAM approach, which may fail in environments with a low amount of visual and geometrical features, they propose elliptical distance correction models for data fusion.

In the work of [166] INSs are aided by two complementary perception sensors (camera and LiDAR) in order to provide accurate relative navigation in GNSS-denied environments. The overall localization is based on EKF. LiDAR measurements are used to accurately estimate the vertical position.

Localization of images/point clouds with models. Several methods consider the challenging problem of photographies upon 2.5 or 3D models, which can belong to official sources such as national mapping agencies (digital elevation models, building models, *etc.*) or to processing such as SfM or scene reconstruction [65,167,79]. They generally consist of projecting one of the engaged data into the representation space of the other, and their main advantages rely on the robustness of the models facing challenging localization problems such as illumination. Boniardi et al. [168] introduce a SLAM system for 2D robot localization in architectural floor plans, starting from a coarse estimation of pose. Their approach consists of exploiting the information encoded in the CAD drawings (reference data) as well as the observations of the real world by means of on-board 2D laser scanner. The latter allows one to online produce a map, represented as a pose-graph. Then, a scan-to-map-matching method based on a generalized ICP (GICP) framework is used to fit the generated pose-graph onto the floor plan from the CAD drawings and compute the trajectory.

The common registration of sparse 3D point clouds obtained with SLAM with geo-referenced 3D building models is proposed by [169]. A similar problem is addressed in [170] by developing an ICP-based method to match 3D point clouds from a mobile mapping system with a 3D geographic database, by taking into account the non-linearity of the trajectory deformations in time. In a similar approach, [171] introduces a monocular SLAM that exploits both 3D building models and a digital elevation model (DEM) to achieve an accurate and robust real-time geo-localization. The proposed solution is based on a constrained bundle adjustment, where the DEM is exploited to correct the camera trajectory.

Simultaneously, geo-referenced objects like facades [172], road surfaces [173], road marks (*e.g.* zebra-crossings) and/or road signs [174,175,153,78] are also applied as a landmark database to compute position information through image registering techniques. Also, Schlichting and Brenner [176] use 3D features such as pole-like objects and planes, measured by a laser scanner to improve accuracy and reliability of localization process. These objects are detected automatically from 3D data acquired by an embedded laser scanner, and then matched to the geo-referenced landmark database data.

Localization coupling at least four modalities. Localization approaches based solely on GNSS, IMU, DMI or their fusion cannot always ensure a precise localization solution. As has already been proven, combining more modalities yields greater precision, and rich sensor systems are becoming more and more widespread.

For urban search and rescue (USAR) missions, [177] one develops a multimodal data fusion system for state estimation of an unmanned ground vehicle, which needs to cope with indoor–outdoor transitions. An EKF is used in the way to combine four modalities: IMU, tracks odometry (DMI), visual odometry (camera), and ICP-based localization (LiDAR).

In the work of [178] an autonomous vehicle localization approach based on loosely-coupled GNSS/IMU/DMI is addressed. In order to improve the performance of classical UKF-based approach, which can be affected by drifts due to the temporary GNSS signal failure, a LiDAR-based lateral error correction is proposed. Thus, the longitudinal curbs estimated in real-time from collected LiDAR scans are used to compute the vehicle's lateral distance. The same distance is also computed using the corresponding curbs extracted from an existing route network definition file (RNDF) map. Finally, the difference of these two distances corresponds to the lateral error of the autonomous vehicle to be considered.

Learning of a modality for localization. Classically, the approaches of localization that rely on reference datasets (*e.g.* content-based image retrieval approaches presented in

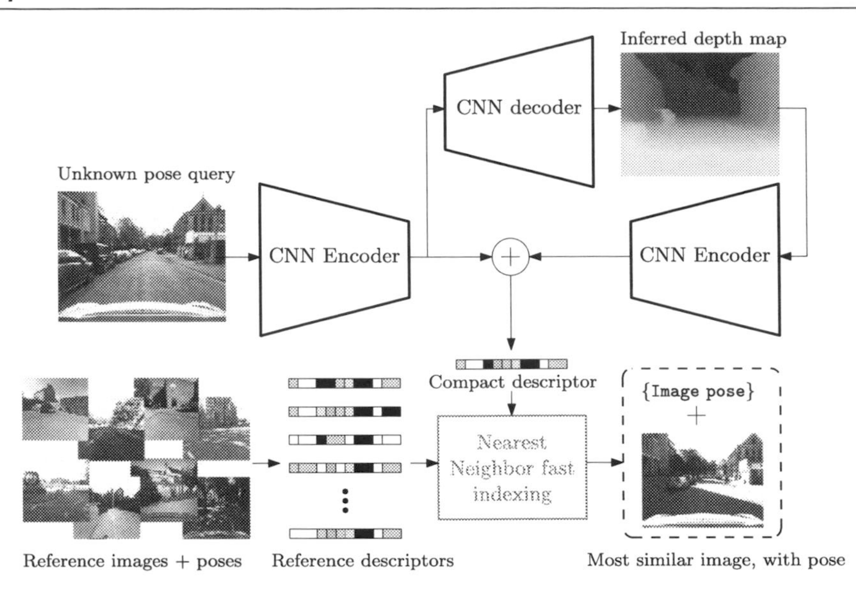

Figure 8.15 : Concept of image-based localization by inferring the geometry (depth map) related to the image query, from a multimodal and geo-localized reference dataset [181].

Sect. 8.3.1.3) compare and match data *of the same nature* between the input to localize and the dataset, for example a query photography and a set of geo-localized images. Under these conditions, such approaches generally under-exploit the datasets made available recently, which include several modalities to map the territories (see Sect. 8.3.2.2), because of their non-availability of the system of localization. Very recently, with the deployment of deep learning, several works have proposed to learn the modality missing in the input, to better exploit the reference dataset where this modality exists. For example, the work in [179] casts the missing information learning problem as a domain adaptation problem. In [180], the authors train a deep neural network to "hallucinate" features from a depth map only present during the training process in the reference dataset; here this work is dedicated to object detection in images. Similarly, but for localization, the approach of [181] increases the localization accuracy by inferring the depth map related to the query image, by benefiting from scene geometry information during training; the concept is illustrated in Fig. 8.15. In the same spirit, some approaches, based on generative adversarial networks (GANs), exploit appearance transfer in adverse conditions (such as day/night and day/snow) to generate new information, which improves the localization process [182].

8.3.2.4 Graph-based localization

When considering the trajectory of a robot in robotics, one general concept consists in setting up a graph being a meta representation able to optimize the pose of the robot within the trajectory. A node in the graph represents a pose of the robot (or some associated features) and an edge the spatial constraints (inter-nodal constraints) between two nodes. This trend corresponds to the family of graph-based SLAM in robotics [183]. One part of the overall problem, usually called *front-end*, is to create the graph by identifying nodes and constraints based on sensor data, while *back-end* refers to the step of finding or correcting the optimal configuration of the poses within the whole trajectory.

As already cited in Sects. 8.3.1.3 on image-based localization and 8.3.2.3 on multimodal localization, ORB-SLAM [86] performs visual SLAM by creating and managing a graph at front-end, aiming at providing a compact representation of the 3D point map generated during the SLAM. Here, each node of the graph, named a co-visibility graph, represents a keyframe, stored because it s characteristic of the environment, and an edge exists if two keyframes have observations of the same map points. The use of the co-visibility graph allows for matching, tracking and mapping focused in a local co-visible area, independently of the global map, making time operation stable in growing environments. In addition, ORB-SLAM performs a pose-graph optimization in order to correct loop closing.

8.3.3 Discussion

A first observation of the analyses of the literature on localization highlights the huge development of approaches involving a visual scene analysis these last years, because of several factors, including the development of portable cameras and LiDAR at a low cost, of topical scenarios of use where other classical modalities may not be permanently available (*e.g.* urban canyon, indoor localization, *etc.*) and because of the arrival of cutting-edge techniques of pattern recognition (deep learning). The researches dealing with the methods making it possible to replace or reinforce GNSS receiver with an alternative on-board sensors are numerous. Under these conditions, the perception information (from camera, radar, LiDAR sensors, *etc.*), if used for localization, may become a fitting solution for environments rich in features and objects. For example, [108] achieved a sub-meter navigation accuracy, when GNSS and LiDAR are used as aiding systems in order to periodically correct an INS-based trajectory of unmanned ground vehicle (UGV) in outdoor and indoor environments. In the same vein, [184] shows that the automatic image aiding significantly improves the attitude errors of a UAV system equipped with a low-cost GNSS/IMU systems. However, it does not significantly affect the positioning accuracy of such a system.

For indoor localization applications, systems that fuse INS with visual data seem to be most widespread. To the contrary, for outdoor environments, the tendency is to combine more modalities to ensure system reliability and continuous absolute pose when GNSS fails, or within unfavorable weather conditions *e.g.* snow, that currently challenge image-based techniques. Thus, as reported in [177], combining multiple modalities, if a system is well designed it yields greater precision and allows one to get pose information, while systems using fewer modalities fail completely. Recently, a number of filters and architectures have been also implemented and tested in order to provide a localization technique suitable especially for autonomous driving. As a general rule, the tightly coupled system is preferable over the loosely-coupled one when considering accuracy criterion. For example, the tests carried out in real urban scenarios by [146] have shown that tightly-coupled GNSS/INS provides better estimates of the vehicle position and attitude, with respect to a commercial GNSS module, loosely integrated with an inertial sensor. More in-depth analyses of the state-of-the-art localization techniques reveal that the accuracy achieved highly depends on the information sources utilized as well as on the specificities of the environments chosen to perform the test (*e.g.* indoor, outdoor like urban downtown, highways, tunnels or open-sky areas). Moreover, the quality metrics used to judge the overall performance and robustness of localization system, differ between one approach to another one, which makes them often incomparable. Not forgetting that it may be also due to the difficulty to having a ground truth to carry out such analyses (*e.g.* Vicon motion capture system for indoor, or theodolite tracker for outdoor localization). Further research works should focus on the common assessment of the most promising localization techniques, with regard to their specific field of application. In the last few years, some standard benchmarking (see Sect. 8.3.2.2) frameworks have helped to compare system performance. For example, the KITTI visual odometer dataset has already been used by several researchers [185–189] to validate their approach, especially VO, VIO, visual SLAM (VSLAM) or visual–inertial SLAM (VI-SLAM).

One recent trend concerns image analysis in challenging appearance conditions (night/day, season, temporal gap, *etc.*), in particular localization and classification, where the visual-based algorithms tend to fail [54]. Deep learning provides promising results in this sense, but it requires the availability of datasets integrating these particular conditions that are only beginning to appear (*e.g. RobotCar Seasons*). In this context, the help of other modalities remains very important.

In addition to this aspect, another promising trend concerns the learning of missing modalities for localization, which gives the possibility to exploit rich geo-referenced datasets with multimodal contents, while the system to localize is not similarly provided. On-going research, relying on deep learning again, provides promising results (Sect. 8.3.2.3) that should open the way to the exploitation of a larger field of datasets, from a given system to localize.

Finally, in the SLAM context, loop closing methods based on deep learning continue to emerge [190,191]. As reported by [192], they are more robust to environmental changes compared with the appearance-based methods, *i.e.* the BoW technique. However, it remains very challenging to design a neural network architecture that can run in real-time in a VI-SLAM or VSLAM system.

8.4 Multimodal Localization for Embedded Systems

Sect. 8.2 describes sensors used to address localization and navigation problem whereas Sect. 8.3 presents the state of the art on general localization and the different methods for heterogeneous sensor fusion. A multimodal embedded system consists both of involved embedded platform with on-board sensing and all external localization system/resources that enable to refine a reliable and accurate position. This section mainly focuses on embedded systems and the corresponding computing architectures, while examining the link to global localization.

First, a view on the application domains is provided in Sect. 8.4.1; then cutting-edge, embedded computing architectures are described in Sect. 8.4.2, and examples of full embedded systems, including sensors and computing architectures, are given in Sect. 8.4.3, in which different examples illustrate constraints and requirements for targeted markets.

8.4.1 Application Domain and Hardware Constraints

An embedded system consists of a mix of software/middleware/hardware modules that must achieve specific functions (real-time, cyber-physical, reactive, . . .). Depending on the environment considered and the application domain, these systems are facing various issues (criticality, real-time and resources).

In the transport domain, an embedded system is by nature transportable, *i.e.* despite the system being constrained by hardware, the power source or the battery capacity is not the main constraint (automotive, train, avionics). These systems mainly target the real-time and the safety constraints with different levels of confidence depending on the means of transport. Furthermore, as the processing board is embedded inside the vehicle, the form factor is less constraining.

For casual electronic consumers and users, an embedded system must fit in a pocket and is able to run several hours before the battery pack needs to be recharged. The device is in some way generalist but has to support audio, video, network functions, it can also be dedicated to gaming as a hand-held console. In the latter case, it is not fully optimized in terms of energy efficiency, it performs well and is generalist.

With the dawn of the internet of things (IoT) domain, energy efficiency criteria become decisive for the design of a compact, frugal, ubiquitous and efficient embedded system. In fact, a constellation of connected objects has emerged the last years that make collecting of data, access to information, screen mirroring, automated localization, *etc.* possible. Wearable devices are now driving the market. Thus, the goal of such devices is to migrate functionalities through different specialized devices (*e.g.* smart watches, health band, remote control, virtual personal assistant, smart headset, *etc.*)

To conclude, as mentioned in Sect. 8.2, there is an increasingly heterogeneous range of sensors available to perform the localization step. Some of them can be mixed up and integrated into the same system for different domains. As they are generating massive amounts of mostly unstructured data, a large diversity of processing has to be done to structure these data and produce enhanced output for the use of technologies like artificial intelligence (AI). Thus, the demand for heterogeneous computing is getting higher, and embedded systems have to cope with different workloads without changing the underlying architecture. Moreover, the processing must be fast and low-latency in order to provide localization in "real-time". The following describes the diversity of cutting-edge technologies involved in these embedded computing architecture.

8.4.2 Embedded Computing Architectures

When application constraints are translated in design constraints, system designers face a wide range of hardware resources to build up a relevant and efficient embedded system. This section focuses on the devices/components choice and their integration. It is divided into two subsections, dealing with design constraints and different hardware components in an embedded system on-chip (SoC).

8.4.2.1 SoC constraints

Today, according to [193], generalist microprocessors have reached their limit regarding energy efficiency. The manufacture process in semiconductor IC device fabrication is now 7 nm in 2018 and it will reach 5 nm in 2020 [194]. However, CMOS scaling no longer provides efficiency gains proportional to the increase in transistor density. Thus, nowadays industrial and academics works and studies are more and more focusing on the design of specialized accelerators. When designing a system, architects have to choose between any two of these three criteria: performance, energy efficiency and ability to perform generalist process [195]. Thus, devices such as general-purpose computing on graphics processing units (GPGPUs) and the last generation of field-programmable gate arrays (FPGAs) became attractive for designers, since they combine both a general purpose part and a specialized part dedicated to an application domain providing a good tradeoff as regards performance/energy efficiency.

Generalist versus specialized capability. The design challenge lies in the way all the selected computation chips, such as application-specific integrated circuit (ASIC), SoC or system in a box, *etc.*, and sensors coexist so that, in our case, the localization and tracking functions are realized. Sensors can settle next the processing part in a centralized manner or they can be distributed on different parts of the system depending on the processing performed. Therefore, in the latter approach, resulting data are pushed to the processing part. A SoC is the perfect candidate to deal with. In fact, SoC is an optimized IC that integrates all the processing components of a system on the same silicon/die. These components may compute general purpose tasks, image/audio process, networks interface communication tasks, graphical/rendering process, chip management tasks, *etc.*

Power consumption. In terms of power consumption, a SoC consumes from few Watts to a dozen of Watts. For example, let us consider a cutting-edge smart phone such as the Nokia 7 from HMD Global, 2 day battery life is claimed by the manufacturer (with a built-in 3800 mAh battery). The activity consists of 5 hours of usage including games, video streaming, phone calls, short message service (SMS), data exchange internet browsing and different applications (social networks, news, GPS, and music). Measurements are done with normal parameters and artificial lighting. The device remains in standby mode all night long. The battery life varies from 14.5 hours to up to 723 hours for video playback resp. in standby mode. It represents an intensity of 5.5 to 260 mA, and a power consumption of 0.13 to 9.67 Watts, considered in this study. A smart phone such as Nokia 7 plus, Huawei P20, Samsung S8, consumes power within a range from 0.85 to 5.6 Watts.

If now considering dedicated hardware components that bring about functionality such as a visual processing unit, Movidius Myriad 2 ICs claimed a power consumption of less than 1 W. Other customized ICs targeting visual–inertial odometry [196] intended to fit in a more compact system such as a nano/micro-aerial vehicle, announcing an average power budget of only 24 mWatts.

Performance and energy metrics. Traditionally, an IC provider always gives typical performance metrics to illustrate the computing capabilities of their chip. Millions of instructions per second (MIPS) or instructions per cycle (IPC) are examples, but this figure is directly linked to the instruction set-architecture (ISA), thus discriminant. Nowadays, we have trillion of floating point operations per second (TFLOPS) for a given data width, giga pixels per second for video, intellectual property (IP), *etc.* In 2016, NVIDIA introduces a new measurement of performance, the deep learning tera operations per second (DL FLOPS) computing figure; it refers to the ability of processing an image classification (AlexNet) using deep CNN (a.k.a. inference). A smart phone manufacturer communicates on the capability of recognizing a number of images per seconds.

These metrics are sometimes controversial because of processing kernel and architecture dependency. They represent a kind of theoretical peak performance not achievable for real-world applications. Thus, some benchmark providers propose typical kernel implementation that can be implemented on candidate architecture and sort them by performance and efficiency in terms of energy (BDti [197], Spec2006 [198] *etc.*). In fact, power and energy aspects consider Watt and Joule metrics. More realistic metrics compute this efficiency, *i.e.* operation per second in power consumption and/or energy delay product (*e.g.* TOPS/Watt), *i.e.* time processing x power consumed depending on executed kernels or applications. A novel metric has been introduced to measure the efficiency of CNN milliJoule/inference or fps/inference per second.

8.4.2.2 IP modules for SoC

As mentioned earlier, a SoC consists of IP modules communicating together on the same die to achieve several functionalities. When application time becomes a hard constraint and acceleration on critical tasks is required, some customized IPs are used to reduce processing time for audio, video, image, radio frequency tasks, *etc.* Thus, IPs such as digital signal processors (DSPs), graphic processing units (GPUs), image signal processors (ISPs), video encoding or vision processing units (VC/VPUs) are implemented to meet critical deadlines. Recently, the neural processing unit (NPU) has been introduced in patent by [199]. Huawei resp. Apple announced the integration of industrial NPU chips Kirin resp. A11 Bionic in their SoC, whereas Qualcomm came up with its neural processing engine, a flexible combination of CPU, DSP and GPU. Huawei's NPU is presented in [200].

DSP. Several solutions are available, such as:

- Cadence [201] proposes its solution, called Tensilica. It allows one to design customizable DSP by extending the instruction set and conjointly implementing corresponding dedicated hardware logic unit. Obviously, the compiling toolchain is also provided. Tensilica DSPs are used in many products such as the Hololens with 300 custom instructions to the DSPs in order to speed up to render real-time augmented reality or Amazon Alexa device for high fidelity (HiFi) DSP. This solution also suits vision processing and convolutional neural net vision recognition systems.
- The CEVA Platform XM6 [202] is the last generation of DSPs optimized for vision and deep learning targeted machine vision and advanced driver-assistance systems (ADASs) market. It is a 8 way very long instruction word (VLIW) processor able to process 128 multiply accumulate (MAC) operations per cycle (16×16 bit). It runs at 1.5 GHz (process node 26 nm). Additionally, 512 to 4096 MACs (8×8 bit) are also available with the Ceva deep neural network (CDNN) engine depending on the targeted market (IoT, smart phone, augmented reality/virtual reality (AR/VR), drones, automotive). Specific memory load (scatter and gather pattern) enables efficient SLAM and depth mapping. The device authorizes 12.5 TOPS for the most advanced configuration.

- Qualcomm Hexagon [203]: The Snapdragon 845's Hexagon 685 DSP is claimed to be a machine learning accelerator able to process large data vector (2×1024 bit). DSP multi-threading enables multiple offload sessions (*i.e.*, accelerate concurrent tasks concurrently for audio, camera, CV).

GPU. These power efficient IPs are dedicated to image/video/rendering processing and targeted tablets, smart phones, wearable, AR/VR, IoT, smart TV, automotive devices. In the case of ARM-G76, the architecture consists of unified shader, dual texture unit, a blender and tile access, depth and stencil, manager and memory. The 20 shader cores include three execution engines (with eight execution lanes for each), a total 480 execution lanes. The execution units have int8 dot product support for improved machine learning performance.

The Adreno architecture is the GPU solution provided by Qualcomm to provide efficient SoC. The last generation is able to support software features such as OpenGL ES 3.2, OpenCL 2.0, DirectX 12.

Image signal processing. One may mention:

- ARM Mali C71: designed to perform low-level process on images, it is also tailored for computer-vision algorithms. This device provides expanded capabilities such as wide dynamic range and multioutput support. Its safety capability (automotive safety integrity level D (ASIL D) and safety integrity level 3 (SIL3) standards-compliant) make it the perfect choice for automotive domain.
- Qualcomm Spectra: 280 ISPs have a parallax-based depth-sensing system that estimates the relative distance of objects from a two-lens perspective (passive stereovision and active IR). It also supports iris and facial scanning for security. It is designed for VR/AR experiences by achieving sub-16 ms motion to photon latency for head and body tracking, with 6 degrees of freedom and a simultaneous location and mapping (SLAM) system (accelerated heterogeneously) that models and tracks the environment around the wearer. Sub-16 ms latency allows for single frame responsiveness at 60 Hz, which is crucial to comfort in particularly intense games and applications. Spectra 280 can go up to 2k $\times$ 2k at 120 frames per second (compatibility with Google, Vive, and Oculus headsets).

Video encoding/decoding unit. Video processor gathers together all scalable IPs able to encode/decode several video streams with different definitions (HD, 4K, 8K, *etc.*), scheme (H264, H265, *etc.*) and data rate (30, 60, 120 fps, *etc.*).

8.4.2.3 SoC

MultiProcessor SoC. The most common generalist IP such as a multiprocessor platform is able to cope with high-level applications and take advantage from application, task/process and thread level parallelism. During the last ten years, the ARM company became the leader of embedded processors and reached 95% of mobile market share since Intel left in 2016. They propose the most power efficient solutions (microcontroller unit, for a wide range of market from IoT world, wearable devices, ADAS to high-end industrial computing [204]. Provided as a license core, ARM products are integrated in a lot of cutting-edge SoCs (Ti, Qualcomm, ATMEL, *etc.*). In fact, thanks to the emergence of wearables devices, ARM processors have taken the lead on the mobile market. Today, their bigLittle architecture is the reference to sustain different types of workloads while keeping the power consumption low. It consists of a combination of small power efficient and large performant flexible cores (for example 4 cortex A55 + 4 cortex A75). The Cortex-A75 is 64 bit out of order 3-way superscalar processor with 15 stages pipeline, *i.e.* able to execute up to three instructions in parallel per clock cycle. It owns seven execution units, two load/stores, two NEON and Floating Point Unit (FPU), a branch, and two integer cores. A specific 8 bit integer format has been introduced for a small dot product to implement neural network operations in a better way. The cortex A55 core is an in-order processor with eight stages pipeline. Today, ARM proposes a new technology to use a combination of impair number of cores (*e.g.* 1A75 + 3A55). It should be noted that the RISC V architecture has emerged in recent years. RISC-V is a free and open-source ISA out of U.C. Berkeley [205]. More than 100 companies are supporting the initiative. Keywords of this new ISA are longevity, portability, and reliability, since a frozen ISA (less than 50 instructions) is provided and extensions are well categorized for a total of 200 instructions (multiply and divide, atomic, single precision floating point, double precision floating point, compressed instructions).

GPGPU. The GPGPU device breaks into the embedded world with the NVIDIA Tegra architecture. The last NVidia Tegra device, called Xavier [206], consists of 9 billion transistors. The SoC consists of:

- 8-core CPU Carmel ARM64 10-wide superscalar with functional safety features plus parity and error-correcting code (ECC) memory suitable for autonomous driving.
- Programmable vision accelerators (PVA) for processing computer-vision tasks: filtering and detection algorithm. Two identical instances are implemented that can work independently or in lockstep mode. Each unit consists of an ARM cortex R5 processor, a direct memory access (DMA) Unit, two memory unit plus two vector processing units (seven ways VLIW). Customizable logic is also available. Operations can be performed on 32×8, 16×16 or 8×32 bit data vector.
- ISP achieving 1,5 GPix/s. (native high dynamic range (HDR)) and multimedia core.
- A new 512-core Volta GPU. This architecture has been designed for the learning machine market and is optimized for inference over a training process. It consists of eight stream multiprocessors with individual 128 KB L1 cache and a shared 512 MB L2 cache. There are also 512 compute unified device architecture (CUDA) tensor cores.

- 8K HDR video processors.
- A deep learning accelerator (DLA) able to achieve 5,7 DL TOPS (FP16) with a configuration/block, input activation and filters weigh, a convolutional core, a post-processing unit and interface with the memories (standard dynamic random access memory (SDRAM) and internal RAM).

DRIVE Xavier SoC puts more processing power to work using less energy, delivering 30 trillion operations per second while consuming just 30 Watts.

8.4.2.4 FPGA

Traditionally, FPGA devices were used only for prototyping and/or make a proof of concept for hardware accelerators before the taping out *i.e.* resolution of the cycle of design for integrated circuits (ASICs), when the photomask of the circuit has been fully created and is sent to the manufacturer for production. FPGA authorizes the device reprogramming of hardware logics cells to perform different functionalities. It can take advantage of fine-grain data parallelism by enabling the concurrent execution of different data processing chain. These last years, FPGA device and their attractiveness for customers has changed since these devices consists of both a hardware logic part and a more generalist software processing system part available on the same silicon. Thus, it is possible to design high efficient hardware components and make easier data exchange with more generalist processing such as multiprocessor platform in the same device. Nowadays, in Xilinx latest product, a.k.a. Zynq Ultrascale+component, 4 ARM A53 plus 2 ARM cortex R5 cores settle next to the logic part. Furthermore, Intel company recently acquires Altera FPGA Manufacturer [207] and proposes the Arria 10 device. Other manufacturers such as Microsemi or Lattice semi only provide FPGA with logic cells.

8.4.2.5 ASIC

ASICs are optimized in terms of performance, power consumption and occupied area (energy efficiency). We consider here chips that are dedicated to new advanced functions.

Holographic chip: intel holographic processing unit [208]. This chip is a custom multiprocessor (used as a co-processor) called the holographic processing unit, or HPU. It is in charge of the integration of all embedded sensors (IMU, custom ToF depth sensor, head-tracking cameras, IR camera) through several interfaces: a mobile industry processor interface (MIPI), a camera system interface/display serial interface (CSI/DSI), an inter-integrated circuit (I2C), and protocol-control information (PCI). This Taiwan Semiconductor Manufacturing Company (TSMC)-fabricated 28 nm co-processor has 65 M logic gates and occupies 144 mm^2. It consists of 24 Tensilica DSP cores. It has around 65 million logic gates, 8MB of static random access memory (SRAM), and an additional layer of 1GB of low-power double data rate type 3 RAM (DDR3 RAM). HPU offers a trillion of calculations per second. Claimed as low power, it consumes 10 W for handle gesture and environment sensing.

Vision processing unit (VPU): intel Movidius Myriad [209]. This type of IPs appeared recently due to the huge demand for devices able to process a large amount of image data while extracting more valuable content such as spatial and temporal data over one or several video streams. In the last years, the IP has received new interfaces to collect data coming from other sensors such as accelerometers, gyroscope, magnetometer, ... They are described as a combination of machine vision with ISP. In this scope, targeted applications are pose estimation, VIO for navigation but also gesture/eye tracking and recognition. The Intel Myriad 2 device MA2x5x combines image signal processing with vision processing. Imaging/vision hardware accelerators can be pipelined avoiding unnecessary data exchange to memory; VLIW processors are provided to support and accelerate vision algorithm. Myriad 2 VPUs offer trillions of floating-point operations per second (TeraFLOPS) of performance within a nominal 1 Watt power envelope. Heterogeneous, high throughput, multicore architecture based on 12 VLIW 128-bit vector SHAVE Processors optimized for machine vision, configurable hardware accelerators for image and vision processing, with line-buffers enabling zero local memory access ISP mode, 2×32-bit RISC processors and programmable interconnect. Interfaces: 12 Lanes MIPI, 1.5 Gbps per lane configurable as CSI-2 or DSI, 1 Gbit ethernet and USB3. The chip occupies 76 mm^2 in a 28 nm high performance computer (HPC) process node. Based on this VPU, [210] designs an eye-of-things board and demonstrates the possibility of having a tiny state-of-the-art machine learning inference (using only 1.1 W) as compared to the call of Google cloud vision application programming interface (API) when performing face emotion recognition with a deep neural network (DNN).

CNN chip: Orlando ST IC [211]. Following the example of Qualcomm and others, in 2017, ST proposed an ultra-low power SoC capable of accelerating deep convolutional neural networks algorithms. An Orlando chip consists of a Cortex-M4 Microcontroller Unit (MCU) and 128 KB of memory, eight programmable clusters each containing two 32-bit DSPs, and four SRAM banks, each bank offering four modules of 2×64 KB each. Coupled with this highly efficient core is an image and CNN co-processor (called the neural processing unit or NPU) that integrates eight convolutional accelerators (CA), among other things. In large networks, engineers can use multiple SoCs and connect them using a 6 Gbps four-lane link. The 34.1 square mm chip uses a 28 nm full depleted silicon on insulator (FDSOI) process node. It runs at near 200 MHz at 0.575 V and consumes only 41 mW when executing AlexNet at 10 fps. Under these considerations, the SoC reaches a peak of 2,9 TOPS/Watt and 130 fps/watt (440 fps/Watt when using 16 CAs).

Table 8.4: System On Chip comparison chart.

Product Name	Qualcomm Snapdragon 845	Huawei Kirin 980	NVIDIA Drive Xavier
Year	2017	2018	2018
Process technology	10 nm FinFet	7 nm	12 nm FinFet TSMC
Transistor nb.	4,3 billions	6,5 billions	9 billions
Area	91 mm^2	100 mm^2	350 mm^2
nb. CPU cores	8	8	8
CPU Arch.	4 A75-like/ 4 A55-like	ARM 4 A76 / 4 A55	Carmel ARM64 8 core
GPU Arch.	Adreno 630	Mali G76 MP10	Volta 512 cores
ISP Arch	1×Spectra 280	new dual arch.	1×
DSP Arch	Hexa 685	in 2×NPU	PVA
Deep Learning Arch	mix of CPU/GPU/DSP	in 2×NPU	DL Acc.
Computing Power	NA	NA	30 TOPS
Total Chips	1	1	1 Xavier
System Mem.	LPDDR4	LPDDR4	16GB 256-bit LPDDR4
Graph. Mem.	29.8 GB/s.	31.8 GB/s.	137 GB/s.
TDP	est. 4.4 W	est. 3.5 W	30 W

Navigation chip [196]. A Navion chip is a customized ICs targeting visual–inertial odometry (see Sect. 8.3.2.3) and is intended to fit in a more compact system such as a nano/micro-aerial vehicle and virtual/augmented reality on portable devices. The chip uses inertial measurements and mono/stereo images to estimate the drone's trajectory and a 3D map of the environment. Several optimizations are performed to minimize chip power and footprint for low-power applications, while maintaining accuracy. The authors announced an average power budget of only 24 mWatts while processing from 28 to 171 fps. The die process is 65 nm CMOS, the chip occupies 20 square feet mm presenting a fully integrated VIO implementation.

8.4.2.6 Discussion

We have presented the different modules involved in embedded computing architectures. ARM-based multiprocessors on a chip are nowadays the most widely used general purpose processors due to their energy efficiency. Now considering the trend in IC design for localization application, traditional combinations such as CPU/DSP/GPU or CPU/GPU have reached their limit for machine learning and computer vision. Since most of the time the learning part of an artificial neural network (ANN) is performed offline, SoC providers design architecture optimized for the inference part of classification through NN. At the same time, the emergence of these specialized co-processors or ASICs demonstrates the need for new architecture capable of merging data from heterogeneous sensors efficiently. Table 8.4 gives the main features of cutting-edge smart phones and automotive SoCs. It also illustrates the difference between these platforms: computing capabilities, memory bandwidth and power consumption.

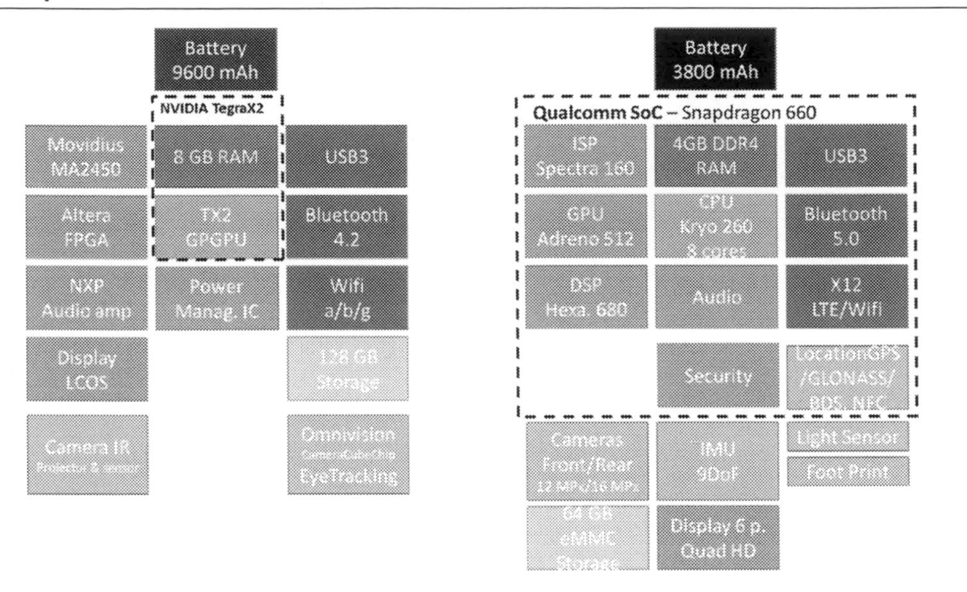

Figure 8.16 : Block diagram of Magic Leap One (left) and Xiaomi Mi8 (right) functionalities.

Since an automotive SoC aggregates a lot of heterogeneous sensors, it must support a wide range of processing, explaining the factor 10 in power consumption as compared to smart phone SoC.

8.4.3 Multimodal Localization in State-of-the-Art Embedded Systems

This section presents the multimodal localization techniques that are currently implemented in state-of-the-art embedded systems. These systems range from devices with a high level of integration constraints such as smart phones and smart glasses to larger systems such as autonomous driving vehicles with lower integration constraints but higher safety requirements.

8.4.3.1 Example of embedded SoC for multimodal localization

This section describes two up-to-date embedded system to illustrate what such a SoC is made of. Fig. 8.16 illustrates the two block diagram of a Xiaomi Mi8 smart phone and Motion Leap One SoCs.

8.4.3.2 Smart phones

Smart phones embed more and more sensors that can be used for localization and tracking such as cameras, depth sensors, and IMUs. The two companies that dominate the software environment in the smart phone ecosystem, namely Google and Apple, are relentlessly working to leverage these sensors to turn smart phones into devices that aim at either adding information to the scene, in the case of AR devices, or changing the perception of the current scene, in the case of VR devices. The architecture of the Xiaomi Mi 8 smart phone is built around a Qualcomm Snapdragon 845 Platform [212]:

- Qualcomm Spectra 280 ISP,
- Kryo 385 Gold (4 A75-like at 2,8 Ghz) and silver (4 A55 at 1,8 Ghz) (In house built-in ARMv8 ISA),
- Hexagon 685 DSP,
- Adreno 630 GPU (240 GFlops),
- 6GB of RAM with 256GB,
- X20 Long Term Evolution (LTE) (1,2Gb/s upstream /150 Mb/s downstream), WiFi 802.11a/b/g/n/ac/ad, Bluetooth 5, GNSS (GPS,Glonass,Beidou,Galileo,QZSS),
- Rear camera: Dual Sony IMX363 12 Mpx f/1.8 1,4 μm + Samsung S5K3M3 12 Mpx f/2.4 1,0 μm,
- Front camera: Samsung S5K3T1 20 Mpx f/2 1.8 μm,
- AMOLEDisplay 1080×2248 definition (Full HD+ (FHD+)).

Qualcomm also announces an AI engine using ISP, Kryo cores, DSP and adreno plus dedicated Software Development Kit (SDK).

Google ARCore. ARCore works on Android devices and focuses on three functionalities. Motion tracking uses the phone's camera to track points of interest in the room coupled to IMU sensor data, in order to determine the 6D pose of the phone as it moves. Environmental understanding detects horizontal surfaces using the same points of interests used for motion tracking. Light estimation leverages the ambient light sensors in order to observe the ambient light in the environment and light the virtual objects in ways that match their surroundings, making their appearance more realistic [213].

Apple ARKit. ARKit works on iOS devices and, similar to ARCore, uses the combination of the devices cameras and IMU to localize and track the device in space and to detect horizontal surfaces in order to be able to place virtual objects on them [214].

8.4.3.3 Smart glasses

Smart glasses are wearable devices that can be used as AR or VR devices. Due to their small form factor, smart glasses present very strict embedded requirements in terms of weight, space occupation, and energy consumption. This section studies different cases of AR and VR smart glasses and highlights implementations of localization algorithms for multiple needs of scene understanding.

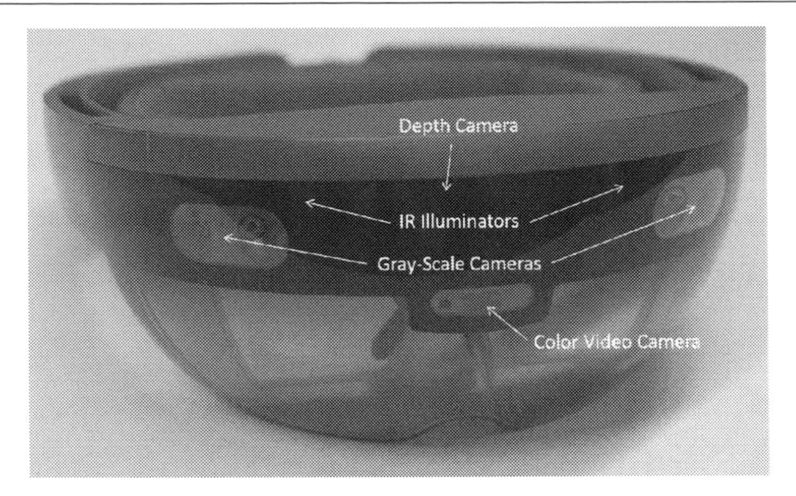

Figure 8.17 : Microsoft Hololens sensors.

Microsoft Hololens. Microsoft introduced the development edition of their Hololens AR device in 2016. The Hololens uses multiple sensors in order to localize the device and understand the scene. These sensors include an IMU, four "environmental understanding" gray scale cameras, a depth camera, short- and long-throw infrared illuminators, as well as a 2 megapixel video camera, a four microphone array, and an ambient light sensor as depicted in Fig. 8.17 [215]. The Hololens embeds processing into the head-mounted device with an Atom x5-Z8500 generalist platform plus a custom-made Microsoft holographic processing unit (HPU) as the main processor for 6D localization and tracking. The HPU has available a thermal sensor, four cameras and an IMU.

Magic Leap One. Magic Leap One is another stand alone AR headset introduced in August 2018 by Magic Leap. It uses an infrared projector and camera combination for 6D localization and tracking and also includes IR eye trackers to track the gaze of the user, a feature that is not available in the Hololens [216]. To cope with the extremely tight space constraints of the headset device, Magic Leap One embeds the processing power into a deported, small factor, computing device that is tethered to the headset with a cable. This device, called lightpack, incorporates a Nvidia Parker SoC with 2 Denver 2.0 64-bit cores and 4 ARM Cortex A57 64-bit cores as well as an Nvidia Pascal GPU with 256 CUDA cores [217].

StarVR one. StarVR one is a VR headset aimed at professional and industrial markets. Its specificity is its very large field of view (210 degree horizontal, 130 degree vertical) and high frame rate tracking (90 Hz) which makes the virtual reality experience immersive. However, in order to achieve these characteristics, StarVR uses embedded active markers on the headset, coupled with an external optical tracking system such as the OptiTrack V120:Trio. It also needs to be tethered to a desktop computer with at least an Intel core i7-7700 or an AMD Ryzen 7 2700X CPU as well as a Nivida GeForce GTX1080 GPU [218].

8.4.3.4 Autonomous mobile robots

Mobile industrial robots are often categorized into two families: automatic guided vehicles (AGVs) and autonomous mobile robots (AMRs). AMR are more autonomous and capable as compared to AGV. The latter are mainly able to go from point A to point B. The former uses SLAM and a map of the building to localize. In terms of perception, they are based on monocular or stereo vision, sonar, LiDAR ToF, and structure light sensors. The market players are Swisslog (KUKA), Omron Adept, Clearpath Robotics, Vecna, Mobile Industrial Robots, SMP Robotics, *etc*. Swisslog (KUKA) and Omron Adept are the leaders of the autonomous mobile robots market. Kuka AMRs provide an autonomous graph-based navigation with safety laser scanners and wheel sensors for SLAM. Navigation software collects all sensors data to give a fine localization for a precise determination of the vehicle position relative to the object or within an environment, fine positioning for increased positioning accuracy and relative positioning through computer aided design (CAD)-based object recognition and tracking, *e.g.* for picking up loads. The robot can sense its position and orientation in the space with millimeter accuracy. Kuka also proposes a 3D perception stereo camera sensor with different baseline distances (65 mm or 160 mm) that provides a depth image calculation for short or long range (<3 m). The image can be displayed in monochrome or in color. The sensor system is built around a NVIDIA Tegra K1 SoC, an IMU, a stereocamera. Fleet management is provided when several AMRs are used. Finally, the whole software suit runs on an embedded industrial PC.

8.4.3.5 Unmanned aerial vehicles

UAVs, commonly known as drones, are aircraft without a human pilot aboard. These range from large military aircrafts such as the General Atomics MQ-9 Reaper to small factor drones aimed at the civilian market. This section deals with localization techniques for the latter category.

Drones use the same sensors as the smart glasses discussed in Sect. 8.4.3.3 for localizations. However, on top of these sensors, drones embed GNSS sensors in order to cover large outdoor areas. Also, some drones include barometers to measure the altitude of the drone and inject this data into their multimodal localization techniques [219]. Reference [154] performed a benchmark comparison of monocular visual–inertial odometry algorithms for a flying robot (see Sect. 8.3.2.1). These algorithms have implemented Meng5 modalities on several architectures (Laptop, Intel NUC, Intel-based small size board, ODROID platform). Using EuRoC MAV datasets [135] containing camera data captured at 20 Hz and synchronized with an IMU at 200 Hz, results are compared with ground-truth positioning given by a laser and Vicon motion capture system. Thus, one noticed that accuracy and robustness can be improved with additional computation. When combining hardware platform like the Up Board or ODROID and SVO+MSF methods, it gives the most efficient performance with a significant sacrifice in accuracy and robustness. VINS-Mono has the best results with non-constrained computation on all hardware platform. ROVIO appears to be the best tradeoff.

8.4.3.6 Autonomous driving vehicles

Autonomous driving (AD) is an important topic of research and development for all major car makers. It is constituted of five levels ranging from driver assistance (level 1) to complete automation (level 5) where absolutely no human attention is required. Most OEMs are currently developing vehicles at level 2 autonomous driving, which is partial automation. At this level, the vehicle can assist with steering or acceleration functions and allow the drivers to disengage from some of their tasks. The driver must always be ready to take control of the vehicle and is still responsible for most safety-critical functions and all monitoring of the environment. Starting at level 3, conditional automation, the vehicle itself controls all monitoring of the environment (using sensors like cameras, LiDAR, *etc.*). The driver's attention is still critical at this level, but the driver can disengage from safety-critical functions like braking and leave it to the technology when conditions are safe. At level 4, high automation, the vehicle would first notify the driver when conditions are safe, and only then does the driver switch the vehicle into autonomous driving mode.

Starting AD level 3, precise localization techniques become crucial for the operation of autonomous vehicles. Current research in localization for AD is focused on a combination of image-based algorithms for lane, road, or curb detection, combined with map matching. These techniques range from low-cost approaches that use data fusion from cameras, GNSS, and IMU [220] to higher cost approaches that also use data from 2D and 3D LiDARs [221]. These techniques, which require well maintained roads with lane markings and corresponding maps, can be applied to AD levels up to level 4. However, for AD level 5, which should be able to pilot a vehicle even in rough terrains such as dirt roads, these techniques come short. Localization techniques based on detailed features descriptors (SIFT, SURF, *etc.*) or semantically segmented images [222] are required to achieve this level of autonomous driving.

Among others, the Volkswagen Group proposes its Audi A8, a level 3 ready vehicle but limited to level 2 for safety reasons. The advanced driver assistance system platform goes by the name of zFas and is constantly evolving. It is of PC motherboard size and consists of a data fusion unit involving a raw processing subsystem that collects cameras data (MobileEyeQ3 SoC) and a safe and secure host subsystem that manages the vehicle (Auryx SoC). A NVIDIA Tegra K1 is also used for graphical computing. A lot of sensors are connected to this zFas:

- twelve ultrasonic sensors on the front, sides and rear,
- four 360° cameras on the front, rear and exterior mirrors,
- a front camera on the top edge of the windshield,
- four mid-range radars at the vehicle's corners,
- a long-range radar on the front,
- an infrared camera (night vision assist) on the front,
- and a laser scanner on the front.

Finally, the localization techniques using on-board sensing systems mentioned below must be combined with vehicle-to-vehicle (V2V) and vehicle-to-infrastructure (V2I) systems. Reference [223] summarized the potentials and limitations of each technique on the implementation on autonomous vehicles. Thus, according to the authors, LiDAR-based techniques are the best in terms of accuracy when mixed with other sensors but this approach is expensive. The V2V and V2I approaches are dependent from the networks' quality of service and must take the security issue into account. Furthermore, the environment in which these networks are deployed is crucial.

8.4.4 Discussion

This section first describes the computational architectures involved in embedded systems dedicated to localization applications. Generalist computer chips remain essential for performing the least specific tasks. Despite this, dedicated processing modules appeared on the market during year 2018. They allow for both the aggregation of data from heterogeneous sensors and for tasks such as neural network classification. More generally, each IC manufacturer or hard/soft IP provider offers its deep learning module, which most often accelerates the processing of so-called convolutional neural networks (CNNs). Embedded systems obviously only provide the inference part of the processing (not the learning one).

Fig. 8.18 presents a classification of a smart phone, a headset mounted, of drone devices and of an autonomous vehicle based on power consumption, integration density versus required localization accuracy and algorithmic complexity.

Obviously, the more sensors are needed, the more complex is the processing to perform data integration and fusion steps. This figure also shows the problem of safety in autonomous vehicles. On the one hand, the sensors are diverse and varied and data fusion must produce very accurate and reliable locations. On the other hand, the determinism of applications and hardware resources must be guaranteed for the obvious reasons of responsibility in the case of an accident and compliance with automotive standards. Thus, the number of integrated circuits and therefore the size and number of electronic cards increase accordingly.

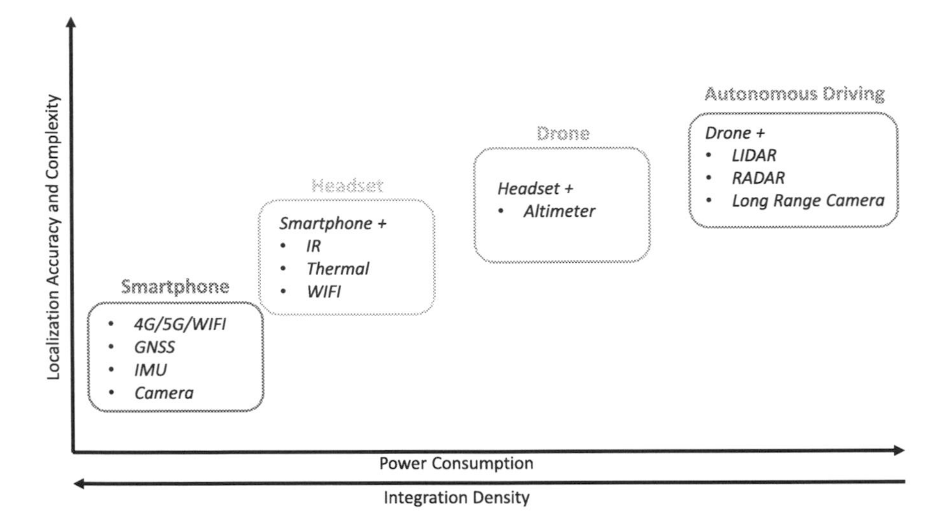

Figure 8.18 : A classification of embedded systems.

Beyond those embedded systems, "out-of-board" localization is another part of the problem. A smart phone and/or headset can take advantage of different techniques: radiofrequency signals more or less greedy in terms of power consumption (NFC, WiFi received signal strength indicator (RSSI), 5G, Precise GPS, ...), external high frequency camera networks, *etc*. Considering an autonomous mobile robot and smart factory context, the need of accuracy and autonomy is combined with real-time constraints, *i.e.* to be able to rely on a dynamic/responsive global localization system and interactions support with other robot/vehicle (infrastructure specific equipment, like vehicle-to-infrastructure for example).

Dealing with the accuracy of embedded localization systems, many study results demonstrate the feasibility and, when required, the effectiveness of their solution.

For a smart phone localization application, it is difficult to obtain precise error estimation because of the lack of reliable ground truth. Moreover, the feasibility is often the main goal and not a precise accuracy.

Concerning the micro aerial vehicle, [154] estimated trajectory alignment to the ground truth and then computed the RMSE position error over the aligned trajectory. The complexity of the datasets varies in terms of trajectory length, light dynamics, and illumination conditions. The best results are obtained for VINS mono with loop closure with an average of 0.12 meters in absolute error translation when considering Up Board (64 bit Intel Atom CPU). This error reaches 0.16 meters for VINS mono on ODROID XU4 (Samsung Exynos 5422 system in an ARM big.Little architecture 4xARM A7 at 1.5 GHz and 4xARM A15 at 2.0 GHz). Reference [196] proposes their Navion chip solution. The first step deals with the migration from a PC machine; then to an ARM Cortex 15, and finally to Xilinx Kintex-7 XC7K355T FPGA) before targeting the ASIC process node [224]. Thus, the greediest modules in terms of processing time have been accelerated on FPGA (front-end, linear solver, linearization and marginalization). Jumping from the initial baseline on PC CPU to FPGA, the power consumption decreases from 28.2 W to 1.46 W. On EuRoC datasets, these optimizations only increases estimation error by 0.029 m, while sustaining an equivalent fps throughput. Globally, the estimation error remains under 0.25 meters. The Navion chip consumes 24 mW when switching to the ASIC 65 nm process node.

For automotive issues, [223] published a survey of the state-of-the-art localization techniques for autonomous vehicle applications. These results include a combination of on-board sensor-based and V2X techniques (V2V/V2I). The following localization techniques are considered relevant from a performance point of view. Reference [225] proposes a combination of localizing ground penetrating radar (LGPR), IMU and GPS reached a 0.04 m estimation error (RMSE). Reference [226] combined LiDAR, GPS, IMU, achieving an accuracy of 0.017 m in lat. and 0.033 m in long (RMSE). Reference [227] associates a camera and IMU for an accuracy of 0.14 m in lat. and 0.19 m in long (RMSE). Reference [228] considers on-board UHF antenna and reached an accuracy up to 0.03 m, lat. Finally, when considering a 5G communications device, [229] achieved an accuracy of 99% below 0.2 m at 100 MHz.

The section also shows that at the cost of hardware restrictions and algorithmic adaptation optimization, an acceptable accuracy can be achieved by the embedded systems. When combining with external positioning systems, a very precise location is even possible. In any case, the design challenge will lie in the implementation and synchronization of all available localization sources available.

8.5 Application Domains

This section presents different application domains where multimodal localization represents a crucial step for the robustness of the system. For each domain, multiple case studies are examined, highlighting the different types of sensors, the localization methods, as well as the computing architectures and the overall embedded system requirements and specifications.

Sect. 8.5.1 presents the scene mapping and cartography applications. Sect. 8.5.2 analyzes applications of pedestrian localization both in indoor and outdoor environments. Sect. 8.5.3 showcases the autonomous navigation application domain. Finally, Sect. 8.5.4 deals with two applications of mixed reality distributed between personal assistance and construction industry.

8.5.1 Scene Mapping

Currently, the mapping and cartography market is growing ever larger with the focus to cover areas quickly and to produce accurate and detailed 2D or 3D maps. Drones and UAV mapping are a powerful new tool that allows one to inspect hard-to-reach areas. They are useful for a wide range of activities by making possible real-time damage assessment and revealing defects, especially checking for the smallest issues, such as cracks, leaks and weak spots. Thus, unmanned aerial vehicles equipped with the adequate equipment are widely used instead of people for civil engineering inspection (*e.g.* viaducts), earthwork inspection (*e.g.* mining, quarries, carriers, rock retaining walls), power networks inspection, crime scene investigations, building inspections of roofs, archeology [230,231], chimneys, towers, stadiums, wind turbines, as well as most recently aircraft-inspection missions [232,233]. We revisit here some application domains, especially UAV-based aircraft-inspection system (Sect. 8.5.1.1), quite similar to the solutions for wind-turbine inspection, as well as an example of mapping drones (Sect. 8.5.1.2) that allows one to collect accurate spatial data. The latter can be exploited for accurate topographic mapping, pre-built and as-built surveying, drone-derived DEM creating, *etc.*

8.5.1.1 Aircraft inspection

Aviation quality is the key factor to ensure safe and sustainable air transport infrastructure. In order to reduce aircraft downtime and human errors, aircraft inspection by drones is a recent industry practice. Such a system allows one to accelerate and facilitate visual inspections, which are necessary to detect defects on the aircraft skin. This work is fundamental to decide if the plane is allowed to fly again. Typically, traditional aircraft visual inspection is performed from the ground or using a telescopic platform, which could last up to one day. In contrast, the UAVs inspection systems through acquired images perform quickly the same task and improve operator safety. The precise localization and obstacle detection for such systems are crucial to maintain safe distance to the parked plane and guarantee the anti-collision of the whole inspection system, while allowing close inspections if necessary. Furthermore, the pose estimation may be used by captured images, which are immediately geo-referenced to the aircraft structure, and then processed to localize and measure aircraft surface target damage. By using a drone, the operator can in real-time map, visualize, and get damage reporting for the inspected plane, and then flag visible defects for closer examination.

The UAVs are suitable to perform more accurate and low-cost inspections thanks to the high technology on which they are based, specifically LiDAR technology and machine learning. The first to have introduced this inspection technique are EasyJet and unmanned-aircraft developer Blue Bear Systems Research in 2015, by inspecting an Airbus A320 using the BlueBear's Riser drone. Since then the research has increasingly focused on this subject.

Figure 8.19 : Donecle aircraft-inspection-drone system [234].

Collaborating with UK company Output42, which provides software for 3D modeling and damage reporting, they developed the aircraft-inspection-drone version which is named remote automated plane inspection and dissemination (RAPID). The goal of this collaboration, known as a maintenance, repair and overhaul (MRO) drone, is to provide a drone-based inspection system.

Other aircraft-inspection-drone systems have been developed by AFI KLM E&M in partnership with start-up Donecle (Fig. 8.19). Their automated drone inspects regulatory stickers on the aircraft fuselage. Its localization in relation to the aircraft is computed thanks to LiDAR-based positioning technology and an uploaded digital model of the aircraft. To verify the quality of the stickers, an on-board intelligent camera captures high-resolution images in order to compare them to the reference database. Then, all defects on these technical markings are detected and located.

Also, the Testia subsidiary of Airbus will make available, in the fourth quarter of the year 2018, their own drone-based inspection system. This automated drone is equipped with an on-board camera to capture images of the aircraft surface, and LiDAR sensors for obstacle detection. Once visual damage on the jetliner's surface is localized and measured, it can be compared with the aircraft digital mock-up.

8.5.1.2 SenseFly eBee classic

The eBee Classic is one of the SenseFly UAV Systems. As illustrated in Fig. 8.20, it is a professional mapping drone, characterized by its light weight (only 0.7 kg), so it is considered one of the lightest UAV mapping systems on the market. The eBee UAV's maximum flight duration is 45 minutes, during which it captures images at a range of 1–10 km^2 for a single flight and with an accuracy of up to 5 cm.

Figure 8.20 : The SenseFly eBee Classic professional mapping drone.

It is equipped by a *SenseFly S.O.D.A*, which is the first device designed for professional photogrammetry of drones. Its principle is to capture exceptionally sharp aerial RGB images in a range of light conditions, permitting the user to produce detailed and lively orthomosaics and very accurate digital surface models. Plus *Sequoia* is the smallest, lightest multispectral Parrot sensor ever released. Principally, the eBee drone contains four 1.2 megapixel cameras, one 16 megapixel RGB camera with rolling shutter, four spectral sensors, GPS, IMU and a magnetometer. Maps and digital elevation models can be created using aerial images taken by the 16 megapixel RGB camera with a resolution of 3 cm. These maps and models are up to 5 cm accurate. SenseFly has also built the intuitive software "*eMotion 2*," which allows one to plan, simulate, follow and control the UAV trajectory, before and during the flight. Thanks to the artificial intelligence given to the SenseFly automatic pilot, IMU and GPS data are constantly analyzed, as well as all other aspects of flight. All system output data are recorded by the eBee automatic pilot, including measurement files and flight trajectory images, and they can be used for other applications. Furthermore, the data are directly compatible with Postflight Terra 3D-EB software, which offers the possibility of automatically processing geo-referenced orthophotomosaics and digital elevation models (DEMs) with an accuracy of 5 cm (relative accuracy). In order to improve accuracy, ground control points can be used.

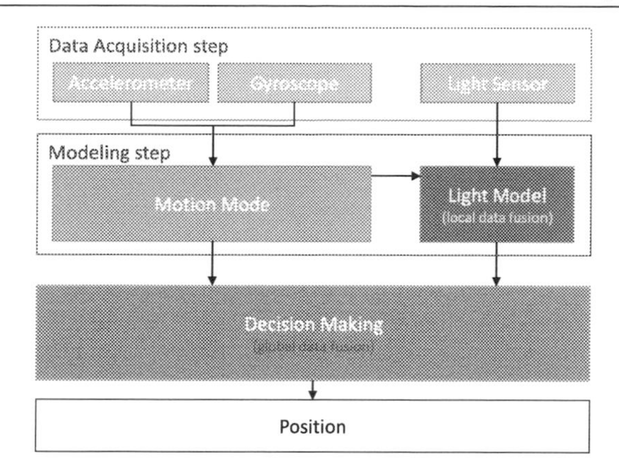

Figure 8.21 : A global system architecture of the indoor localization system discussed in [235].

8.5.2 *Pedestrian Localization*

Pedestrian localization has many applications ranging from commercial cases where stores use the location of their clients in order to push notifications on their smart phones relevant to the precise location in the store, to vital emergency cases where victims of accidents or crimes need to be precisely located. In all cases, the pedestrian localization needs to be precise and in real-time in order to be leveraged for such application.

This section examines two use cases of embedded multimodal localization used to get a real-time precise pedestrian location, whether in an indoor or in an outdoor setting.

8.5.2.1 Indoor localization in large-scale buildings

Localization systems have various application areas, including critical and constraining ones that require the use of rather embedded systems. For example, in [235], an indoor localization system is proposed to help pedestrians to locate themselves in large-scale buildings like hospitals. Implemented on an *ASUS Z00Ld* smart phone, this system is based on inertial sensors (gyroscope and accelerometer) as a dead reckoning system and a light sensor embedded in the smart phone. As most of the systems are based on the combination of different sensors and data fusion, the proposed system adopts the same general system architecture as shown in Fig. 8.21. It proceeds through the data acquisition step where the data is resampled and filtered in order to be exploited for the end of processing. Then comes the modeling step. In this case, this operation is illustrated by two models: the motion model and the light model. The motion model mainly concerns three functional components: the step detection, the heading direction estimation and the positioning. The step detection is managed by the accelerometer, in particular the accelerometer's z-axis, which is easily affected by the smart phone vibration caused by the pedestrian movement. For the second functional component it is handled by the gyroscope, the heading direction estimation is done thanks to the z-axis gyroscope data. Finally, for the positioning the authors use the PDR system (see Sect. 8.3.1.1) and the footpath approach by fusing locally the different pieces of information having been produced in the components before. The final position resulting from this step is used after making a decision regarding the location of the pedestrian. On the other hand, the light model aims to modify the pedestrian's position and correct the step length information; it is based on the light sensor and the step detection data fusion. This is a second local data fusion. It is followed by the decision making phase which represents the global data fusion process. The final correct and accurate position is calculated and produced in this step.

8.5.2.2 Precise localization of mobile devices in unknown environments

Other applications have taken advantage of the camera associated with the IMU in a smart phone to locate it in different environments under different conditions, and track it. As in [16], a localization system based on a loose IMU-camera coupling was proposed. As this is a loose coupling, the inertial data, coming from the IMU, and the visual data are processed differently. First, the IMU measurements are used to estimate the position and orientation of the smart phone. After modeling the IMU measurements and filtering them, they are used for the fusion filter prediction step used in this work, which is the EKF, relying on the fourth-order Runge–Kutta method. Then comes the camera data (image) processing. First of all, the features are detected using the ORB detector [236] because of its invariance to illumination and viewpoint changes as well as its detection rapidity. These features are then tracked by the two KLT algorithms [237,238] followed by detect-then-track, in order to allow the pose to be restored using epipolar geometry with a RANSAC method [239] for removing outliers.

8.5.3 Automotive Navigation

When talking about autonomous navigation for the general public, people most often think of the concept of autonomous driving and the changes that this evolution will bring about in their daily lives on the road. However, in reality the sector most concerned is the industrial world. The difficult tasks and the transport of materials/packages/heavy loads in the warehouse are nowadays carried out by a platform/robot that is either fixed or mobile. For the latter, the notion of autonomous navigation takes on its full meaning, to transport loads from point A to point B but also to be able to adapt to the risks of avoiding obstacles and/or operators. Other application domains such as tactical robots for the military or other purposes are also under development [240].

Autonomous navigation requires precise and robust localization in order to complete a reliable mission in complex scenarios. In fact, accurate and real-time localization of autonomous

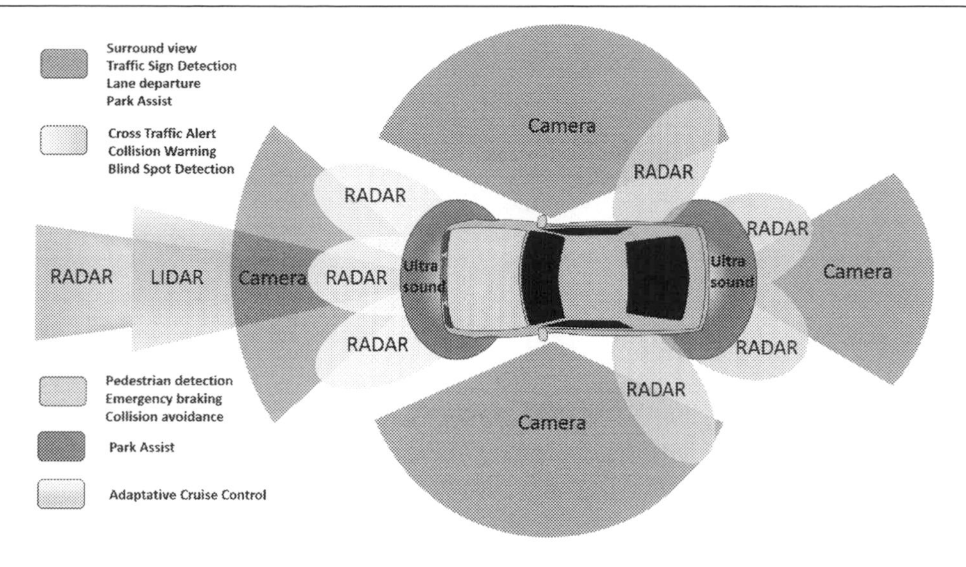

Figure 8.22 : Overview of integrated sensors for ADAS [243].

vehicles/robots is necessary for both decision making and path planning. This can only be achieved with multimodal localization in order to guarantee, not only precision, but also robustness of localization. This subsection explores the case of an embedded implementation for autonomous driving and smart factories.

8.5.3.1 Autonomous driving

Since the 2008 automotive crisis, the automotive market has recovered and 100M vehicles should be produced by 2019. [241] claimed that by 2045, 70% of all vehicles sold will integrate autonomous capabilities. By 2050, 5% of sold vehicle could be a level 5 autonomous car and personal cars could be replaced by "shared cars". These figures illustrate the global trends for car manufacturer. Thus, all companies and especially General Motors, Ford, Volkswagen, Daimler, Renault, BMW, Honda have developed their own autonomous model with different levels of maturity (from the prototype-based platform to a more advanced platform). In the next 20 years, those cars will integrate more than 30 sensors just for ADAS. Different sensors technologies are involved (see Fig. 8.22) and the number of cameras could go up to ten. The evolution of embedded computing resources follow the same path. Today, 50 electronic control units (ECUs) are built-in in a car performing door control, engine control, electric power steering control, human–machine interface (HMI), a power control module, *etc.* The processing architecture will evolve to integrate all the collected data and provide feedback on the actuators efficiently. Data management, bandwidth management, cross domain and flexibility will be the topics of concern of tomorrow. Finally, safety in the car and for road users (cyclist, pedestrians, *etc.*) is one of the major concerns for the car manufacturer. Thus, some sensor providers, such as MobileEye, propose cutting-edge products for perception awareness. Their latest vision-based device, called MobileEye Shield [242], includes several services: pedestrian and cyclist collision warning, forward collision warning, headway monitoring warning, lane departure warning and speed limit indicator. It should be noted that MobileEye, Intel and BMW work together inside a consortium; another hint of the necessity for IC founder, sensor and car manufacturer to collaborate to design efficient devices.

8.5.3.2 Smart factory

The concept of the industry 4.0 defines the next-gen factory. After mechanization, mass production assembly lines and computer and automation steps, today the fourth step of the industrial revolution deals with cyber physical systems (CPSs) that allow for the emergence of a digital, flexible simulation/logistics, and energy-efficient factory. Powerful and autonomous embedded systems are increasingly connected wirelessly to each other and to the internet. This leads to the convergence of the physical world and the virtual world (cyberspace) in the form of CPS.

For this purpose, autonomous mobile robots (AMR) become essential as a smart, mobile, autonomous and cooperative (robot/human) machine. Such robots have already emerged in warehouses such as Amazon facilities for example. Additionally to traditional robot system providers (Omron tech, Swiss Log, *etc.*), other actors develop their own technologies: Amazon Robotics, Fenwick Robotics, *etc.*

From on-board PC boards, systems are now evolving to a more compact and energy-efficient embedded architecture that enables SLAM capacity with LiDAR, use of stereocamera, IMU, IR and so on. Of course, interactions with local networks through radiofrequency media are also available.

8.5.4 Mixed Reality

Mixed reality (MR) is a term referring to a live direct or indirect view of a physical, real-world environment whose elements are augmented by computer-generated sensory input, such as sound, graphics, labels or 3D (animated) models. By way of simplification, some works speak of *augmented reality* instead, although the view of the surrounding scene is not direct (projection on a screen). Nowadays, the potential applications are numerous and concern various domains, for instance:

- medical, personal aid: "virtual x-ray" vision of the patient during surgery, assistance to disabled persons, *etc.*;

- engineering, construction and maintenance: overlaying dedicated data (sensor data, professional data, GIS information, service instruction diagrams, *etc.*) over large machinery, construction work sites or infrastructures;
- navigation and discovery: overlaying locations of friends and landmarks, visualizing routes;
- games, magic books, augmented archaeological sites and museums for tourism, multi-party videoconferencing around a virtual table, remote collaboration on a virtual model, *etc.*

MR requires the control of several main steps to ensure a robust system, where the localization step is central:

- the localization of the device, by precise pose estimation, to be able to register real and virtual information in detail, which can be potentially in 3D;
- tracking, to be able to maintain the right position and orientation of the virtual information despite the motion of the device, added to a potential relocalization in case of drift;
- registration, which concerns the alignment of real and virtual information.

We illustrate the exploitation of embedded systems for MR, with a focus on the key step of localization, through two domains presented in the two following sections.

8.5.4.1 Virtual cane system for visually impaired individuals

In the context of person assistance, an application particularly designed for visually impaired people is a real-time virtual blind cane system [244,245]. The solutions mentioned have similar characteristics: they combine three sensors in a loosely-coupled way, a camera, an IMU and a laser. Implementation [245] is done on a CMOS image sensor (CMOS camera) and it is implemented on a *Xilinx ZYNQ-7000* FPGA device. In this work, the FPGA accelerates the laser stripe detection by processing the images captured by the camera, using the centroid method. The distance of obstacles with respect to the camera is calculated by the triangulation using the laser stripe center position and the camera parameters. Thus, inertial data are integrated at this step, using the Kalman filter algorithm; the pose of the system is calculated and used to determine the distance to obstacles with respect to the user's body. The use of FPGA acceleration makes the system reactive in real-time. The H2020 INSPEX project [246] proposes to integrate a device plugged on a white cane (see Fig. 8.23) for visually impaired and blind (VIB) people, to detect an obstacle in agreement with the person's height, to provide audio feedback about harmful obstacles, to improve their mobility confidence and to reduce injuries. The device is planned to offer a "safety cocoon" to its user. The proposal goes beyond electronic white canes such as [247]. Their software platform is able to co-integrate different sensors, short- and long-range LiDAR [248], RF UWB radar [249] and ultrasonic information.

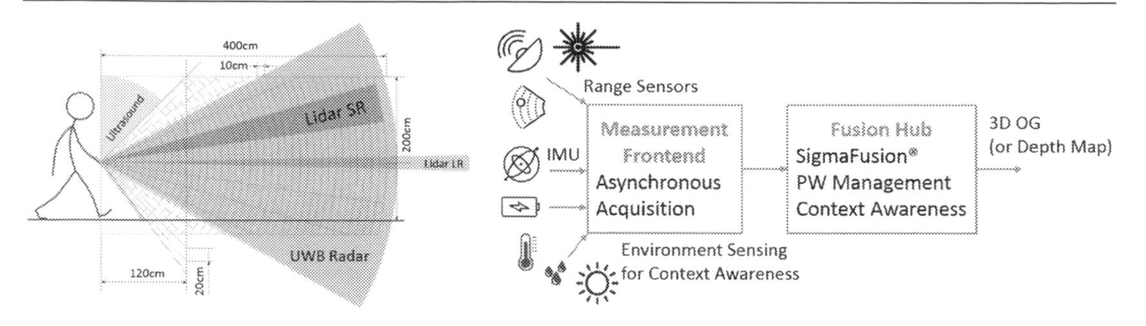

Figure 8.23 : Overview of the real-time virtual blind cane system proposed in [246].

8.5.4.2 Engineering, construction and maintenance

Mixed reality on embedded systems has seen the emergence of several industrial applications impacting different major sectors like smart cities, aeronautics, rail or naval sectors. For a smart city for example, they have a large panel of business-specific declinations, such as local authorities, urban planning, network managers, construction and public works markets and transport infrastructure and operators. Linked to the development of building information modeling (BIM) in the construction sector, several service providers, specialized in project management assistance in the follow-up of construction sites, have become involved to jointly exploit SLAM and BIM software tools (for example Autodesk Revit, Graphisoft ARCHICAD, Bentley Microstation solution) to offer enhanced visualization solutions in situ. Many companies in the construction sector are now involved; one may mention Microsoft in partnership with Trimble (positional technology, 3D modeling) and AECOM (one of the world's largest engineering firms) to bring HoloLens AR to their projects.

Another example is DAQRI, an American company developing products based on an augmented reality wearable technology product, DAQRI smart glasses, designed for the industry. This device becomes a wireless IoT device that connects with Wi-Fi and Bluetooth. High performance optical sensors include a wide-angle tracking camera, a depth-sensing camera, and an HD color camera for taking photos and videos. Advanced, hardware-accelerated computer-vision technology combines SLAM and sparse mapping technology to deliver persistent and reliable position tracking across large spaces. Users can "pin" an MR model to any point and it will remain in the right spatial position and orientation, regardless of user movement over large distances, such as an entire manufacturing facility or construction site.

A last example concerns the European research project LARA ("LBS augmented reality assistive system for utilities infrastructure management through Galileo and EGNOS", called H2020 – Galileo 1st Call) [250], dedicated to MR visualization of underground infrastructures and networks on mobile devices, based on very fine GPS positioning. A hand-held, low-cost,

Figure 8.24 : Illustration of the European project LARA mobile device (photography from [250]).

mobile device is targeted, with integration of state-of-the-art technologies in the domain of positioning and sensors (GNSS), computer graphics, AR and 3D GIS geo-databases [251, 252]. Fig. 8.24 illustrates the mobile device targeted in the project.

8.6 Conclusion

This chapter explored the state of the art of multimodal localization for embedded systems. In order to study the different requirements of multimodal localization in the embedded computing world, it started, see Sect. 8.2, by exploring the multiple inputs that are currently used in such embedded systems, namely, the different positioning systems and perception sensors. This exploration showed a plethora of heterogeneous data which have to be processed, analyzed and potentially combined with different strategies of fusion. The most widespread sensor is the GNSS, but we observed that several scenarios from today, *e.g.* indoor localization or urban navigation, make it often unusable. Many other sensors can be employed and combined successfully, as highlighted in Sect. 8.3.

In particular, the literature on localization highlights the huge development and relevance of approaches involving an analysis of the scene based on cameras and LiDAR these last years, because of the development of low-cost sensors. For indoor localization applications, systems that fuse INS with visual data seem to be the most widespread. To the contrary, for outdoor environments, the tendency is to combine more modalities to ensure system reliability and continuous absolute pose when GNSS fails, having the urban-canyoning effect or under unfavorable weather conditions, *e.g.* snow, which currently challenge image-based techniques. More generally speaking, an in-depth evaluation of multimodal localization solutions is very difficult to conduct globally. Evaluation already performed reveals that the accuracy achieved highly depends on the information sources utilized as well as on specificities of the environments chosen to perform the tests, making the rich panel of proposals difficult to compare because often evaluated on different benchmarks. The quality metrics used to judge the overall performance and robustness of localization system often differ between one approach and another, which makes them often incomparable, in addition to the fact that it may also be due to the difficulty of having a ground truth to judge of the precision of the localization obtained. Recent trends in deep learning newly address the problems of visual-based localization in challenging appearance conditions and of the learning of missing modalities for localization. These researches are on-going; they promise to further improve the quality of localization under even severe conditions, while relying on increasingly varied geo-referenced datasets.

The relevancy of different methods of localization was then analyzed in Sect. 8.4 under consideration of the prism of embedded systems constraints, whether relative to application domains or to hardware factors. In order to achieve this goal, the different types of embedded computing architectures were studied and multiple types of embedded systems with heterogeneous needs of localization were exposed. This led to a categorization of the different requirements of localization, depending on the types of embedded systems and the corresponding sensors used in these systems. This section has highlighted the interest and viability of embedded systems for multimodal localization. Nowadays, ARM-based multiprocessors on a chip are the most widely used general purpose processors due to their energy efficiency. Generalist computer chips remain essential for performing the least specific tasks, but various dedicated processing modules appeared on the market during the year. They allow for both the aggregation of data from heterogeneous sensors and tasks such as neural network classifications and computer-vision processing. Obviously, the more sensors are needed, the more complex is the processing to perform data integration and fusion steps, *i.e.* an embedded platform with safety features will require powerful computing architecture and additional reliability resources and, by the way, larger and more distributed solutions. We also observed that at the cost of hardware restrictions and algorithmic adaptation optimization, an acceptable accuracy can be achieved by the embedded systems.

Finally, in Sect. 8.5 we have presented a representative palette of the various application domains where embedded localization is either indispensable or where it is a tool that opens the way to new services. For several use cases, we have exhibited the heterogeneous input, the multimodal localization methods, the computing hardware, and the embedded constraints relative to each of these application domains.

References

[1] G. Artese, A. Trecroci, Calibration of a low cost MEMS INS sensor for an integrated navigation system, The International Archives of the Photogrammetry, Remote Sensing and Spatial Information Sciences (2008) 877–882.

[2] Q. Cai, G. Yang, N. Song, Y. Liu, Systematic calibration for ultra-high accuracy inertial measurement units, Sensors 16 (6) (2016) 940.

[3] B. Bidikar, G.S. Rao, L. Ganesh, M.S. Kumar, Satellite clock error and orbital solution error estimation for precise navigation applications, Positioning 5 (01) (2014) 22.
[4] B.J. Madeja, M. Kubíček, Development of High Resolution RGB Camera, Master's thesis, Brno University of Technology, Faculty of Electrical Engineering and Communication, Brno, Czech Republic, 2017.
[5] L. Millet, et al., A 5500fps 85GOPS/W 3D stacked BSI vision chip based on parallel in-focal-plane acquisition and processing, in: IEEE Symposium on VLSI Circuits, Honolulu, HI, 2018, pp. 245–246.
[6] L. Lee, P. Hui, Interaction methods for smart glasses: a survey, IEEE Access 6 (2018) 28712–28732, https://doi.org/10.1109/ACCESS.2018.2831081.
[7] Nancy Owano, Meta Glasses to Place Virtual Reality Worlds at Fingertips (w/ Video), Tech. rep., PHYS.ORG, 2013, https://phys.org/pdf288096606.pdf.
[8] E. Mueggler, B. Huber, D. Scaramuzza, Event-based, 6-DOF pose tracking for high-speed maneuvers, in: Intelligent Robots and Systems, 2014 IEEE/RSJ International Conference on, IROS 2014, IEEE, 2014, pp. 2761–2768.
[9] PROPHESEE, PROPHESEE reveals ONBOARD, the most advanced event-based reference system combining for the first time Prophesee's patented vision sensor and AI algorithms, https://www.prophesee.ai/onboard/, 2018.
[10] A. Group, Perimeter protection devices around machinery, https://blaxtair.com/the-blaxtair-pedestrian-machinery-anti-collision-camera, 2018.
[11] F. Alhwarin, A. Ferrein, I. Scholl, IR stereo kinect: improving depth images by combining structured light with IR stereo, in: D.-N. Pham, S.-B. Park (Eds.), PRICAI 2014: Trends in Artificial Intelligence, Springer International Publishing, Cham, 2014, pp. 409–421.
[12] H.F. Durrant-Whyte, Sensor models and multisensor integration, The International Journal of Robotics Research 7 (6) (1988) 97 113.
[13] W. Elmenreich, Sensor Fusion in Time-Triggered Systems, Ph.D. thesis, Institut fur Technische Informatik, Vienna University of Technology, Vienna, Austria, 2002.
[14] B. Chandrasekaran, S. Gangadhar, J.M. Conrad, A survey of multisensor fusion techniques, architectures and methodologies, in: SoutheastCon, 2017, IEEE, 2017, pp. 1–8.
[15] R.R. Brooks, S.S. Iyengar, Multi-sensor Fusion: Fundamentals and Applications with Software, Prentice-Hall, Inc., 1998.
[16] W. Fang, L. Zheng, H. Deng, A motion tracking method by combining the F and camera in mobile devices, in: Sensing Technology, 2016 10th International Conference on, ICST, IEEE, 2016, pp. 1–6.
[17] G. Zhou, J. Ye, W. Ren, T. Wang, Z. Li, On-board inertial-assisted visual odometer on an embedded system, in: Robotics and Automation, 2014 IEEE International Conference on, ICRA, IEEE, 2014, pp. 2602–2608.
[18] J. Nikolic, J. Rehder, M. Burri, P. Gohl, S. Leutenegger, P.T. Furgale, R. Siegwart, A synchronized visual–inertial sensor system with FPGA pre-processing for accurate real-time SLAM, in: Robotics and Automation, 2014 IEEE International Conference on, ICRA, IEEE, 2014, pp. 431–437.
[19] Z. Zhang, S. Liu, G. Tsai, H. Hu, C.-C. Chu, F. Zheng, PIRVS: An Advanced Visual–Inertial SLAM System with Flexible Sensor Fusion and Hardware Co-design, 2018, pp. 1–7.
[20] S. Shen, N. Michael, V. Kumar, Tightly-coupled monocular visual–inertial fusion for autonomous flight of rotorcraft MAVs, in: Robotics and Automation, 2015 IEEE International Conference on, ICRA, IEEE, 2015, pp. 5303–5310.
[21] M. Hwangbo, J.-S. Kim, T. Kanade, Gyro-aided feature tracking for a moving camera: fusion, auto-calibration and GPU implementation, The International Journal of Robotics Research 30 (14) (2011) 1755–1774.
[22] H. Durrant-Whyte, Introduction to Decentralised Data Fusion, Australian Centre for Field Robotics, The University of Sydney NSW, Australia, 2006, https://www.acfr.usyd.edu.au/pdfs/training/decentDataFusion/notes.pdf.
[23] A. Puengnim, L. Patino-Studencka, J. Thielecke, G. Rohmer, Precise positioning for virtually synchronized pseudolite system, in: Indoor Positioning and Indoor Navigation, 2013 International Conference on, IPIN, IEEE, 2013, pp. 1–8.

[24] R. Xu, W. Chen, Y. Xu, S. Ji, A new indoor positioning system architecture using GPS signals, Sensors 15 (5) (2015) 10074–10087.

[25] C. Medina, J.C. Segura, A. De la Torre, Ultrasound indoor positioning system based on a low-power wireless sensor network providing sub-centimeter accuracy, Sensors 13 (3) (2013) 3501–3526.

[26] D. Ruiz, E. García, J. Ureña, D. de Diego, D. Gualda, J.C. García, Extensive ultrasonic local positioning system for navigating with mobile robots, in: Positioning Navigation and Communication, 2013 10th Workshop on, WPNC, IEEE, 2013, pp. 1–6.

[27] J.H. Oh, D. Kim, B.H. Lee, An indoor localization system for mobile robots using an active infrared positioning sensor, Journal of Industrial and Intelligent Information 2 (1) (2014) 35–38.

[28] D. Hauschildt, N. Kirchhof, Advances in thermal infrared localization: challenges and solutions, in: Indoor Positioning and Indoor Navigation, 2010 International Conference on, IPIN, IEEE, 2010, pp. 1–8.

[29] H.-H. Liu, W.-H. Lo, C.-C. Tseng, H.-Y. Shin, A WiFi-based weighted screening method for indoor positioning systems, Wireless Personal Communications 79 (1) (2014) 611–627.

[30] Z.-A. Deng, G. Wang, D. Qin, Z. Na, Y. Cui, J. Chen, Continuous indoor positioning fusing WiFi, smartphone sensors and landmarks, Sensors 16 (9) (2016) 1427.

[31] S. Lee, B. Koo, M. Jin, C. Park, M.J. Lee, S. Kim, Range-free indoor positioning system using smartphone with bluetooth capability, in: Position, Location and Navigation Symposium, PLANS 2014, 2014 IEEE/ION, IEEE, 2014, pp. 657–662.

[32] Y. Zhuang, J. Yang, Y. Li, L. Qi, N. El-Sheimy, Smartphone-based indoor localization with bluetooth low energy beacons, Sensors 16 (5) (2016) 596.

[33] S. Gezici, Z. Tian, G.B. Giannakis, H. Kobayashi, A.F. Molisch, H.V. Poor, Z. Sahinoglu, Localization via ultra-wideband radios: a look at positioning aspects for future sensor networks, IEEE Signal Processing Magazine 22 (4) (2005) 70–84.

[34] S.-H. Fang, C.-H. Wang, T.-Y. Huang, C.-H. Yang, Y.-S. Chen, An enhanced zigbee indoor positioning system with an ensemble approach, IEEE Communications Letters 16 (4) (2012) 564–567.

[35] N. Samama, Global positioning: technologies and performance, in: Wiley Survival Guides in Engineering and Science, Wiley-Interscience, 2008, https://hal.archives-ouvertes.fr/hal-01398442.

[36] R.E. Hattach, J. Erfanian, Ngmn 5g White Paper, Next Generation Mobile Networks Alliance, 2015.

[37] E.J. Lefferts, F.L. Markley, M.D. Shuster, Kalman filtering for spacecraft attitude estimation, Journal of Guidance, Control, and Dynamics 5 (5) (1982) 417–429.

[38] P.M. Kachmar, L. Wood, Space navigation applications, Journal of the Institute of Navigation 42 (1995) 187–234, https://doi.org/10.1002/j.2161-4296.1995.tb02335.x.

[39] S.K. Park, Y.S. Suh, A zero velocity detection algorithm using inertial sensors for pedestrian navigation systems, Sensors 10 (10) (2010) 9163–9178.

[40] A. Peng, L. Zheng, W. Zhou, C. Yan, Y. Wang, X. Ruan, B. Tang, H. Shi, H. Lu, H. Zheng, Foot-mounted indoor pedestrian positioning system based on low-cost inertial senors, in: Proceedings of the 2016 International Conference on Indoor Positioning and Indoor Navigation, IPIN, Alcalá de Henares, Spain, 2016, pp. 4–7.

[41] P. Chen, Y. Kuang, X. Chen, A UWB/improved PDR integration algorithm applied to dynamic indoor positioning for pedestrians, Sensors 17 (9) (2017) 2065.

[42] M. Kourogi, T. Kurata, A method of pedestrian dead reckoning for smartphones using frequency domain analysis on patterns of acceleration and angular velocity, in: 2014 IEEE/ION Position, Location and Navigation Symposium, PLANS 2014, 2014, pp. 164–168.

[43] F. Gu, K. Khoshelham, J. Shang, F. Yu, Z. Wei, Robust and accurate smartphone-based step counting for indoor localization, IEEE Sensors Journal 17 (11) (2017) 3453–3460.

[44] B. Wang, X. Liu, B. Yu, R. Jia, X. Gan, Pedestrian dead reckoning based on motion mode recognition using a smartphone, Sensors 18 (6) (2018) 1811, https://doi.org/10.3390/s18061811, http://www.mdpi.com/1424-8220/18/6/1811.

[45] Sunghee Choi, Jinhyun Do, Boyeon Hwang, Jangmyung Lee, Static attitude control for underwater robots using multiple ballast tanks, IEEJ Transactions on Electrical and Electronic Engineering 9 (S1) (2014) S49–S55.
[46] A. Rossi, M. Pasquali, M. Pastore, Performance analysis of an inertial navigation algorithm with DVL auto-calibration for underwater vehicle, in: 2014 DGON Inertial Sensors and Systems, ISS, 2014, pp. 1–19.
[47] A. Filippeschi, N. Schmitz, M. Miezal, G. Bleser, E. Ruffaldi, D. Stricker, Survey of motion tracking methods based on inertial sensors: a focus on upper limb human motion, Sensors 17 (6) (2017) 1257.
[48] A. Rietdorf, C. Daub, P. Loef, Precise positioning in real-time using navigation satellites and telecommunication, in: Proceedings of the 3rd Workshop on Positioning, Navigation and Communication, WPNC'06, 2006, pp. 123–128.
[49] Y. Morales, T. Tsubouchi, DGPS, RTK-GPS and StarFire DGPS Performance Under Tree Shading Environments, 2007, pp. 519–524.
[50] B. Li, J. Zhang, P. Mumford, A.G. Dempster, How good is assisted GPS?, in: Proceedings of the International GNSS Society Symposium, IGNSS 2011, 2011, Citeseer.
[51] J. Brejcha, M. Cadik, GeoPose3K: mountain landscape dataset for camera pose estimation in outdoor environments, Image and Vision Computing 66 (2017) 1–14, https://doi.org/10.1016/j.imavis.2017.05.009, http://linkinghub.elsevier.com/retrieve/pii/S0262885617300963.
[52] A.R. Zamir, A. Hakeem, L. Van Gool, M. Shah, R. Szeliski, Large-Scale Visual Geo-Localization, 2016.
[53] N. Piasco, D. Sidibé, C. Demonceaux, V. Gouet-Brunet, A survey on visual-based localization: on the benefit of heterogeneous data, Pattern Recognition 74 (2018) 90–109, https://doi.org/10.1016/j.patcog.2017.09.013.
[54] T. Sattler, W. Maddern, A. Torii, J. Sivic, T. Pajdla, M. Pollefeys, M. Okutomi, Benchmarking 6DOF urban visual localization in changing conditions, in: IEEE Conference on Computer Vision and Pattern Recognition, CVPR 2018, 2018.
[55] E. Garcia-Fidalgo, A. Ortiz, Vision-based topological mapping and localization methods: a survey, Robotics and Autonomous Systems 64 (2015) 1–20, https://doi.org/10.1016/j.robot.2014.11.009.
[56] K. Mikolajczyk, C. Schmid, Scale & affine invariant interest point detectors, International Journal of Computer Vision 60 (1) (2004) 63–86.
[57] D.G. Lowe, Distinctive image features from scale-invariant keypoints, International Journal of Computer Vision 60 (2) (2004) 91–110.
[58] N. Dalal, B. Triggs, Histograms of oriented gradients for human detection, in: Proceedings of the IEEE Conference on Computer Vision and Pattern Recognition, vol. 1, CVPR, IEEE, 2005, pp. 886–893.
[59] C. McManus, B. Upcroft, P. Newmann, Scene signatures: localised and point-less features for localisation, in: Robotics: Science and Systems X, University of California, Berkeley, CA, 2014, https://eprints.qut.edu.au/76158/.
[60] B. Morago, G. Bui, Y. Duan, 2D matching using repetitive and salient features in architectural images, IEEE Transactions on Image Processing 7149 (2016) 1–12, https://doi.org/10.1109/TIP.2016.2598612.
[61] H.J. Kim, E. Dunn, J.-M. Frahm, Predicting good features for image geo-localization using per-bundle VLAD, in: Proceedings of the IEEE International Conference on Computer Vision, ICCV, 11-18-Dece, 2015, pp. 1170–1178.
[62] J. Matas, O. Chum, M. Urban, T. Pajdla, Robust wide-baseline stereo from maximally stable extremal regions, Image and Vision Computing 22 (10) (2004) 761–767.
[63] M. Paulin, M. Douze, Z. Harchaoui, J. Mairal, F. Perronnin, C. Schmid, Local convolutional features with unsupervised training for image retrieval, in: Proceedings of the IEEE International Conference on Computer Vision, ICCV, 11-18-Dece, 2015, pp. 91–99.
[64] J.L. Schönberger, H. Hardmeier, T. Sattler, M. Pollefeys, Comparative Evaluation of Hand-Crafted and Learned Local Features, January 2017, pp. 6959–6968.
[65] C. Arth, C. Pirchheim, J. Ventura, D. Schmalstieg, V. Lepetit, Instant outdoor localization and SLAM initialization from 2.5D maps, IEEE Transactions on Visualization and Computer Graphics 21 (11) (2015) 1309–1318, https://doi.org/10.1109/TVCG.2015.2459772.

[66] B.C. Russell, J. Sivic, J. Ponce, H. Dessales, Automatic alignment of paintings and photographs depicting a 3D scene, in: Proceedings of the IEEE International Conference on Computer Vision Workshop, ICCVW, 2011.
[67] A. Oliva, A. Torralba, Modeling the shape of the scene: a holistic representation of the spatial envelope, International Journal of Computer Vision 42 (3) (2001) 145–175, https://doi.org/10.1023/A:1011139631724.
[68] C. Azzi, D. Asmar, A. Fakih, J. Zelek, Filtering 3D keypoints using GIST for accurate image-based localization, in: British Machine Vision Conference, No. 2, BMVC, 2016, pp. 1–12.
[69] X. Wan, J. Liu, H. Yan, G.L.K. Morgan, Illumination-invariant image matching for autonomous UAV localisation based on optical sensing, ISPRS Journal of Photogrammetry and Remote Sensing 119 (2016) 198–213, https://doi.org/10.1016/j.isprsjprs.2016.05.016.
[70] A. Babenko, A. Slesarev, A. Chigorin, V. Lempitsky, Neural codes for image retrieval, in: Proceedings of the IEEE European Conference on Computer Vision, ECCV, 2014, pp. 584–599.
[71] R. Arandjelović, P. Gronat, A. Torii, T. Pajdla, J. Sivic, NetVLAD: CNN architecture for weakly supervised place recognition, IEEE Transactions on Pattern Analysis and Machine Intelligence 40 (6) (2018) 1437–1451, https://doi.org/10.1109/TPAMI.2017.2711011.
[72] A. Gordo, J. Almazán, J. Revaud, D. Larlus, End-to-end learning of deep visual representations for image retrieval, International Journal of Computer Vision 124 (2) (2017) 237–254, https://doi.org/10.1007/s11263-017-1016-8.
[73] H.J. Kim, E. Dunn, J. Frahm, Learned contextual feature reweighting for image geo-localization, in: 2017 IEEE Conference on Computer Vision and Pattern Recognition, CVPR, 2017, pp. 3251–3260.
[74] F. Radenović, G. Tolias, O. Chum, CNN image retrieval learns from BoW: unsupervised fine-tuning with hard examples, in: B. Leibe, J. Matas, N. Sebe, M. Welling (Eds.), Computer Vision, ECCV 2016, Springer International Publishing, Cham, 2016, pp. 3–20.
[75] Q. Liu, R. Li, H. Hu, D. Gu, Extracting semantic information from visual data: a survey, Robotics 5 (1) (2016) 8, https://doi.org/10.3390/robotics5010008, http://www.mdpi.com/2218-6581/5/1/8/htm.
[76] I. Kostavelis, A. Gasteratos, Semantic mapping for mobile robotics tasks: a survey, Robotics and Autonomous Systems 66 (2015) 86–103, https://doi.org/10.1016/j.robot.2014.12.006.
[77] Y. LeCun, Y. Bengio, G. Hinton, Deep learning, Nature 521 (7553) (2015) 436.
[78] X. Qu, B. Soheilian, N. Paparoditis, Landmark based localization in urban environment, in: Geospatial Computer Vision, ISPRS Journal of Photogrammetry and Remote Sensing 140 (2018) 90–103, https://doi.org/10.1016/j.isprsjprs.2017.09.010, http://www.sciencedirect.com/science/article/pii/S0924271617302228.
[79] O. Saurer, G. Baatz, K. Köser, L. Ladicky, M. Pollefeys, Image based geo-localization in the alps, International Journal of Computer Vision 116 (3) (2016) 213–225, https://doi.org/10.1007/s11263-015-0830-0.
[80] N. Sünderhauf, S. Shirazi, F. Dayoub, B. Upcroft, M.J. Milford, On the performance of ConvNet features for place recognition, in: 2015 IEEE/RSJ International Conference on Intelligent Robots and Systems, IROS, 2015-Decem, 2015, pp. 4297–4304.
[81] T. Weyand, I. Kostrikov, J. Philbin, Planet – photo geolocation with convolutional neural networks, in: ECCV, vol. 9912, 2016, pp. 37–55.
[82] M. Maimone, Y. Cheng, L. Matthies, Two years of visual odometry on the mars exploration rovers: field reports, Journal of Field Robotics 24 (3) (2007) 169–186, https://doi.org/10.1002/rob.v24:3.
[83] C. Forster, M. Pizzoli, D. Scaramuzza, SVO: fast semi-direct monocular visual odometry, in: 2014 IEEE International Conference on Robotics and Automation, ICRA, 2014, pp. 15–22.
[84] C. Cadena, L. Carlone, H. Carrillo, Y. Latif, D. Scaramuzza, J. Neira, I. Reid, J. Leonard, Past, present, and future of simultaneous localization and mapping: towards the robust-perception age, IEEE Transactions on Robotics 32 (6) (2016) 1309–1332.
[85] G. Klein, D. Murray, Parallel tracking and mapping for small ar workspaces, in: Proceedings of the 2007 6th IEEE and ACM International Symposium on Mixed and Augmented Reality, ISMAR '07, IEEE Computer Society, Washington, DC, USA, 2007, pp. 1–10.

[86] R. Mur-Artal, J.M.M. Montiel, J.D. Tardos, ORB-SLAM: a versatile and accurate monocular SLAM system, IEEE Transactions on Robotics 31 (5) (2015) 1147–1163.
[87] R. Mur-Artal, J.D. Tardós, ORB-SLAM2: an open-source SLAM system for monocular, stereo, and RGB-D cameras, IEEE Transactions on Robotics 33 (2017) 1255–1262.
[88] T. Owen, Three-Dimensional Computer Vision: A Geometric Viewpoint by Olivier Faugeras, The MIT Press, London, UK, ISBN 0-262-06158-9, 1993, 663 pages incl index, vol. 12, Cambridge University Press, 1994.
[89] A. Irschara, C. Zach, J. Frahm, H. Bischof, From structure-from-motion point clouds to fast location recognition, in: 2009 IEEE Computer Society Conference on Computer Vision and Pattern Recognition Workshops, CVPR Workshops 2009, 2009 IEEE Computer Society Conference on Computer Vision and Pattern Recognition, IEEE Computer Society, 2009, pp. 2599–2606.
[90] T. Sattler, B. Leibe, L. Kobbelt, Efficient & effective prioritized matching for large-scale image-based localization, IEEE Transactions on Pattern Analysis and Machine Intelligence 39 (9) (2017) 1744–1756.
[91] W. Wan, Z. Liu, K. Di, B. Wang, J. Zhou, A cross-site visual localization method for yutu rover, The International Archives of the Photogrammetry, Remote Sensing and Spatial Information Sciences XL-4 (May 2014) 279–284, https://doi.org/10.5194/isprsarchives-XL-4-279-2014, http://www.int-arch-photogramm-remote-sens-spatial-inf-sci.net/XL-4/279/2014/.
[92] W. Fröstner, B.P. Wrobel, Photogrammetric Computer Vision: Statistics, Geometry, Orientation and Reconstruction, first edition, Springer Publishing Company, Incorporated, 2016.
[93] X. Qu, B. Soheilian, E. Habets, N. Paparoditis, Evaluation of SIFT and SURF for vision based localization, The International Archives of the Photogrammetry, Remote Sensing and Spatial Information Sciences 41 (July 2016) 685–692, https://doi.org/10.5194/isprsarchives-XLI-B3-685-2016.
[94] D.P. Paudel, A. Habed, C. Demonceaux, P. Vasseur, Robust and optimal sum-of-squares-based point-to-plane registration of image sets and structured scenes, in: Proceedings of the IEEE International Conference on Computer Vision, ICCV, 2015, pp. 2048–2056.
[95] Ceyhun Celik, Hasan Sakir Bilge, Content based image retrieval with sparse representations and local feature descriptors: a comparative study, Pattern Recognition 68 (2017) 1–13, https://doi.org/10.1016/j.patcog.2017.03.006.
[96] L. Zheng, Y. Yang, Q. Tian, SIFT meets CNN: a decade survey of instance retrieval, IEEE Transactions on Pattern Analysis and Machine Intelligence 40 (5) (2018) 1224–1244, https://doi.org/10.1109/TPAMI.2017.2709749.
[97] Y. Jing, M. Covell, D. Tsai, J.M. Rehg, Learning query-specific distance functions for large-scale web image search, IEEE Transactions on Multimedia 15 (2013) 2022–2034.
[98] Y. Song, X. Chen, X. Wang, Y. Zhang, J. Li, 6-DOF image localization from massive geo-tagged reference images, IEEE Transactions on Multimedia 18 (8) (2016) 1542–1554.
[99] A. Kendall, M. Grimes, R. Cipolla, Posenet: a convolutional network for real-time 6-dof camera relocalization, in: Proceedings of the IEEE International Conference on Computer Vision, 2015, pp. 2938–2946.
[100] A. Kendall, R. Cipolla, et al., Geometric loss functions for camera pose regression with deep learning, in: Proc. CVPR, vol. 3, 2017, p. 8.
[101] J. Shotton, B. Glocker, C. Zach, S. Izadi, A. Criminisi, A. Fitzgibbon, Scene coordinate regression forests for camera relocalization in RGB-D images, in: Proceedings of the IEEE Conference on Computer Vision and Pattern Recognition, CVPR, 2013, pp. 2930–2937.
[102] J. Valentin, A. Fitzgibbon, M. Nießner, J. Shotton, P.H.S. Torr, Exploiting uncertainty in regression forests for accurate camera relocalization, in: Proceedings of the IEEE Conference on Computer Vision and Pattern Recognition, CVPR, 2015, pp. 4400–4408.
[103] P.J. Besl, N.D. McKay, A method for registration of 3-D shapes, IEEE Transactions on Pattern Analysis and Machine Intelligence 14 (2) (1992) 239–256, https://doi.org/10.1109/34.121791.
[104] F. Lu, E. Milios, Robot pose estimation in unknown environments by matching 2D range scans, Journal of Intelligent & Robotic Systems 18 (3) (1997) 249–275, https://doi.org/10.1023/A:1007957421070.

[105] M. Tomono, A scan matching method using euclidean invariant signature for global localization and map building, in: IEEE International Conference on Robotics and Automation, 2004, Proceedings, vol. 1, ICRA '04, 2004, pp. 866–871.
[106] G. Weiß, E. von Puttkamer, A Map Based on Laserscans Without Geometric Interpretation, PIOS Press, 1995, pp. 403–407.
[107] J.W.K.-W. Jorg, E. von Puttkamer, APR-Global Scan Matching Using Anchor Point Relationships, IOS Press, 2000, p. 471.
[108] Y. Gao, S. Liu, M.M. Atia, A. Noureldin, INS/GPS/LiDAR integrated navigation system for urban and indoor environments using hybrid scan matching algorithm, Sensors 15 (9) (2015) 23286–23302, https://doi.org/10.3390/s150923286, http://www.mdpi.com/1424-8220/15/9/23286.
[109] M. Bosse, J. Roberts, Histogram matching and global initialization for laser-only SLAM in large unstructured environments, in: Proceedings 2007 IEEE International Conference on Robotics and Automation, 2007, pp. 4820–4826.
[110] E.B. Olson, Real-time correlative scan matching, in: IEEE Intern. Conf. on Robotics and Automation, ICRA, 2009, pp. 4387–4393.
[111] Jaromir Konecny, M. Prauzek, J. Hlavica, SLAM algorithm based on cross-correlation scan matching, in: Proceedings of the 8th International Conference on Signal Processing Systems, ICSPS 2016, ACM, New York, NY, USA, 2016, pp. 89–93, http://doi.acm.org/10.1145/3015166.3015184.
[112] A. Censi, L. Iocchi, G. Grisetti, Scan matching in the hough domain, in: Proceedings of the 2005 IEEE International Conference on Robotics and Automation, 2005, pp. 2739–2744.
[113] P. Biber, W. Strasser, The normal distributions transform: a new approach to laser scan matching, in: Proceedings 2003 IEEE/RSJ International Conference on Intelligent Robots and Systems (Cat. No. 03CH37453), vol. 3, IROS 2003, 2003, pp. 2743–2748.
[114] A. Burguera, Y. González, G. Oliver, On the use of likelihood fields to perform sonar scan matching localization, Autonomous Robots 26 (4) (2009) 203–222, https://doi.org/10.1007/s10514-009-9108-0.
[115] S.G. Francesco Amigoni, M. Gini, Scan matching without odometry information, in: First International Conference on Informatics in Control, Automation and Robotics, ICINCO2004, 2004, pp. 349–352.
[116] R.B. Rusu, N. Blodow, M. Beetz, Fast point feature histograms (FPFH) for 3D registration, in: 2009 IEEE International Conference on Robotics and Automation, 2009, pp. 3212–3217.
[117] A. Mian, M. Bennamoun, R. Owens, On the repeatability and quality of keypoints for local feature-based 3D object retrieval from cluttered scenes, International Journal of Computer Vision 89 (2) (2010) 348–361.
[118] M. Poreba, F. Goulette, A robust linear feature-based procedure for automated registration of point clouds, Sensors 15 (1) (2015) 1435–1457, https://doi.org/10.3390/s150101435.
[119] X. Yuan, C.-X. Zhao, Z.-M. Tang, LiDAR scan-matching for mobile robot localization, Information Technology Journal 9 (2010) 27–33.
[120] D. Wujanz, S. Schaller, F. Gielsdorf, L. Gründig, Plane-based registration of several thousand laser scans on standard hardware, The International Archives of the Photogrammetry, Remote Sensing and Spatial Information Sciences XLII-2 (2018) 1207–1212, https://doi.org/10.5194/isprs-archives-XLII-2-1207-2018, https://www.int-arch-photogramm-remote-sens-spatial-inf-sci.net/XLII-2/1207/2018/.
[121] Y.-T. Wang, C.-C. Peng, A.A. Ravankar, A. Ravankar, A single LiDAR-based feature fusion indoor localization algorithm, Sensors 18 (4) (2018) 1294, https://doi.org/10.3390/s18041294, http://www.mdpi.com/1424-8220/18/4/1294.
[122] R.W. Wolcott, R.M. Eustice, Robust LiDAR localization using multiresolution gaussian mixture maps for autonomous driving, The International Journal of Robotics Research 36 (3) (2017) 292–319.
[123] C. Bruns, LiDAR-Based Vehicle Localization in an Autonomous Valet Parking Scenario, Ph.D. thesis, The Ohio State University, 2016.
[124] R.E. Kalman, et al., A new approach to linear filtering and prediction problems, Journal of Basic Engineering 82 (1) (1960) 35–45.
[125] S.J. Julier, J.K. Uhlmann, New extension of the Kalman filter to nonlinear systems, in: International Society for Optics and Photonics, AeroSense'97, 1997, pp. 182–193.

[126] S. Julier, J. Uhlmann, H.F. Durrant-Whyte, A new method for the nonlinear transformation of means and covariances in filters and estimators, IEEE Transactions on Automatic Control 45 (3) (2000) 477–482.
[127] M.S. Arulampalam, S. Maskell, N. Gordon, T. Clapp, A tutorial on particle filters for online nonlinear/non-gaussian bayesian tracking, IEEE Transactions on Signal Processing 50 (2) (2002) 174–188.
[128] J. Carpenter, P. Clifford, P. Fearnhead, Improved particle filter for nonlinear problems, IEE Proceedings. Radar, Sonar and Navigation 146 (1) (1999) 2–7.
[129] M. Corey, The Kalman Filter and Related Algorithms: A Literature Review, 2011.
[130] H. Fang, N. Tian, Y. Wang, M. Zhou, M.A. Haile, Nonlinear bayesian estimation: from Kalman filtering to a broader horizon, IEEE/CAA Journal of Automatica Sinica 5 (2) (2018) 401–417, https://doi.org/10.1109/JAS.2017.7510808.
[131] M. Menze, A. Geiger, Object scene flow for autonomous vehicles, in: IEEE International Conference on Robotics and Automation, ICRA, IEEE, 2015, pp. 3061–3070.
[132] G. Mattyus, S. Wang, S. Fidler, R. Urtasun, Enhancing road maps by parsing aerial images around the world, in: Proceedings of the 2015 IEEE International Conference on Computer Vision, ICCV, ICCV '15, IEEE Computer Society, Washington, DC, USA, 2015, pp. 1689–1697.
[133] W. Maddern, G. Pascoe, C. Linegar, P. Newman, 1 year, 1000 km: the Oxford RobotCar dataset, The International Journal of Robotics Research 36 (1) (2017) 3–15.
[134] S. Wang, M. Bai, G. Máttyus, H. Chu, W. Luo, B. Yang, J. Liang, J. Cheverie, S. Fidler, R. Urtasun, Torontocity: seeing the world with a million eyes, in: Proceedings of the IEEE International Conference on Computer Vision, ICCV, 2017, pp. 3028–3036.
[135] M. Burri, J. Nikolic, P. Gohl, T. Schneider, J. Rehder, S. Omari, M.W. Achtelik, R. Siegwart, The EuRoC micro aerial vehicle datasets, The International Journal of Robotics Research 35 (10) (2016) 1157–1163, https://doi.org/10.1177/0278364915620033.
[136] J. Sturm, N. Engelhard, F. Endres, W. Burgard, D. Cremers, A benchmark for the evaluation of RGB-D SLAM systems, in: Proc. of the International Conference on Intelligent Robot Systems, IROS, 2012.
[137] J. Chen, G. Ou, A. Peng, L. Zheng, J. Shi, An INS/WiFi indoor localization system based on the weighted least squares, Sensors 18 (2018) 1458.
[138] Y. Zhuang, N. El-Sheimy, Tightly-coupled integration of WiFi and MEMS sensors on handheld devices for indoor pedestrian navigation, IEEE Sensors Journal 16 (1) (2016) 224–234, https://doi.org/10.1109/JSEN.2015.2477444.
[139] J. Sasiadek, P. Hartana, GPS/INS sensor fusion for accurate positioning and navigation based on Kalman filtering, in: 7th IFAC Symposium on Cost-Oriented Automation, COA 2004, Gatineau, Québec, Canada, 6-9 June 2004, IFAC Proceedings Volumes 37 (5) (2004) 115–120, https://doi.org/10.1016/S1474-6670(17)32353-4, http://www.sciencedirect.com/science/article/pii/S1474667017323534, 2004.
[140] D.J. Kim, M.K. Kim, K.S. Lee, H.G. Park, M.H. Lee, Localization system of autonomous vehicle via Kalman filtering, in: 2011 11th International Conference on Control, Automation and Systems, 2011, pp. 934–937.
[141] R. Toledo-Moreo, M.A. Zamora-Izquierdo, B. Ubeda-Minarro, A.F. Gomez-Skarmeta, High-integrity IMM-EKF-based road vehicle navigation with low-cost GPS/SBAS/INS, IEEE Transactions on Intelligent Transportation Systems 8 (3) (2007) 491–511, https://doi.org/10.1109/TITS.2007.902642.
[142] P. Zhang, J. Gu, E.E. Milios, P. Huynh, Navigation with IMU/GPS/digital compass with unscented Kalman filter, in: IEEE International Conference Mechatronics and Automation, vol. 3, 2005, pp. 1497–1502.
[143] M. Enkhtur, S. Yun Cho, K.-H. Kim, Modified unscented Kalman filter for a multirate INS/GPS integrated navigation system, ETRI Journal 35 (5) (2013) 943–946, https://doi.org/10.4218/etrij.13.0212.0540.
[144] F. Caron, M. Davy, E. Duflos, P. Vanheeghe, Particle filtering for multisensor data fusion with switching observation models: application to land vehicle positioning, IEEE Transactions on Signal Processing 55 (6) (2007) 2703–2719, https://doi.org/10.1109/TSP.2007.893914.
[145] D. Titterton, J. Weston, J. Weston, Strapdown Inertial Navigation Technology, I. of Electrical Engineers, A.I. of Aeronautics and Astronautics, Electromagnetics and Radar Series, Institution of Engineering and Technology, 2004.

[146] G. Falco, M. Pini, G. Marucco, Loose and tight GNSS/INS integrations: comparison of performance assessed in real urban scenarios, Sensors 17 (2) (2017) 255.
[147] A. Nemra, N. Aouf, Robust INS/GPS sensor fusion for UAV localization using SDRE nonlinear filtering, IEEE Sensors Journal 10 (4) (2010) 789–798, https://doi.org/10.1109/JSEN.2009.2034730.
[148] S. Wang, Z. Deng, G. Yin, An accurate GPS-IMU/DR data fusion method for driverless car based on a set of predictive models and grid constraints, Sensors 16 (3) (2016) 280, https://doi.org/10.3390/s16030280.
[149] Y. Zhao, GPS/IMU Integrated System for Land Vehicle Navigation Based on MEMS, Ph.D. thesis, KTH Royal Institute of Technology, 2011.
[150] A. Ali Broumandan, G. Lachapelle, Spoofing detection using GNSS/INS/odometer coupling for vehicular navigation, Sensors 18 (5) (2018) 789–798, https://doi.org/10.3390/s18051305, http://www.mdpi.com/1424-8220/18/5/1305.
[151] M. Aftatah, A. Lahrech, A. Abounada, A. Soulhi, GPS/INS/Odometer data fusion for land vehicle localization in GPS denied environment, Modern Applied Science 11 (1) (2017), https://doi.org/10.5539/mas.v11n1p62.
[152] R. Mur-Artal, J.D. Tardós, Visual–inertial monocular SLAM with map reuse, IEEE Robotics and Automation Letters 2 (2) (2017) 796–803.
[153] L. Wei, B. Soheilian, V. Gouet-Brunet, Augmenting vehicle localization accuracy with cameras and 3D road infrastructure database, in: ECCV Workshop on Computer Vision in Vehicle Technology, Zurich, Switzerland, in: Lecture Notes in Computer Science, vol. 8925, 2014, pp. 194–208.
[154] J.A. Delmerico, D. Scaramuzza, A benchmark comparison of monocular visual–inertial odometry algorithms for flying robots, in: 2018 IEEE International Conference on Robotics and Automation, ICRA 2018, Brisbane, Australia, May 21-25, 2018, 2018, pp. 2502–2509.
[155] A.I. Mourikis, S.I. Roumeliotis, A multi-state constraint Kalman filter for vision-aided inertial navigation, in: Proc. IEEE Int. Conf. on Robotics and Automation, 2007, pp. 10–14.
[156] S. Leutenegger, S. Lynen, M. Bosse, R. Siegwart, P. Furgale, Keyframe-based visual–inertial odometry using nonlinear optimization, The International Journal of Robotics Research 34 (3) (2015) 314–334.
[157] M. Bloesch, S. Omari, M. Hutter, R. Siegwart, Robust visual inertial odometry using a direct EKF-based approach, in: 2015 IEEE/RSJ International Conference on Intelligent Robots and Systems, IROS, 2015, pp. 298–304.
[158] T. Qin, P. Li, S. Shen, VINS-Mono: a robust and versatile monocular visual–inertial state estimator, IEEE Transactions on Robotics 34 (4) (2018) 1004–1020, https://doi.org/10.1109/TRO.2018.2853729.
[159] S. Lynen, M.W. Achtelik, S. Weiss, M. Chli, R. Siegwart, A robust and modular multi-sensor fusion approach applied to MAV navigation, in: 2013 IEEE/RSJ International Conference on Intelligent Robots and Systems, 2013, pp. 3923–3929.
[160] M. Kaess, H. Johannsson, R. Roberts, V. Ila, J.J. Leonard, F. Dellaert, iSAM2: incremental smoothing and mapping using the Bayes tree, The International Journal of Robotics Research 31 (2) (2012) 216–235, https://doi.org/10.1177/0278364911430419.
[161] V. Creuze, Monocular odometry for underwater vehicles with online estimation of the scale factor, in: IFAC 2017 IFAC 2017 World Congress, Toulouse, France, 2017, https://hal-lirmm.ccsd.cnrs.fr/lirmm-01567463.
[162] J.-Y. Bouguet, OpenCV Document, in: Intel, Microprocessor Research Labs. 1, 2000.
[163] F. Shkurti, I. Rekleitis, M. Scaccia, G. Dudek, State estimation of an underwater robot using visual and inertial information, in: 2011 IEEE/RSJ International Conference on Intelligent Robots and Systems, 2011, pp. 5054–5060.
[164] A. Burguera, F. Bonin-Font, G. Oliver, Trajectory-based visual localization in underwater surveying missions, Sensors 15 (1) (2015) 1708–1735.
[165] I. Nikolov, C. Madsen, LiDAR-based 2D localization and mapping system using elliptical distance correction models for UAV wind turbine blade inspection, in: Proceedings of the 12th International Joint Conference on Computer Vision, Imaging and Computer Graphics Theory and Applications, SCITEPRESS Digital Library, 2017.

[166] S. Yun, Y.J. Lee, S. Sung, IMU/Vision/LiDAR integrated navigation system in GNSS denied environments, in: Aerospace Conference, 2013 IEEE, IEEE, 2013, pp. 1–10.
[167] G. Pascoe, W. Maddern, P. Newman, Direct visual localisation and calibration for road vehicles in changing city environments, in: Proceedings of the IEEE International Conference on Computer Vision Workshops, 2015, pp. 9–16.
[168] F. Boniardi, T. Caselitz, R. Kümmerle, W. Burgard, Robust LiDAR-based localization in architectural floor plans, in: Intelligent Robots and Systems, 2017 IEEE/RSJ International Conference on, IROS, IEEE, 2017, pp. 3318–3324.
[169] P. Lothe, S. Bourgeois, F. Dekeyser, E. Royer, M. Dhome, Towards geographical referencing of monocular SLAM reconstruction using 3D city models: application to real-time accurate vision-based localization, in: 2009 IEEE Conference on Computer Vision and Pattern Recognition, 2009, pp. 2882–2889.
[170] F. Monnier, B. Vallet, N. Paparoditis, J.-P. Papelard, N. David, Registration of terrestrial mobile laser data on 2D or 3D geographic database by use of a non-rigid ICP approach, ISPRS Annals of Photogrammetry, Remote Sensing and Spatial Information Sciences II-5/W2 (2013) 193–198.
[171] D. Larnaout, S. Bourgeois, V. Gay-Bellile, M. Dhome, Towards bundle adjustment with GIS constraints for online geo-localization of a vehicle in urban center, in: 2012 Second International Conference on 3D Imaging, Modeling, Processing, Visualization Transmission, 2012, pp. 348–355.
[172] C. Cappelle, M.E.B. el Najjar, F. Charpillet, D. Pomorski, Outdoor obstacle detection and localisation with monovision and 3D geographical database, in: 2007 IEEE Intelligent Transportation Systems Conference, 2007, pp. 1102–1107.
[173] W. Burgard, O. Brock, C. Stachniss, Map-Based Precision Vehicle Localization in Urban Environments, MITP, 2008, https://ieeexplore.ieee.org/xpl/articleDetails.jsp?arnumber=6280129.
[174] B.S.O. Tournaire, N. Paparoditis, Towards a subdecimeter georeferencing of ground-based mobile mapping systems in urban areas: matching ground-based and aerial-based imagery using roadmarks, The International Archives of the Photogrammetry, Remote Sensing and Spatial Information Sciences 36 (1) (2006).
[175] M. Schreiber, C. Knöppel, U. Franke, Laneloc: lane marking based localization using highly accurate maps, in: 2013 IEEE Intelligent Vehicles Symposium, IV, 2013, pp. 449–454.
[176] A. Schlichting, C. Brenner, Localization using automotive laser scanners and local pattern matching, in: 2014 IEEE Intelligent Vehicles Symposium Proceedings, 2014, pp. 414–419.
[177] V. Kubelka, L. Oswald, F. Pomerleau, F. Colas, T. Svoboda, M. Reinstein, Robust data fusion of multimodal sensory information for mobile robots, Journal of Field Robotics 32 (4) (2015) 447–473.
[178] X. Meng, H. Wang, B. Liu, A robust vehicle localization approach based on GNSS/IMU/DMI/LiDAR sensor fusion for autonomous vehicles, Sensors 17 (9) (2017) 2140.
[179] W. Li, L. Chen, D. Xu, L. Van Gool, Visual recognition in RGB images and videos by learning from RGB-D data, IEEE Transactions on Pattern Analysis and Machine Intelligence 40 (8) (2018) 2030–2036, http://ieeexplore.ieee.org/document/8000401/.
[180] J. Hoffman, S. Gupta, T. Darrell, Learning with side information through modality hallucination, in: Proceedings of the IEEE Conference on Computer Vision and Pattern Recognition, CVPR, 2016, pp. 826–834, http://ieeexplore.ieee.org/document/7780465/.
[181] N. Piasco, D. Sidibé, V. Gouet-Brunet, C. Demonceaux, Learning scene geometry for visual localization in challenging conditions, in: IEEE International Conference on Robotics and Automation, ICRA 2019, Montreal, Canada, 2019, pp. 1–7.
[182] H. Porav, W. Maddern, P. Newman, Adversarial training for adverse conditions: robust metric localisation using appearance transfer, in: ICRA, 2018, pp. 1011–1018.
[183] P. Agarwal, Robust Graph-Based Localization and Mapping, Ph.D. thesis, Technische Fakultat Albert-Ludwigs-Universitat, Dissertation zur Erlangung des akademischen Grades Doktor der Naturwissenschaften, Freiburg, Germany, 2015.
[184] M. Veth, R. Anderson, F. Webber, M. Nielsen, Tightly-Coupled INS, GPS, and Imaging Sensors for Precision Geolocation, vol. 1, 2008.

[185] H. Chu, H. Mei, M. Bansal, M.R. Walter, Accurate vision-based vehicle localization using satellite imagery, CoRR, arXiv:1510.09171, arXiv:1510.09171, http://arxiv.org/abs/1510.09171.
[186] M.A. Brubaker, A. Geiger, R. Urtasun, Map-based probabilistic visual self-localization, IEEE Transactions on Pattern Analysis and Machine Intelligence 38 (4) (2016) 652–665, https://doi.org/10.1109/TPAMI.2015.2453975.
[187] H. Yin, L. Tang, X. Ding, Y. Wang, R. Xiong, Locnet: Global Localization in 3d Point Clouds for Mobile Vehicles, 2017.
[188] F. Bellavia, M. Fanfani, C. Colombo, Selective visual odometry for accurate auv localization, Autonomous Robots 41 (1) (2017) 133–143, https://doi.org/10.1007/s10514-015-9541-1.
[189] Maxime Ferrera, Julien Moras, et al., The Aqualoc Dataset: Towards Real-Time Underwater Localization From a Visual–Inertial-Pressure Acquisition System, 2018.
[190] X. Gao, T. Zhang, Unsupervised learning to detect loops using deep neural networks for visual slam system, Autonomous Robots 41 (1) (2017) 1–18, https://doi.org/10.1007/s10514-015-9516-2.
[191] R. Arandjelović, P. Gronat, A. Torii, T. Pajdla, J. Sivic, NetVLAD: CNN architecture for weakly supervised place recognition, in: IEEE Conference on Computer Vision and Pattern Recognition, 2016.
[192] C. Chang, H. Zhu, M. Li, S. You, A review of visual–inertial simultaneous localization and mapping from filtering-based and optimization-based perspectives, Robotics 7 (2018) 45, https://doi.org/10.3390/robotics7030045.
[193] H. Esmaeilzadeh, E. Blem, R.St. Amant, K. Sankaralingam, D. Burger, Dark silicon and the end of multicore scaling, in: Proceedings of the 38th Annual International Symposium on Computer Architecture, ISCA '11, ACM, New York, NY, USA, 2011, pp. 365–376, http://doi.acm.org/10.1145/2000064.2000108.
[194] ITRS, International Technology Roadmap for Semiconductors 2.0, Tech. rep., ITRS, 2015, http://www.itrs2.net/itrs-reports.html.
[195] H. Esmaeilzadeh, A. Sampson, L. Ceze, D. Burger, Neural acceleration for general-purpose approximate programs, in: Proceedings of the 2012 45th Annual IEEE/ACM International Symposium on Microarchitecture, MICRO-45, IEEE Computer Society, Washington, DC, USA, 2012, pp. 449–460.
[196] A. Suleiman, Z. Zhang, L. Carlone, S. Karaman, V. Sze, Navion: a fully integrated energy-efficient visual–inertial odometry accelerator for autonomous navigation of nano drones, in: IEEE Symposium on VLSI Circuits, VLSI-Circuits, 2018.
[197] B.D.T. Inc., Benchmark, https://www.bdti.com/, 2018.
[198] S.P.E. Corporation, SPEC CPU2006 analysis papers: guest editor's introduction, https://www.spec.org/cpu2006/publications/SIGARCH-2007-03/, 2018.
[199] P. et al., Neural Processing Unit, patent No.: US 8,655,815 B2 (May 2 2014), https://patents.google.com/patent/US8655815.
[200] A. Frumusanu, Hisilicon kirin 970 - android soc power & performance overview, https://www.anandtech.com/show/12195/hisilicon-kirin-970-power-performance-overview, 2018.
[201] Cadence, Tie language – the fast path to high-performance embedded soc processing, Tech. rep., https://ip.cadence.com/knowledgecenter/resources/know-dip-wp.
[202] CEVA XM6 product note, 5 2017.
[203] Qualcomm Hexagon DSP: an architecture optimized for mobile multimedia and communications, 2017.
[204] ARM, Products processors, https://www.arm.com/products/silicon-ip-cpu, 2018.
[205] A. Waterman, Y. Lee, D.A. Patterson, K. Asanović, The RISC-V Instruction Set Manual, volume i: Base User-Level isa, Tech. Rep. UCB/EECS-2011-62, EECS Department, University of California, Berkeley, May 2011, http://www2.eecs.berkeley.edu/Pubs/TechRpts/2011/EECS-2011-62.html.
[206] M. Demler, XAVIER Simplifies Self-driving Cars, Tech. rep., Microprocessor Report, The Linley Group, Juin 2017, http://www.linleygroup.com/mpr/h/article.php?id=11820.
[207] Intel, Intel SoCs FPGA, https://www.intel.com/content/www/us/en/products/programmable/soc.html, 2018.
[208] R. Furlan, The future of augmented reality: Hololens - Microsoft's ar headset shines despite rough edges [resources_tools and toys], IEEE Spectrum 53 (6) (2016) 21, https://doi.org/10.1109/MSPEC.2016.7473143.

[209] B. Barry, C. Brick, F. Connor, D. Donohoe, D. Moloney, R. Richmond, M. O'Riordan, V. Toma, Always-on vision processing unit for mobile applications, IEEE MICRO 35 (2) (2015) 56–66, https://doi.org/10.1109/MM.2015.10.

[210] O. Deniz, N. Vallez, J.L. Espinosa-Aranda, J.M. Rico-Saavedra, J. Parra-Patino, G. Bueno, D. Moloney, A. Dehghani, A. Dunne, A. Pagani, S. Krauss, R. Reiser, M. Waeny, M. Sorci, T. Llewellynn, C. Fedorczak, T. Larmoire, M. Herbst, A. Seirafi, K. Seirafi, Eyes of things, Sensors 17 (5) (2017) 1173, https://doi.org/10.3390/s17051173, http://www.mdpi.com/1424-8220/17/5/1173.

[211] G. Desoli, V. Tomaselli, E. Plebani, G. Urlini, D. Pau, V. D'Alto, T. Majo, F. De Ambroggi, T. Boesch, S.-p. Singh, E. Guidetti, N. Chawla, The Orlando project: a 28 nm FD-SOI low memory embedded neural network ASIC, in: J. Blanc-Talon, C. Distante, W. Philips, D. Popescu, P. Scheunders (Eds.), Advanced Concepts for Intelligent Vision Systems, Springer International Publishing, Cham, 2016, pp. 217–227.

[212] Qualcomm, Snapdragon 845 mobile platform, https://www.qualcomm.com/products/snapdragon-845-mobile-platform, 2018.

[213] Google, ARCore fundamental concepts, Website, https://developers.google.com/ar/discover/concepts, 2018.

[214] Apple, ARKit, Website, https://developer.apple.com/arkit/, 2018.

[215] I. Spillinger, Microsoft Hololens and Mixed Reality Holographic Processing Unit, Semicon Korea, February 2017.

[216] iFixit, Magic leap one teardown, Website, https://www.ifixit.com/Teardown/Magic+Leap+One+Teardown/112245, August 2018.

[217] M. Leap, Magic leap one specs, Website, https://www.magicleap.com/magic-leap-one, 2018.

[218] StarVR, Starvr one specifications, Website, https://www.starvr.com/products/, 2018.

[219] J. Jansen, Autonomous Localization and Tracking for UAVs Using Kalman Filtering, Master's thesis, Norwegian University of Science and Technology, Trondheim, 2014.

[220] R. Vivacqua, R. Vassallo, F. Martins, A Low Cost Sensors Approach for Accurate Vehicle Localization and Autonomous Driving Application, vol. 17, Multidisciplinary Digital Publishing Institute, 2017, p. 2359.

[221] W. Liang, Y. Zhang, J. Wang, Map-based localization method for autonomous vehicles using 3D-LiDAR, IFAC-PapersOnLine 50 (1) (2017) 276–281.

[222] E. Stenborg, C. Toft, L. Hammarstrand, Long-term visual localization using semantically segmented images, in: Conference: 2018 IEEE International Conference on Robotics and Automation, ICRA, 2018.

[223] S. Kuutti, S. Fallah, K. Katsaros, M. Dianati, F. Mccullough, A. Mouzakitis, A survey of the state-of-the-art localization techniques and their potentials for autonomous vehicle applications, IEEE Internet of Things Journal 5 (2) (2018) 829–846, https://doi.org/10.1109/JIOT.2018.2812300.

[224] Z. Zhang, A. Suleiman, L. Carlone, V. Sze, S. Karaman, Visual–Inertial Odometry on Chip: An Algorithm-and-Hardware Co-Design Approach, 2017.

[225] M. Cornick, J. Koechling, B. Stanley, B. Zhang, Localizing ground penetrating radar: a step toward robust autonomous ground vehicle localization, Journal of Field Robotics 33 (1) (2016) 82–102, https://doi.org/10.1002/rob.21605.

[226] A.Y. Hata, D.F. Wolf, Feature detection for vehicle localization in urban environments using a multilayer lidar, IEEE Transactions on Intelligent Transportation Systems 17 (2) (2016) 420–429, https://doi.org/10.1109/TITS.2015.2477817.

[227] R.W. Wolcott, R.M. Eustice, Visual localization within lidar maps for automated urban driving, in: 2014 IEEE/RSJ International Conference on Intelligent Robots and Systems, 2014, pp. 176–183.

[228] H. Nabil, T. Ditchi, G. Emmanuel, J. Lucas, S. Holé, Electronics RF infrastructure cooperative system for in lane vehicle localization, Sensors 2014 (3) (2014) 598–608.

[229] J.A. del Peral-Rosado, J.A. López-Salcedo, S. Kim, G. Seco-Granados, Feasibility study of 5G-based localization for assisted driving, in: 2016 International Conference on Localization and GNSS, ICL-GNSS, 2016, pp. 1–6.

[230] G. Zhang, Y. Chen, H. Moyes, Optimal 3D reconstruction of caves using small unmanned aerial systems and RGB-D cameras, in: 2018 International Conference on Unmanned Aircraft Systems, ICUAS, 2018, pp. 410–415.

[231] G. Zhang, B. Shang, Y. Chen, H. Moyes, Smartcavedrone: 3D cave mapping using UAVs as robotic co-archaeologists, in: 2017 International Conference on Unmanned Aircraft Systems, ICUAS, 2017, pp. 1052–1057.
[232] R.F. Fernandez, K. Keller, J. Robins, Design of a system for aircraft fuselage inspection, in: 2016 IEEE Systems and Information Engineering Design Conference, 2016, pp. 283–288.
[233] M.N. Graham Warwick, Aircraft inspection drones entering service with airline mros, https://www.mro-network.com/, 2018.
[234] Donecle, Donecle drone, https://fr.wikipedia.org/wiki/Fichier:AFI_05_2017_Donecle_drone_002.jpg, 2018-12-20.
[235] H.-Y. Huang, C.-Y. Hsieh, K.-C. Liu, H.-C. Cheng, S.J. Hsu, C.-T. Chan, Multimodal sensors data fusion for improving indoor pedestrian localization, in: 2018 IEEE International Conference on Applied System Invention, ICASI, IEEE, 2018, pp. 283–286.
[236] E. Rublee, V. Rabaud, K. Konolige, G. Bradski, ORB: An efficient alternative to SIFT or SURF, in: Computer Vision, 2011 IEEE International Conference on, ICCV, IEEE, 2011, pp. 2564–2571.
[237] Carlo Tomasi, Takeo Kanade, Detection and Tracking of Point Features, School of Computer Science, Carnegie Mellon Univ. Pittsburgh, 1991.
[238] J.K. Suhr, Kanade-Lucas-Tomasi (KLT) feature tracker, Computer Vision EEE6503 (2009) 9–18.
[239] S. Choi, T. Kim, W. Yu, Performance evaluation of RANSAC family, in: Proceedings of the British Machine Vision Conference 2009, vol. 24, 2009.
[240] B. Dynamics, Boston dynamics is changing your idea of what robots can do, https://www.bostondynamics.com/, 2018.
[241] Yole, Sensors and Data Management for Autonomous Vehicles Report, Tech. rep., Yole Developpement, October 2015, http://www.yole.fr/.
[242] M. Eye, Mobileye shield +, https://www.mobileye.com/fr-fr/produits/mobileye-shield/, 2018.
[243] G.A. Kedar Chitnis, Roman Staszewski, TI Vision SDK, Optimized Vision Libraries for ADAS Systems, Tech. rep.. Texas Instruments, April 2014, http://www.ti.com/lit/wp/spry260/spry260.pdf.
[244] Q.K. Dang, Y. Chee, D.D. Pham, Y.S. Suh, A virtual blind cane using a line laser-based vision system and an inertial measurement unit, Sensors 16 (1) (2016) 95, https://doi.org/10.3390/s16010095, http://www.mdpi.com/1424-8220/16/1/95.
[245] K. Fan, C. Lyu, Y. Liu, W. Zhou, X. Jiang, P. Li, H. Chen, Hardware implementation of a virtual blind cane on FPGA, in: Real-time Computing and Robotics, 2017 IEEE International Conference on, RCAR, IEEE, 2017, pp. 344–348.
[246] I.H. Project, Inspex- integrated smart spatial exploration system, http://www.inspex-ssi.eu/, 2018.
[247] K.L. Li, Electronic Travel Aids for Blind Guidance – An Industry Landscape Study, ind eng 290, 12 2015.
[248] R. O'Keeffe, S. Gnecchi, S. Buckley, C. O'Murchu, A. Mathewson, S. Lesecq, J. Foucault, Long range lidar characterisation for obstacle detection for use by the visually impaired and blind, in: ECTC, 2018, pp. 533–538.
[249] L. Ouvry, et al., Uwb Radar Sensor Characterization for Obstacle Detection with Application to the Smart White Cane, 04 2018.
[250] Website of the European research project LARA, http://lara-project.eu, 2018.
[251] E. Stylianidis, E. Valari, K. Smagas, A. Pagani, J. Henriques, A. Garca, E. Jimeno, I. Carrillo, P. Patias, C. Georgiadis, A. Kounoudes, K. Michail, LBS augmented reality assistive system for utilities infrastructure management through GALILEO and EGNOS, The International Archives of the Photogrammetry, Remote Sensing and Spatial Information Sciences XLI-B1 (2016) 1179–1185.
[252] R.J.P. Jimenez, E.M. Becerril, R.M. Nor, K. Smagas, E. Valari, E. Stylianidis, Market potential for a location based and augmented reality system for utilities management, in: 22nd International Conference on Virtual System & Multimedia, VSMM'16, Kuala Lumpur, 2016, pp. 1–4.

CHAPTER 9

Self-Supervised Learning from Web Data for Multimodal Retrieval

Raul Gomez[*,†], **Lluis Gomez**[†], **Jaume Gibert**[*], **Dimosthenis Karatzas**[†]

[*]*Eurecat, Centre Tecnològic de Catalunya, Unitat de Tecnologies Audiovisuals, Barcelona, Spain*
[†]*Computer Vision Center, Universitat Autònoma de Barcelona, Barcelona, Spain*

Contents

Multimodal Scene Understanding
https://doi.org/10.1016/B978-0-12-817358-9.00015-9

9.1 Introduction

9.1.1 Annotating Data: A Bottleneck for Training Deep Neural Networks

Large annotated datasets, powerful hardware and deep learning techniques are allowing one to get outstanding machine learning results, not only in traditional classification problems but also in more challenging tasks such as image captioning or language translation. Deep neural networks allow for building pipelines that can learn patterns from any kind of data with impressive results.

Deep learning has two strong requirements: Computation power and tons of data. The computation power requirement is fulfilled by GPUs and other AI specialized hardware, such as TPUs. Moreover, the hardware power is evolving fast without an apparent roof together with deep learning algorithms requirements. The story with the data requirement is different. Despite the existence of large-scale annotated datasets such as ImageNet [1], COCO [2] or Places [3], the lack of data limits the application of deep learning to specific problems where it is difficult or economically non-viable to get proper annotations. Although there exist some tools to facilitate human data annotation, such as the Amazon Mechanical Turk,[1] annotating the tons of data required to train supervised deep learning models is a very expensive and manual task, whose efficiency cannot evolve over time.

9.1.2 Alternatives to Annotated Data

A common strategy to overcome the lack of annotated data is to first train models in generic datasets, as ImageNet, and then fine-tune them to other tasks using smaller, specific datasets [4]. But still we depend on the existence of annotated data to train our models. Another strategy to overcome the insufficiency of data is to use computer graphics techniques to generate artificial data inexpensively. However, while synthetic data has proven to be a valuable source of training data for many applications such as pedestrian detection [5], image semantic segmentation [6] and scene text detection and recognition [7,8], nowadays it is still not easy to generate realistic complex images for some tasks.

[1] https://www.mturk.com.

An alternative to this strategies and a solution to overcome the annotated data requirements of supervised deep learning techniques are not fully supervised techniques. Among them, self-supervised learning exploits multimodal data to learn relations between two or more data modalities using paired instances. Web and social media offer an immense amount of images accompanied by other information such as the image title, description or date. This data is noisy and unstructured but it is free and nearly unlimited. We mentioned that data annotation efficiency does not improve with time. As a contrast, the amount of available multimodal data in the web does. Designing algorithms to learn from web data is an interesting research area as it would disconnect the deep learning evolution from the scaling of human-annotated datasets, given the enormous amount of existing web and social media data. We call this scenario self-supervised learning because it consists in exploiting relations between different modalities (in this case images and text) of multimodal data as supervision.

9.1.3 Exploiting Multimodal Web Data

Lately, web data has been used to build classification datasets, such as in the WebVision Challenge [9] and in this Facebook work [10]. In these works, to build a classification dataset, queries are made to search engines using class names and the retrieved images are labeled with the querying class. In such a configuration the learning is limited to some pre-established classes, thus it could not generalize to new classes. While working with image labels is very convenient for training traditional visual models, the semantics in such a discrete space are very limited in comparison with the richness of human language expressiveness when describing an image. Instead we define here a scenario where, by exploiting distributional semantics in a given text corpus, we can learn from every word associated to an image. As illustrated in Fig. 9.1, by leveraging the richer semantics encoded in the learned embedding space, we can infer previously unseen concepts even though they might not be explicitly present in the training set.

The noisy and unstructured text associated to web images provides information about the image content that we can use to learn visual features. A strategy to do that is to embed the multimodal data (images and text) in the same vectorial space. In this work we represent text using five different state-of-the-art methods and eventually embed images in the learned semantic space by means of a regression CNN. We compare the performance of the different text space configurations under a text-based image retrieval task.

9.2 Related Work

Multimodal image and text embeddings have been lately a very active research area. The possibilities of learning together from different kinds of data have motivated this field of study,

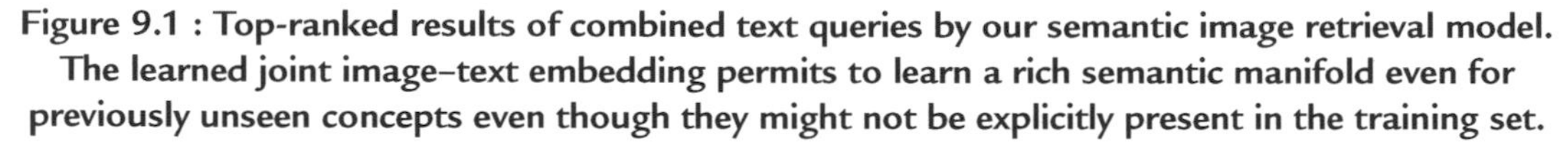
Figure 9.1 : Top-ranked results of combined text queries by our semantic image retrieval model. The learned joint image–text embedding permits to learn a rich semantic manifold even for previously unseen concepts even though they might not be explicitly present in the training set.

where both general and applied research has been done. DeViSE [11] proposes a pipeline that, instead of learning to predict ImageNet classes, learns to infer the Word2Vec [12] representations of their labels. The result is a model that makes semantically relevant predictions even when it makes errors, and generalizes to classes outside of its labeled training set. Gordo & Larlus [13] use captions associated to images to learn a common embedding space for images and text through which they perform semantic image retrieval. They use a *tf-idf*-based BoW representation over the image captions as a semantic similarity measure between images and they train a CNN to minimize a margin loss based on the distances of triplets of query-similar-dissimilar images. Gomez, Patel et al. [14,15] use LDA [16] to extract topic probabilities from a bunch of Wikipedia articles and train a CNN to embed their associated images in the same topic space. Wang et al. [17] propose a method to learn a joint embedding of images and text for image-to-text and text-to-image retrieval, by training a neural net to embed in the same space Word2Vec [12] text representations and CNN extracted features.

Other than semantic retrieval, joint image–text embeddings have also been used in more specific applications. Patel et al. [18] use LDA [16] to learn a joint image–text embedding and generate contextualized lexicons for images using only visual information. Gordo et al. [19] embed word images in a semantic space relying in the graph taxonomy provided by Word-

Net [20] to perform text recognition. In a more specific application, Salvador et al. [21] propose a joint embedding of food images and their recipes to identify ingredients, using Word2Vec [12] and LSTM representations to encode ingredient names and cooking instructions and a CNN to extract visual features from the associated images. Exploiting Instagram publications related to #Barcelona, Gomez et al. [22] learn relations between words, images and Barcelona neighborhoods to study which words and visual features tourists and locals relate with each neighborhood.

The robustness against noisy data has also been addressed by the community, though usually in an implicit way. Patrini et al. [23] address the problem of training a deep neural network with label noise with a loss correction approach and Xiau et al. [24] propose a method to train a network with a limited number of clean labels and millions of noisy labels. Fu et al. [25] propose an image tagging method robust to noisy training data and Xu et al. [26] address social image tagging correction and completion. Zhang et al. [27] show how label noise affects the CNN training process and its generalization error.

9.2.1 Contributions

The work presented here brings in a performance comparison between five state-of-the-art text embeddings in self-supervised learning, showing results in three different datasets. Furthermore it proves that self-supervised multimodal learning can be applied to web and social media data achieving competitive results in text-based image retrieval compared to pipelines trained with human-annotated data. Finally, a new dataset formed by Instagram images and their associated text is presented: InstaCities1M.

9.3 Multimodal Text–Image Embedding

One of the objectives of this work is to serve as a fair comparative of different text embeddings methods when learning from web and social media data. Therefore we design a pipeline to test the different methods under the same conditions, where the text embedding is a module that can be replaced by any text representation.

The proposed pipeline is as follows: First, we train the text embedding model on a dataset composed of pairs of images and correlated texts (I, x). Second, we use the text embedding model to generate vectorial representations of those texts. Given a text instance x, we denote its embedding by $\phi(x) \in \mathbb{R}^D$. Third, we train a CNN to regress those text embeddings directly from the correlated images. Given an image I, its representation in the embedding space is denoted by $\psi(I) \in \mathbb{R}^D$. Thereby the CNN learns to embed images in the vectorial

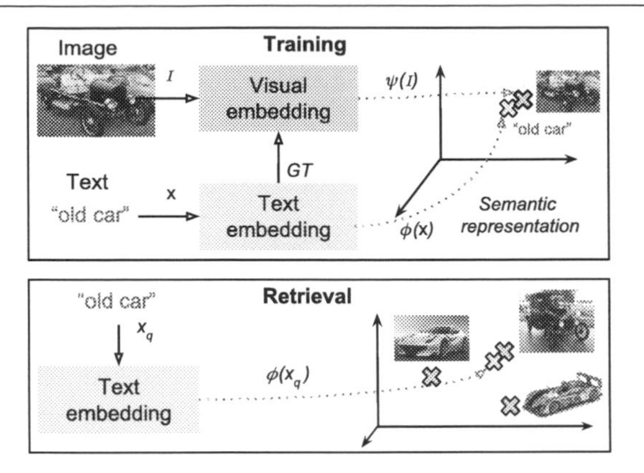

Figure 9.2 : Pipeline of the visual embedding model training and the image retrieval by text.

space defined by the text embedding model. The trained CNN model is used to generate visual embeddings for the test set images. Fig. 9.2 shows a diagram of the visual embedding training pipeline and the retrieval procedure.

In the image retrieval stage the vectorial representation in the joint text–image space of the querying text is computed using the text embedding model. Image queries can also be handled by using the visual embedding model instead of the text embedding model to generate the query representation. Furthermore, we can generate complex queries combining different query representations applying algebra in the joint text–image space. To retrieve the most semantically similar image I_R to a query x_q, we compute the cosine similarity of its vectorial representation $\phi(x_q)$ with the visual embeddings of the test set images $\psi(I_T)$, and retrieve the nearest image in the joint text–image space:

$$\arg\min_{I_T \in \text{Test}} \frac{\langle \phi(x_q), \psi(I_T) \rangle}{||\phi(x_q)|| \cdot ||\psi(I_T)||}. \tag{9.1}$$

State-of-the-art text embedding methods trained on large text corpus are very good generating representations of text in a vector space where semantically similar concepts fall close to each other. The proposed pipeline leverages the semantic structure of these text embedding spaces training a visual embedding model that generates vectorial representations of images in the same space, mapping semantically similar images close to each other, and also close to texts correlated to the image content. Note that the proposed joint text–image embedding can be extended to other tasks besides image retrieval, such as image annotation, tagging or captioning.

A CNN is trained to regress text embeddings from the correlated images minimizing a sigmoid cross-entropy loss. This loss is used to minimize distances between the text and image embeddings. Let $\{(I_n, x_n)\}_{n=1:N}$ be a batch of image–text pairs. If $\sigma(\cdot)$ is the component-wise sigmoid function, we denote $p_n = \sigma(\phi(x_n))$ and $\hat{p}_n = \sigma(\psi(I_n))$. Note $p_n, \hat{p}_n \in \mathbb{R}^D$ where D is the dimensionality of the joint embedding space. Let the loss be

$$L = -\frac{1}{ND}\sum_{n=1}^{N}\sum_{d=1}^{D}[\, p_{nd}\log \hat{p}_{nd} + (1 - p_{nd})\log(1 - \hat{p}_{nd})\,]. \tag{9.2}$$

The GoogleNet architecture [28] is used, customizing the last layer to regress a vector of the same dimensionality as the text embedding. We train with a Stochastic Gradient Descent optimizer with a learning rate of $1e^{-3}$, multiplied by 0.1 every 100k iterations, and a momentum of 0.9. The batch size is set to 120 and random cropping and mirroring are used as online data augmentation. With these settings the CNN trainings converge after around 300K–500K iterations. We use the Caffe [29] framework and initialize with the ImageNet [1] trained model to make the training faster. Notice that, despite initializing with a model trained with human-annotated data, this does not denote a dependence on annotated data, since the resulting model can generalize to much more concepts than the ImageNet classes. We trained one model from scratch obtaining similar results, although more training iterations were needed. Cross entropy loss is not usually used for regression problems, where mean square error loss is often used. We chose cross entropy loss empirically, since it was the one providing a stable training and better performance. Although cross entropy loss tends to be considered a loss for classification, it is also suitable for regression problems: despite this loss will not be zero when the regression solution matches the ground truth, it will always be minimum compared to other solutions.

9.4 Text Embeddings

Text vectorization methods are diverse in terms of architecture and the text structure they are designed to deal with. Some methods are oriented to vectorize individual words and others to vectorize full texts or paragraphs. In this work we consider the top-performing text embeddings and test them in our pipeline to evaluate their performance when learning from web and social media data. Here we explain briefly the main characteristics of each text embedding method used.

LDA [16]. Latent Dirichlet allocation learns latent topics from a collection of text documents and maps words to a vector of probabilities of those topics. It can describe a document by assigning topic distributions to it, which in turn have word distributions assigned. An advantage of this method is that it gives interpretable topics.

GloVe [30]. It is a count-based model. It learns the vectors by essentially doing dimensionality reduction on the co-occurrence counts matrix. Training is performed on aggregated global word-word co-occurrence statistics from a corpus.

Word2Vec [12]. Learns representations for words based on their context using a single hidden layer feed-forward neural network. It has two variants: In the CBOW (Continuous Bag of Word) approach, the neural network is trained to predict a word given as input its surrounding context (surrounding words). In the Skip-gram model, opposite to the CBOW model, the neural network is trained to predict a word context given that word as an input. In this work we use the most extended and efficient CBOW approach.

Doc2Vec [31]. Extends the Word2Vec idea to documents, being able to create a numeric representation for them, regardless of their length. Extending Word2Vec CBOW model, it adds another input vector to the input context, which is the paragraph identifier. When training the word vectors, the document vector is trained as well, and at the end it holds a numeric representation of the whole document. As with Word2Vec, in this work we use the CBOW approach.

FastText [32]. It is an extension of Word2Vec which treats each word as composed of character ngrams, learning representations for ngrams instead of words. The idea is to take into account and exploit the morphology of words. Each word is split in n-grams which are all inputted separately to the model, which can be trained using the CBOW or the skip-gram approach. The vector for each word is made of the sum of its character n grams, so it can generate embeddings for out of vocabulary words. By exploiting words morphology, FastText tries to generate better embeddings for rare words, assuming their character ngrams are shared with other words. It also allows one to generate embeddings for out of vocabulary words. To train FastText we use the originally proposed and most extended skigram approach.

To the best of our knowledge, this is the first time these text embeddings are trained from scratch on the same corpus and evaluated under the image retrieval by text task. We used Gensim[2] implementations of LDA, Word2Vec, FastText and Doc2Vec and the GloVe implementation by Maciej Kula.[3] While LDA and Doc2Vec can generate embeddings for documents, Word2Vec, GloVe and FastText only generate word embeddings. To get documents embeddings from these methods, we consider two standard strategies: First, computing the document embedding as the mean embedding of its words. Second, computing a *tf-idf* weighted mean of the words in the document. For all embeddings a dimensionality of 400 has been used. The value has been selected because is the one used in the Doc2Vec paper [31], which compares Doc2Vec with other text embedding methods, and it is enough to get optimum performances of Word2Vec, FastText and GloVe, as [12,32,30] show, respectively. For LDA a dimensionality of 200 has also been considered.

---

[2] https://radimrehurek.com/gensim.

[3] https://github.com/maciejkula/glove-python.

Figure 9.3 : Examples of InstaCities1M dataset images.

9.5 Benchmarks

In this section we present the datasets used in this work and show some examples of their images and their associated text.

9.5.1 InstaCities1M

A dataset formed by Instagram images associated with one of the 10 most populated English speaking cities all over the world (in the images captions one of the names of these cities appears). It contains 100K images for each city, which makes a total of 1M images, split in 800K training images, 50K validation images and 150K test images. The interest of this dataset is that is formed by recent social media data. The text associated with the images is the description and the hashtags written by the photo up-loaders, so it is the kind of free available data that would be very interesting to be able to learn from. Fig. 9.3 shows some examples of InstaCities1M images and their associated text. The InstaCities1M dataset is available on https://gombru.github.io/2018/08/01/InstaCities1M/.

9.5.2 WebVision

The Webvision dataset [33] contains more than 2.4 million images crawled from the Flickr website and Google Images search. The same 1000 concepts as the ILSVRC 2012 dataset [1] are used for querying images. The textual information accompanying those images (caption, user tags and description) is provided. The validation set, which is used as test in this work, contains 50K images. Fig. 9.4 shows some examples of WebVision images and their associated text.

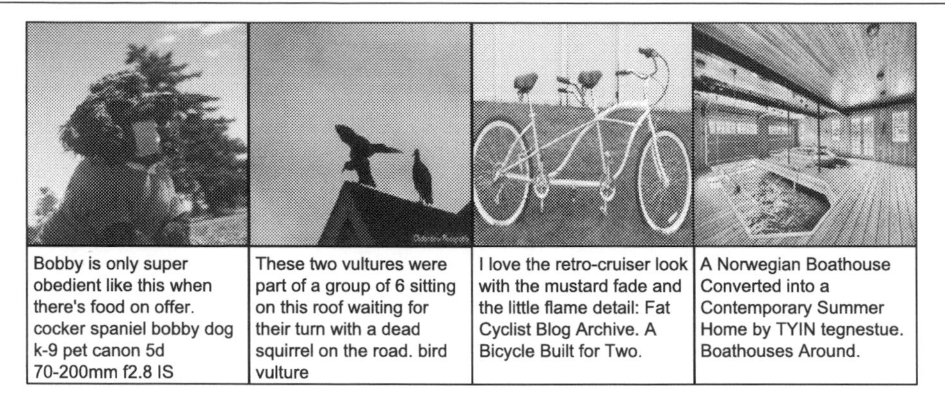

Figure 9.4 : Examples of WebVision dataset images.

Figure 9.5 : Examples of MIRFlickr dataset images.

9.5.3 MIRFlickr

The MIRFlickr dataset [34] contains 25,000 images collected from Flickr, annotated using 24 predefined semantic concepts. Fourteen of those concepts are divided in two categories: 1) strong correlation concepts and 2) weak correlation concepts. The correlation between an image and a concept is strong if the concept appears in the image predominantly. For differentiation, we denote strong correlation concepts by a suffix "*". Finally, considering strong and weak concepts separately, we get 38 concepts in total. All images in the dataset are annotated by at least one of those concepts. Additionally, all images have associated tags collected from Flickr. Following the experimental protocol in [35–38] tags that appear less than 20 times are first removed and then instances without tags or annotations are removed. Fig. 9.5 shows some examples of MIRFlickr images and their associated text.

Table 9.1: Queries for the retrieval experiments on InstaCities1M and WebVision datasets.

	Simple	Complex
Urban	car, skyline, bike	yellow+car, skyline+night, bike+park
Weather	sunrise, snow, rain	sunrise+beach, snow+ski, rain+umbrella
Food	ice-cream, cake, pizza	ice-cream+beach, chocolate+cake, pizza+wine
People	woman, man, kid	woman+bag, man+boat, kid+dog

9.6 Retrieval on InstaCities1M and WebVision Datasets

In this section we perform image retrieval experiments in the InstaCites1M and the WebVision datasets, comparing the performance of the different text embeddings in our pipeline. We analyze the performance of each text embedding, present an error analysis of our pipeline and show qualitative retrieval results of both image by text retrieval and image retrieval using multimodal queries.

9.6.1 Experiment Setup

To evaluate the learned joint embeddings, we define a set of textual queries and check visually if the TOP-5 retrieved images contain the querying concept. We define 24 different queries. Half of them are single word queries and the other half two word queries. They have been selected to cover a wide area of semantic concepts that are usually present in web and social media data. Both simple and complex queries are divided in four different categories: Urban, weather, food and people. Queries are listed in Table 9.1. For complex queries, only images containing both querying concepts are considered correct.

9.6.2 Results and Conclusions

Tables 9.2 and 9.3 show the mean Precision at 5 for InstaCities1M and WebVision datasets and transfer learning between those datasets. To compute transfer learning results, we train the model with one dataset and test with the other. Table 9.4 shows the mean precision at 5 for InstaCities1M with introduced additional noise and of a model trained with mean square error loss. The noise is introduced by changing the indicated % of captions to random captions from the training set. Figs. 9.1, 9.6 and 9.7 show the first retrieved images for some complex textual queries. Fig. 9.7 also shows results for non-object queries, proving that our pipeline works beyond traditional instance-level retrieval. Figs. 9.8 and 9.9 show that retrieval also works with multimodal queries combining an image and text.

For complex queries, where we demand two concepts to appear in the retrieved images, we obtain good results for those queries where the concepts tend to appear together. For instance,

Table 9.2: Performance on InstaCities1M and WebVision. First column shows the mean P@5 for all the queries, second for the simple queries and third for complex queries.

Text embedding	InstaCities1M			WebVision		
Queries	All	S	C	All	S	C
LDA 200	0.40	0.73	0.07	0.11	0.18	0.03
LDA 400	0.37	0.68	0.05	0.14	0.18	0.10
Word2Vec mean	0.46	0.71	**0.20**	0.37	0.57	0.17
Word2Vec tf-idf	0.41	0.63	0.18	**0.41**	0.58	0.23
Doc2Vec	0.22	0.25	0.18	0.22	0.17	**0.27**
GloVe	0.41	0.72	0.10	0.36	**0.60**	0.12
GloVe tf-idf	**0.47**	**0.82**	0.12	0.39	0.57	0.22
FastText tf-idf	0.31	0.50	0.12	0.37	0.60	0.13

Table 9.3: Performance on transfer learning. First column shows the mean P@5 for all the queries, second for the simple queries and third for complex queries.

Text embedding	Train: WebVision Test: InstaCities			Train: InstaCities Test: WebVision		
Queries	All	S	C	All	S	C
LDA 200	0.14	0.25	0.03	0.33	0.55	0.12
LDA 400	0.17	0.25	0.08	0.24	0.39	0.10
Word2Vec mean	0.41	**0.63**	0.18	0.33	0.52	0.15
Word2Vec tf-idf	**0.42**	0.57	**0.27**	0.32	0.50	0.13
Doc2Vec	0.27	0.40	0.15	0.24	0.33	0.15
GloVe	0.36	0.58	0.15	0.29	0.53	0.05
GloVe tf-idf	0.39	0.57	0.22	**0.51**	**0.75**	**0.27**
FastText tf-idf	0.39	0.57	0.22	0.18	0.33	0.03

Table 9.4: Performance on InstaCities1M using GloVe tf-idf introducing noise by changing the indicated % of captions by random captions from the training set.

Experiment	InstaCities1M		
Queries	All	S	C
Without introduced noise	0.47	0.82	0.12
10% introduced noise	0.25	0.43	0.07
20% introduced noise	0.18	0.32	0.05
30% introduced noise	0.15	0.25	0.05

Figure 9.6 : First retrieved images for complex queries with Word2Vec on InstaCites1M.

Figure 9.7 : First retrieved images for complex queries (left), city related complex queries (top-right) and non-object queries (bottom-right) with Word2Vec on InstaCites1M.

Figure 9.8 : First retrieved images for multimodal queries (concepts are added or removed to bias the results) with Word2Vec on WebVision.

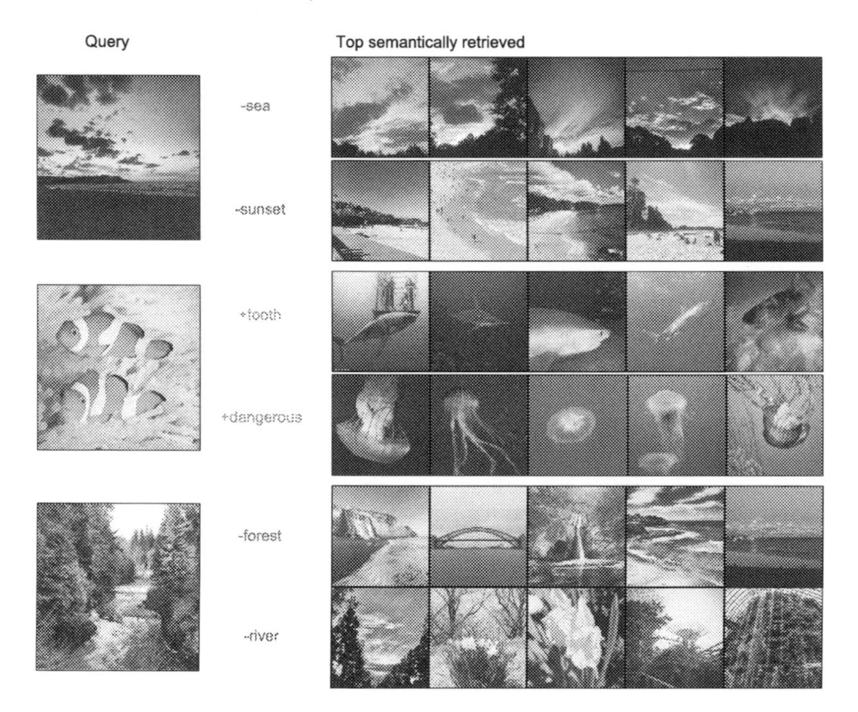

Figure 9.9 : First retrieved images for multimodal complex queries with Word2Vec on WebVision.

we generally retrieve correct images for “skyline + night” and for “bike + park”, but we do not retrieve images for “dog + kid”. When failing with this complex queries, usually images where only one of the two querying concepts appears are retrieved. Fig. 9.10 shows that in some cases images corresponding to semantic concepts between the two querying concepts

Figure 9.10 : First retrieved images for simple (left and right columns) and complex weighted queries with Word2Vec on InstaCites1M.

are retrieved. That proves that the common embedding space that has been learned has a semantic structure. The performance is generally better in InstaCities1M than in WebVision. The reason is that the queries are closer to the kind of images people tend to post in Instagram than to the ImageNet classes. However, the results on transfer learning show that WebVision is a better dataset to train than InstaCities1M. That is because WebVision has more images than InstaCities1M (2.4M training images vs. 800k training images) and shows that the learned models are robust, general and scalable: Having more data, even if it is not specifically related with the target task, allows for learning embedding models that perform better in that task. Results show that all the tested text embeddings methods work quite well for simple queries. However, LDA fails when is trained in WebVision. That is because LDA learns latent topics with semantic sense from the training data. Every WebVision image is associated to one of the 1000 ImageNet classes, which influences a lot the topics learning. As a result, the embedding fails when the queries are not related to those classes. The top-performing methods are GloVe when training with InstaCities1M and Word2Vec when training with WebVision, but the difference between their performance is small. FastText achieves a good performance on WebVision but a bad performance on InstaCities1M compared to the other methods. An explanation is that, while social media data contains more colloquial vocabulary, WebVision contains domain specific and diverse vocabulary, and since FastText learns representations for character ngrams, is more suitable to learn representations from corpus that are morphologically rich. Doc2Vec does not work well in any database. That is because it is oriented to deal with larger texts than the ones we find accompanying images in web and social media. For the word embedding methods Word2Vec and GloVe, the results computing the text representation as the mean or as the *tf-idf* weighted mean of the words embeddings are similar.

The overall conclusion of the performance comparison between text embeddings in this experiment is that word-level text embeddings such as Word2Vec and GloVe perform better than document level text embeddings (LDA, Doc2Vec) and character ngrams level text embeddings (FastText). The reason is that captions associated to images in social media tend to be quite concise, so averaging the word-level embeddings of a caption still gives us an informative representation that allows us to take profit of the rich semantic space learned by this kind of embeddings. The fact that this semantic space is quite sparse allows us to perform arithmetic between embeddings in it, and also to be able to learn from those representations averaged over caption's words. The introduction of additional artificial noise deteriorates the results heavily. This indicates that, despite the proposed learning pipeline can learn powerful visual features from web and social media data with its inherent noise, reducing it may lead to huge performance improvements.

9.6.3 Error Analysis

Remarkable sources of errors are listed and explained in this section.

9.6.3.1 Visual features confusion

Errors due to the confusion between visually similar objects may occur, for instance retrieving images of a quiche when querying "pizza". Those errors could be avoided using more data and higher dimensional representations, since the problem is the lack of training data to learn visual features that generalize to unseen samples.

9.6.3.2 Errors from the dataset statistics

An important source of errors is due to dataset statistics. As an example, the WebVision dataset contains a class which is "snow leopard" and it has many images of that concept. The word "snow" appears frequently in the images correlated descriptions, so the net learns to embed together the word "snow" and the visual features of a "snow leopard". There are many more images of "snow leopard" than of "snow", therefore, when we query "snow" we get snow leopard images. Fig. 9.11 shows this error and how we can use complex multimodal queries to bias the results.

9.6.3.3 Words with different meanings or uses

Words with different meanings or words that people use in different scenarios introduce unexpected behaviors. For instance when we query "woman + bag" in the InstaCities1M dataset

Figure 9.11 : First retrieved images for text queries using Word2Vec on WebVision. Concepts are removed to bias the results.

we usually retrieve images of pink bags. The reason is that people tend to write “woman” in an image caption when pink stuff appears. Those are considered errors in our evaluation, but inferring which images people relate with certain words in social media can be a very interesting research.

9.7 Retrieval in the MIRFlickr Dataset

To compare the performance of our pipeline to other types of image retrieval by text systems we use the MIRFlickr dataset, which is typically used to train and evaluate image retrieval systems. The objective is to prove the quality of the multimodal embeddings learned solely with web data comparing them to supervised methods.

9.7.1 Experiment Setup

We consider three different experiments: 1) Using as queries the tags accompanying the query images and computing the MAP of all the queries. Here a retrieved image is considered correct if it shares at least one tag with the query image. For this experiment, the splits used are 5% queries set and 95% training and retrieval set, as defined in [36,38]. 2) Using as queries the class names. Here a retrieved image is considered correct if it is tagged with the query concept. For this experiment, the splits used are 50% training and 50% retrieval set, as defined

Table 9.5: MAP on the image by text retrieval task on MIRFlickr as defined in [36,38].

Method	Train	map
LDA 200	InstaCites1M	0.736
LDA 400	WebVision	0.627
Word2Vec tf-idf	InstaCites1M	0.720
Word2Vec tf-idf	WebVision	0.738
GloVe tf-idf	InstaCites1M	**0.756**
GloVe tf-idf	WebVision	0.737
FastText tf-idf	InstaCities1M	0.677
FastText tf-idf	WebVision	0.734
Word2Vec tf-idf	MIRFlickr	0.867
GloVe tf-idf	MIRFlickr	**0.883**
DCH [36]	MIRFlickr	0.813
LSRH [37]	MIRFlickr	0.768
CSDH [38]	MIRFlickr	0.764
SePH [35]	MIRFlickr	0.735
SCM [39]	MIRFlickr	0.631
CMFH [40]	MIRFlickr	0.594
CRH [41]	MIRFlickr	0.581
KSH-CV [42]	MIRFlickr	0.571

Table 9.6: MAP on the image by text retrieval task on MIRFlickr as defined in [43].

Method	Train	map
GloVe tf-idf	InstaCites1M	**0.57**
GloVe tf-idf	MIRFlickr	**0.73**
MML [43]	MIRFlickr	0.63
InfR [43]	MIRFlickr	0.60
SBOW [43]	MIRFlickr	0.59
SLKL [43]	MIRFlickr	0.55
MLKL [43]	MIRFlickr	0.56

in [44]. 3) Same as experiment 1 but using the MIRFlickr train-test split proposed in Zhang et al. [43].

9.7.2 Results and Conclusions

Tables 9.5 and 9.6 show the results for the experiments 1 and 3, respectively. We see that our pipeline trained with web and social media data in a multimodal self-supervised fashion achieves competitive results. When trained with the target dataset, our pipeline outperforms the other methods. Table 9.7 shows results for the experiment 2. Our pipeline with the GloVe

Table 9.7: AP scores for 38 semantic concepts and MAP on MIRFlickr. Underlined numbers compare our method trained with InstaCities and other methods trained with the target dataset.

Method	GloVe tf-idf	MMSHL [44]	SCM [39]	GloVe tf-idf
Train	MIRFlickr			InstaCities
animals	0.775	0.382	0.353	0.707
baby	0.337	0.126	0.127	0.264
baby*	0.627	0.086	0.086	0.492
bird	0.556	0.169	0.163	0.483
bird*	0.603	0.178	0.163	0.680
car	0.603	0.297	0.256	0.450
car*	0.908	0.420	0.315	0.858
female	0.693	0.537	0.514	0.481
female*	0.770	0.494	0.466	0.527
lake	0.403	0.194	0.182	0.230
sea	0.720	0.469	0.498	0.565
sea*	0.859	0.242	0.166	0.731
tree	0.727	0.423	0.339	0.398
tree*	0.894	0.423	0.339	0.506
clouds	0.792	0.739	0.698	0.613
clouds*	0.884	0.658	0.598	0.710
dog	0.800	0.195	0.167	0.760
dog*	0.901	0.238	0.228	0.865
sky	0.900	0.817	0.797	0.809
structures	0.850	0.741	0.708	0.703
sunset	0.601	0.596	0.563	0.590
transport	0.650	0.394	0.368	0.287
water	0.759	0.545	0.508	0.555
flower	0.715	0.433	0.386	0.645
flower*	0.870	0.504	0.411	0.818
food	0.712	0.419	0.355	0.683
indoor	0.806	0.677	0.659	0.304
plant_life	0.846	0.734	0.703	0.564
portrait	0.825	0.616	0.524	0.474
portrait*	0.841	0.613	0.520	0.483
river	0.436	0.163	0.156	0.304
river*	0.497	0.134	0.142	0.326
male	0.666	0.475	0.469	0.330
male*	0.743	0.376	0.341	0.338
night	0.589	0.564	0.538	0.542
night*	0.804	0.414	0.420	0.720
people	0.910	0.738	0.715	0.640
people*	0.945	0.677	0.648	0.658
MAP	0.738	0.451	0.415	0.555

tf-idf text embedding trained with InstaCites1M outperforms state-of-the-art methods in most of the classes and in MAP. If we train with the target dataset, results are improved significantly. Notice that despite being applied here to the classes and tags existing in MIRFlickr, our pipeline is generic and has learned to produce joint image and text embeddings for many more semantic concepts, as seen in the qualitative examples.

9.8 Comparing the Image and Text Embeddings

In this section we analyze the semantic quality of the learned joint embedding spaces showing how the CNN has learned to embed images in them.

9.8.1 Experiment Setup

To evaluate how the CNN has learned to map images to the text embedding space and the semantic quality of that space, we perform the following experiment: We build random image pairs from the MIRFlickr dataset and we compute the cosine similarity between both their image and their text embeddings. In Fig. 9.12 we plot the images embeddings distance vs. the text embedding distance of 20,000 random image pairs. If the CNN has learned correctly to map images to the text embedding space, the distances between the embeddings of the images and the texts of a pair should be similar, and points in the plot should fall around the identity line $y = x$. Also, if the learned space has a semantic structure, both the distance between images embeddings and the distance between texts embeddings should be smaller for those pairs sharing more tags: The plot points' color reflects the number of common tags of the image pair, so pairs sharing more tags should be closer to the origin of the axis.

As an example, take a dog image with the tag "dog", a cat image with the tag "cat" and one of a scarab with the tag "scarab". If the text embedding has been learned correctly, the distance between the projections of dog and scarab tags in the text embedding space should be bigger than the one between dog and cat tags, but smaller than the one between other pairs not related at all. If the CNN has correctly learned to embed the images of those animals in the text embedding space, the distance between the dog and the cat image embeddings should be similar than the one between their tags embeddings (and the same for any pair). So the point given by the pair should fall in the identity line. Furthermore, that distance should be closer to the coordinates origin than the point given by the dog and scarab pair, which should also fall in the identity line and nearer to the coordinates origin that another pair that has no relation at all.

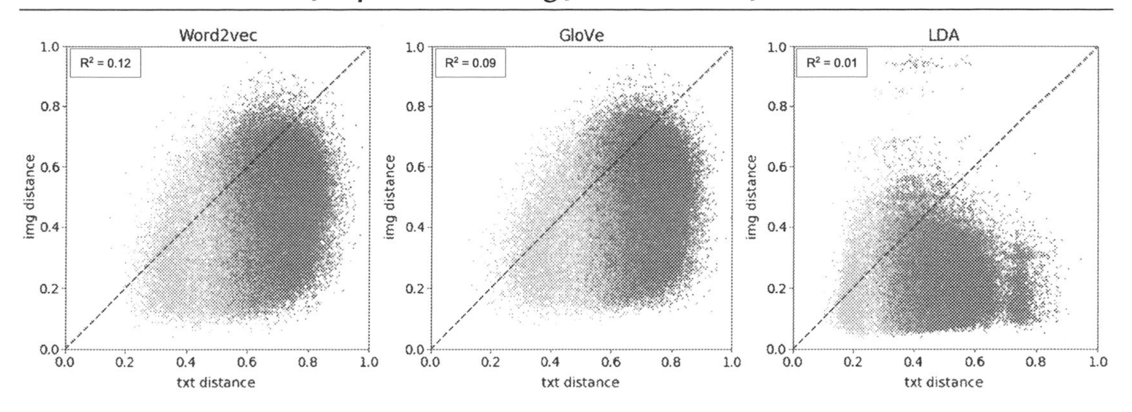

Figure 9.12 : Text embeddings distance (X) vs. the images embedding distance (Y) of different random image pairs for LDA, Word2Vec and GloVe embeddings trained with InstaCities1M. Distances have been normalized between [0, 1]. Points are red if the pair does not share any tag, orange if it shares one, light orange if it shares two, yellow if it shares three and green if it shares more. R^2 is the coefficient of determination of images and texts distances.

9.8.2 Results and Conclusions

The plots in Fig. 9.12 for both the Word2Vec and the GloVe embeddings show a similar shape. The resulting blob is elongated along the $y = x$ direction, which proves that both image and text embeddings tend to provide similar distances for an image pair. The blob is thinner and closer to the identity line when the distances are smaller (so when the image pairs are related), which means that the embeddings can provide a valid distance for semantic concepts that are close enough (dog, cat), but fails inferring distances between weak related concepts (car, skateboard). The colors of the points in the plots show that the space learned has a semantic structure. Points corresponding to pairs having more tags in common are closer to the coordinates origin and have smaller distances between the image and the text embedding. From the colors it can also be deduced that the CNN is good inferring distances for related images pairs: there are just a few images having more than three tags in common with image embedding distance bigger than 0.6, while there are many images with bigger distances that do not have tags in common. However, the visual embedding sometimes fails and infers small distances for image pairs that are not related, as those image pairs having no tags in common and an image embedding distance below 0.2.

The plot of the LDA embedding shows that the learned joint embedding is not so good in terms of the CNN images mapping to the text embedding space nor in terms of the space semantic structure. The blob does not follow the identity line direction that much, which means that the CNN and the LDA are not inferring similar distances for images and texts of pairs. The points colors show that the CNN is inferring smaller distances for more similar image pairs only when the pairs are very related.

The coefficient of determination R^2 shown at each graph measures the proportion of the variance in a dependent variable that is predicted by linear regression and a predictor variable. In this case, it can be interpreted as a measure of in how far image distances can be predicted from text distances and, therefore, of how well the visual embedding has learned to map images to the joint image–text space. It ratifies our plots' visual inspection, proving that visual embeddings trained with Word2Vec and GloVe representations have learned a much more accurate mapping than LDA, and shows that Word2Vec is better in terms of that mapping.

9.9 Visualizing CNN Activation Maps

We have proved that, using only social media data, state-of-the-art CNNs can be trained in a self-supervised way to learn powerful visual features, capable to discriminate among a huge variety of scenes: from objects to outdoor scenes, abstract concepts or specific buildings. In this experiment we visualize the images from the InstaCities1M retrieval set that generated the highest activations in some CNN units, using the GoogleNet trained from scratch with InstaCites1M and GloVe tf-idf text embedding as self-supervision. We also show the regions of the images that activated most the selected units. To generate those activations maps we used *deconvnet*, proposed by Zeiler et al. [45] and the Caffe implementation presented in [46]. Fig. 9.13 shows the results of a selection of neurons in the *pool5* layer of our model. We can notice that network units are selective to specific buildings, such as the Golden Gate Bridge, objects such as guitars, drums or lights to identify concert scenes, or even basketball t-shirts.

9.10 Visualizing the Learned Semantic Space with t-SNE

In this section we use the t-SNE dimensionality reduction method to reduce the dimensionality of the joint embedding space to 2 dimensions and we show images in that space to visualize its semantic structure.

9.10.1 Dimensionality Reduction with t-SNE

Inspired by A. Karpathy's work,[4] who uses t-SNE to visualize CNN layer features, we use t-SNE[5] [47] to visualize the learned joint visual and textual embedding. t-SNE is a non-linear dimensionality reduction method, which we use on our 400 dimensional embeddings to produce 2 dimensional embeddings. For each one of the given 400 dimensional visual or textual

[4] https://cs.stanford.edu/people/karpathy/cnnembed/.

[5] https://github.com/lvdmaaten/bhtsne/.

Figure 9.13 : Top-5 activations for five units in pool5 layer of GoogleNet model trained from scratch with InstaCities1M using GloVe tf-idf as self-supervision and their activation maps.

embeddings, t-SNE computes a 2 dimensional embedding arranging elements that have similar representations nearby, providing a way to visualize the learned joint image–text space and analyze qualitatively its semantic structure.

9.10.2 Visualizing Both Image and Text Embeddings

As we have learned a joint image and text embedding space, we can apply t-SNE to both modalities of embeddings at once. We apply t-SNE to a set formed by the visual embeddings of the images in test set of InstaCities1M and the text embeddings of the selected querying terms (Table 9.1). In this experiment, we use the Word2Vec model trained on InstaCities1M dataset.

9.10.3 Showing Images at the Embedding Locations

First, we set a canvas with predefined dimensions (2000×2000 pixels). Then we normalize the 2 dimensional embeddings given by t-SNE to fit in the canvas size. Finally, we visualize images at their embedding locations, setting their top-left corner at their embedding location and resizing them to 50×50 pixels. For text embeddings, we use an image containing its words as their representations in the canvas. To get an interpretable visualization avoiding images overlaps, if two images share any pixel in the output figure we omit one of them (prioritizing word images). Therefore, images surrounding word images are not necessary top retrieval results for that word, but they are the nearest images of the ones being represented in the figure.

9.10.4 Semantic Space Inspection

The joint embedding 2 dimensional visualization in Fig. 9.14 shows the semantic structure of the learned space. It shows semantic clusters that the joint embedding has learned in a self-supervised way from the data distribution, corresponding to different kind of images people tend to post on Instagram. For instance, the figure shows a cluster for food images, a cluster for sport images, a cluster for sunrise images, or a cluster for animal images. It also shows that images of people are very numerous, and that the joint embedding groups them correctly. It can also be appreciated how images we might consider noise, such as images with logos or text, are clustered together. The majority of those images are far from the semantic clusters, isolated and near the figure edges. That is because the joint embedding has not been able to find semantic relations between these images and the rest, so it assigns to them embeddings that have no relation with the others. When computing t-SNE, as the objective is to place similar images nearby, this images without semantic relations are set far

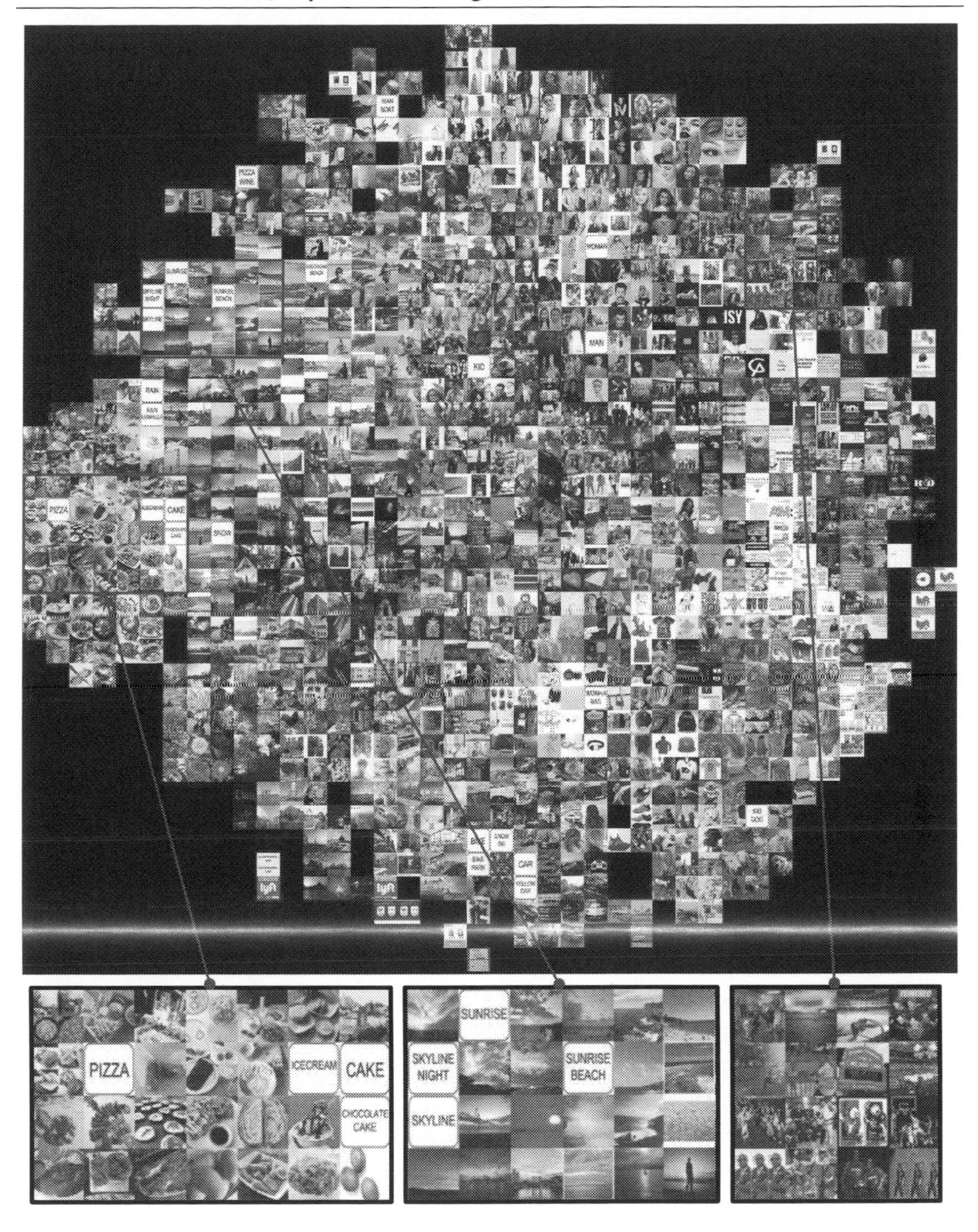

Figure 9.14 : Visualization (2000 x 2000 px) of the joint embedding with Word2Vec on InstaCities1M dataset.

from the others. Therefore, we can conclude that the pipeline is quite robust to social media noise. More t-SNE visualizations of the learned joint embeddings are available in https://gombru.github.io/2018/08/01/learning_from_web_data.

9.11 Conclusions

In this work we learn a joint visual and textual embedding using web and social media data and we benchmark state-of-the-art text embeddings in the image retrieval by text task, concluding that GloVe and Word2Vec are the best ones for this data, having a similar performance and competitive performances over supervised methods in the image retrieval by text task. We show that our models go beyond instance-level image retrieval to semantic retrieval and that can handle multiple concepts queries and also multimodal queries, composed of a visual query and a text modifier to bias the results. We clearly outperform the state of the art in the MIRFlickr dataset when training in the target data. The code used in this work is available on https://github.com/gombru/LearnFromWebData.

Acknowledgments

This work was supported by the Doctorats Industrials program from the Generalitat de Catalunya, the Spanish project TIN2017-89779-P, the H2020 Marie Skłodowska-Curie actions of the European Union, grant agreement No. 712949 (TECNIOspring PLUS), and the Agency for Business Competitiveness of the Government of Catalonia (ACCIO).

References

[1] J. Deng, W. Dong, R. Socher, L.-J. Li, K. Li, L. Fei-Fei, ImageNet: a large-scale hierarchical image database, in: CVPR, 2009.
[2] T.Y. Lin, M. Maire, S. Belongie, J. Hays, P. Perona, D. Ramanan, P. Dollar, C.L. Zitnick, Microsoft COCO: Common Objects in Context, Lect. Notes Comput. Sci., 2014.
[3] B. Zhou, A. Lapedriza, A. Khosla, A. Oliva, A. Torralba, Places: a 10 million image database for scene recognition, in: TPAMI, 2017.
[4] J. Yosinski, J. Clune, Y. Bengio, H. Lipson, How transferable are features in deep neural networks?, in: NIPS, 2014.
[5] J. Mar, V. David, D. Ger, M.L. Antonio, Learning appearance in virtual scenarios for pedestrian detection, in: CVPR, 2010.
[6] G. Ros, L. Sellart, J. Materzynska, D. Vazquez, A.M. Lopez, The SYNTHIA dataset: a large collection of synthetic images for semantic segmentation of urban scenes, in: CVPR, 2016.
[7] T.Q. Phan, P. Shivakumara, S. Tian, C.L. Tan, Recognizing text with perspective distortion in natural scenes, in: ICCV, 2013.
[8] A. Gupta, A. Vedaldi, A. Zisserman, Synthetic data for text localisation in natural images, in: CVPR, 2016.
[9] W. Li, L. Wang, W. Li, E. Agustsson, J. Berent, A. Gupta, R. Sukthankar, L. Van Gool, WebVision challenge: visual learning and understanding with web data, arXiv:1705.05640, 2017.
[10] D. Mahajan, R. Girshick, V. Ramanathan, K. He, M. Paluri, Y. Li, A. Bharambe, L. Der Van, M. Facebook, Exploring the limits of weakly supervised pretraining, in: ECCV, 2018.

[11] M. Norouzi, T. Mikolov, S. Bengio, Y. Singer, J. Shlens, A. Frome, G.S. Corrado, J. Dean, Zero-shot learning by convex combination of semantic embeddings, in: NIPS, 2013.
[12] T. Mikolov, G. Corrado, K. Chen, J. Dean, Efficient estimation of word representations in vector space, in: ICLR, 2013.
[13] A. Gordo, D. Larlus, Beyond instance-level image retrieval: leveraging captions to learn a global visual representation for semantic retrieval, in: CVPR, 2017.
[14] L. Gomez, Y. Patel, M. Rusiñol, D. Karatzas, C.V. Jawahar, Self-supervised learning of visual features through embedding images into text topic spaces, in: CVPR, 2017.
[15] Y. Patel, L. Gomez, R. Gomez, M. Rusiñol, D. Karatzas, C.V. Jawahar, TextTopicNet – self-supervised learning of visual features through embedding images on semantic text spaces, arXiv:1705.08631, 2018.
[16] D.M. Blei, A.Y. Ng, M.I. Jordan, Latent Dirichlet allocation, Journal of Machine Learning Research 3 (2003) 993–1022.
[17] L. Wang, Y. Li, S. Lazebnik, Learning deep structure-preserving image–text embeddings, in: CVPR, 2016.
[18] Y. Patel, L. Gomez, M. Rusiñol, D. Karatzas, Dynamic lexicon generation for natural scene images, in: ECCV, 2016.
[19] A. Gordo, J. Almazan, N. Murray, F. Perronin, LEWIS: latent embeddings for word images and their semantics, in: ICCV, 2015.
[20] Princeton University, WordNet, 2010.
[21] A. Salvador, N. Hynes, Y. Aytar, J. Marin, F. Ofli, I. Weber, A. Torralba, Learning cross-modal embeddings for cooking recipes and food images, in: CVPR, 2017.
[22] R. Gomez, L. Gomez, J. Gibert, D. Karatzas, Learning from #Barcelona Instagram data what locals and tourists post about its neighbourhoods, in: ECCV Workshops, 2018.
[23] G. Patrini, A. Rozza, A. Menon, R. Nock, L. Qu, Making deep neural networks robust to label noise: a loss correction approach, in: CVPR, 2016.
[24] T. Xiao, T. Xia, Y. Yang, C. Huang, X. Wang, Learning from massive noisy labeled data for image classification, in: CVPR, 2015.
[25] J. Fu, Y. Wu, T. Mei, J. Wang, H. Lu, Y. Rui, Relaxing from vocabulary: robust weakly-supervised deep learning for vocabulary-free image tagging, in: ICCV, 2015.
[26] X. Xu, L. He, H. Lu, A. Shimada, R.I. Taniguchi, Non-linear matrix completion for social image tagging, in: IEEE Access, 2017.
[27] M. Melucci, Relevance feedback algorithms inspired by quantum detection, IEEE Transactions on Knowledge and Data Engineering 28 (4) (2016) 1022–1034.
[28] C. Szegedy, V. Vanhoucke, S. Ioffe, J. Shlens, Rethinking the inception architecture for computer vision, in: CVPR, 2016.
[29] Y. Jia, E. Shelhamer, J. Donahue, S. Karayev, J. Long, R. Girshick, S. Guadarrama, T. Darrell, Caffe: convolutional architecture for fast feature embedding, arXiv:1408.5093, 2014.
[30] J. Pennington, R. Socher, C. Manning, Glove: global vectors for word representation, in: EMNLP, 2014.
[31] Q.V. Le, T. Mikolov, Distributed representations of sentences and documents, in: NIPS, 2014.
[32] P. Bojanowski, E. Grave, A. Joulin, T. Mikolov, Enriching word vectors with subword information, arXiv: 1607.04606, 2016.
[33] W. Li, L. Wang, W. Li, E. Agustsson, L. Van Gool, WebVision database: visual learning and understanding from web data, arXiv:1708.02862, 2017.
[34] M.J. Huiskes, M.S. Lew, The MIR flickr retrieval evaluation, in: ACM Int. Conf. Multimed. Inf. Retr., 2008.
[35] Z. Lin, G. Ding, M. Hu, J. Wang, Semantics-preserving hashing for cross-view retrieval, in: CVPR, 2015.
[36] X. Xu, F. Shen, Y. Yang, H.T. Shen, X. Li, Learning discriminative binary codes for large-scale cross-modal retrieval, IEEE Transactions on Image Processing 26 (5) (2017) 2494–2507.
[37] K. Li, G.J. Qi, J. Ye, K.A. Hua, Linear subspace ranking hashing for cross-modal retrieval, IEEE Transactions on Pattern Analysis and Machine Intelligence 39 (9) (2017) 1825–1838.
[38] L. Liu, Z. Lin, L. Shao, F. Shen, G. Ding, J. Han, Sequential discrete hashing for scalable cross-modality similarity retrieval, IEEE Transactions on Image Processing 26 (1) (2017) 107–118.

[39] D. Zhang, W.-j. Li, Large-scale supervised multimodal hashing with semantic correlation maximization, in: AAAI, 2014.
[40] G. Ding, Y. Guo, J. Zhou, Collective matrix factorization hashing for multimodal data, in: CVPR, 2014.
[41] Y. Zhen, D.-Y. Yeung, Co-regularized hashing for multimodal data, in: NIPS, 2012.
[42] J. Zhou, G. Ding, Y. Guo, Q. Liu, X. Dong, Kernel-based supervised hashing for cross-view similarity search, in: IEEE Int. Conf. Multimed. Expo, 2014.
[43] X. Zhang, X. Zhang, X. Li, Z. Li, S. Wang, Classify social image by integrating multi-modal content, in: Multimed. Tools Appl., Springer US, 2018.
[44] J. Wang, G. Li, A multi-modal hashing learning framework for automatic image annotation, in: Int. Conf. Data Sci. Cybersp., 2017.
[45] M.D. Zeiler, R. Fergus, Visualizing and understanding convolutional networks, in: ECCV, Springer, Cham, 2014.
[46] J. Yosinski, J. Clune, A. Nguyen, T. Fuchs, H. Lipson, Understanding neural networks through deep visualization, arXiv:1506.06579.
[47] L. Van Der Maaten, A. Courville, R. Fergus, C. Manning, Accelerating t-SNE using tree-based algorithms, Journal of Machine Learning Research 15 (2014) 3221–3245.

CHAPTER 10

3D Urban Scene Reconstruction and Interpretation from Multisensor Imagery

Hai Huang, Andreas Kuhn, Mario Michelini, Matthias Schmitz, Helmut Mayer

Institute for Applied Computer Science, Bundeswehr University Munich, Neubiberg, Germany

Contents

Multimodal Scene Understanding
https://doi.org/10.1016/B978-0-12-817358-9.00016-0

10.1 Introduction

Related work on 3D (urban) scene reconstruction goes back to the 1970s and 1980s. The Ascona workshop in 1995 [1] gave a good overview of the state reached by the middle of the 1990s. Since then, numerous developments have occurred.

This paper deals with the complete chain from pose estimation via dense 3D reconstruction, scene classification, scene and building decomposition to building modeling of levels of detail 2 (LoD2) and 3. Because the related work for all these different areas is rather disparate, we decided to include it in the respective sections. In the following we start with the description of our work on pose estimation suitable for unordered image sets possibly containing wide-baseline connections whose closing is essential to avoid that the image sets are split into disjoint partitions.

10.2 Pose Estimation for Wide-Baseline Image Sets

Pose estimation was in the focus of computer vision in the 1990s with the book of Hartley and Zisserman [2] summing up the theoretical findings. Arguably, random sample consensus—RANSAC [3] and the five-point algorithm [4] introduced the two most important ingredients to make fully automatic pose estimation without approximate values for the pose as in classical photogrammetric setups [5] a reality.

By now classical work such as by Pollefeys et al. [6] was based on (linear) image sequences. Later, a focus have been (extremely) large sets of images, so-called "community photo collections" [7]. While the feasibility has been shown in [8], particularly, [9] and later [10] have demonstrated how to deal with millions of or even one hundred million images.

Compared to the latter, our focus is rather different. We only deal with thousands of images, but we want to link as many of them as possible even if they are connected by a large baseline in comparison to the distance of the scene, as is needed when combining images from the ground with images from unmanned aerial vehicles (UAVs). This implies rather different viewing angles on the scene and, thus, makes the matching of points difficult.

Our approach requires no additional information except (an approximate) camera calibration. The latter can often be obtained from the meta-data of the images. The presentation in the following is split in two parts: In Sects. 10.2.1 and 10.2.2 means for pose estimation suitable for wide baselines are given, assuming that the overlap between the images is known. Part one basically follows [11,12], but with a couple of extensions and improvements added over the years. The second part in Sect. 10.2.3 consists of the capability to determine the overlap and, thus, to estimate the pose for the images even when the ordering of the images is unknown following [13,14].

10.2.1 Pose Estimation for Wide-Baseline Pairs and Triplets

As detailed below in Sect. 10.2.2, the basic building block for the determination of the poses for an image block by merging are triplets. While one could derive the poses for triplets directly from matches in three images, we start pose estimation with the smallest possible image sets, namely pairs. By this means, the complexity for matching in three images can be reduced by constraining it to the areas around the epipolar lines estimated for the pairs.

Because we want to deal with rather large baselines, we decided to employ affine least-squares matching [15,16] as the basic means for matching. "Affine" means that a patch is rotated by two angles (rotation and sheer), different scales are employed for the two basic axes and two translations (in x- and y-direction) are used. This enables a much better adaptation of two patches than the usually employed translations in combination with one rotation and one scale. What is more, least-squares matching does not only produce very accurate results, but additionally also provides estimates for the precision in the form of covariance matrices.

While the mathematically correct model for matching a planar patch would be the eight parameter projective transformation, we found empirically that it cannot be reliably estimated from the typical patch size of 13×13 pixels. The latter has also been determined empirically and gives a good balance between sufficient information content and not being too large so that disturbances, e.g., due to occlusions or non-planarity, do not affect the matching too adversely.

Unfortunately, least-squares matching is computationally very demanding and, thus, has to be limited to regions where it is likely to be successful. To this end, first points are detected by means of the scale invariant feature transform—SIFT [17]—in images reduced to a resolution of about 200×150 pixels. The points are then matched via (normalized) cross-correlation, because we found that the results of matching by cross-correlation do not degrade as abruptly as for matching by means of SIFT. Pairs with a correlation coefficient higher than an empirically determined low threshold of 0.5 are then input to least-squares matching, the results of which are only accepted if a high threshold of 0.95 is exceeded.

The resulting point pairs are input to the estimation of the relative pose of the pairs based on RANSAC and the five-point algorithm. We have extended this by a variant of the expectation maximization (EM) algorithm where we vary between expectation in the form of the determination of the inliers and maximization realized by means of robust bundle adjustment [18].

Once the relative pose has been determined for the pairs, the matching is repeated on double the resolution for triplets, reducing the search space considerably by means of the epipolar lines derived from the essential matrices estimated for the pairs. Pose estimation employs one "master" image for which the relative pose is determined to the other two "slave" images as above with RANSAC and the five-point algorithm for the same five points in the "master image". To link the pairs, relative scale is computed by means of triangulating the five points and computing the relative distances in both pairs. The median of the ratio of the distances is taken as the relative scale, i.e., relative length of the baselines. This basic result for triplets is refined on again the double resolution, projecting points in three images via the trifocal tensors derived from the relative poses of the triplets.

Additionally to the above approach for pairs, we have developed means to estimate also the poses for pairs with extreme baselines [19]. It is based on distorting the images by means of a large set of different projective transformations, detecting and matching points in the transformed images. While the results demonstrate that it is possible to estimate the pose for these extreme configurations, the computational complexity is extremely high, making the approach only useful if not more than a couple of image pairs need to be matched. In [20] an approach has been presented addressing the problem that often the pose is precisely determined only for parts of an image, as not enough points can be matched on difficult areas with strong perspective distortions such as roads seen from ground level. While [20] demonstrates the feasibility, it is limited to specific setups (roads seen from ground level in the lower part of the image).

10.2.2 Hierarchical Merging of Triplets

The derived triplets are the basis for the determination of the poses for the whole image set. The approach is described in detail in [21]. At the core is the hierarchical merging of image sets starting with triplets as the minimum sets.

While in principle 3D points alone or also combinations of 3D points and camera poses can be used to align the coordinate system of an image set to another image set, here the poses for two images included in both image sets (e.g., images 2 and 3 in the two triplets consisting of images 1, 2, 3 and 2, 3, 4, respectively) are used to compute a 3D Euclidean transformation. Even though the poses for one image are enough to compute the relative translation and rotation, the relative scale of both coordinate systems can only be computed using additional information such as 3D points seen in two or more images in both sets. As individual 3D points can be wrong due to matching errors particularly for wide baselines, we opted to use the poses of an additional image included in both sets (i.e., image 3 additionally to image 2 in the above example) as accumulator for the relative distances of the 3D points and the relative scale of both sets is computed as the ratio between the lengths of the baselines (of images 2 and 3 in the above example) in both sets.

Because two images included in both sets are used, a merge of two image sets with n_1 and n_2 images leads to a combined set of images consisting of $n_1 + n_2 - 2$ images.

While in earlier work [11] we had employed a serial extension by one triplet at a time, this was replaced by a hierarchical approach in [21]. Here, sets of approximately the same size are merged. This has two advantages: First, because we have found that after merging bundle adjustment is a must to obtain a reliable result by staying close to an optimal solution, the hierarchical approach avoids numerous bundle adjustments for large sets only slightly smaller than the complete set (in the serial case one has to adjust the complete set, the set minus one image, the set minus two images, etc.). Second, when the sets are of significantly different size, the employed robust bundle adjustment (cf. Sect. 10.2.1) tends to throw out the points from the smaller set, which due to a lower redundancy are less precise and thus more likely regarded as outliers.

In [21] it is also shown how one should reduce the points for merging if one wants to obtain a higher efficiency. The only way found to reduce the number of points which avoids a bias in the result is by randomly deleting 3D points. Intuitive means such as keeping points seen in as many images as possible also in combination with a policy to keep points well spread over the images give a better precision, but it was found that it can lead to a severe bias.

Recently, in [14] it has been shown that a search for maximum matching in the graph describing the connections between the images leads to a higher degree of parallelization and, therefore, a higher efficiency.

10.2.3 Automatic Determination of Overlap

The approach introduced in Sects. 10.2.1 and 10.2.2 assumes that the relation, i.e., overlap, between the images is known. Determining the overlap for larger image sets is a difficult problem, because basically all images have to be compared to all other. The least-squares matching approach of Sect. 10.2.1 can deal with large baselines, but its complexity is way too high to be used for all possible pairs.

Thus, a multistage approach has been devised [13,14], which first matches SIFT points in image pairs highly efficiently. The number of matches in each pair is the basis to restrict the approach introduced in Sect. 10.2.1 to a small subset of promising pairs and triplets. This allows one to estimate the pose for image sets containing essential wide-baseline connections which are the only means to connect certain subsets, but at the same time limits the computational complexity.

The relationships between images are modeled by an undirected weighted graph $G_I = (V_I, E_I)$, where a node $v \in V_I$ represents an image and an edge $(v_i, v_j) \in E_I$ connects two overlapping images corresponding to the nodes v_i and v_j. The edges are weighted using the Jaccard distance [22] with correspondences determined by matching binary SIFT descriptors. The latter are obtained by embedding SIFT descriptors [17] from continuous space $\mathbb{R}^{128}$ into Hamming space $\mathbb{H}^{128} = \{0, 1\}^{128}$ based on orthants. By this means, a compact representation of the descriptors as bit vectors is achieved resulting in a very fast comparison using the Hamming distance followed by the distance ratio test [17].

The graph G_I allows for a straightforward modeling if pairs are used for linking. However, it lacks descriptiveness in the case of triplet-based linking requiring higher-order relationships. In addition, the employed hierarchical merging of triplets (cf. Sect. 10.2.2) uses pairs to propagate the geometry (i.e., linkable triplets must have two images in common) and this constraint cannot be modeled by G_I.

Therefore, we employ the line graph $L(G_I) = (V_L, E_L)$ of the graph G_I to describe the linking of images. It has as set of nodes V_L the edges of G_I. An edge $(v_i, v_j) \in E_L$ exists iff the incident nodes v_i and v_j, corresponding to edges in E_I, have exactly one node $v \in V_I$ in common. Hence, V_L contains nodes corresponding to pairs of overlapping images, where two nodes are adjacent if the pairs have an image in common. Because of this, the linking of triplets using pairs for geometry propagation corresponds to a traversal through $L(G_I)$. An edge in $L(G_I)$ corresponds to a triplet and its weighting is based on the lowest quality of the three pairs of the triplet, where the quality of a pair is defined as the number of correspondences weighted by the roundness [23] of the corresponding reconstructed 3D points.

For a direct modeling of images, we extend $L(G_I)$ to explicitly represent the images using a second node type leading to the so-called *linking graph* $G = (V_I \cup V_L, E)$. It comprises *pair nodes* $n_p \in V_L$ corresponding to image pairs and *image nodes* $n_i \in V_I$ corresponding to images. An image and a pair node are adjacent if the pair node contains the image corresponding to the image node. Because the image nodes are only required for a complete modeling, no edges exist between them.

The linking graph completely describes the image linking, but can contain links of varying quality, e.g., due to critical camera configurations [24], as well as redundant links which are either less suitable or not required for pose estimation. Thus, we determine a linking subgraph (LSG) containing only essential links by searching for a terminal minimum Steiner tree [25] in the linking graph.

However, due to the tree-like structure, potential image loops in the LSG are not closed. At this stage, we can obtain additional information in the form of approximate poses very fast by hierarchically merging the triplets contained in the LSG without bundle adjustment. By this means we are able to efficiently search for image pairs which can be used to close loops by restricting the search space using a Euclidean neighborhood as well as the view direction

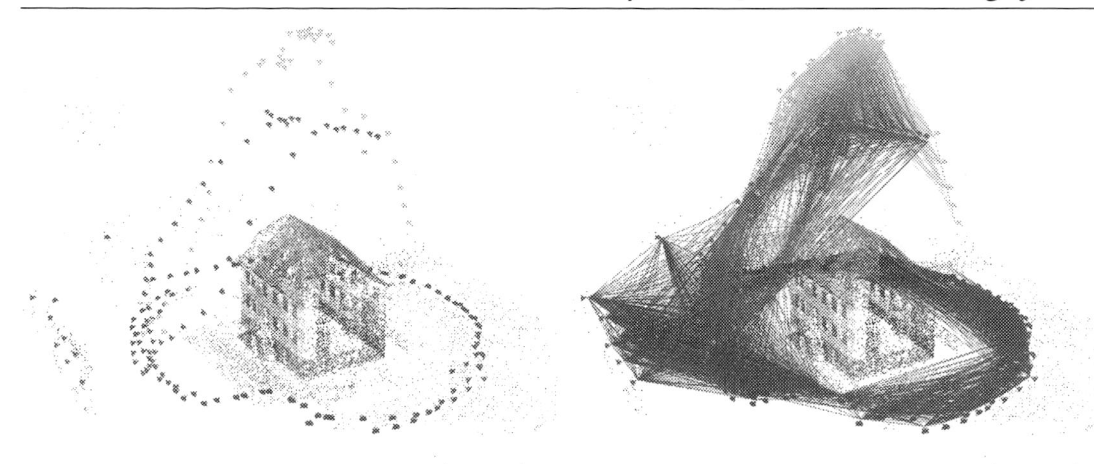

Figure 10.1 : Pose estimation for Building 1. Camera poses are presented as pyramids and the colors symbolize a different camera type/calibration. The links between the poses visualize detected overlap between the respective images.

difference between images. In addition, we determine the length of a potential image loop in the form of the graph distance between its outermost image nodes and restrict it as well.

For an efficient pose estimation the linking graph is iteratively constructed using minimum spanning trees of G_I as well as its subgraph LSG determined until all images are suitably linked or the further search is not meaningful anymore. As the terminal Steiner problem has been shown to be NP-complete [25], an approximation [26] is used for the determination of the LSG. After closing the image loops, hierarchical merging of triplets (cf. Sect. 10.2.2) is employed to determine the actual poses.

Fig. 10.1 shows the poses (left) as well as the links between them (right) for our running example scene Building 1.

10.3 Dense 3D Reconstruction

Accurately estimated camera poses for each image are the basis for the generation of dense depth maps considering geometric epipolar constraints. For surface reconstruction from the depth maps a multiplicity of methods has been developed in the recent decades.

In particular, volumetric methods based on local optimization have demonstrated their potential for 3D Reconstruction on a larger scale. The basic idea of extracting an iso-surface from voxels numerically occupied by propagated truncated signed distance functions was first presented by Curless and Levoy [27]. Goesele et al. [28] showed their adaptation to 3D

Reconstruction from Multi-View Stereo (MVS)-based depth maps while Sagawa et al. [29] extended the fusion to varying voxel sizes. Finally, Fuhrmann and Goesele [30] employed variable surface qualities for large-scale volumetric reconstruction from MVS images.

Instead of linear distance functions used in the latter methods, we employ an alternative probabilistic distance function for varying voxel resolutions. An additional filtering step allows for outlier removal in challenging configurations from noisy depth maps. The filtering is based on the idea of free-space constraints, which was originally proposed by Merrell et al. [31]. In contrast to the image-based consistency checks, we propose a probabilistic filtering in the volumetric 3D space.

10.3.1 Dense Depth Map Generation and Uncertainty Estimation

We use semi-global matching (SGM) [32] for MVS matching as it allows for an efficient but still effective processing even for high-resolution images. In our pipeline, stereo pairs that have a sufficient overlap determined from the sparse 3D point cloud generated by pose estimation are selected from the entire image set. Subsequently, the pairwise estimated stereo depth maps are fused for each image to compensate for noise and to filter outliers. In addition, SGM makes use of peak filtering by means of depth map clustering.

In general, the optimization scheme employed in SGM allows for a dense reconstruction of disparities even of weakly textured areas. Because of the reasonably lower quality for weakly textured areas, we additionally employ a pixelwise uncertainty estimation of disparities as post-processing [33,20]. It has been shown that besides the lack of texturedness especially the orientation of the surface relative to the cameras' line-of-sight influences the disparity uncertainty. This can be traced back to fronto-parallel prior assumptions embedded in the SGM optimization framework. Therefore, we classify the disparity uncertainties to consider their quality subsequently in the 3D surface reconstruction (cf. Fig. 10.2).

While the estimation of surface orientation (e.g. the normal vector) cannot be done in a stable way on disparity maps, we found a feature based on the local Total Variation (TV) of disparities to work well for uncertainty estimation. More precisely, we determine the TV in an iteratively growing pixel neighborhood (window) until it exceeds a given threshold. We define the resulting window size as a TV class and use its uncertainty for the estimation of the spatial error. To assign a disparity uncertainty to the TV class, we make use of stereo ground truth data [34] and learn a Gaussian uniform mixture function representing noise and outliers with an EM approach for each TV class. The finally employed disparity uncertainty range from a quarter of a pixel to a couple of pixels demonstrates the span of uncertainty of individual disparities which strongly influences the uncertainty of the 3D point.

Figure 10.2 : From left to right: 1. Input image, 2. Disparity map estimated by SGM and 3. Uncertainty map derived from the disparity map. The disparity map represents distances from light gray (near) to dark gray (far). The uncertainty map shows uncertainty classes of the disparity map from red (high uncertainty) to blue (low uncertainty). Especially slanted surfaces and low-texture areas are classified as having a relatively higher uncertainty.

10.3.2 3D Uncertainty Propagation and 3D Reconstruction

From dense disparity maps with pixelwise disparity uncertainty we estimate the 3D coordinates and the corresponding 3D error functions [35,20]. To this end, we use well-known analytic error propagation [36]. Beside the disparity uncertainty, the 3D error depends on focal length, baseline length, image resolution and distance of the 3D point to the camera.

For an unrestricted 3D reconstruction from 3D point clouds with varying quality, especially volumetric approaches have shown their potential for 3D reconstruction on a large scale [20]. For a volumetric representation, the 3D space is discretized into voxels in the surrounding of a 3D point. Depending on the spatial distance to the 3D point, individual voxels are assigned values that describe their likeliness to be behind or in front of a surface (cf. Fig. 10.3). We use probabilistic Gaussian cumulative distribution functions (CDFs) for the estimation of the voxel values [35,20]. For measurements from multiple cameras, the individual values are fused using a sound theoretical model employing binary Bayes theory. The final volumetric space can be transformed to 3D point clouds or triangle meshes by a Gaussian regression considering neighboring voxel values.

The volumetric reconstruction is implemented runtime and memory efficient by using octree data structures. For the generation of 3D models from images captured at varying distances it is inevitable to propagate the depths into multiple octree levels. To this end, the voxel size (cf. Fig. 10.3) corresponding to the octree depth is estimated for each measurement. More precisely, individual 3D points are included in the octree on varying levels which are selected considering the scale implied by the 3D error derived from the disparity uncertainty and the camera configuration. At this point, our TV prior functions act as a local regularization term and points on lower octree depths are filtered by points with relatively lower uncertainty.

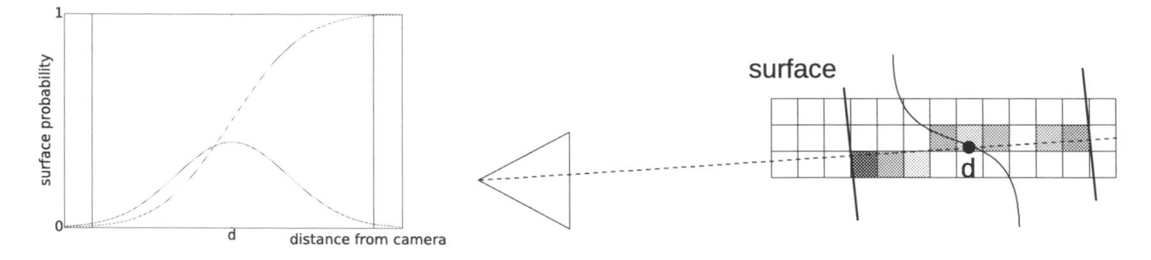

Figure 10.3 : The accuracy of a 3D point on the line-of-sight can be represented by a Gaussian distribution (blue). The integral over the function (green) describes the probability of the actual point being behind a surface. We use this function to assign voxels on the line-of-sight a surface probability (right). Blue voxels have a low probability while red voxels have a high probability of being behind a surface.

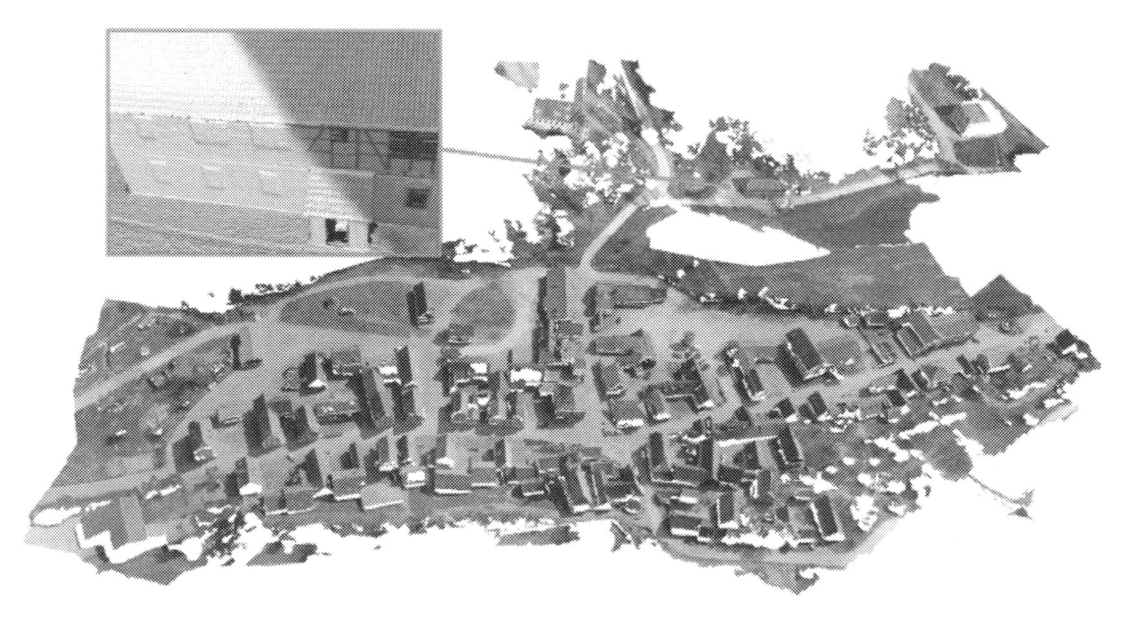

Figure 10.4 : Village Bonnland reconstructed from hundreds of high-resolution images. Employing terrestrial images capturing the scene at close distance our method allows for the reconstruction of important details (zoomed area). The reconstruction space was split in hundreds of subspaces and processed in parallel on a cluster system in a couple of hours. Due to the local optimization employed in the surface reconstruction neighboring surface parts are consistent even though they have been processed independently.

We have also demonstrated that the employed local optimization allows for the reconstruction of very large scenes by means of space division of the entire 3D reconstruction space into smaller subsets [37] (cf. Fig. 10.4). To this end, the space is incrementally split until the number of 3D points is below a given number in each subset. While a globally optimized surface would lack in consistent scalability, our local optimization paradigm avoids complex divide and conquer strategies for fusing neighboring subspaces.

10.4 Scene Classification

We propose a robust and efficient analytical approach for automatic urban scene classification based on imagery and elevation (2.5D) data. Extending the approach described in [38], relative features for both color and geometry, i.e., color coherence, relative height and 3D point coplanarity, are proposed to deal with object diversity and scene complexity. The classification is conducted using a random forest (RF) [39] classifier. Feature extraction and classification can be performed in parallel on independent partitions of the scene, speeding up the processing significantly.

10.4.1 Relative Features

Replacing absolute by relative features to obtain more stable classifications is not a novel idea. E.g., relative heights of buildings and trees in relation to the ground derived based on an estimated DTM (digital terrain model) have become standard and are even considered the most significant features [40] for urban scene classification. Previous work, however, suffers from (1) the heterogeneous appearance of objects of the same class, e.g., diverse sizes and types of buildings as well as different materials/colors of streets, and (2) the similarity of features of objects from different classes, e.g., heights of buildings and trees. Shadows and variable lighting conditions during data acquisition make the situation even more difficult. The challenge is to devise more intra-class stable and, at the same time, more inter-class discriminative features concerning both color and geometry information, which is one of the goals of this work.

10.4.1.1 Color coherence

In urban scene classification changing lighting conditions usually cannot be avoided and may considerably influence the results. For instance, shadows of buildings and trees are cast on the streets, lawns and roofs. These regions are often miss-classified because of their substantially different color values. Color coherence allows for intra-class stability by dealing with objects under various lighting conditions. It measures the difference between two colors with a single value and has been proposed for image segmentation in [41]. We employ the $L^*a^*b^*$ color space, which has an independent channel (L^*) for lightness/luminance, i.e., it can deal with various illumination conditions. The other two channels a^* and b^* represent red/green and yellow/blue, respectively. We assume that (1) the a^* channel is sensitive to vegetation

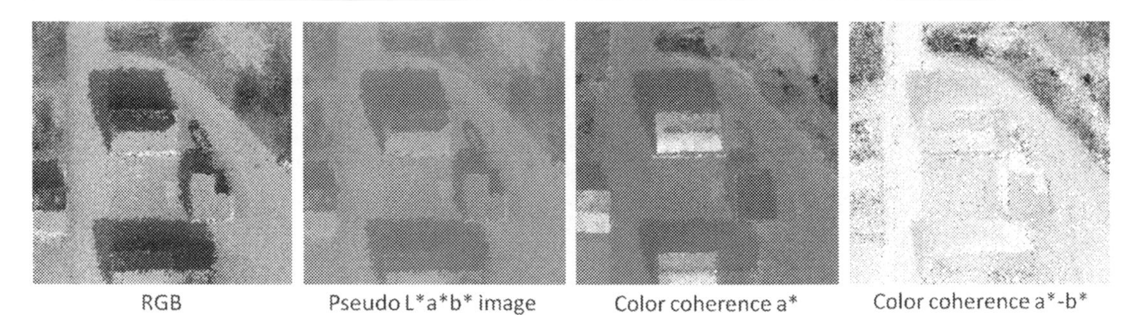

Figure 10.5 : Color coherence concerning vegetation for an RGB image.

objects, i.e., trees and lawn, and (2) the b^* channel can help to analyze objects lying in the shadow. This is because Rayleigh scattering, i.e., diffuse sky radiation (scattered solar radiation), makes the sky blue and the sun itself is more yellowish. Thus, as the bright side of a roof takes more direct sun light, it has a more yellow tone, while the dark side of a roof reflects more sky radiation and, therefore, has a more blue tone.

Color coherence is a relative feature concerning a predefined color. In contrast to [41], we do not measure the direct color distance between two arbitrary objects, but the distance from the current data point to a given object class. As the "reference" class we choose vegetation because of its relatively invariant appearance concerning color for both RGB and multispectral images. Fig. 10.5 (right) shows a map of color coherence concerning vegetation for $a^* - b^*$, where the vegetation areas are dark while all other objects are lighter, i.e., have a significant difference to vegetation. Fig. 10.5 also shows that in comparison with only the a^* channel (center right), subtracting the b^* channel gives a much better result concerning the influence of shadows (right).

10.4.1.2 Definition of neighborhood

The spatial neighborhood relationship is employed in many approaches to integrate contextual information for more plausible results. The definition of neighborhood shared by most related work consists of the data points around the current point. For a square image grid, the range of neighbors is a simple $(2m + 1) * (2m + 1)$ matrix with m the order of neighbors (Fig. 10.6, left). While this is a general setup widely used, it only works well under the assumption that the meaningful neighbors are isotropically distributed or there is no direction-related information to consider. Another tricky problem is the order/size of the neighborhood: A larger window is desirable to include more contextual information, but it also implies a more time-consuming processing. The latter becomes particularly critical when graphical models are used as higher-order neighbors mean an exponentially increased computational effort.

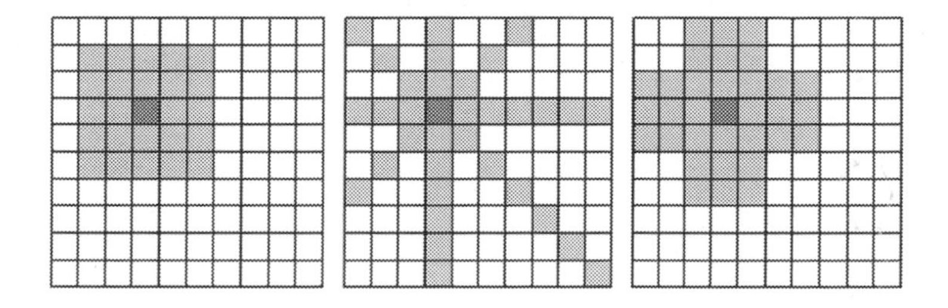

Figure 10.6 : Definitions of neighborhood: Conventional definition with second-order neighbors (left) and two radial patterns (center and right).

Figure 10.7 : Employed neighborhoods for relative height for flat (left) and undulating terrain (right).

Inspired by [32], radial neighborhood patterns (Fig. 10.6, center and right) are proposed. The lengths of the "beams" can be set to "infinity" [32] to reach the boundary of the data and cross as many different objects as possible, while the size of the whole population of neighbors is still acceptable. The length as well as the thickness of the "beams" (Fig. 10.6, right) can be adapted to the object characteristics, e.g., height information in undulating terrain (cf. Sect. 10.4.1.3), where a finite setup is appropriate.

10.4.1.3 Relative height

Relative height is defined as the elevation difference between the current data point and the estimated local ground level. By this means, the differences between classes become more discriminative and stable. A radial neighborhood pattern (cf. above) is employed for relative height computation. For scenes with flat terrain, for instance Fig. 10.7 (left), an "infinite" length of the "beams" is used to gather as much information as possible by including all available ground data.

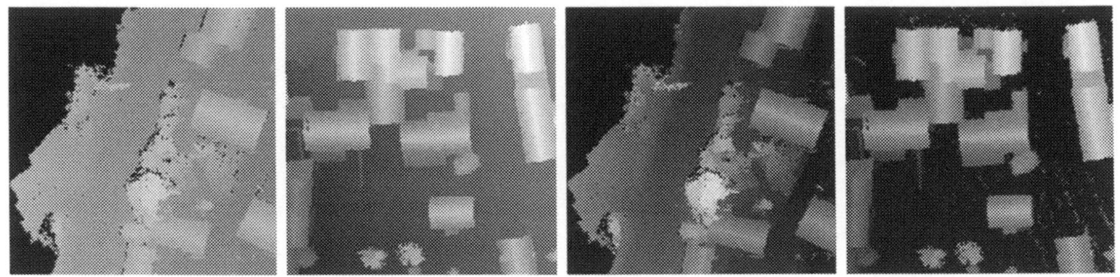

Figure 10.8 : Absolute and relative height for valley and hillside.

In contrast, a limited length is mandatory for scenarios with larger height differences like for undulating terrain (Fig. 10.7, right). On hillsides or in valleys, the ground level of one building can be similar or even higher than the roof of another building in the neighborhood. This implies that the absolute height from the training data is not a valid prior for buildings any more. In this case, the range of neighbors is set slightly larger than the average building length.

As shown in Fig. 10.8, the height maps for valley and hillside areas are enhanced using the relative height. What is more important, the height values of different objects, e.g., roofs and ground, are directly comparable in the relative height maps. That is, a classifier trained on one dataset can be applied to both areas.

10.4.1.4 Coplanarity of 3D points

Coplanarity measures how well the current point and its neighbors form a plane. A common plane for the given point and its neighbors is estimated using RANSAC [3] and their coplanarity is quantified by calculating the percentage of inliers to the estimated plane. Coplanarity is employed as a feature to infer the probability of the current point being a part of a planar object. In this work, it was found to be very effective to distinguish trees from other objects, especially roofs, which might have very similar heights in most European urban areas. Fig. 10.9 presents a coplanarity map calculated from the direct neighbors of each pixel.

In summary, relative features integrate both local and contextual information. The definition of "context" is extended from a geometric neighborhood to a more semantic description of environment (ground height) and class (vegetation). In comparison to conventional features with absolute values, the proposed relative features are more discriminative between different classes, and at the same time more stable for objects of the same class.

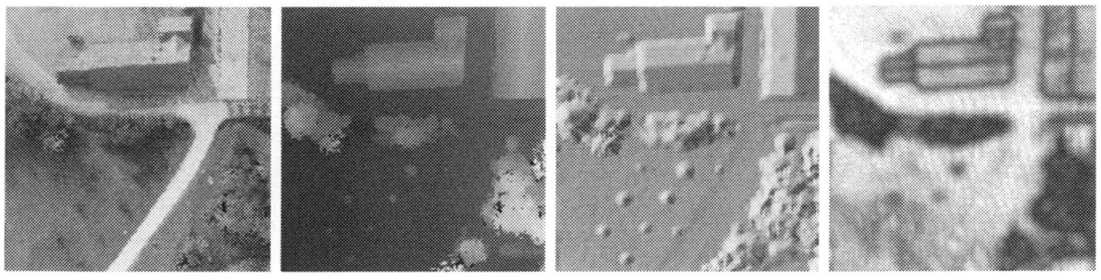

Figure 10.9 : Coplanarity map.

10.4.2 Classification and Results

A standard RF classifier is employed for the classification. The calculation of the features and the classification with the trained classifier are both implemented for parallel processing splitting the dataset into independent partitions. Because the proposed features are robust for various scenarios, we empirically found that a generally trained classifier can be directly applied to all partitions from the same dataset and provide reasonable results without additional local or incremental training. The computation time can be reduced considerably because the partitions can be distributed. As long as the partitions have a reasonable size, i.e., are large enough to contain whole major objects like buildings and road segments with full width, the division will only marginally deteriorate the results.

10.4.2.1 Post-processing

Post-processing with blob filters is conducted to correct local errors based on semantic consistency constraints: e.g., the regions of roads and buildings should be homogeneous without small spots of tree or lawn inside. On the other hand, gaps are allowed for trees and lawn. For instance, ground or lawn may be visible in a gap between trees and trees may be found in the middle of a lawn.

The post-processing corrects also errors caused by data artifacts and improves the plausibility of the results. The size of the blobs to be filtered is determined based on the empirically derived object size of each class and has to be adapted to varying data resolution and quality.

10.4.2.2 Results for Bonnland

The Bonnland dataset (cf. also Sect. 10.2 and Fig. 10.10) includes buildings in the valley and on the hillside. The height of the hillside ground level can be greater than that of some building roofs in the valley. Accordingly, a classifier trained with absolute height values for ground

Figure 10.10 : Bonnland data with undulating terrain: The ground height (solid line) of a hillside building can be higher than the roof top (dashed line) of a building in the valley.

Figure 10.11 : Classification of Bonnland data with classes ground (gray), building (red), high vegetation (green) and low vegetation (blue).

and buildings will not work in this area. The proposed relative feature for height, by contrast, is stable in the undulating terrain without need to train the classifier locally. Fig. 10.11 presents the classification result for the whole dataset.

We use a reduced version of the reconstructed 3D point cloud (cf. Sect. 10.2.2; about 8.8 million out of over 1 billion points), which is rasterized into a 2.5D point cloud with a resolution of 0.2 meters. Fig. 10.11 presents the whole Bonnland dataset and the classification result. We have defined four object classes, i.e., ground (gray), building (red), high vegetation/tree (green) and low vegetation/lawn (blue).

10.5 Scene and Building Decomposition

The parsing of complex scenes including buildings is one of the main challenges towards fully automatic city model reconstruction and the automatic reconstruction of larger urban areas. We assume that a reasonable decomposition, subdividing the whole scene as well as heterogeneous buildings into regular components is the key to tackle this.

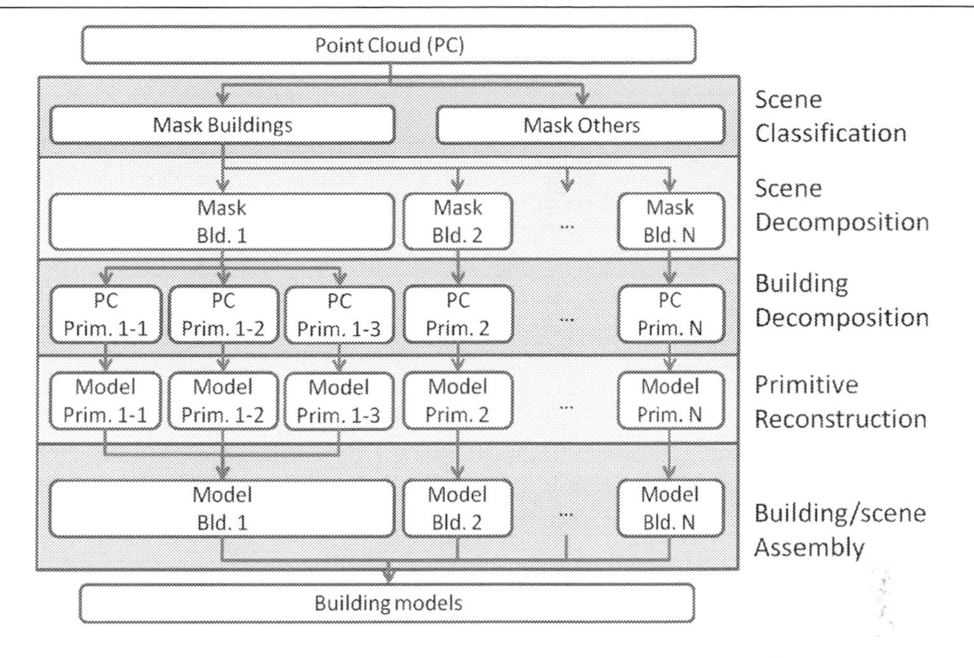

Figure 10.12 : Workflow of building reconstruction with decomposition and assembly.

Fig. 10.12 shows a workflow which links the approaches for scene classification [38] and primitive-based reconstruction [42,43] by

- preprocessing/decomposition—a reliable decomposition of the whole scene as well as of complex buildings into regular primitives precedes model reconstruction—and
- post-processing/assembly of primitives—the assembly of the primitives performs a true model merging in CAD (computer aided design) style—

to complete the pipeline.

10.5.1 Scene Decomposition

The goal of scene decomposition is to extract individual buildings from the building mask derived by scene classification (cf. Sect. 10.4). As the separation of buildings in the building mask may be adversely affected by data as well as labeling errors (cf. Fig. 10.13, red dashed contour), a mathematical morphological "opening" operation is conducted, as shown in Fig. 10.13, to remove trivial connections for a better isolation of the buildings and to eliminate small outliers. A disk-shaped structuring element is employed with a radius of 1 meter.

We start to detect individual buildings via connected components (CCs). The input building mask, as presented in Fig. 10.13, is segmented into rectangular tiles accordingly. Overlap

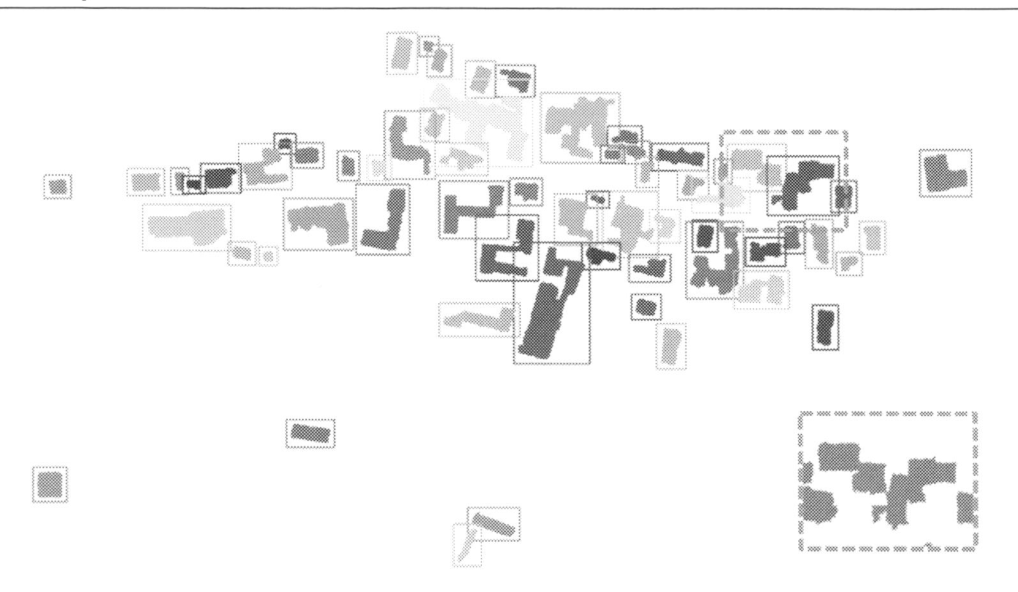

Figure 10.13 : The building mask is decomposed into connected components. Mathematical morphological (opening) is used for a better separation of adjacent buildings.

between tiles is allowed to make sure that the buildings are completely included in the tiles. Please note that global coordinates and the height (for undulating areas) are tagged to each tile for the final model assembly.

Due to the complexity of the scene and the errors of the classification, however, it cannot be guaranteed that each CC contains exactly one individual building. That is, as Fig. 10.14 shows, after decomposition, a tile may contain a single detached building, a complex building (a building consisting of multiple building components), or even multiple buildings which are closely adjacent to each other. The last case is often found in densely inhabited areas.

In this work we name both, complex buildings and adjacent building groups, "building complexes". Further parsing of building complexes is described in the following.

10.5.2 Building Decomposition

Building decomposition works on the results of scene decomposition. It further divides building complexes into standard building components—primitives, which are employed in the following building model reconstruction. Although our approach presented in [43] allows one to model multiple building components by means of the "birth" and "death" jumps in a reversible jump Markov chain Monte Carlo (RJMCMC) framework, a preceding decomposition makes the statistical search more efficient and robust.

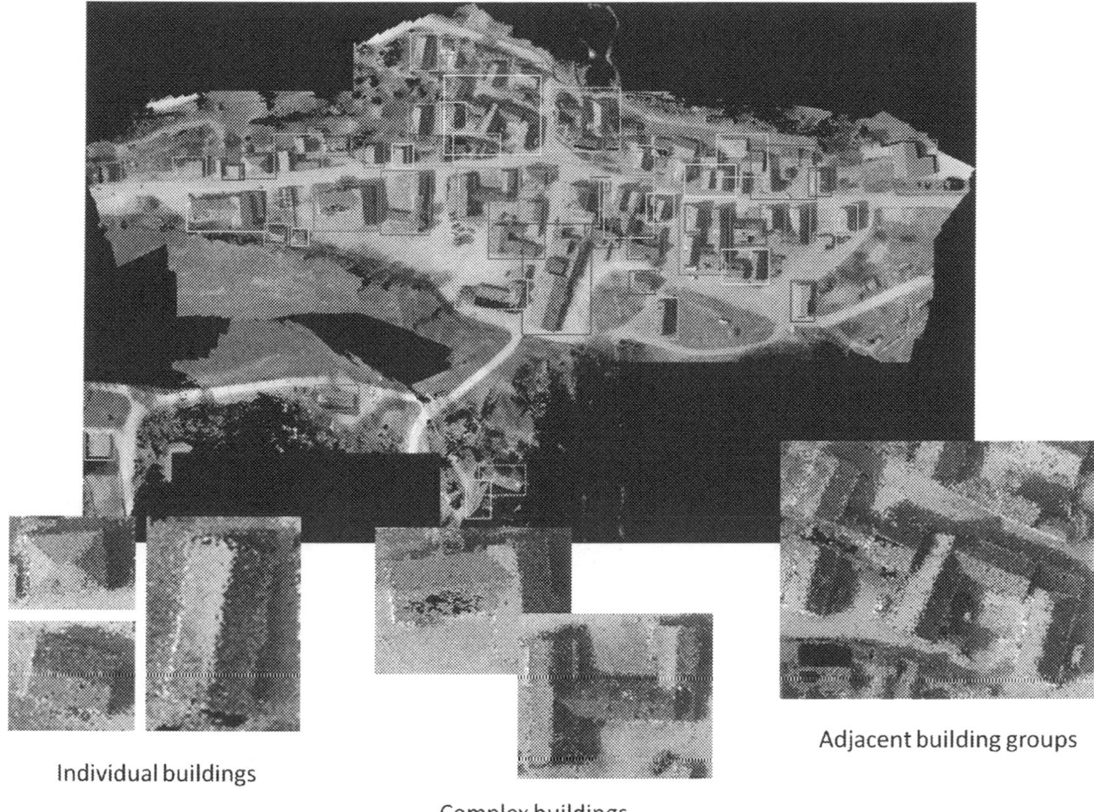

Figure 10.14 : The results of scene decomposition may contain individual buildings, complex buildings as well as building groups consisting of adjacent buildings.

We employ a combined bottom-up and top-down scheme for building decomposition, which uses 3D geometry parsing based on a predefined primitive library. Conventional building decomposition is conducted bottom-up based on 2D footprints that are either already available [44,45] or derived from the data [46]. The performance of footprint-based decomposition is limited where 3D geometric information, e.g., different heights or roof types of buildings, has to be considered [47]. 3D geometric parsing, however, cannot be conducted on 3D building models, because the latter do not yet exist. The best decomposition is found by means of statistical model selection and optimization.

Please note that the decomposition and the following building modeling have to share a common construction principle to guarantee a consistent work pipeline. Two basic strategies for building (footprint) decomposition can be differentiated based on one key difference: if the components of a building are allowed to overlap [48,42] or not [44,45]. Different definitions

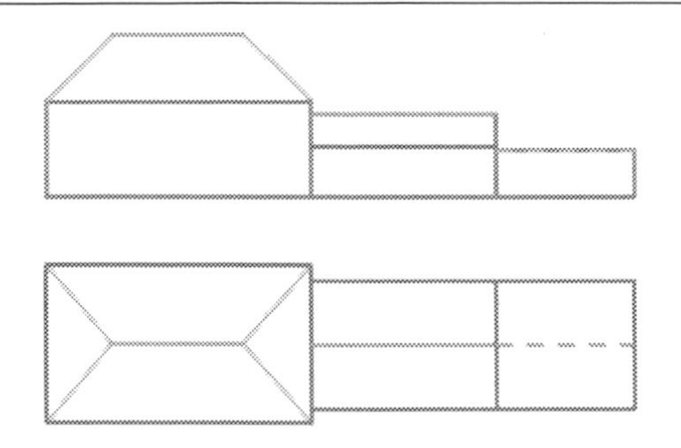

Figure 10.15 : Ridge lines of hipped (left), gable (center) and flat (right) roofs. For flat roofs, central lines (dashed) are used to describe the height of the building.

are correspondingly employed for the primitives. In [43], we have demonstrated that the first concept allowing for overlap fits better to generative modeling because of its flexibility and most importantly, the potential to generate complete and plausible models.

10.5.2.1 Ridge extraction

From the 3D point cloud, we employ bottom-up methods to extract ridge lines, which can be considered as key geometric features of buildings. As shown in Fig. 10.15, we define the edges on the roof as: (1) Horizontal ridge line (red), which connects two apexes of the roof, (2) diagonal ridge line (green) connecting one apex and one eave corner, and (3) eave line (blue), which links two eave corners. The roof contour consisting of eave lines can be used as approximation of the building footprint when the roof overhang is ignored.

For non-flat roofs, ridge lines can be determined as the intersection lines of the individual roof planes as also demonstrated in current work [49,50]. Planes of a building complex are detected from the 3D point cloud by a RANSAC-based approach. All intersection lines are filtered by means of the "Relation Matrix" [51] to separate ridge lines from other intersection lines as well as to determine the type of ridge lines, i.e., horizontal or diagonal ridges. "Central lines" are used for flat roofs, which do not have any ridge. In comparison with building skeletons, central lines of roofs are 3D. That is, as shown in Fig. 10.15 (right), they have a height value and in the case of adjacent flat roofs they can differentiate multiple building components by means of their heights.

Ridge lines have the following advantages for the decomposition of complicated buildings:

- Full 3D information: different heights can be used to distinguish different building components even though they have an identical width (cf. Fig. 10.15, bottom).
- High accuracy and reliability: The ridge lines are calculated by the intersection of robustly determined planes, i.e., they are actually derived from all inliers of the involved planes.

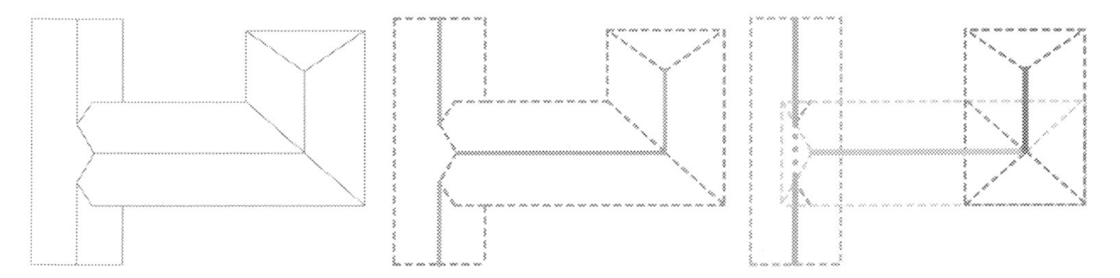

Figure 10.16 : Building decomposition: Determination of a combination of primitives (right) derived from ridge lines (center, red) to represent the underlying model (left).

- The horizontal ridges consist of straight line segments, which can be regarded as ridge lines of individual primitives and, thus, can guide the decomposition.

10.5.2.2 Primitive-based building decomposition

A statistical primitive-based decomposition guided by the ridges is illustrated by Fig. 10.16. The detected horizontal ridge lines (Fig. 10.16, center, red) consist of straight line segments (Fig. 10.16, right, bold). The primitives selected from a predefined library [52] have a rectangular contour as well as a single straight horizontal ridge line (except for flat roof and shed roof). The end points of the segments are determined by intersection with diagonal ridges or the boundary of the building mask.

The horizontal ridges guide the decomposition: The most appropriate primitives (cf. Fig. 10.16, right) are statistically selected with the goals to fit (1) to the already extracted diagonal ridges and (2) to the rest of the edges, without conflicts with the known planes and the boundary of the building mask.

The building decomposition does not deal with actual building models, because they do not yet exist. The goal of the decomposition is to find an optimal combination of primitives to approximate the underlying model (Fig. 10.16, left). That is, the decomposition determines the number and types of primitives as well as the way of their combination (Fig. 10.16, right). Additionally, building decomposition also results in initial values of parameters for the primitives (cf. Sect. 10.6.1).

10.6 Building Modeling

10.6.1 Primitive Selection and Optimization

We propose a statistical building modeling based on generative primitive models [43]. The primitives are defined with parameters (cf. Fig. 10.17):

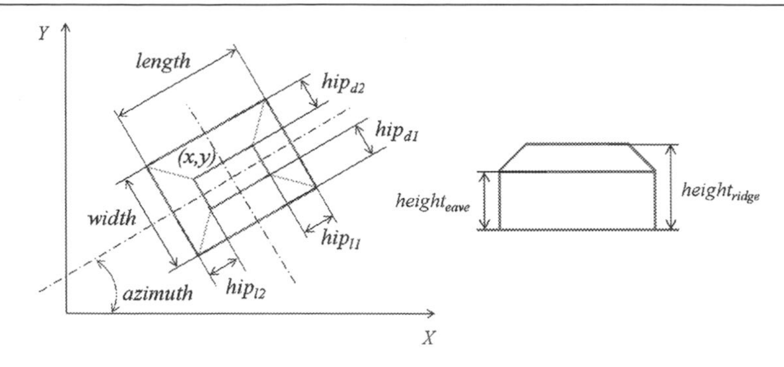

Figure 10.17 : Parameters of a primitive for generative modeling.

$$\theta \in \Theta; \Theta = \{\mathcal{P}, \mathcal{C}, \mathcal{S}\}, \tag{10.1}$$

where the parameter space Θ consists of position parameters $\mathcal{P} = \{x, y, azimuth\}$, contour parameters $\mathcal{C} = \{length, width\}$ (rectangular footprint), and $\mathcal{S}$ containing shape parameters: Ridge/eave height and the depth parameters of hips.

The maximum a posteriori (MAP) estimate of Θ is employed to find the optimal model fitting the data:

$$\widehat{\Theta}_{MAP} = \underset{\Theta}{argmax}\left\{\frac{L(\mathcal{D}|\Theta)p(\Theta)}{P(\mathcal{D})}\right\} = \underset{\Theta}{argmax}\left\{L(\mathcal{D}|\Theta)p(\Theta)\right\}, \tag{10.2}$$

where $L(\mathcal{D}|\Theta)$ is the likelihood function presenting the goodness of fit of the model to the data $\mathcal{D}$ and $p(\Theta)$ presents the prior for Θ, which is derived from empirical knowledge and incrementally improved during the reconstruction. That is, the parameter values of the already found building components or of adjacent buildings are used to update the priors. $P(\mathcal{D})$ is the marginal probability, which is regarded as constant in the optimization as it does not depend on Θ.

Reversible rump Markov chain Monte Carlo (RJMCMC) is used for the statistical search of the parameters, allowing for an efficient exploration of the high-dimensional (determined by the number of parameters) search space. The reversible jumps allow one to switch between different search spaces, i.e., different types of primitives. Model selection is integrated in the transition kernel of reversible jumps to guide the search of the optimal primitive type. Multiple hypothetical models are generated via statistical sampling of the primitive type as well as the corresponding parameters. The final model is the verified candidate with the best goodness of fit to the data.

As mentioned above in Sect. 10.5.2.2, certain initial values of the parameters can be derived from the primitive-based building decomposition. The following parameters can be determined directly from nothing but the horizontal ridges (cf. Sect. 10.5.2.1):

- Coordinates of the centroid (x and y) from the center of the ridge line.
- Orientation (*azimuth*): corresponds to orientation of the ridge line.
- Ridge height (z_2) of the roof.

Additional parameters can be determined by taking the building footprints into account, which either stem from given building masks or can be derived from the diagonal ridges (cf. Sect. 10.5.2.1):

- Length and width derived from orientation and footprint.
- Depths of hips (hip_{l1}, hip_{l2}, hip_{d2}, and hip_{d2}) are the longitudinal and radial distances from the end points of the horizontal ridge to the boundary of the building mask.

Known initial values can significantly improve the performance of the statistical search. They are, to a certain extent, reliable and specific, as the ridge lines can be precisely determined (cf. Sect. 10.5.2.1).

Fig. 10.18 presents the reconstruction result of the running example Building 1 (top) and demonstrates the robustness of the proposed modeling approach against data flaws (bottom). 3D point clouds from image matching may contain flaws because of the quality and coverage of the images, homogeneous texture of the objects (i.e., points cannot be matched), occlusion, etc. They lead not only to false colors or incorrect positions of points, but also to gaps (missing points) in the objects. While conventional bottom-up methods may encounter insurmountable difficulties in this case, resulting in irregular and/or incomplete building components, the proposed method guarantees plausible results despite such flaws.

10.6.2 Primitive Assembly

If a building consists of multiple primitives, the reconstructed primitives are assembled into a single model in two consecutive steps: (1) Joint parametric adjustment and (2) geometrical model merging.

The "joint parametric adjustment" helps to remove trivial conflicts between primitives and compensates for small deviations (cf. Fig. 10.19, top left), which can occur during the reconstruction driven by a stochastic process. In the adjustment, the change of each side of a primitive is proportional to its size, i.e., footprint area. The parameters of all building components are jointly adjusted using two rules [43]:

- Rule 1: The intersection angles of the primitives are jointly regularized to 0° or 90° if the deviation is less than a given threshold.
- Rule 2: Heights of flat roofs or ridge- and eave-heights of other roofs are adjusted to one value if the deviation is less than a given threshold,

where the thresholds are determined according to data resolution and quality.

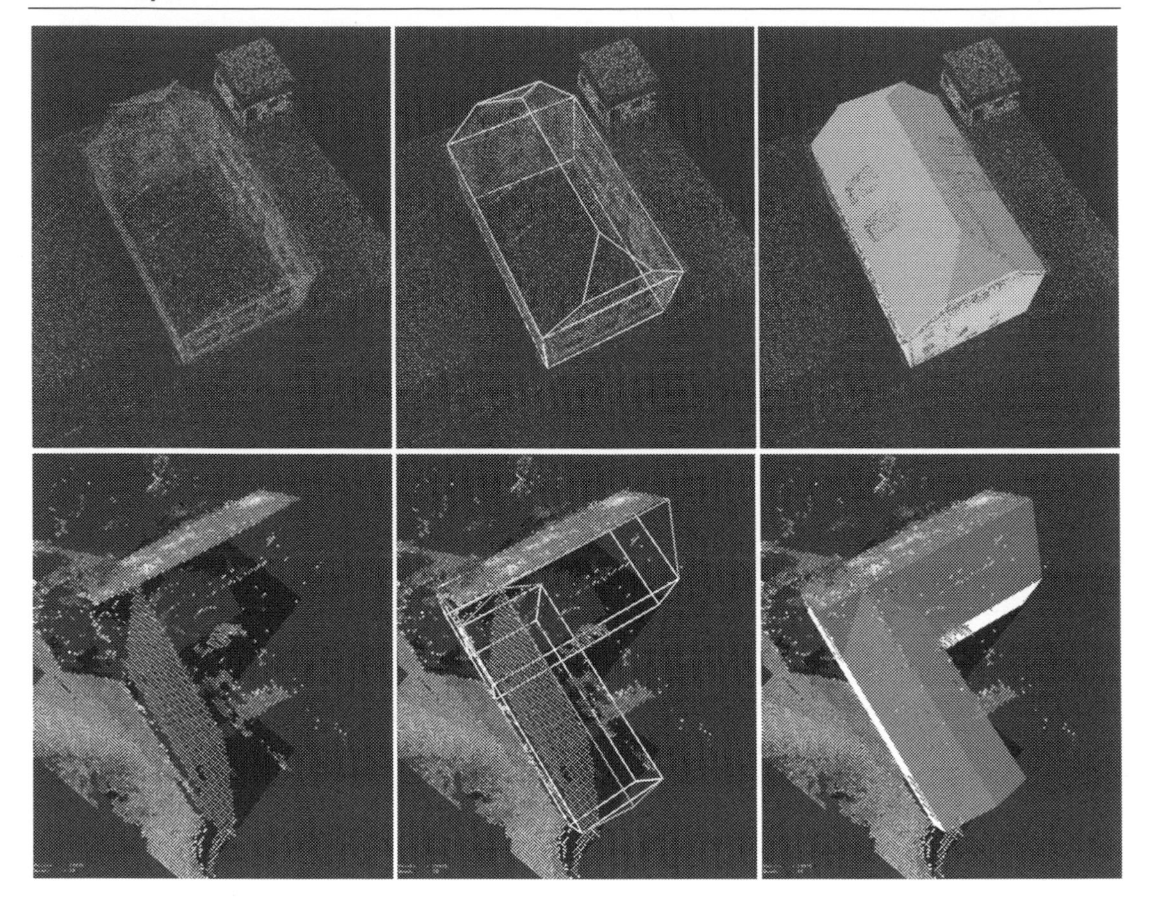

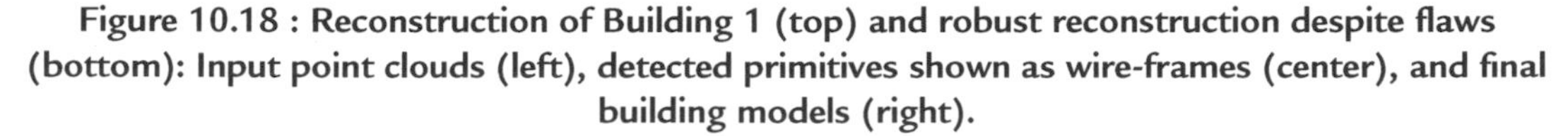

Figure 10.18 : Reconstruction of Building 1 (top) and robust reconstruction despite flaws (bottom): Input point clouds (left), detected primitives shown as wire-frames (center), and final building models (right).

The parameter adjustment, however, cannot deal with all mismatching positions of the primitives. The mismatching is the result of deviations caused by stochastic processes and the uncertainty of the data, which in principle cannot be corrected in this step. Therefore, a further geometrical adjustment is required. "Geometrical model merging" generates the final single model of a building complex. Similar to [43], we conduct a simple vertex-shifting to correct the geometrical mismatching and all the primitives are matched to each other (Fig. 10.19, bottom left). The primitives are originally generated as Boundary Representation (B-Rep) models, as shown in Fig. 10.19 (center), and simply placed together as separate models which overlap. Although in the rendered model (left) the intersecting part is hidden and does not affect the appearance, the model is ontologically not a single "subject" and geometrically not watertight. To fix this, our model merging employs Constructive Solid Geometry (CSG) mod-

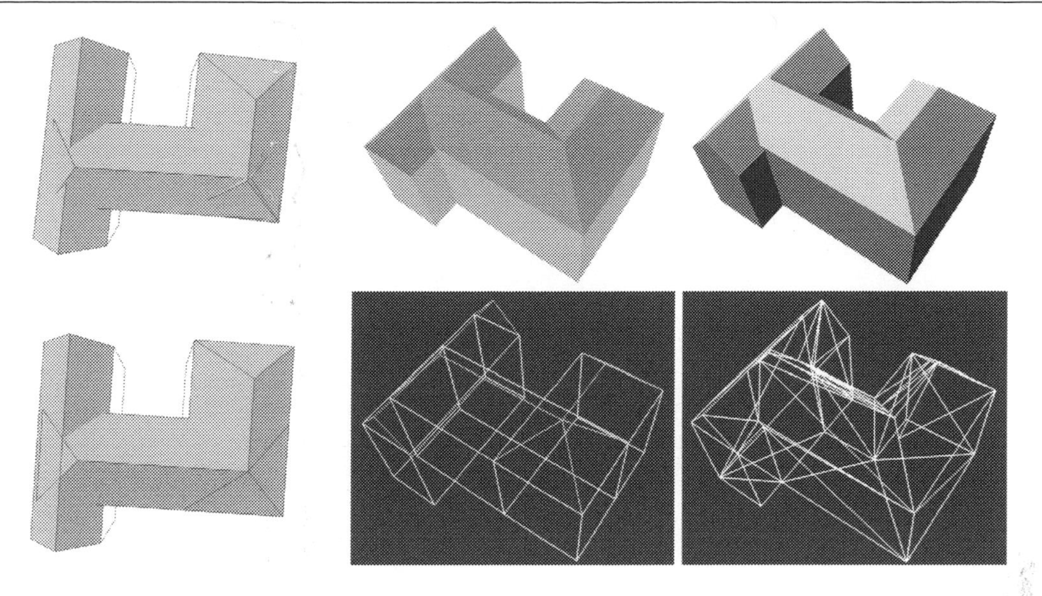

Figure 10.19 : Primitive assembly from individual primitives (left) to a watertight CSG model (right).

eling. The B-Rep primitives are first converted into CSG models and merged with a "union" operation into a single solid body (cf. Fig. 10.19, right). The latter is then converted back to a single and watertight B-Rep model, i.e., the final model.

10.6.3 LoD2 Models

We have performed experiments on the Bonnland dataset (cf. Sect. 10.4), representing a complete and typical central European village with a mixture of detached buildings and building complexes, a church, as well as a small castle on a hill. The 3D point cloud has been reconstructed from UAV imagery [33] and covers about 0.12 square kilometers of undulating terrain. We use a reduced and rasterized version with a resolution of 0.2 meters.

The building mask provided by scene classification (cf. Sect. 10.4) is decomposed into 62 connected components and corresponding data tiles (cf. Fig. 10.14), which are processed in parallel. The reconstructed building models are assembled in the global coordinate system. Bird's-eye views of input point cloud (top) and the reconstructed model (bottom) are presented in Fig. 10.20. Along with the watertight building models a mesh model is generated from the non-building points to model the ground.

The runtime of building modeling for 33 detached buildings and 29 building complexes, which consist of 112 primitives, is about 14 minutes on a laptop with a four cores/eight

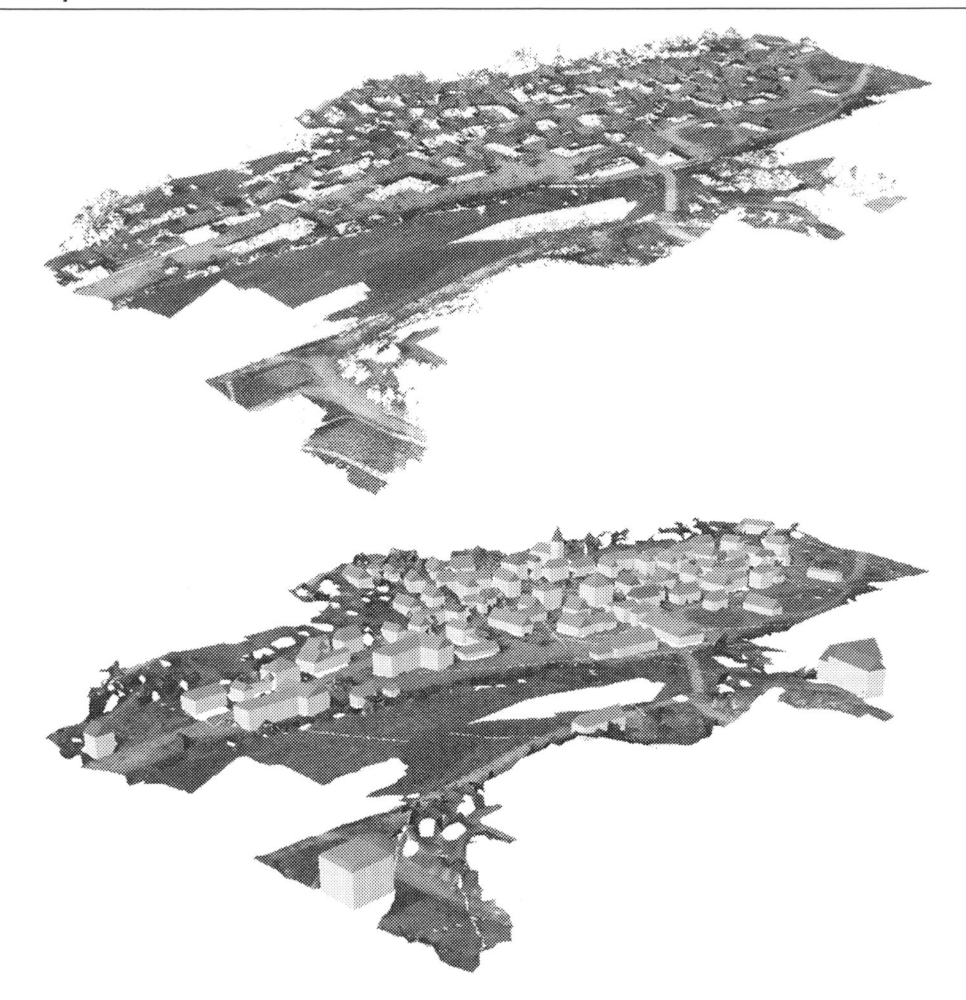

Figure 10.20 : Building reconstruction for Bonnland: Input point cloud (top) and reconstructed models (bottom).

threads CPU at 2.3 GHz. Except for buildings with large flaws in the data (cf. Fig. 10.18, bottom) or occlusions, where the average deviation does not reflect the reconstruction accuracy, the reconstruction error of the majority of the buildings is less than half of the resolution, i.e., 10 centimeters.

10.6.4 Detection of Facade Elements

As the original input to the described workflow are images and because semantic analysis by Convolutional Neural Networks (ConvNets) has made significant progress in recent years,

Figure 10.21 : Projection of different images onto the same virtual fronto-parallel plane.

it seems promising to use this technique for the detection of facade elements. To reduce the necessary computational complexity when using a ConvNet, we restrict the image regions to the facades.

The building primitives define planar elements for roofs and facades. Once the optimum primitives have been determined, the facade planes can be derived in the form of polygons defined by vertices. The corresponding regions of a facade can then be extracted from the images and projected via a planar homography onto the same virtual fronto-parallel plane. Assuming that the facade including all elements, such as windows and doors, is almost planar, the projections from all images should have a similar position on the virtual plane. This reduces the search space for our ConvNet to a limited two-dimensional space. Fig. 10.21 shows the projections of three different views of one facade onto its corresponding virtual plane.

An adapted ConvNet [53] is employed to detect the facade elements in the images (cf. Fig. 10.22). The network is based on AlexNet [54], which was pretrained on the ImageNet dataset [55] and is extended by a set of convolutional (Conv) and deconvolutional (DeConv) layers to achieve pixelwise classification.

Rectified facades from the eTRIMS dataset [56] were used for fine-tuning the network to classify facade images into the four classes building, door, wall, and other. The results of the ConvNet for individual images are combined by summation of the output of the last fully connected (FC) layer and application of the softmax function to the sum. A rectangle is fitted to each window and door and, finally, the facade elements are projected back into the original 3D model which allows one to generate the LOD3 model (cf. next section). In Fig. 10.23 the process is visualized using one of several images.

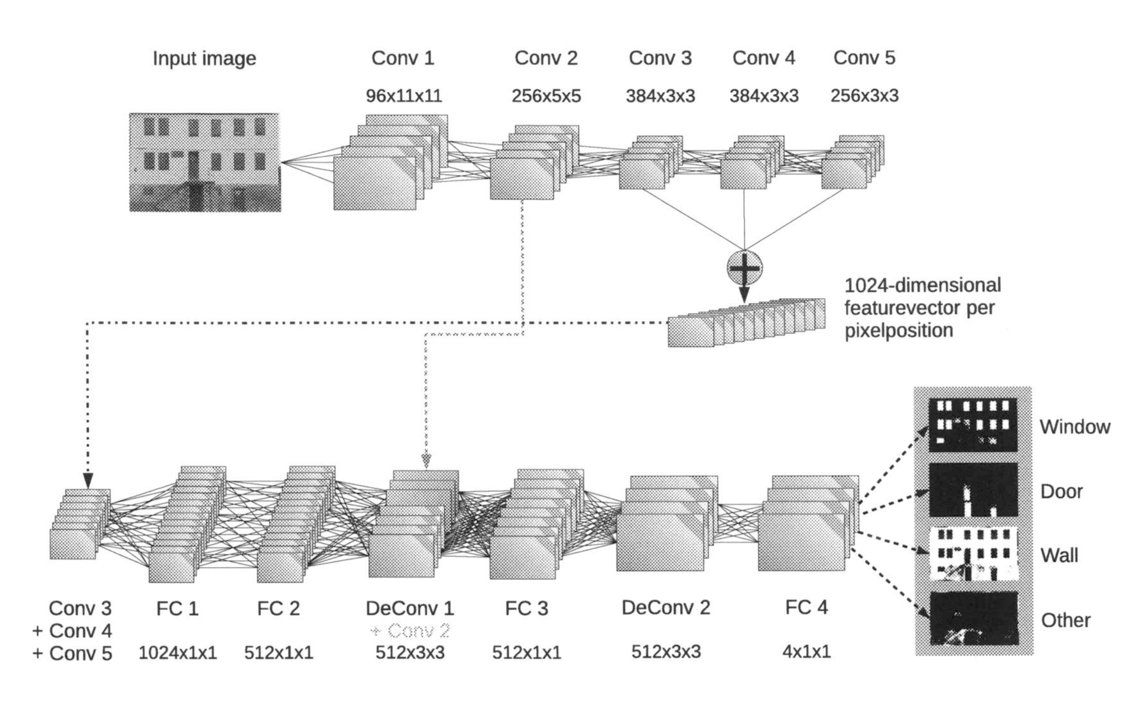

Figure 10.22 : Architecture of the Convolutional Neural Network (ConvNet).

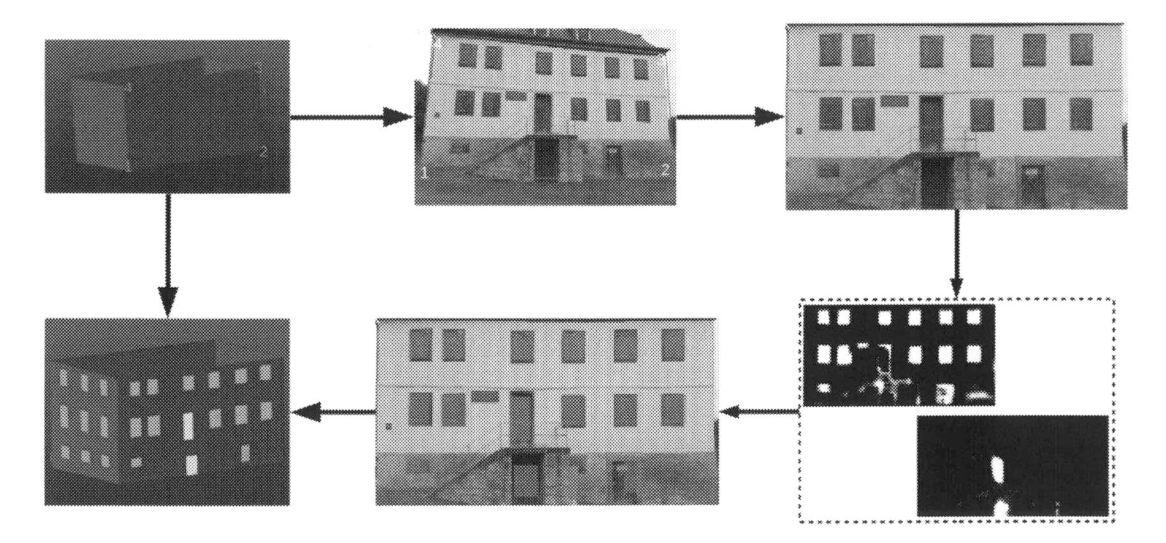

Figure 10.23 : Detection of windows and doors with a ConvNet on a planar facade projected into one image.

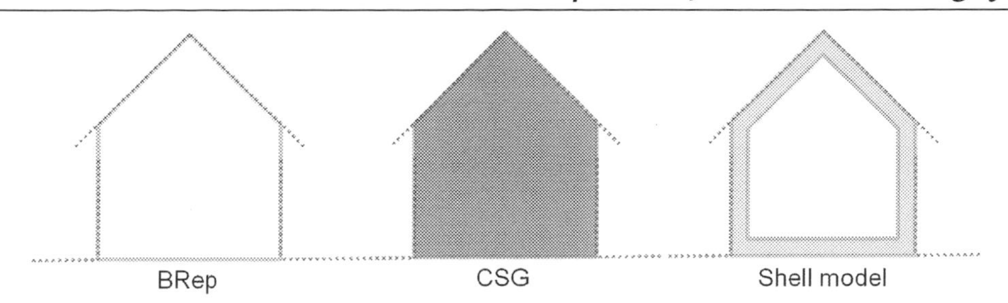

Figure 10.24 : Comparison of shell model (right) with BRep (left) and CSG (center) models.

10.6.5 Shell Model

We propose the so-called "shell model"—a hybrid model combining elements of CSG and BRep—for the modeling of buildings in LoD3. The motivations for the "shell model" are:

- 3D Measurement data (from almost all acquisition technologies) reveal only the surface instead of the solid body of the objects.
- Data with deviations form no perfect surface but a layer with a certain thickness representing the data noise.

A shell is, therefore, considered a more reasonable and practical geometrical assumption (model) compared to a simple surface or a solid body. The concept of the shell model is presented in Fig. 10.24. It consists of parallel inner- and outer-layers and the solid body defined by them.

The shell model is specifically designed for our application, in which the images from both airborne and terrestrial cameras are fused and the reconstructed point cloud (cf. Sect. 10.3) is available for both roofs and facades. In comparison with the modeling approaches for either roof or facade data (Fig. 10.25, top), the advantages of the shell model can be summarized as follows:

1. Roof as well as facades can both be precisely modeled. In roof-based modeling, the walls are approximated by extruding the boundary of the roof (Fig. 10.25, top left). The model based on facades does not deal with the roof overhang (Fig. 10.25, top right).
2. Roof and facades are inherently assembled. No additional effort is required for the adjustment and adaption of roof and facade planes.
3. Windows and doors can be modeled quite naturally as openings.

Fig. 10.26 presents the reconstruction of Building 1 using the shell model. The shell model (center) is initially used for building detection from the input point cloud (left) with inner

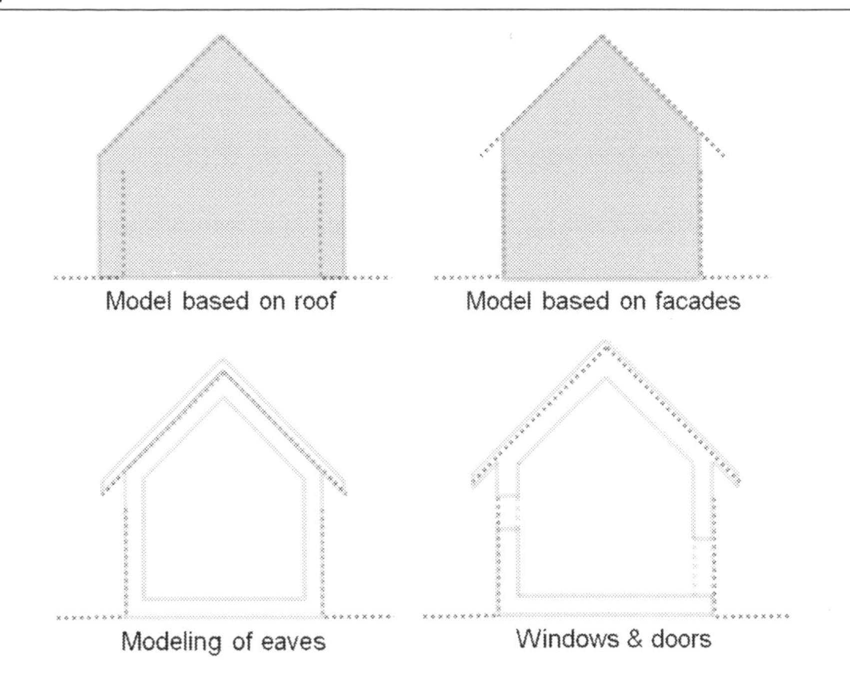

Figure 10.25 : Top—modeling for roof (extrusion to the ground) and facades (no overhang considered). On the bottom it is shown how eaves and windows as well as doors can be represented by the shell model.

Figure 10.26 : LoD3 modeling of Building 1: Input point cloud (left), shell model used for building detection (center) and the final model (right) with windows and door modeled as openings.

(red) and outer (blue) layers. The model of best fit is then assumed to be the layer (green) between them. The final model (right) is generated with integrated windows as well as door and a given thickness of the walls.

10.7 Conclusion and Future Work

We have presented an approach for 3D urban scene reconstruction and interpretation based on the fusion of airborne and terrestrial images. It is one step forward towards a complete and fully automatic pipeline for large-scale urban reconstruction. The main contributions of this work can be summarized as follows:

- Fusion of images from different platforms (terrestrial, UAV) by means of pose estimation and 3D reconstruction of the scene.
- Combination of color and 3D geometric information for scene classification.
- Primitive-based model decomposition and assembly for the parsing of complex scenes as well as buildings.
- ConvNets for the detection of windows and doors on facades.
- Automatic modeling of buildings at LoD2 and LoD3.

We are aware that many challenges remain, for instance, the reliable parsing and decomposition of building-blocks in densely inhabited urban areas with an intricate neighborhood of individual but similar buildings. Many public and commercial buildings also have special shapes that cannot be represented by the introduced rectangular primitives. Concerning future work we, thus, consider to extend the primitives with a more flexible geometry. Further building elements such as balconies, dormers and chimneys could be modeled with (variants of) the given primitives. Deep neural networks can be extended to fully utilize the available multimodal data (e.g., color and depth information in the scope of this paper) for not only the scene classification but also for the detection of facade elements. Reference [57] demonstrates a start of our exploration.

References

[1] A. Grün, O. Kübler, P. Agouris (Eds.), Automatic Extraction of Man-Made Objects from Aerial and Space Images, 1995.

[2] R. Hartley, A. Zisserman, Multiple View Geometry in Computer Vision, second edition, Cambridge University Press, Cambridge, UK, 2003.

[3] M. Fischler, R. Bolles, Random sample consensus: a paradigm for model fitting with applications to image analysis and automated cartography, Communications of the ACM 24 (6) (1981) 381–395.

[4] D. Nistér, An efficient solution to the five-point relative pose problem, IEEE Transactions on Pattern Analysis and Machine Intelligence 26 (6) (2004) 756–770.

[5] C. McGlone (Ed.), Manual of Photogrammetry, sixth edition, American Society of Photogrammetry and Remote Sensing, Bethesda, USA, 2013.

[6] M. Pollefeys, L. Van Gool, M. Vergauwen, F. Verbiest, K. Cornelis, J. Tops, Visual modeling with a hand-held camera, International Journal of Computer Vision 59 (3) (2004) 207–232.

[7] M. Goesele, J. Ackermann, S. Fuhrmann, R. Klowsky, F. Langguth, P. Muecke, M. Ritz, Scene reconstruction from community photo collections, IEEE Computer 43 (6) (2010) 48–53.

[8] S. Agarwal, N. Snavely, I. Simon, S. Seitz, R. Szeliski, Building Rome in a day, in: IEEE International Conference on Computer Vision, 2009, pp. 72–79.

[9] J.-M. Frahm, D. Gallup, T. Johnson, R. Raguram, C. Wu, Y.-H. Jen, E. Dunn, B. Clipp, S. Lazebnik, M. Pollefeys, Building Rome on a cloudless day, in: Eleventh European Conference on Computer Vision, vol. IV, 2010, pp. 368–381.
[10] J. Heinly, J. Schönberger, E. Dunn, J.-M. Frahm, Reconstructing the world in six days (as captured by the Yahoo 100 million image dataset), in: IEEE Conference on Computer Vision and Pattern Recognition, 2015, pp. 3287–3295.
[11] J. Bartelsen, H. Mayer, H. Hirschmüller, A. Kuhn, M. Michelini, Orientation and dense reconstruction from unordered wide baseline image sets, Photogrammetrie—Fernerkundung—Geoinformation 4 (12) (2012) 421–432.
[12] H. Mayer, J. Bartelsen, H. Hirschmüller, A. Kuhn, Dense 3D reconstruction from wide baseline image sets, in: F. Dellaert, J.-M. Frahm, M. Pollefeys, L. Leal-Taixé, B. Rosenhahn (Eds.), Outdoor and Large-Scale Real-World Scene Analysis, Springer Berlin Heidelberg, Berlin, Heidelberg, 2012, pp. 285–304.
[13] M. Michelini, H. Mayer, Efficient wide baseline structure from motion, ISPRS Annals of the Photogrammetry, Remote Sensing and Spatial Information Sciences III-3 (2016) 99–106.
[14] M. Michelini, Automatische Kameraposeschätzung für komplexe Bildmengen, Dissertation, Universität der Bundeswehr München, 2018.
[15] W. Förstner, On the geometric precision of digital correlation, International Archives of Photogrammetry and Remote Sensing 24 (3) (1982) 176–189.
[16] A. Grün, Adaptive least squares correlation: a powerful image matching technique, South African Journal of Photogrammetry, Remote Sensing and Cartography 14 (3) (1985) 175–187.
[17] D. Lowe, Distinctive image features from scale-invariant keypoints, International Journal of Computer Vision 60 (2) (2004) 91–110.
[18] W. Förstner, B. Wrobel (Eds.), Photogrammetric Computer Vision—Statistics, Geometry, Orientation and Reconstruction, Springer-Verlag, Berlin, Germany, 2016.
[19] L. Roth, A. Kuhn, H. Mayer, Wide-baseline image matching with projective view synthesis and calibrated geometric verification, PFG—Journal of Photogrammetry, Remote Sensing and Geoinformation Science 85 (2) (2017) 85–95.
[20] A. Kuhn, H. Hirschmüller, D. Scharstein, H. Mayer, A TV prior for high-quality scalable multi-view stereo reconstruction, International Journal of Computer Vision 124 (1) (2017) 2–17.
[21] H. Mayer, Efficient hierarchical triplet merging for camera pose estimation, in: German Conference on Pattern Recognition, Springer-Verlag, Berlin, Germany, 2014, pp. 399–409.
[22] M. Levandowsky, D. Winter, Distance between sets, Nature 234 (5) (1971) 34–35.
[23] C. Beder, R. Steffen, Determining an initial image pair for fixing the scale of a 3D reconstruction from an image sequence, in: Pattern Recognition, DAGM 2006, Springer-Verlag, Berlin, Germany, 2006, pp. 657–666.
[24] M. Michelini, H. Mayer, Detection of critical camera configurations for structure from motion, The International Archives of the Photogrammetry, Remote Sensing and Spatial Information Sciences XL-3/W1 (2014) 73–78.
[25] G. Lin, G. Xue, On the terminal Steiner tree problem, Information Processing Letters 84 (2) (2002) 103–107.
[26] Y. Chen, An improved approximation algorithm for the terminal Steiner tree problem, in: Computational Science and Its Applications, ICCSA 2011, in: Lecture Notes in Computer Science, Springer-Verlag, Berlin, Germany, 2011, pp. 141–151.
[27] B. Curless, M. Levoy, Volumetric method for building complex models from range images, in: SIGGRAPH, ACM Transactions on Graphics, 1996.
[28] M. Goesele, B. Curless, S. Seitz, Multi-view stereo revisited, in: IEEE Conference on Computer Vision and Pattern Recognition, vol. 2, 2006, pp. 2402–2409.
[29] R. Sagawa, K. Nishino, K. Ikeuchi, Adaptively merging large-scale range data with reflectance properties, IEEE Transactions on Pattern Analysis and Machine Intelligence 27 (3) (2005) 392–405.
[30] S. Fuhrmann, M. Goesele, Fusion of depth maps with multiple scales, in: SIGGRAPH Asia, ACM Transactions on Graphics, 2011.

[31] P. Merell, A. Akbarzadeh, L. Wang, P. Mordohai, J.-M. Frahn, R. Yang, D. Nistér, M. Pollefeys, Real-time visibility-based fusion of depth maps, in: IEEE Conference on Computer Vision and Pattern Recognition, 2007, pp. 1–8.

[32] H. Hirschmüller, Stereo processing by semiglobal matching and mutual information, IEEE Transactions on Pattern Analysis and Machine Intelligence 30 (2) (2008) 328–341.

[33] A. Kuhn, H. Mayer, H. Hirschmüller, D. Scharstein, A TV prior for high-quality local multi-view stereo reconstruction, in: 2nd International Conference on 3D Vision, 3DV, 2014, pp. 65–72.

[34] D. Scharstein, H. Hirschmüller, Y. Kitajima, G. Krathwohl, N. Nesic, X. Wang, P. Westling, High-resolution stereo datasets with subpixel-accurate ground truth, in: German Conference on Pattern Recognition, Springer, 2014.

[35] A. Kuhn, H. Hirschmüller, H. Mayer, Multi-resolution range data fusion for multi-view stereo reconstruction, in: German Conference on Pattern Recognition, Springer-Verlag, Berlin, Germany, 2013, pp. 41–50.

[36] N. Molton, M. Brady, Practical structure and motion from stereo when motion is unconstrained, International Journal of Computer Vision 39 (1) (2000) 5–23.

[37] A. Kuhn, H. Mayer, Incremental division of very large point clouds for scalable 3D surface reconstruction, in: IEEE International Conference on Computer Vision Workshops, ICCVW, 2015, pp. 157–165.

[38] H. Huang, H. Mayer, Robust and efficient urban scene classification using relative features, in: 23rd ACM SIGSPATIAL International Conference on Advances in Geographic Information Systems, GIS '15, ACM, New York, NY, USA, 2015, pp. 81:1–81:4.

[39] L. Breiman, Random forests, Machine Learning 45 (1) (2001) 5–32.

[40] L. Guo, N. Chehata, C. Mallet, S. Boukir, Relevance of airborne lidar and multispectral image data for urban scene classification using random forests, ISPRS Journal of Photogrammetry and Remote Sensing 66 (1) (2011) 56–66.

[41] H. Huang, H. Jiang, C. Brenner, H. Mayer, Object-level segmentation of RGBD data, ISPRS Annals of the Photogrammetry, Remote Sensing and Spatial Information Sciences II-3 (2014) 73–78.

[42] H. Huang, C. Brenner, M. Sester, 3D building roof reconstruction from point clouds via generative models, in: 19th ACM SIGSPATIAL International Conference on Advances in Geographic Information Systems, GIS, 1-4 November, ACM Press, Chicago, IL, USA, 2011, pp. 16–24.

[43] H. Huang, C. Brenner, M. Sester, A generative statistical approach to automatic 3D building roof reconstruction from laser scanning data, ISPRS Journal of Photogrammetry and Remote Sensing 79 (2013) 29–43.

[44] F. Lafarge, X. Descombes, J. Zerubia, M. Pierrot-Deseilligny, Structural approach for building reconstruction from a single DSM, IEEE Transactions on Pattern Analysis and Machine Intelligence 32 (1) (2010) 135–147.

[45] M. Kada, L. McKinley, 3D building reconstruction from LiDAR based on a cell decomposition approach, The International Archives of the Photogrammetry, Remote Sensing and Spatial Information Sciences 38(3/W4) (2009) 47–52.

[46] T. Partovi, H. Huang, T. Krauß, H. Mayer, P. Reinartz, Statistical building roof reconstruction from worldview-2 stereo imagery, The International Archives of the Photogrammetry, Remote Sensing and Spatial Information Sciences XL(3/W2) (2015) 161–167.

[47] H. Huang, H. Mayer, Towards automatic large-scale 3d building reconstruction: primitive decomposition and assembly, in: Societal Geo-Innovation, the 20th AGILE Conference on Geographic Information Science, in: Lecture Notes in Geoinformation and Cartography, Springer, 2017.

[48] C. Brenner, N. Haala, Erfassung von 3D Stadtmodellen, Photogrammetrie—Fernerkundung—Geoinformation 2 (2000) 109–117.

[49] W. Nguatem, H. Mayer, Contiguous patch segmentation in pointclouds, in: German Conference on Pattern Recognition, 2016, pp. 131–142.

[50] W. Nguatem, H. Mayer, Modeling urban scenes from pointclouds, in: International Conference on Computer Vision, 2017, pp. 3837–3846.

[51] H. Huang, C. Brenner, Rule-based roof plane detection and segmentation from laser point clouds, in: Joint Urban Remote Sensing Event, JURSE, 2011, 11-13 April, IEEE, Munich, Germany, 2011, pp. 293–296.

[52] H. Huang, B. Kieler, M. Sester, Urban building usage labeling by geometric and context analyses of the footprint data, in: 26th International Cartographic Conference, ICC, International Cartographic Association (ICA), Dresden, Germany, 2013.
[53] M. Schmitz, H. Mayer, A convolutional network for semantic facade segmentation and interpretation, The International Archives of the Photogrammetry, Remote Sensing and Spatial Information Sciences XLI-B3 (2016) 709–715.
[54] A. Krizhevsky, I. Sutskever, G.E. Hinton, Imagenet classification with deep convolutional neural networks, in: Advances in Neural Information Processing Systems, 2012, pp. 1097–1105.
[55] J. Deng, W. Dong, R. Socher, L.-J. Li, K. Li, L. Fei-Fei, ImageNet: a large-scale hierarchical image database, in: IEEE Conference on Computer Vision and Pattern Recognition, 2009, pp. 248–255.
[56] F. Korč, W. Förstner, eTRIMS Image Database for Interpreting Images of Man-Made Scenes, Technical report TR-IGG-P-2009-01, Dept. of Photogrammetry, University of Bonn, 2009, http://www.ipb.uni-bonn.de/projects/etrims_db/.
[57] W. Zhang, H. Huang, M. Schmitz, X. Sun, H. Wang, H. Mayer, Effective fusion of multi-modal remote sensing data in a fully convolutional network for semantic labeling, Remote Sensing 10 (1) (2018) 52.

CHAPTER 11

Decision Fusion of Remote-Sensing Data for Land Cover Classification

Arnaud Le Bris[*], **Nesrine Chehata**[*,†], **Walid Ouerghemmi**[*,¶], **Cyril Wendl**[*,‡], **Tristan Postadjian**[*], **Anne Puissant**[§], **Clément Mallet**[*]

[*]*Univ. Paris-Est, LASTIG STRUDEL, IGN, ENSG, Saint-Mande, France* [†]*EA G&E Bordeaux INP, Université Bordeaux Montaigne, Pessac, France* [‡]*Student at Ecole Polytechnique Fédérale de Lausanne (EPFL), Lausanne, Switzerland* [§]*CNRS UMR 7362 LIVE-Université de Strasbourg, Strasbourg, France* [¶]*Aix-Marseille Université, CNRS ESPACE UMR 7300, Aix-en-Provence, France*

Contents

Multimodal Scene Understanding
https://doi.org/10.1016/B978-0-12-817358-9.00017-2

11.1 Introduction

The last years have witnessed the emergence of a large variety of new sensors with various characteristics. The possibility to collect different kinds of observations over the same area has considerably increased: remote sensing can now be considered generically multimodal [1]. Those sensors can use different modalities (radar, Lidar or optical). They can be air-borne or satellite-borne. Even for the same modality, they can exhibit very distinct characteristics: for instance, optical sensors show a large range of spectral configurations (number of spectral bands, position and width of the spectral bands), spatial resolutions, coverage and, for spaceborne sensors, revisit times (i.e., minimum delay between two possible consecutive acquisitions over the same area, thus conditioning the possibility to capture genuine time series). As a consequence, combining remote-sensing data with different characteristics is a standard remote-sensing problem that has been extensively investigated in the literature [2]. The overall aim consists in fusing multisensor information as a means of combining the respective advantages of each sensor. Complementary observations can thus be exploited for land cover mapping purposes, which is a core remote-sensing application and the necessary input for a large number of public policies and environmental models. Combining existing sensors can mitigate limitations of any one particular sensor for various land cover issues [3,4].

This chapter specifically focuses on the fusion of one data type, exhibiting a *very high spatial resolution*, with another one, exhibiting a *lower spatial resolution but enhanced complementary characteristics*. Indeed, very high spatial resolution (VHR) multispectral imagery enables an accurate spatial delineation of objects and a possible use of texture information for enhanced class discrimination [5]. On the other hand, sensors with lower spatial resolutions offer enhanced spectral or temporal information, making it possible to consider richer land cover semantics. To illustrate this problem, two use cases will be considered, accompanied by two methodological contributions:

- The first one considers the fusion of very high spatial resolution multispectral imagery with lower spatial resolution hyperspectral imagery for detailed urban land cover classification. Hyperspectral imagery provides an accurate spectral information but generally with a low geometric precision. Hence, it can provide finer land cover semantics classification results, while a VHR image helps to retrieve the geometric contours of such classes. Combining both data sources helps to reach better accuracy scores at the highest spatial resolution of both datasets.
- The second one integrates very high spatial resolution multispectral/monodate SPOT 6/7 satellite imagery (classified with a convolutional neural network, CNN), and Sentinel-2 time series (classified with a random forest). The final aim is urban footprint detection. In this case, VHR imagery provides texture information and fine object delineation, while the Sentinel-2 time series gives access to better contextual information.

In both cases, the land cover fusion scheme targets benefiting from the complementary characteristics of these multimodal sources.

Existing data fusion approaches will be analyzed in Sect. 11.1.1. From this review, existing methods will be discussed, and a fusion strategy elaborated (Sect. 11.1.2). This proposed framework will then be presented in detail (Sect. 11.2), before being applied to the two above-mentioned use cases (Sects. 11.3 and 11.4, respectively).

11.1.1 Review of the Main Data Fusion Methods

Fusion of heterogeneous data sources have been widely investigated in the remote-sensing literature (e.g., [6–9]). Fusion can be carried out at three different levels [10]:

- The observation level: early fusion.
- The attribute/feature level: intermediate fusion.
- The decision level: late fusion.

11.1.1.1 Early fusion – fusion at the observation level

Fusion can be achieved at the observation level, i.e., through the direct joint analysis of the pixel values (with or without calibration procedures). For that purpose, pan-sharpening is a well-known technique that integrates the geometric details of a high resolution panchromatic image and the color information of a low spatial resolution multispectral (or hyperspectral) image to produce a high spatial resolution multispectral (or hyperspectral) image. Pan-sharpening methods usually use the panchromatic image to replace the high frequency part in the low resolution image [11]. Other fusion algorithms have been proposed to merge multispectral (or hyperspectral) and panchromatic (but also multispectral and hyperspectral) images to combine complementary characteristics in terms of spatial and spectral resolutions [12–14]). A review of such methods can be found in [14]. Eventually, super-resolution is another approach relying on early fusion of several sensors [15].

11.1.1.2 Intermediate fusion – fusion at the attribute/feature level

Data sources can be merged at the feature level. Features (spectral indices, texture-based, etc.) are computed for each source separately or for both of them and fed into the same classifier through a unique feature set [16]. Examples of remote-sensing pipelines involving fusion at the attribute level can be found in [17–22]. For instance, [18] proposed a conditional random field (CRF) model for building detection using InSAR and orthoimage features. Reference [20] merged Lidar and optical aerial image features for forest stand extraction: the proposed approach involves several steps (segmentation, classification, and regularization) and fusion is performed at each of them. Improvement is noticed for each step. More recently, deep convolutional neural networks were used to perform data fusion at the attribute level [23]. Reference [22] applied for instance deep forests to Lidar and hyperspectral imagery features. A detailed review can be found in [24]. Several datasets and challenges have been released in the last decades, under the aegis of the IEEE GRSS society [25–27].

11.1.1.3 Late fusion – fusion at the decision level

Late decision fusion happens after the classification process: the outputs of multiple independent classifiers are combined in order to provide a more reliable decision. Such classification results can be either label maps or class membership probability maps. Various late decision fusion methods have been proposed. Most of them can be divided into different categories: consensus rules (majority voting), probabilistic approaches (Bayesian fusion), credibilist or evidential ones, and possibilist ones.

The probabilistic, evidential, and credibilist decision fusion approaches are generic and can be applied to different fusion problems. They only require class "membership" measures (probabilities or belief masses depending on the approach) for each source and for each class, or at least a confidence measure for each source.

Possibilist methods use fuzzy logic-based fusion rules [28–30]. They require one to define weights [31] in order to better deal with the uncertainty of the different sources. Such generic approaches have been applied to remote-sensing data [32].

Evidential approaches are a generalization of probabilistic ones. They include the well-known Dempster–Shafer fusion rule [33]. In remote sensing, this rule has often been used to merge classifications or alarm detection results. For instance, [34–36] applied the Dempster–Shafer rule to combine several different supervised building detectors based on different remote-sensing modalities (optical, Lidar, radar). This rule was also used by [37] for the fusion of different road obstacle detectors in the context of intelligent vehicle development. Reference [38] used the Dempster–Shafer rule for the fusion of urban footprints detected at different dates out of satellite archival images. References [39,40] applied this rule to detecting changes in an unsupervised classification context.

Another evidential fusion rule is the Yager rule. It was applied by [41] to combine different road obstacle detectors for vehicle navigation. Other evidential rules have been proposed more recently: the Dezert–Smarandache rule may be mentioned [42]. Other new rules have also been proposed by [43,44]: they extend Dempster–Shafer rules in order to achieve a better management of the conflict between sources. However efficient in some cases, Dempster–Shafer remains a theoretical complex framework that does not easily apply when dealing with heterogeneous and multiple data.

Another important issue consists in defining the input of the different fusion rules. Different situations can be taken into account depending on whether a global confidence measure is affected for each source, or, conversely, whether class posterior probabilities for a given source are directly available for each pixel (directly provided by the initial classifier). In the former situation, global confidence measures affected to each source are calculated from the confusion matrices (indeed, a confusion matrix provides the probability of an object labeled class A by one source to belong in fact to class B). Validation data are thus necessary to calculate these weights. Furthermore, for evidential methods and especially Dempster–Shafer methods, uncertainty classes (i.e. unions of original classes) must necessarily be defined. Hence, belief masses associated to these classes must be computed. If a global confidence is associated to a source, some solutions have been proposed. For instance, we have Appriou's method [45], or more recently the method presented in [46,47]. If a class membership measure is affected to each pixel for each source and for each class, it is also possible to derive belief masses for these uncertainty (union) classes. Finally, [38] propose an alternative way to integrate these two kinds of information.

Last but not least, the last category of late fusion approaches consists in supervised learning. They automatically learn from training examples the best way to merge input sources. Per-source posterior class probabilities are concatenated and considered as a feature vector. This feature vector is then provided as an input to a classifier which is trained to perform the best possible fusion. Thus, they can be considered as being at the interplay between late and intermediate fusion (classifiers previously applied to each source can then be considered as a kind of feature generators, then referred to as auto-context classification [48]). Such approaches have been used with different classifiers: random forest [49,50], Adaboost [50], support vector machines [51,52]. These supervised learning-based methods enable good results but require a sufficient amount of training data to model the classes and avoid over-fitting (especially for deep learning).

In addition, it must here be said that most fusion methods mentioned in this section generate measures to assess the conflict between two sources.

Late approaches have often been applied to remote-sensing fusion problems [53–56,32,51, 57–61]. Fusion methods operating at the decision level can be applied in two situations, considering if they try to merge multiple classifiers applied to either the same data source, or to multiple sources. For instance, [53] combined neuronal and statistical maximum likelihood classifiers using several consensus theory rules (i.e., majority voting, complete agreement) to classify multispectral and hyperspectral images. References [57,58] merged posterior probabilities from maximum likelihood classifications of optical images with prior information about classes derived out of, respectively, digital terrain models or digital surface models, as well as information from existing land cover databases. Reference [55] investigated the fusion of multitemporal thematic mapper images, using decision fusion-based methods (i.e., joint likelihood and weighted majority fusion). A characterization of the spatial organization of SAR image elements is investigated by [54] merging the responses of multiple low-level detectors applied to the same image within a Dempster–Shafer scheme. Reference [32] investigated the use of fuzzy decision rules to combine the classification results of a conjugate gradient neural network and a fuzzy classifier over an IKONOS image. Reference [61] combined convolutional neural networks and random forest classifiers using a multiplication Bayesian scheme.

11.1.2 Discussion and Proposal of a Strategy

Fusion can be performed at different levels. Fusion at the observation level, e.g., pan or multisharpening, is limited to specific situations where it has a real physical meaning (e.g., hyperspectral and multispectral images acquired simultaneously). It is not generic enough.

Fusion at the observation level or at the decision level is more generic and also eligible to the present fusion problem. Two main issues remain: (i) for both, the spatial scale of analysis and, subsequently, the interpolation process; (ii) for feature-based approaches, the ability to correctly handle the various data sources in the decision process. In the case of imbalanced feature sets, supervised techniques such as random forests or support vector machines, even with feature selection strategies, may prefer the data source generating the larger number of attributes. Thus, the process will not fully benefit from the advantages of all datasets.

As a consequence, late fusion strategy is adopted in this chapter. Indeed, contrary to intermediate level fusion methods, it makes it possible to initially process each input data source independently through specific optimal methods. Moreover, it even enables one to use already existing results from available operational land cover classification services, as long as they provide class membership confidence measures. Besides, especially in this last situation, it can be used without any ground truth (training) information, contrary to intermediate level fusion methods.

Most existing decision fusion methods do not explicitly take into account the fact that input data sources have different spatial resolutions, and thus do not explicitly deal with both semantic and spatial uncertainties. Spatial uncertainty handling here consists in removing classification noise, and enforcing that the classification result follows as closely as possible the natural borders in the original images. Such a task can be cast in the form of a smoothing problem. Local smoothing methods exist: majority voting, Gaussian and bilateral filtering [62], as well as probabilistic relaxation [63] are possible. The majority vote can be used in particular when a segmentation of the area is available: the major class is assigned to the segment. The vote can also be weighted by class probabilities of the different pixels. The probabilistic relaxation is another local smoothing method that aims at homogenizing probabilities of a pixel according to its neighbors. It is an iterative algorithm in which the class probability at each pixel is updated at each iteration in order to have it closer to the probabilities of its neighbors.

However, these local smoothing methods are generally outperformed by global regularization strategies [64,20]. Global regularization methods consider the whole image by connecting each pixel to its neighbors. They traditionally adopt Markov random fields (MRFs): the labels at different locations are not considered to be independent and the global solution can be retrieved with the simple knowledge of the close neighborhood for each pixel. The optimal configuration of labels is retrieved when finding the maximum *a posteriori* over the entire field [65,64]. The problem is therefore considered as a minimization procedure of a global energy over the whole image. Despite a simple neighborhood encoding (pairwise relations are often preferred), the optimization procedure propagates over large distances. Global regularization is often considered as a post-processing step within a classification process. It has been associated to late fusion in recent works, as for instance in [66,67].

As a consequence, so as to benefit from the complementarity of a very high spatial resolution sensor with another one, exhibiting lower spatial resolution but enhanced complementary characteristics, the proposed fusion framework involves (i) fusion at decision level, (ii) associated to global regularization. It mostly relies on existing state-of-the-art methods, but combines them in order to cope both with semantic and spatial uncertainties. Besides, it is flexible enough to integrate several fusion rules and be applied to various use cases.

11.2 Proposed Framework

A late fusion framework is proposed in order to benefit both from low spatial resolution data (but enhanced spectrally or temporally) and very high spatial resolution multispectral mono-

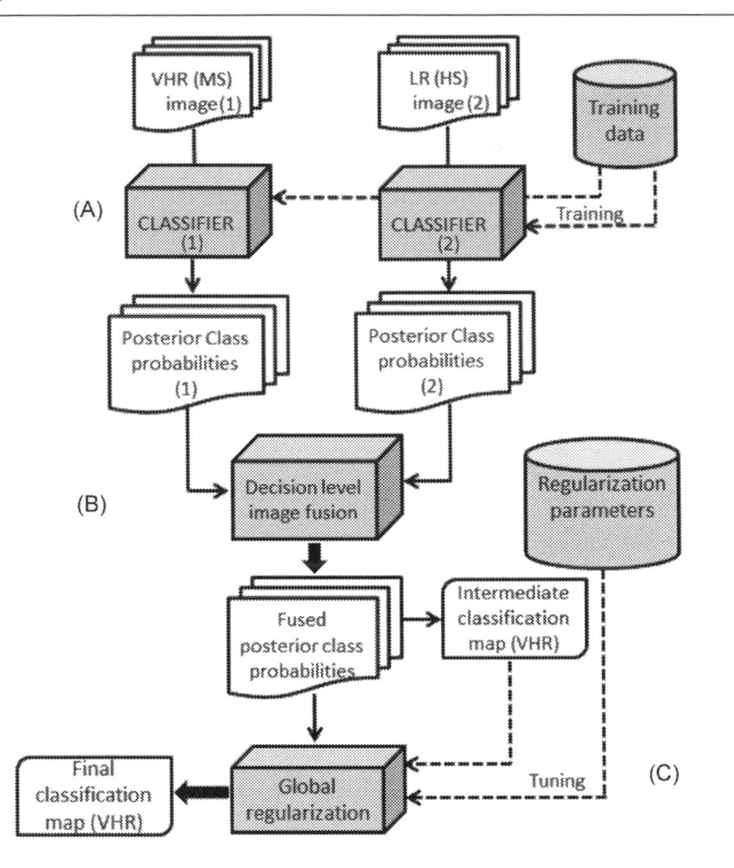

Figure 11.1 : Proposed generic framework.

date data. It aims at dealing both with semantic and spatial uncertainties. It consists in three main steps, presented in Fig. 11.1.

(A) **Classification of each original source and generation of the posterior class probabilities:** the two input data sources are first automatically labeled, independently, by specific adapted processes. At the end, a posterior class probability map is generated for each of them. Here, predictions have already been computed and are considered as granted.

(B) **Per pixel fusion of the posterior probabilities at the decision level:** a per-pixel decision fusion is applied to these maps so as to combine them into a more accurate decision at the highest spatial resolution. Different fusion rules are considered. This step aims at dealing with semantic uncertainties between sources. A conflict measure between sources can also be generated at this step (Sect. 11.2.1).

(C) **Global regularization:** The final classification map is retrieved as a result of a global regularization step of the merged class membership probability map obtained at the previous step. It allows one to deal with spatial uncertainties between both sources, to reduce remaining noise and to take into account the contrast information of the original image and subsequently to follow as closely as possible the natural borders in images (Sect. 11.2.2).

11.2.1 Fusion Rules

11.2.1.1 Fuzzy rules

The first tested fusion approach is based on fuzzy rules [30]. Fuzzy rule theory states that a fuzzy set A in a reference set of classes $\mathcal{L}$ is a set of ordered pairs:

$$A = [(c, P_A^{(c)}(x)|c \in \mathcal{L})], \tag{11.1}$$

where the membership probability of A in $\mathcal{P}$ is given by $P_A^{(c)} : \mathcal{L} \to [0, 1]$. The measure of conflict $(1 - K)$ between two sources is given as [68]

$$K = \sup_{c \in \mathcal{L}} \min(P_A^{(c)}(x), P_B^{(c)}(x)). \tag{11.2}$$

In order to account for the fact that fuzzy sets with a strong fuzziness possibly hold unreliable information, each fuzzy set i is weighted according to a pointwise confidence measure w_i [32]:

$$w_i = \frac{\sum_{k=0, k \neq i}^{n} H_{\alpha QE}(P_k)}{(n-1)\sum_{k=0}^{n} H_{\alpha QE}(P_k)}. \tag{11.3}$$

n is the number of sources and $H_{\alpha QE}$ is a fuzziness measure called α-quadratic entropy (QE) [31]. Each fuzzy set i is weighted by the fuzziness degree of all other fuzzy sets (*i.e.*, classifications). If the fuzziness degree of the other sets is high, the weight of a given source i will be high too. For all the following fusion rules, the fuzzy sets have been weighted as $\tilde{P}_A^{(c)}(x) = w_A \cdot P_A^{(c)}(x)$, $\tilde{P}_B^{(c)}(x) = w_B \cdot P_B^{(c)}(x)$, $P_A^{(c)}(x)$ and $P_B^{(c)}(x)$ being the original membership probabilities and w_A, w_B their corresponding pointwise measures.

In further experiments, all fuzzy rules have been tested as input using the probabilities weighted by the pointwise measure.

The following fusion rules based on fuzzy logic were considered:

1. **Minimum rule (Min)** as the intersection of two fuzzy sets P_A and P_B, given by the minimum of their membership probabilities (conjunctive behavior):

$$\begin{aligned} \forall c \in \mathcal{L} \quad (P_A \cap P_B)^{(c)}(x) &= P_{fusion}^{(c)}(x) \\ &= \min\left(P_A^{(c)}(x), P_B^{(c)}(x)\right). \end{aligned} \tag{11.4}$$

2. **Maximum rule (Max)** as the union between the two fuzzy sets P_A and P_B, given by the maximum of their membership probabilities (disjunctive behavior):

$$\forall c \in \mathcal{L} \quad (P_A \cup P_B)^{(c)}(x) = P^{(c)}_{fusion}(x) = \max\left(P^{(c)}_A(x), P^{(c)}_B(x)\right). \tag{11.5}$$

3. **Compromise operator**:

$$P^{(c)}_{fusion}(x) = \begin{cases} \max\left(T_1, \min\left(T_2, (1-K)\right)\right) \\ \quad \text{if}(1-K) \neq 1, \\ \max\left(P^{(c)}_A(x), P^{(c)}_B(x)\right) \\ \quad \text{if } (1-K) = 1. \end{cases} \tag{11.6}$$

$T_1 = \frac{\min\left(P^{(c)}_A(x), P^{(c)}_B(x)\right)}{K}$, $T_2 = \max\left(P^{(c)}_A(x), P^{(c)}_B(x)\right)$.
It can be noticed that the operator behavior is conjunctive when the conflict between A and B is low ($1 - K \approx 0$), and disjunctive when the conflict is high ($1 - K \approx 1$). When the conflict is partial, the operator behaves in a compromise way [68].

4. **Compromise modified**: Since a pure compromise fusion rule would favor T_1, [69] proposed to measure the intra-class conflict as $f_c = \text{abs}(C_{\text{best1}} - C_{\text{best2}})$ with $C_{\text{best1}} = \text{argmax}_{c \in \mathcal{L}} P^{(c)}_S(x)$ and $C_{\text{best2}} = \text{argmax}_{c \in \mathcal{L} \setminus C_{\text{best1}}} P^{(c)}_S(x)$. They set a conflict threshold t_c (e.g., $t_c = 0.25$ for experiments in Sect. 11.3) to be used as follows:

Algorithm 1 Compromise rule according to [69]

if $f_c < t_c$ **then** ▷ existence of intra-membership conflict
  $P^{(c)}_{fusion} = \max\left(P^{(c)}_A(x), P^{(c)}_B(x)\right)$
else ▷ no intra-membership conflict
  $P^{(c)}_{fusion}$ following Eq. (11.6)
end if

5. **Prioritized operators** (referred to as Prior 1 for Eq. (11.7) and Prior 2 for Eq. (11.8)):

$$P^{(c)}_{fusion}(x) = \max\left(P^{(c)}_A(x), \min(P^{(c)}_B(x), K)\right), \tag{11.7}$$

$$P^{(c)}_{fusion}(x) = \min\left(P^{(c)}_A(x), \max\left(P^{(c)}_B(x), (1-K)\right)\right). \tag{11.8}$$

If the conflict between A and B is high (*i.e.*, $K \approx 0$), only $P^{(c)}_A(x)$ is taken into account (prioritized) and $P^{(c)}_B(x)$ is considered as a specific piece of information. Thus, the fusion result depends on the order of the sources A and B.

6. An **accuracy dependent** (AD) operator [32], integrating both local and global confidence measurements:

$$P^{(c)}_{fusion}(x) = \max\left(\min\left(w_i.P_i^{(c)}(x), f_i^{(c)}(x)\right), i \in [1, n]\right), \tag{11.9}$$

where $f_i^{(c)}$ is the global confidence of source i regarding class c, P_i is a class membership of source i, and w_i is a normalization factor (see Eq. (11.3)). This operator ensures that only reliable sources are taken into consideration for each class, via the predefined coefficients $f_i^{(c)}$. The idea seems interesting. Nevertheless, the final result depends on the reliability of the classifier and also on the availability of ground truth data, which is mandatory to generate the $f_i^{(c)}$ term.

11.2.1.2 Bayesian combination and majority vote

A straightforward approach is to sum or multiply the input class membership probabilities as a Bayesian sum ($\sim$ majority vote) or product [70]:

$$P^{(c)}_{fusion\_sum}(x) = P_A^{(c)}(x) + P_B^{(c)}(x), \tag{11.10}$$

$$P^{(c)}_{fusion\_product}(x) = P_A^{(c)}(x) \times P_B^{(c)}(x). \tag{11.11}$$

Those rules will be referred to as **Bayesian sum** and **Bayesian product**, respectively.

11.2.1.3 Margin-based rules

The aim of these rules is to take into account the confidence, measured by the classification margin, of each source. The classification margin is defined for each pixel x and each source s as the difference between the two highest class probabilities:

$$margin^{(s)}(x) = P_s^{(C_{\text{best1}})}(x) - P_s^{(C_{\text{best2}})}(x), \tag{11.12}$$

with $C^{(s)}_{\text{best1}}(x) = \text{argmax}_{c \in \mathcal{L}}\, P_s^{(c)}(x)$ and $C^{(s)}_{\text{best2}}(x) = \text{argmax}_{c \in \mathcal{L} \setminus \{C^{(s)}_{\text{best1}}(x)\}}\, P_s^{(c)}(x)$.

Fusion can then be carried out preferring for each pixel the most confident source, i.e. the one with the highest margin. This fusion rule (referred to as **margin-Max**) selects, for each pixel, the source for which the margin between the two highest probabilities is the highest, with sources $\mathcal{S} = A, B$ and the classes $\mathcal{L} = \{c_i\}_{1 \leq i \leq n}$:

$$\forall x, \forall c \in \mathcal{L} \;\; P^{(c)}_{fusion}(x) = P^{(c)}_{S_{best}}(x), \tag{11.13}$$

where $S_{best} = \text{argmax}_{\mathcal{S} \in \mathcal{C}}\, margin^{(s)}(x)$.

The classifier confidence information provided by the margin can also be used to weight the class probabilities of each source in the Bayesian sum and product (respectively, **margin Bayesian sum weighted**, **margin Bayesian product weighted**):

$$P^{(c)}_{fusion\_sum}(x) = \frac{P_A^{(c)}(x) \cdot margin^{(A)}(x) + P^{B(c)}(x) \cdot margin_B^{(c)}(x)}{margin^{(A)}(x) + margin^{(B)}(x)}, \tag{11.14}$$

$$P^{(c)}_{fusion\_product}(x) = \left(P_A^{(c)}(x)\right)^{\frac{margin^{(A)}(x)}{margin^{(A)}(x)+margin^{(B)}(x)}} \times \left(P_B^{(c)}(x)\right)^{\frac{margin^{(B)}(x)}{margin^{(A)}(x)+margin^{(B)}(x)}}. \tag{11.15}$$

11.2.1.4 Dempster–Shafer evidence theory

According to the Dempster–Shafer (DS) formalism, an information from a source s for a class c can be given as a mass function $m_c | m_c \in [0, 1]$ [33]. Dempster–Shafer's evidence theory rule assumes simple classes $c \in \mathcal{L}$ as well as composed classes, which were hence limited to two simple classes at most [69].

Masses associated to each simple class are directly the class membership probabilities:

$$m_s^{(c)}(x) = P_s^{(x)}. \tag{11.16}$$

For mixed classes, $\forall c_1, c_2 \in \mathcal{L}$, $\forall$ pixel x and $\forall s \in \mathcal{S}$, two versions were tested, denoted DSV_1 and DSV_2, respectively:

$$\begin{aligned} m_s^{(c_1 \cup c_2)}(x) = & \left(P_s^{(c_1)}(x) + P_s^{(c_2)}(x)\right) \\ & \times \left(1 - \max\left(P_s^{(c_1)}(x), P_s^{(c_2)}(x)\right)\right) \\ & + \min\left(P_s^{(c_1)}(x), P_s^{(c_2)}\right), \end{aligned} \tag{11.17}$$

$$\begin{aligned} m_s^{(c_1 \cup c_2)}(x) = & \frac{1}{2}\left(P_s^{(c_1)}(x) + P_s^{(c_2)}(x)\right) \\ & \times \left(1 - \max\left(P_s^{(c_1)}(x), P_s^{(c_2)}(x)\right)\right) \\ & + \min\left(P_s^{(c_1)}(x), P_s^{(c_2)}\right). \end{aligned} \tag{11.18}$$

This leads to a mass $m_s^{(c_1 \cup c_2)}(x) \in [0, 1]$, being 1 if $P_s^{(c_1)} = 0$, $P_s^{(c_2)} = 1$ or $P_s^{(c_2)} = 0$, $P_s^{(c_1)} = 1$. In both versions, all masses are normalized such that $\sum_{c \in \mathcal{L}}^{(s)} m_c^{(s)}(x) = 1$.

The fusion rule is based on the following conflict measure between two sources A and B:

$$K(x) = \sum_{\substack{c,d \in \mathcal{L}' \\ c \cap d \neq \emptyset}}^{(s)} = m_A^{(c)}(x) m_B^{(c)}(x), \tag{11.19}$$

$c, d \in \mathcal{L}$ being mixed classes with $c \cap d = \emptyset$.

The fusion is performed by

$$m^{(c)}_{fusion}(x) = \frac{1}{1-K(x)} \sum^{(s)}_{\substack{c_1,c_2 \in \mathcal{L}' \\ c_1 \cap c_2 = c}} m^{(c)}_A(x) m^{(c)}_B(x). \tag{11.20}$$

11.2.1.5 Supervised fusion rules: learning based approaches

In addition to the previous standard fusion rules, learning-based supervised methods were also tested [51,52,23]. Such methods consist in learning how to best merge both sources (based on a ground truth). A classifier is trained to label feature vectors corresponding to the concatenation of class membership measures from both sources. Thus, such a strategy can be considered at the interplay between late and intermediate fusion (classifiers applied independently to each source can then be considered as a kind of feature generators). It is similar to *auto-context* approaches. A drawback stems from the fact that they require a significant amount of reference data.

In next experiments, two classifiers were considered for supervised fusion: random forests (RFs) [71] and support vector machines (SVMs) [72] with a linear or a radial basis function (rbf) kernel.

11.2.2 Global Regularization

After fusion rules have been applied at the pixel level, a spatial regularization of the obtained classification map is performed. This regularization here aims at dealing with spatial uncertainties between both sources. Indeed, the fusion result still contains noisy patches, especially in transition areas between neighboring classes. Besides, considering original image information, such a regularization also enables one to preserve real-world contours more accurately.

A global regularization strategy is adopted [66]. The problem is expressed using an energetic graphical model and solved as a min-cut problem. Indeed, such a formulation has been used successfully for many purposes related to image processing in the last years [73].

11.2.2.1 Model formulation(s)

The problem is formulated in terms of an energy E that has to be minimized over the whole image I in order to retrieve a labeling C of the entire image which corresponds to a minimum of E. As commonly adopted in the literature, E consists of two terms, one related to the data fidelity, and one to prior spatial knowledge, setting constraints on class transitions between the different pairs of neighboring pixels N. Several options were considered for the different energy terms. We have

$$E(P_{fusion}, C_{fusion}, C) = \sum_{x \in I} E_{data}(C(x)) + \lambda \sum_{\substack{x,y \in N \\ x \neq y}} E_{reg}(C(x), C(y)), \tag{11.21}$$

where

$$\begin{aligned}
&E_{data}\big(C(x)\big) = f\Big(P_{fusion}\big(C(x)\big)\Big),\\
&E_{reg}\big(C(x) = C(y)\big)\\
&\quad = g\Big(P_{fusion}\big(C(x)\big), P_{fusion}\big(C(y)\big), C_{fusion}(x), C_{fusion}(y), I(x), I(y)\Big),\\
&E_{reg}\big(C(x) \neq C(y)\big)\\
&\quad = h\Big(P_{fusion}\big(C(x)\big), P_{fusion}\big(C(y)\big), C_{fusion}(x), C_{fusion}(y), I(x), I(y)\Big).
\end{aligned}$$

The final label map corresponds to the configuration C which minimizes E over I.

The **data term** is a fit-to-data attachment term. It relies on the probability distribution P_{fusion}, defined by a function $f(.)$. The function f ensures that if the probability for a pixel x to belong to class $C(x)$ is close to 1, E_{data} will be small. It will not impact the total energy E. Conversely, if the probability for a pixel x to belong to class $C(x)$ is low, E_{data} will be near its maximum and will penalize such a configuration. The following options were tested:

$$\text{Option 1: } f(t) = -\log(t), \tag{11.22}$$

$$\text{Option 2: } f(t) = 1 - t. \tag{11.23}$$

Earlier experiments [20] verified that *Option 2* tends to smooth the classification map more than *Option 2*. Thus, the data term will be selected among these options depending on the targeted application and on the input data. *Option 1* will be selected to keep small regions as long as they are relevant according to class probabilities, while *Option 2* will be used to obtain smoother maps with wider flat areas [20].

The **regularization term** E_{reg} defines the interactions between a pixel x and its eight neighbors, setting a constraint to smooth the initial classification map. Several options were also considered. They were all based on an enhanced Potts model [64] in order to guarantee smoother label changes.

- A simple Potts model is defined by

$$\begin{aligned}
&E_{reg}(C(x) = C(y)) = 0,\\
&E_{reg}(C(x) \neq C(y)) = 1.
\end{aligned} \tag{11.24}$$

- This model can be modified to integrate an image contrast constraint. It accounts for the fact that label changes should be less strongly penalized in high-contrast areas of the image:

$$\begin{aligned} &E_{reg}(C(x)=C(y))=0, \\ &E_{reg}(C(x)\neq C(y))=(1-\gamma)+\gamma V(x,y,\varepsilon). \end{aligned} \tag{11.25}$$

$\gamma=0$ yields a pure Potts model, while $\gamma=1$ puts all weight on the contrast term. The contrast component accounts for the fact that label changes need to be penalized in high frequencies areas. $V(x,y,\varepsilon)$ is the term that integrates the contrast of the image (as defined below).

- Another model was proposed:

$$\begin{aligned} &E_{reg}(C(x)=C(y))=0, \\ &E_{reg}(C(x)\neq C(y))=(1-\gamma).\left(1-P_{fusion}(C_{fusion}(x))^{\beta}\right)+\gamma V(x,y,\varepsilon). \end{aligned} \tag{11.26}$$

This re-writing of the Potts model regarding the regularization term handles more efficiently the smoothing procedure. Indeed, when $C(x)\neq C(y)$, $E_{reg}(C(x)\neq C(y))$ becomes a function of P_{fusion} and V. If $P_{fusion}(C_{fusion}(x))$ is close to 1, decision fusion gives a high confidence for x to belong to the class C_{fusion}. Then, E_{reg} becomes only dependent on V, which will decide whether the configuration C_{fusion} is favored or not. Conversely, if $P_{fusion}(C_{fusion}(x))$ is close to 0, E_{reg} is high, and the configuration C_{fusion} is prone to be rejected. A Potts model is obtained if $\gamma=0$ and $\beta=0$.

The term $V(x,y,\varepsilon)$ for image contrast is the same for all regularization term formulations. It is based on [66,74] and defined by

$$V(x,y)=\frac{1}{dim}\sum_{i\in[0,dim]}V_i(x,y)^{\varepsilon}, \tag{11.27}$$

$V_i(x,y)=\exp\left(\frac{-\left(I_i(x)-I_i(y)\right)^2}{2\cdot MeanGrad(I)}\right)$ and $MeanGrad(I)=\frac{1}{Card(N)}\sum_{x,y\in N;x\neq y}\left(I_i(x)-I_i(y)\right)^2$.
dim is the dimension of the image I, $I_i(x)$ is the intensity for x in band i in the multispectral image I; $\varepsilon\in[0,\infty[$.

11.2.2.2 Optimization

Once the energy E has been defined, it has to be minimized in order to get a labeling configuration C (*i.e.*, a classification map) of the entire image, which corresponds to a minimum of the energy E. This model can be expressed as a graphical model and solved as a min-cut problem [73,75].

The graph-cut algorithm employed here is the quadratic pseudo-Boolean optimization (QPBO)[1] [75,76]. QPBO is a classical graph-cut method that builds a probabilistic graph where each pixel is a node. The minimization is computed by finding the minimal cut. Contrary to several standard graph-cut methods for which the pairwise term E_{regul} can only consider the two configurations $(C(x) = C(y))$ and $(C(x) \neq C(y))$, QPBO enables one to integrate more constraints, defining a pairwise term E_{regul} differently according to these four configurations $(C(x) = 0, C(y) = 0)$, $(C(x) = 0, C(y) = 1)$, $(C(x) = 1, C(y) = 0)$ and $(C(x) = 1, C(y) = 1)$.

QPBO performs binary classification. Extension to the multiclass problem is performed using an α-expansion routine [73]. Each label α is visited in turn and a binary labeling is solved between that label and all others, thus flipping the labels of some pixels to α. These expansion steps are iterated until convergence and at the end the algorithm returns a labeling C of the entire image which corresponds to a minimum of the energy E.

11.2.2.3 Parameter tuning

The regularization term E (Eq. (11.21)) is controlled by up to four parameters, depending on the retained formulation for E_{reg}: λ, γ, β and ε. Each of them is attached to a particular sub-term of E.

- $\lambda \in [0, \infty[$ is a trade-off parameter between the terms E_{data} and E_{reg}. The more λ increases the more is the regularization effect. The choice of this parameter will depend on the distribution of the decision fusion map to be optimized.
- $\gamma \in [0, 1]$ is a trade-off parameter between the basic energy model and the rectified model, integrating the contrast measure.
- $\varepsilon \in [0, \infty[$ is a parameter controlling the influence of the contrast measure in the energy term.
- $\beta \in [0, \infty[$ is a trade-off parameter between the smoothing criterion and the importance of C_{fusion} in the model. If β is high, the smoothing criterion is predominant and the model approximates a Potts model. On the opposite, if β is low, the model will tend to follow the classification given by C_{fusion}.

A simple Potts model can be obtained using the following parameterizations:

- $\gamma = 0$ and $\beta \rightarrow +\infty$,
- $\gamma = 1$ and $\beta \rightarrow +\infty$ and $\varepsilon = 0$.

A greedy way for parameter optimization was presented in [66]. The value λ_{opt} maximizing the classification result for a Potts model is assumed to be the same value as the one maximizing the fusion model classification result. Hence, λ_{opt} is first computed using a simple Potts

---

[1] For an implementation, see http://pub.ist.ac.at/~vnk/software/QPBO-v1.4.src.zip.

model. Then, with λ_{opt} and $\gamma = 0$, β_{opt} is found. Similarly, using λ_{opt} and $\gamma = 1$, the value ϵ_{opt} is computed. Lastly, the trade-off parameter γ_{opt} maximizing results of the model is chosen in the $[0, 1]$ interval. The process is iterated, optimizing parameters in the same order at each iteration according to the current parameter set.

Such a strategy can be performed by quantitative cross-validation when sufficient reference data is available. Otherwise, it can be empirically performed by qualitative (visual) evaluation of the results. The set of parameters yielding the nicest and smoothest possible result while following the real object contours can thus be identified. This solution is relevant when regularization also targets improving the visual quality and the interpretability of classification results in operational contexts.

In practice, a set of parameters defined for a classification problem and a decision fusion rule is stable enough to be used in other, similar situations.

11.3 Use Case #1: Hyperspectral and Very High Resolution Multispectral Imagery for Urban Material Discrimination

11.3.1 Introduction

This first use case concerns the joint use of hyperspectral and very high resolution (VHR) multispectral imagery for fine urban land cover classification. Indeed, several applications require fine-grained knowledge about urban land cover and especially urban material maps [77,78]. As no geodatabases contain such information, remote-sensing techniques are urgently required.

Mapping urban environments requires VHR optical images. Indeed, such a spatial resolution is necessary to individualize and precisely delineate urban objects and to consider sharper geometrical details (e.g., [79,80]). However, VHR sensors have generally a poor spectral configuration (usually four bands, blue–green–red–near infrared), limiting their ability to discriminate fine classes [81–84], compared to superspectral or hyperspectral (HS) sensors. Unfortunately, the latter generally exhibit a lower spatial resolution. To overcome the weaknesses of both sensors, HS and VHR multispectral (MS) images can be jointly integrated to benefit from their complementary characteristics and subsequently efficiently separate the classes of interest. Thus, the fusion of such sensors should enhance the classification performance at the highest spatial resolution.

It here may be recalled that early fusion (at observation level) *i.e.*, image sharpening [14], could be applied within this context. However, late fusion is more generic and still valid even for images not acquired simultaneously and processed by specific land cover labeling approaches.

11.3.2 Fusion Process

As mentioned earlier, the method is based on three main steps:

1. **Classification of HS and MS images and generation of the posterior class probabilities:** the two images were classified independently. A SVM classifier with a radial basis function (rbf) kernel [72] was used. The posterior class probabilities were retrieved with the Platt technique [85]. The SVM classifier was used as a baseline, since it was shown to provide good results for this kind of data. However, other supervised classifiers could be used (e.g., random forest) as well as specific methods dedicated to HS imagery such as spectral unmixing [86] (endmember abundances would substitute standard class probabilities).
2. **Fusion at the decision level:** A decision fusion was applied to these posterior class probability maps to combine them at the highest resolution. Different fusion rules listed in Sect. 11.2.1 were tested: fuzzy decision rules (Min, Max, compromise, prioritized, accuracy dependent), Bayesian combination (sum and product based rules), evidence theory (Dempster–Shafer rule), margin theory (margin-Max rule).
3. **Final optimization:** This last step consists in performing a global regularization of the classification map obtained at previous step so as to deal with spatial uncertainties between both sources. The graphical model introduced in Sect. 11.2.2 was used: *Option 1* ($f(t) = -\log(t)$) was retained for the data term, while the contrast sensitive regularization term was formulated following Eq. (11.26).
   The parameters differ from the Potts configuration, which over-smooths the decision fusion classification: $\gamma = 0.5$, $\beta = 1$, and $\varepsilon = 1$. For λ, two configurations were tuned, depending on the decision rule: $\lambda = 0.1$ for the Min and Dempster–Shafer rules, and $\lambda = 10$ for the compromise rule.

11.3.3 Datasets

Experiments were performed over three datasets captured over the cities of Pavia (Italy), and Toulouse (France; see Fig. 11.2). For all datasets, a SVM classifier was trained using 50 samples per class extracted from the images.

Concerning Pavia city (Italy), two datasets called "Pavia University" and "Pavia Center" were used. They are free datasets widely used by the hyperspectral community and available on line.[2] Initially captured by a ROSIS hyperspectral sensor, these datasets have, respectively, 103 and 102 spectral bands from 430 to 860 nm. Pavia University is a 335×605 pixels image, Pavia Center is a 715×1096 pixels image, and both have a GSD of 1.3 m. Both

[2] Available at http://www.ehu.eus/ccwintco/index.php/Hyperspectral_Remote_Sensing_Scenes.

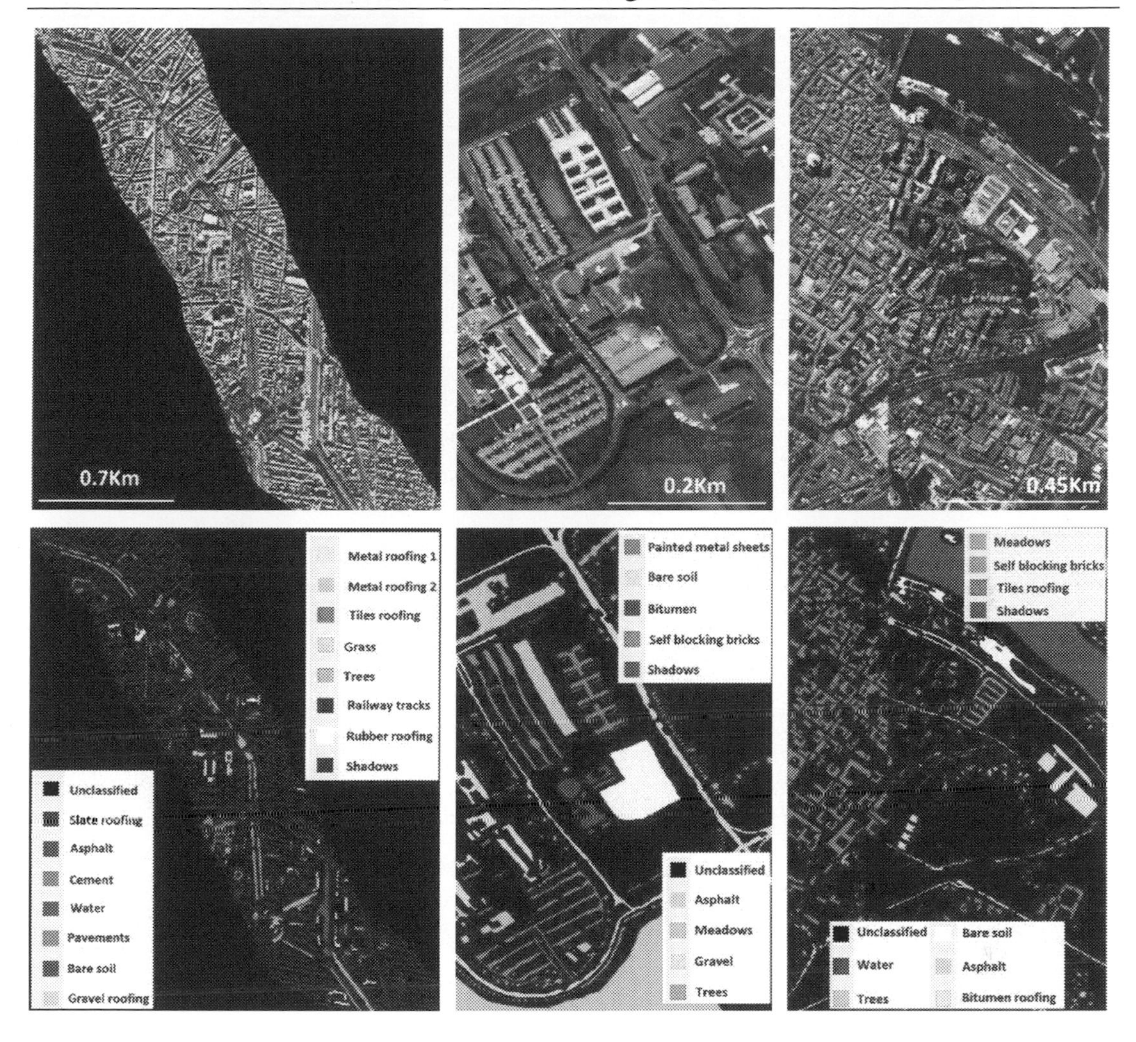

Figure 11.2 : Datasets and corresponding ground truth with labels. From left to right, Toulouse Center, Pavia University, and Pavia Center.

scenes are composed of nine land cover classes (Fig. 11.2): *Asphalt, Meadows, Gravel, Trees, Painted Metal Sheets, Bare Soil, Bitumen, Self-Blocking Bricks, Shadows* for Pavia University and *Water, Trees, Meadows, Self-Blocking Bricks, Bare soil, Asphalt, Bitumen roofing, Tiles roofing, Shadows* for Pavia Center. MS images were generated for a Pleiades satellite spectral configuration (limited to three bands, red–green–blue), with a GSD of 1.3 m, while HS images were resampled at a lower spatial resolution of 7.8 m and at the full original spectral range (*i.e.*, 103 and 102 bands), so that their pan-sharpening ratio would be the same as for the Toulouse dataset.

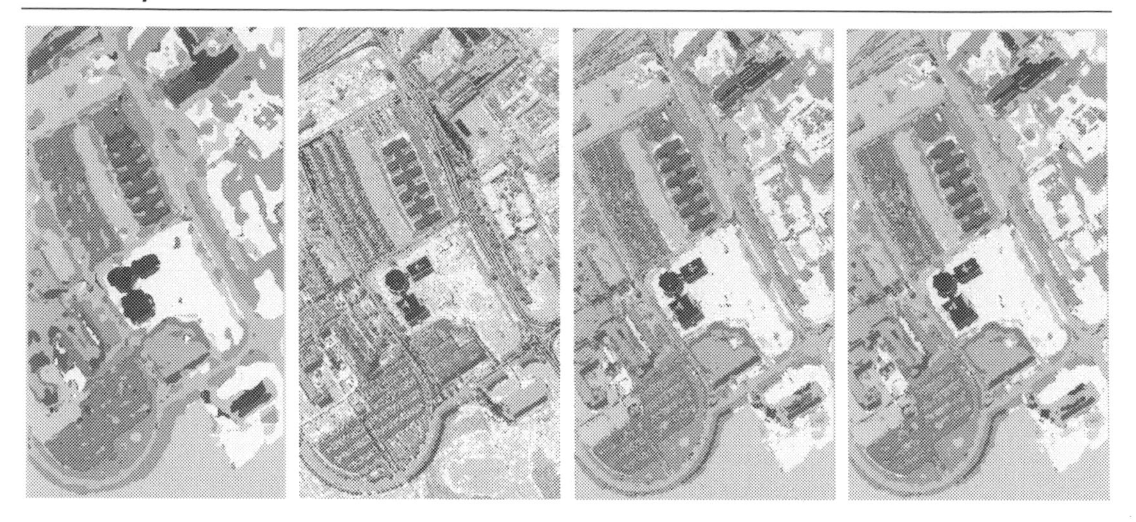

Figure 11.3 : Pavia University classification results using the best decision fusion rule. From left to right, SVM classification of HS image, SVM classification of MS image, classification fusion by Min rule, global classification regularization.

The third dataset is called "Toulouse Center" (France). It was captured over the city Toulouse in 2012 by Hyspex sensors [87]. It has 405 spectral bands ranging from 400 to 2500 nm, and an initial GSD of 1.6 m. Its associated land cover is composed of 15 classes (Fig. 11.2): *Slate roofing, Asphalt, Cement, Water, Pavements, Bare soil, Gravel roofing, Metal roofing 1, Metal roofing 2, Tiles roofing, Grass, Trees, Railway tracks, Rubber roofing, Shadows*. MS and HS images were created for the fusion purpose; a MS image using Pleiades satellite spectral configuration (four bands, red (R)–green (G)–blue (B)–near infrared (NIR)), with a GSD of 1.6 m, and a HS image which is a resampled version of the original image at a spatial resolution of 8 m [88].

11.3.4 Results and Discussion

11.3.4.1 Source comparison

The MS image is characterized by a high spatial resolution and few bands, while the HS one has a low spatial resolution and a hundred(s) of bands. As expected, the SVM classifier applied over these images led to:

- sharp object delineation in the MS image due to its good spatial resolution, but also a lot of artifacts (see Figs. 11.3, 11.4, and 11.5);
- a good discrimination of the different classes in the HS image. However, blurry object delineation is also noticed, due to its low spatial resolution (see Figs. 11.3, 11.4, and 11.5).

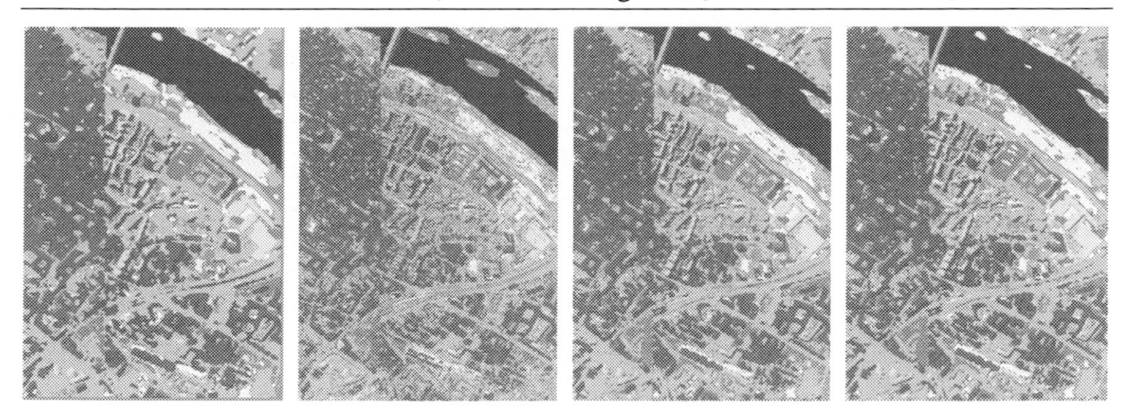

Figure 11.4 : Pavia Center classification results using the best decision fusion rule. From left to right, SVM classification of HS image, SVM classification of MS image, classification fusion by Min rule, global classification regularization.

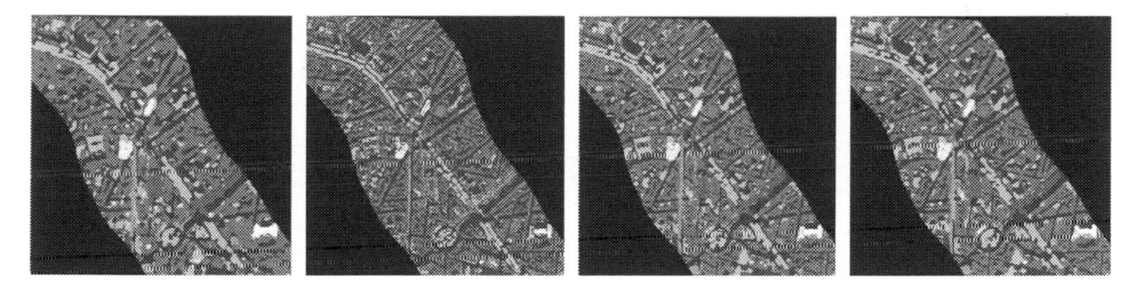

Figure 11.5 : Toulouse Center classification results using the best decision fusion rule. From left to right, SVM classification of HS image, SVM classification of MS image, classification fusion by Min rule, global classification regularization.

The corresponding classification accuracies are listed in Table 11.2: better results are retrieved using the HS image.

11.3.4.2 Decision fusion classification

10 different decision fusion rules were first tested and compared over the three datasets. The quantitative results provided in Table 11.1 lead us to consider the compromise, Bayesian product, margin-Max and Dempster–Shafer rules to be the most efficient rules. The comparison must also take into consideration the visual inspection of the results, as ground truth data remains very limited on these datasets. For Pavia University, four of the best accuracies were reached for Min, compromise, Bayesian product, and Dempster–Shafer rules. In practice, the Min/Compromise rules give the most satisfactory rendering, especially regarding the *Self-*

Table 11.1: Classification accuracies (in %) after fusion procedure, 10 fusion rules at decision level are compared. (OA = Overall Accuracy; F-score = mean F-score.)

	Pavia University			Pavia Center			Toulouse Center		
	OA	Kappa	F-score	OA	Kappa	F-score	OA	Kappa	F-score
Max	92.8	90.7	90.6	98.5	97.8	96.0	**75.6**	**62.4**	**69.8**
Min	**96.1**	**94.9**	**95.1**	98.6	98.0	96.3	72.2	58.7	65.8
Compromise	**96.1**	**95.0**	**95.0**	**98.8**	**98.3**	**96.7**	**73.6**	**60.2**	**68.0**
Prior1	94.7	93.1	93.4	98.2	97.5	95.3	71.3	57.7	65.5
Prior2	92.8	90.7	90.6	98.5	97.8	96.0	**75.6**	**62.4**	**69.8**
AD	95.0	93.5	93.5	**99.0**	**98.7**	**97.7**	75.8	58.1	28.3
Sum Bayes	95.0	93.5	93.2	98.7	98.1	96.5	**75.7**	**62.7**	**70.5**
Prod Bayes	**96.6**	**95.5**	**95.6**	**99.0**	**98.6**	**97.2**	74.5	61.4	69.8
Margin-Max	94.0	92.2	92.0	**98.8**	**98.3**	**96.6**	**75.6**	**62.5**	**69.6**
Dempster-Shafer V1	**96.4**	**95.4**	**95.3**	**98.9**	**98.5**	**97.1**	**74.6**	**61.5**	**69.8**

Table 11.2: Classification accuracy of images HS and MS separately, after decision fusion, and after global regularization. For each dataset, results are provided for the fusion rule achieving the best final results after global regularization.

	Image HS classification	Image MS classification	Decision fusion	After regularization
		Pavia University (Min rule)		
OA (%)	94.7	68.8	96.1	**97.0**
Kappa (%)	93.1	61.6	94.9	**96.1**
F-score (%)	93.4	72.8	95.1	**96.3**
		Pavia Center (Dempster-Shafer V1 rule)		
OA (%)	98.2	92.0	98.9	**99.3**
Kappa (%)	97.5	89.0	98.5	**99.0**
F-score (%)	95.3	83.5	97.1	**98.0**
		Toulouse Center (Compromise rule)		
OA (%)	71.2	69.2	73.5	**74.6**
Kappa (%)	57.6	53.8	60.2	**61.5**
F-score (%)	65.4	55.9	68.0	**70.9**

Blocking Bricks class which is a conflicting class (see Fig. 11.3, magenta color class). The two other rules seem to overestimate this class and to a greater extent consider the HS classification map in the fusion process. This explains their higher accuracy (Table 11.1). The Min rule acts in a cautious way when taking the best of the lowest memberships, while the compromise rule acts depending on the degree of conflict between sources. The Bayesian product rule is a good and simple trade-off if the initial classification maps are not highly conflicting. Otherwise, the result will be degraded by wrong information.

Concerning Pavia Center, all the rules seem accurate (Fig. 11.4, e.g., with the Dempster–Shafer rule), with an overall accuracy higher than 98% (Table 11.1). When visually inspecting the results, all rules gave similar good results excepting Prior 1, showing a result guided by the HS classification map rather than the MS one.

The Toulouse dataset is the largest one, with up to 15 classes. This explains the lower accuracies reached for this dataset. The best results were given by the Max, Prior 2, Bayesian sum and Dempster–Shafer rules. In practice, the Max, prior and sum rules seem to overestimate certain classes; especially tile roofing and vegetation. The best qualitative results are given by the Min, compromise and Dempster–Shafer rules. Despite a satisfactory accuracy, the AD rule exhibits many misclassifications regarding tile roofing (*i.e.*, underestimation), metal roofing 1 (*i.e.*, overestimation), and an erroneous detection of the gravel roofing. This is mainly due to the global accuracy measure which is included in the rule and calculated thanks to a limited ground truth data.

However, due to the very limited amount of reference data, the quantitative accuracies do not necessarily transcribe the real potential of the fusion rules. The best ones from a quantitative and practical qualitative point of view are the compromise, the Bayesian product and the Dempster–Shafer rules

In this study, VHR-MS images as well as HS ones at lower resolution were generated from VHR HS original images. Thus, working on such synthetic datasets leads to quite optimistic results, but this is sufficient to assess the different fusion rules. Besides, the fusion method is flexible enough for instance to integrate a specific process to deal with shadows in a diachronic acquisition context.

11.3.4.3 Regularization

Global regularization was applied to enhance the classification results and eliminate the artifacts. Table 11.2 presents the optimization results for the best fusion rules per dataset. Indeed, the optimization procedure permits one to enhance further the classification. Quantitatively, it slightly enhances the decision fusion classification (by 1–2%) but offers a better visual rendering with an elimination of the artifacts, a better decimation of the classes borders, and a regularization of the scattered pixels (Figs. 11.3, 11.4 and 11.5). These optimized maps seem better in modeling the real scene. The optimization effect is more visible over Pavia University and Toulouse Center. Concerning Pavia Center, the decision fusion gives already good results and thus, the optimized maps are only slightly improved (Table 11.2). Results obtained over the Pavia datasets are comparable to other studies (e.g., [17]). For Pavia University; the painted metal sheets are better recovered and no mismatches with the surrounding road are noticeable. The proposed method permits one to extract some bitumen buildings that were difficult to differentiate from roads (*i.e.*, upper right and lower right, Fig. 11.5), even if the gravel buildings could still be better refined. For Pavia center, the global rendering is enhanced with a minimization of the classification artifacts.

11.3.5 Conclusion

Several decision fusion methods were tested and compared. Among the fuzzy rules, the Min and Compromise rules are the most efficient. The Max rule often leads to misclassifications due to the fact it pays more confidence to the highest membership. The prioritized rules favor a source rather than the other. Indeed, the reliability is not ensured, as noticed for Prior 1, which gives confidence to the less reliable source. The AD rule accuracy is too dependent on the ground truth reliability: it gives encouraging results for Pavia datasets, but the accuracy was not sufficient for Toulouse dataset. The Bayesian sum and product rules can be interesting in the case of low conflict between sources, since they give acceptable results over Pavia Center and Toulouse. Concerning the proposed margin-based rule, it performs well over Pavia center, and correctly over Toulouse. However, it is not sufficient over Pavia University. Finally, the Dempster–Shafer rule has homogeneous performance over the three datasets, leading always to interesting results.

Even if the decision fusion enables one to increase the classification accuracy compared to the initial maps, the results remain affected by classification artifacts and unclear borders. The final maps are either guided by one of the initial maps or by both: the final result is, therefore, a better version of the initial maps. The optimization procedure gives encouraging results, with clear borders among the different classes, and artifacts elimination.

The method also has the possibility to integrate other decision rules in a fully tunable way. The optimization model is simple and flexible and could be modified according to the used dataset and the spatial resolution of the data sources. In further work one will investigate the explicit use of conflict measures from the fusion step within the regularization framework. At the moment, the optimization parameter selection is rather manual; some automation could be included, and other contrast measures could be tested to improve the accuracy.

11.4 Use Case #2: Urban Footprint Detection

11.4.1 Introduction

This second use case focuses on the detection of urban areas out of SPOT 6 and Sentinel-2 satellite imagery. Mapping urban areas is important to monitor urban sprawl and soil imperviousness, and to predict their further evolution [89,90]. Remote sensing is highly relevant for such a regular and continuous monitoring over time. Supervised classification approaches using satellite imagery have been extensively studied in order to automate the process of land cover (LC) classification [91–93,38], but often rely only on one sensor.

Urban and peri-urban areas are complex and heterogeneous landscapes containing impervious areas, trees, grass, bare ground, and water [94,95]. "Artificialized areas" can be defined as irreversibly impervious areas, including buildings and roads, but also small enclosed pervious structures such as gardens, backyards, and green public spaces [96]. There is no unique clear definition of the urban area or footprint. It generally corresponds to a simplification of the artificialized area: road networks outside of built areas are then excluded.

This study aims at detecting such area automatically out of multisource remote-sensing data, trying to follow the real-world city boundary contour as closely as possible. Isolated built-up areas should also be retrieved.

The remote-sensing paradigm has drastically changed in the very last years with the advent of new sensors exhibiting enhanced spectral, spatial, swath or revisit period characteristics, making it possible to acquire datasets at country scale in a limited time. SPOT 6/7 and Sentinel-2 are examples of these new sensors. They will be used in this use case. Indeed, on one hand, they are freely available over the whole French territory thanks to the Théia initiative[3] and GEOSUD Equipex.[4] On the other hand, they exhibit complementary characteristics:

- VHR sensors, e.g., SPOT 6/7, enable the delineation of small features and the use of texture information. However, they often do not have enough spectral information to distinguish fine Land cover types.
- Sentinel-2 sensors exhibit more spectral bands coupled with an important revisit frequency but with a rather limited geometric resolution (10–20 m).

Last years have witnessed the advent of deep learning methods, and especially Convolutional Neural Networks (CNNs) [97–99]. Such approaches have shown their superiority compared to standard classification processes. Indeed, thanks to their end-to-end process, CNNs directly learn optimized (spectral and textural) features (convolution filters) for each classification problem as well as the best way to use them (*i.e.*, the classification model). Besides, implicit features directly take into account the context and thus perform a multiscale analysis of the image. As a counterpart, they require a huge amount of training data.

New studies as regards urban footprint detection have been initiated by the advent of Sentinel data. Sentinel-2 optical images exhibit excellent spectral and temporal characteristics. They are perfectly tailored for land cover production. In [91], Sentinel-2 time series are directly classified by a Random Forest for the yearly extraction of 20-class land cover maps.

[3] https://www.theia-land.fr/.

[4] http://ids.equipex-geosud.fr/.

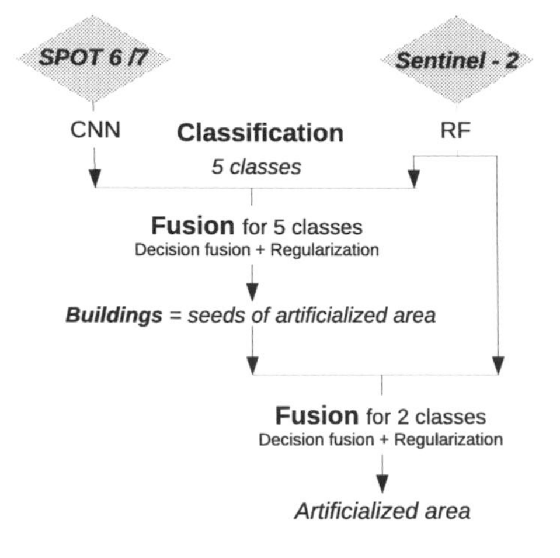

Figure 11.6 : Proposed framework for urban footprint detection.

The method presented in [38] can also be applied to such time series, classifying each date independently before merging the results by a Dempster–Shafer process. It must here be said that, as Sentinel-2 exhibits 10 m GSD for some bands, several studies have tried to use both their radiometric and texture information to detect urban areas [92,100]. However, here, it was decided to focus on Sentinel-2 specificities (enhanced spectral characteristics and time series) and not to exploit the texture information poorer than the one from SPOT 6/7.

To summarize, deep learning approaches are optimal to analyze SPOT 6/7 images. Their spatial resolution makes it possible to try to detect urban elements (e.g. buildings) and to use them to derive urban areas. For Sentinel-2 data, it is more interesting to focus on their specificities (enhanced spectral characteristics and time series). Thus, the fusion of such sources would combine their advantages to reduce spatial and semantic uncertainties. The late fusion scheme proposed in Sect. 11.2 is adapted, considering again that original data have been classified earlier and independently by specific methods. Besides, it can enable the integration of existing land cover maps produced such as Théia's ones based on [91].

11.4.2 Proposed Framework: A Two-Step Urban Footprint Detection

The proposed workflow (Fig. 11.6) consists of three steps.

1. The SPOT 6/7 image and the Sentinel-2 time series are individually classified according to a 5-class nomenclature: *buildings*, *roads*, *water*, *forest*, and *other vegetation*. A membership to each class is provided per pixel.
2. The two classification results are merged at the decision level, aiming at the best detection of building objects.

3. These detected buildings are considered as seeds of urban areas: they are used as prior knowledge for being in an urban area, which is then merged with a binary *urban/non-urban* Sentinel-2 classification within a second fusion still at the decision level.

Both fusions (for the 5-class and the binary classifications) are performed following the scheme presented in Sect. 11.2.

11.4.2.1 Initial classifications

Both sources are classified individually. The Sentinel-2 time series is labeled using a random forest (RF) classifier trained from 50,000 samples per class. RF is used to have a framework similar to the one from [91], of which the LC maps are intended to be available at a national scale. The SPOT 6/7 image is classified using a deep Convolutional Neural Network (CNN) [98], because of its high ability to efficiently exploit context and texture information from VHR image. The training process of the CNN used 10,000 samples per class (from which 10% were kept for cross-validation). Both classifiers produce membership probabilities for the five classes.

11.4.2.2 First regularization

Both fusions (*i.e.*, for the 5-class nomenclature and the *urban/non-urban* one) are performed according to Sect. 11.2, involving a per-pixel decision fusion followed by a spatial regularization.

1. Per pixel fusion: several rules are compared. In addition to the rules tested in Sect. 11.3, supervised ones, consisting in learning from the ground truth how to best merge both sources, are investigated for the 5-class problem. A classifier is trained to label feature vectors corresponding to the concatenation of class membership measures from both sources. Two classifiers are compared for fusion: RF and Support Vector Machine (SVM):
   - RF, with 10,000 training samples per class and 100 trees;
   - SVM with a linear kernel, with 10,000 training samples per class;
   - SVM with a radial basis kernel function (rbf) with a lower number of 500 training samples per class due to practical reasons of higher calculation times. Parameters of the SVM model were optimized using cross-validation.
2. Regularization: In order to smooth the fusion result, still containing noisy labels, and make it follow real-world image contours more accurately, a global regularization is performed. The graphical model introduced in Sect. 11.2.2 is used: *Option 2* ($f(t) = 1 - t$) is retained for the data term, while the contrast-sensitive regularization term is formulated as in Eq. (11.25).

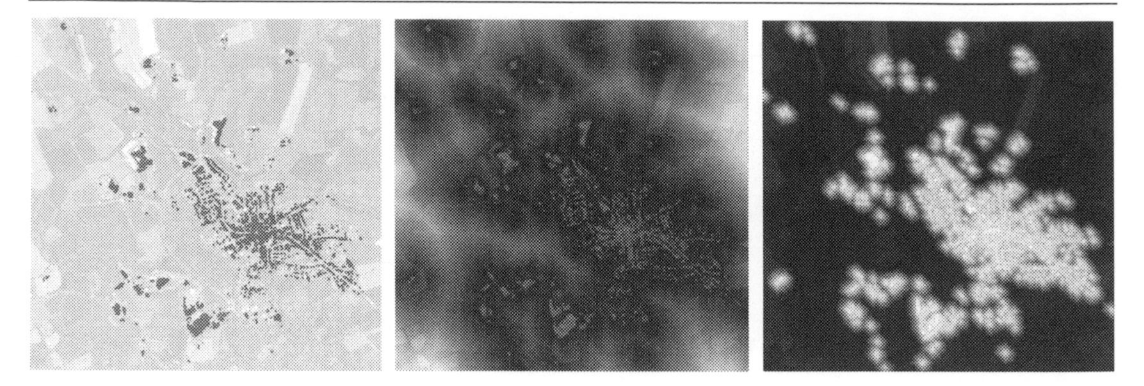

Figure 11.7 : From left to right, buildings detected from the 5-class fusion scheme (red), distance to these buildings (black: low → white: high), prior probability to be in an urban area (black: low → white: high).

For the 5-class fusion, the contrast term is calculated on the SPOT 6/7 image blurred by a Gaussian filter of standard deviation 2 to obtain smoother contours. For the binary classification fusion, it is calculated on the Sentinel-2 image, also blurred by a Gaussian filter of standard deviation 2.

11.4.2.3 Binary classification and fusion

The "*urban/non-urban*" fusion requires one to derive binary class probabilities from results of the previous steps. Buildings from the 5-class fusion result are considered as seeds of urban areas and used to define a prior to be in an urban area (see Fig. 11.7): a linearly decreasing function assigning a probability is applied surrounding all buildings. This probability to be in an urban area starts from 1 and reaches a value of 0 after a distance of 100 m to a building.

Class posterior probabilities from the Sentinel-2 image RF 5-class classification are converted to a binary classification with an *urban area* class (u), and a *non-urban* area class ($\neg u$): $P(u) = P(b) + P(r)$ and $P(\neg u) = 1 - P(u)$, with $P(b)$ and $P(r)$ the class probabilities for buildings and roads, respectively.

Then, the prior probability map to be in an urban area according to previous building detection is merged together with binary class probabilities from the Sentinel-2 RF classifier.

11.4.3 Data

Sentinel-2 offers both a spectral configuration improved over the usual multispectral sensors and has an ability to acquire time series (5 day revisit). In further experiments, only the 10 spectral bands having a 10 or 20 m GSD are used. All are upsampled to 10 m GSD. Six dates

Figure 11.8 : Left: SPOT 6/7 image. Right: the associated ground truth.

(namely, August 15th 2016, January 25th 2017, March 16th 2017, April 12th 2017 and May 25th 2017) are kept. They were retained both because of their low cloud cover and in order to have different seasons/appearances of land cover classes.

SPOT 6/7 includes four spectral bands (red–green–blue–near infrared) pan-sharpened to 1.5 m. A single date (April 16th) is used.

A ground truth of five classes is generated (Fig. 11.8) from available national reference geo databases (training and evaluation: the number of pixels used as training samples is very small compared to the total amount of samples). Buildings, roads and water areas are extracted from IGN's BD Topo®[5]) topographic database, forests from IGN's BD Forêt®[6] database and crops from the Référentiel Parcellaire Graphique[7] of the French Ministry of Agriculture.

Experiments are performed over a test area spanning 648 km^2 in Finistère, North Western France. This study area contains both urban, peri-urban, rural, and natural landscapes.

11.4.4 Results

11.4.4.1 Five-class classifications

For the sake of visibility, the results are shown over a restricted area of 0.64 km^2 (Figs. 11.10, 11.9 and 11.11). There, the original classifications exhibit several errors: the impact of data

[5] http://professionnels.ign.fr/bdtopo.
[6] http://professionnels.ign.fr/bdforet.
[7] http://professionnels.ign.fr/rpg.

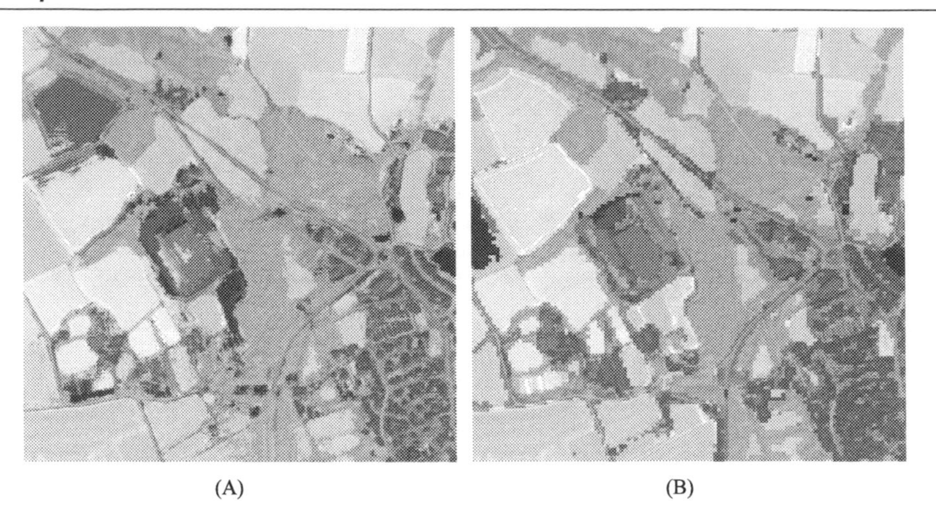

Figure 11.9 : The two initial classifications. (A) SPOT 6 CNN classification. (B) S2 RF classification. The SPOT 6/7 image is superimposed to the classifications.

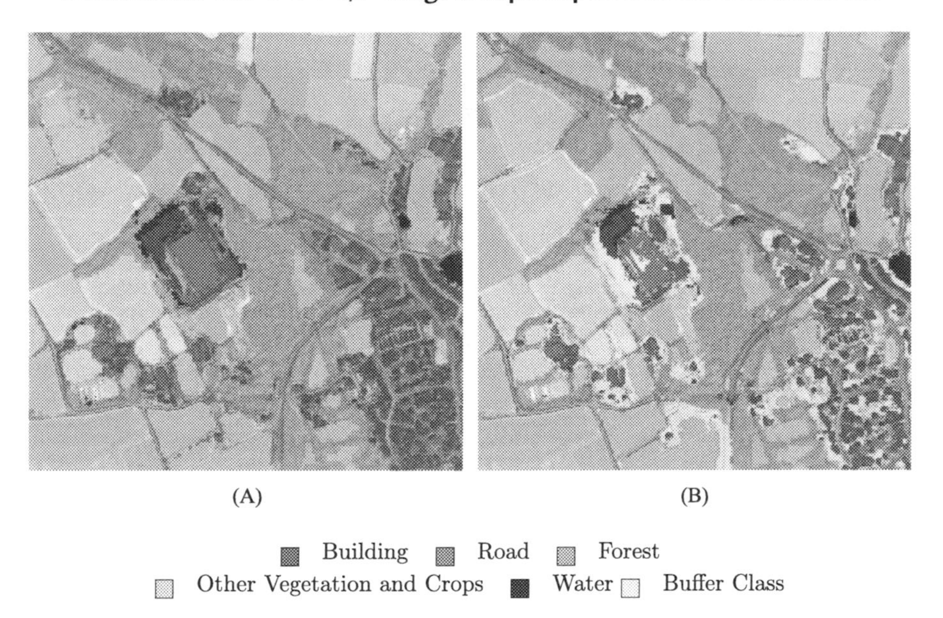

Figure 11.10 : Fusion results before regularization. (A) Min rule. (B) SVM rbf with buffer class.

fusion can be clearly demonstrated. Quantitative evaluation over image is for five tiles of 3000 × 3000 m size, totaling an area of 45 km^2, distributed over the entire 648 km^2 study zone. Each classification is compared to the class labels of the ground truth (Fig. 11.8). Evaluation measures for individual classifications, fusion and regularization, all using five classes, are shown in Table 11.3.

Figure 11.11 : SVM rbf fusion after regularization.

Table 11.3: Accuracy scores (in %) for the first step. OA = overall accuracy, AA = average accuracy, Fm = Mean F-score, Fb = F-score for buildings. Fusion rules are described in Sect. 11.2.1.

Method	Kappa	OA	AA	Fm	Fb
Original classifications					
Sentinel-2	72.0	83.7	81.5	64.6	52.2
SPOT 6	73.3	85.2	70.8	63.4	62.5
Per pixel fusion (before regularization)					
Fuzzy Min	79.1	88.4	84.7	76.7	**73.8**
Fuzzy Max	77.3	87.4	84.2	73.2	70.5
Compromise	78.8	88.2	84.5	76.2	72.5
Prior 1	73.3	85.2	70.8	63.4	62.5
Prior 2	73.3	85.2	70.8	63.4	62.5
Bayesian Sum	78.3	87.9	84.8	74.8	71.9
Bayesian Product	79.1	88.5	85.0	76.7	**73.8**
Dempster–Shafer V1	79.1	88.4	**85.1**	76.5	**73.7**
Dempster-Shafer V2	79.0	88.4	**85.1**	76.4	**73.7**
Margin Maximum	77.7	87.6	84.5	73.4	70.2
Margin Bayesian Product	78.4	88.0	84.7	75.1	72.0
Margin Bayesian Sum	78.0	87.7	84.6	74.0	71.0
RF	81.8	90.0	**90.1**	81.2	**81.6**
SVM linear	80.5	89.3	**88.6**	77.9	**80.4**
SVM rbf	81.0	89.6	**89.1**	79.1	**83.4**
Fusion and regularization					
Fuzzy Min	75.1	85.8	82.3	73.9	**73.8**
SVM rbf	81.4	89.8	**89.3**	79.8	**83.9**

Initial classifications. Original classifications confirm the initial observation that the SPOT 6/7 CNN result tends to preserve small objects. However, some confusions between (bare soil) crops and built-up areas occur. The Sentinel-2 RF classification overall behaves better, but it mixes buildings and roads due to its coarse spatial resolution. The results from the individual classifications on the SPOT 6/7 and Sentinel-2 images are shown on Fig. 11.9. This area was selected to illustrate problems of both classifications and improvements obtained with fusion and regularization (better results are observed in all other areas). Confusion between water and built areas can be noticed. This phenomenon is caused by the fact that the water area database used to generate training samples included ponds at the bottom of careers appearing white and very similar to built-up areas.

Per pixel fusion. This first fusion is performed at the SPOT 6/7 resolution (1.5 m) as it aims at achieving the most fine detection of building objects.

Among the classic fusion rules proposed by [69], the *Min* fuzzy rule produces the best results, following the objects' borders most precisely while producing the least class confusions (Fig. 11.10 and Table 11.3). Considering Fig. 11.10, all rules managed to eliminate the wrongly classified building patch (top-left), preferring the Sentinel-2 classification over the SPOT 6/7 one. The Min fusion rule follows the field contours a bit less smoothly than the other ones. The industrial area (at the center of the displayed area) is still confused with water in both results (such a confusion can be explained by the presence of very white water area training samples).

The RF/SVM supervised fusions initially tend to produce patches of buildings rather than separate buildings due to missing training data (and thus constraints) between buildings and roads (Fig. 11.8). Indeed the ground truth contains gaps between buildings and so the classifiers tend to aggregate individual buildings as there are no training data available around building to prevent the classifier from doing it. Adding an additional sixth buffer class "*around buildings*" helps to refine the contours and to obtain a higher level of detail than just using the Min or Bayes rules. It preserves more details of individual buildings, but can cause some confusion between buildings and this "buffer" class in the center of very wide buildings in industrial areas (see Fig. 11.10). It can also erase small patches of buildings. However, this remains quite an exception and is not a real problem for the final goal of detection of urban areas. The same observations are made on all areas, even those on which the supervised fusion model is not trained. Thus, the result of the supervised fusion using a rbf SVM classifier with buffer class is used in subsequent steps.

Regularization. The parameters are as follows: $\lambda = 10$, $\gamma = 0.7$, $\varepsilon = 50$. They were selected by cross-validation, optimizing visual qualities of the regularization result. The contrast term is calculated from the SPOT 6/7 image: it especially aims at obtaining the finest detection of

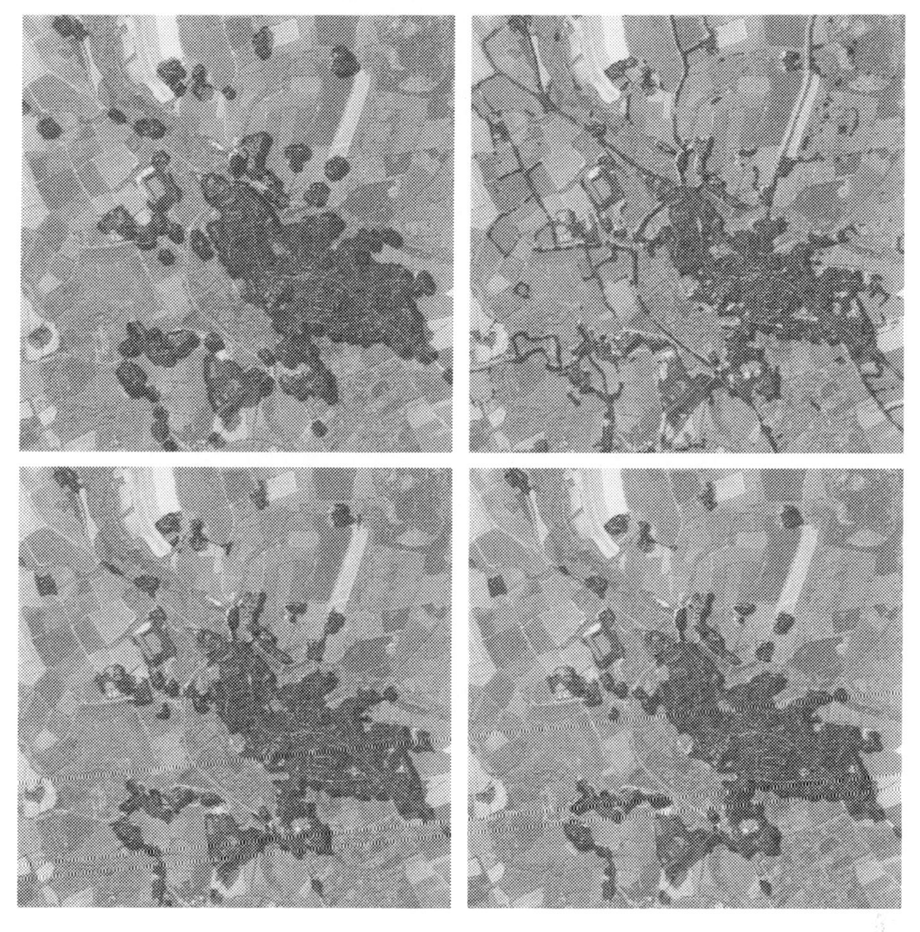

Figure 11.12 : Urban area: input dilated detected building mask (top left), Sentinel-2 classification (top right) and fusion result before (bottom left) and after (bottom right) regularization.

building objects. The result is shown in Fig. 11.11. The regularization performs well, smoothing out small noisy patches and yielding a visually more appealing result. However, accuracy measures of the regularization remain similar to the ones of the fusion (cf. Table 11.3).

11.4.4.2 Urban footprint extraction

Raw urban area maps directly derived using the binary class probabilities are shown in Fig. 11.12. Again, they are first merged using a per pixel fusion rule. The supervised learning based fusion approaches cannot be used here since no training data for *urban/non-urban*

Table 11.4: Accuracy measures: F-score for the buildings class (F-score$_u$), Kappa, overall accuracy (OA) and intersection over union for the buildings class (IoU$_u$).

Classification	Ground Truth	F-Score$_u$ [%]	Kappa [%]	OA [%]	IoU$_u$ [%]
Binary dilated buildings	BD Topo®	**86.7**	**83.2**	**94.5**	**76.5**
	OSO	56.8	50.1	86.8	39.7
	OSM	58.3	51.9	87.3	41.2
Binary Sentinel-2	BD Topo®	65.2	56.4	85.9	48.3
	OSO	63.9	58.3	89.3	46.9
	OSM	52.6	45.5	86.0	35.7
Fusion Min	BD Topo®	79.8	75.2	92.4	66.3
	OSO	66.9	62.2	91.2	50.3
	OSM	62.9	57.6	90.2	45.9
Regularization	BD Topo®	79.7	75.1	92.5	66.2
	OSO	67.4	62.8	91.5	50.9
	OSM	64.4	59.4	90.7	47.4

areas was available. Several rules yield visually similar results; the Min fuzzy fusion rule is eventually chosen.

Global regularization is then performed with the same $(\lambda, \gamma, \varepsilon)$ parameters as for the 5-class regularization, but the image contrast term is now calculated from the Sentinel-2 image. The result obtained is shown in Fig. 11.12. Compared to Sentinel-2 detection, roads and small misclassifications have been removed. Compared to the dilated building mask from previous step, the objects' borders fit better to the true urban area borders.

As no true reference data for urban areas is available, strict evaluation is not possible. However, the detected artificialized area can be compared to binary ground truth maps derived from the following other related databases:

- Dilated BD Topo® ground truth: Buildings of this database used as ground truth for the 5-class classification were dilated by 20 m.
- OpenStreetMap (OSM) data which uses refined Corine land cover data;
- The CESBIO land cover OSO product[8] [91], gathering the classes "*urban diffuse*", "*dense urban*" and "*industrial and commercial zones*".

Strict quantitative evaluation is not possible, but such data can provide some hints: accuracies are provided in Table 11.4. Besides, a visual comparison with the different sources is shown in Fig. 11.13.

[8] http://osr-cesbio.ups-tlse.fr/~oso/.

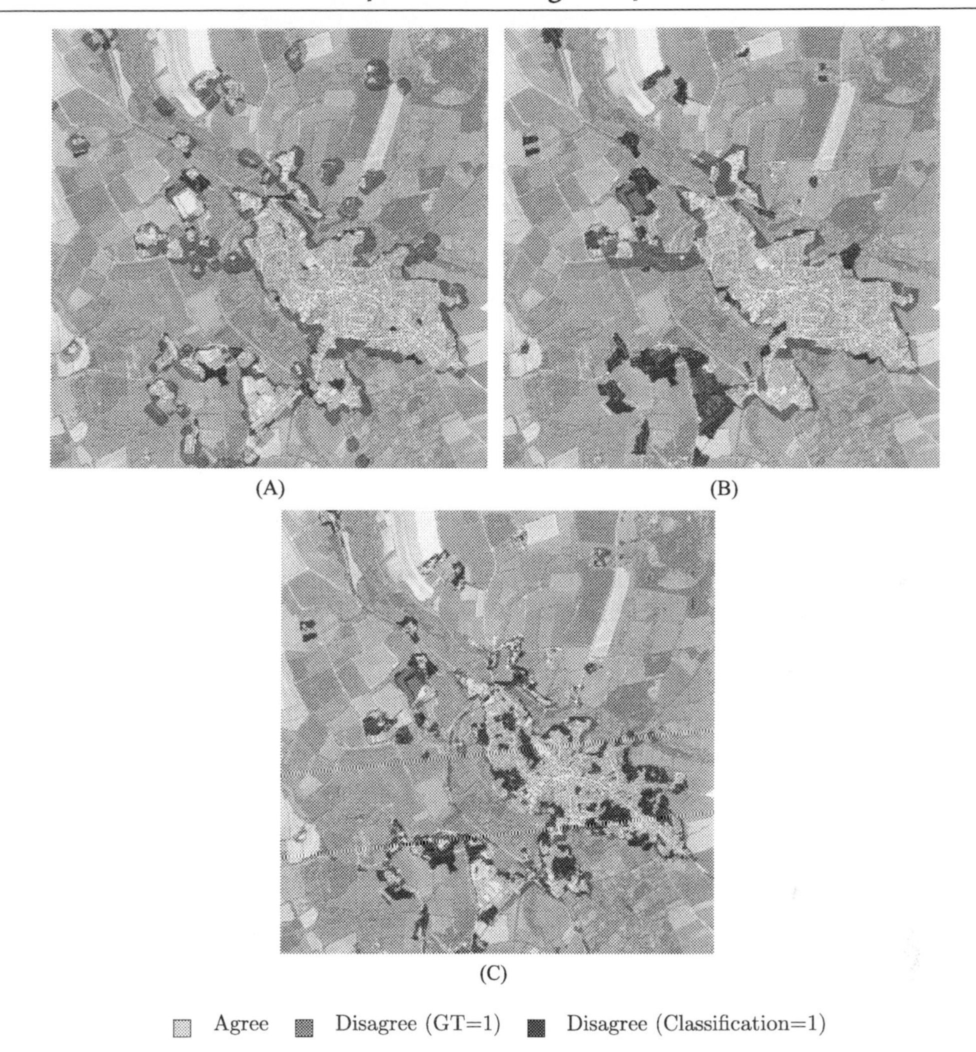

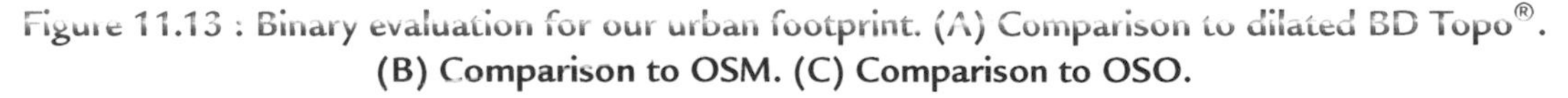

Figure 11.13 : Binary evaluation for our urban footprint. (A) Comparison to dilated BD Topo®. (B) Comparison to OSM. (C) Comparison to OSO.

The following aspects can be underlined:

- The derived OSM ground truth is rather simplified and contains few, rough urbanized patches.
- The derived OSO ground truth is partial, containing gaps of artificial areas within urban patches. This is due to the resolution and methodology used for the classification and the selection of classes considered as *urban*.
- The dilated BD Topo® ground truth is the most truthful, as expected. However, it is also approximate and exceeds the urban area due to the dilatation used to generate it.

The dilated BD Topo® ground truth generally yields the highest agreement with classifications in terms of accuracy measures. The fusion and regularization steps improve the accuracy measures over the individual input classifications with the exception of the binary regularization input, which can be explained by the fact that both have been produced by dilatation.

11.4.5 Conclusion

A framework was proposed to detect urban areas. Sentinel-2 and SPOT 6/7 data were classified individually in five topographic classes. Decision-level fusion and regularization then enable one to obtain a result preserving high geometric details while reducing misclassifications. Results presented in this study were presented on one dataset, but the processing chain was applied to another region (Gironde department in South Western France) exhibiting a different (climatic and topographic) landscape. It led to similar conclusions, showing high generalization potential despite varying behaviors of the initial classifiers.

Traditional fusion methods enable artifact removal but can keep confusion between buildings and roads. In contrast, supervised fusion methods enable an enhanced detection of buildings at the price of a new artificial "*building buffer*" class. Such an introduction comes at the cost of introducing a certain additional amount of semantic uncertainty. However, new class confusions mostly occur between the vegetation and forest classes, which is not a problem here. Buildings could thus be extracted with a higher amount of semantic details.

Second, the urban area can be approximated by fusion and regularization merging the urban / non-urban class membership probabilities of the Sentinel-2 classification and a urban prior measure derived from previously detected buildings. A simple function was used to derive a probability map of urbanized area. Improvements could be made having a more advanced urban membership prior measure, decreasing faster for uncertain buildings.

Furthermore, the promising results of supervised fusion would justify the use of CNNs for fusion. Although such a strategy looks also promising for the identification of artificialized areas, it would have to face missing ground truth data and heterogeneous, user- and application-dependent definitions of such areas.

11.5 Final Outlook and Perspectives

A fusion framework was proposed to merge very high spatial resolution, monodate, multispectral images with time series of images exhibiting an enhanced spectral configuration but with a lower spatial resolution. It mostly relies on existing state-of-the-art methods, combined in order to cope both with semantic and with spatial uncertainties. Besides, it is flexible enough to integrate other fusion rules. It is a late fusion strategy, permitting one to initially process each input data source independently through specific methods, and even to use already calculated results from existing operational land cover classification services. Besides, in this case, it can be used without any ground truth information, contrary to intermediate level fusion methods.

The proposed framework was applied to two different use cases. For each of them, classification results were improved. Several fusion rules were tested. Good results were reached for several of them, but the best results were obtained by the "Minimum" fuzzy rule, by Dempster–Shafer rules, and, when sufficient training data is available, by supervised learning-based methods.

At present, the proposed fusion framework has been applied to only two sources, but it could easily be extended to more sources. Besides, for the moment, it has been tested only for optical images exhibiting different characteristics, but it is generic enough to be applied to input data of different modalities. It has also been used only to merge class membership probability maps from classification results, but it would be interesting to apply it to other kinds of results. Especially for the lower spatial resolution source, it would be relevant to use class abundances obtained from an unmixing process (applied to the hyperspectral case or time series of data).

References

[1] L. Gomez-Chova, D. Tuia, G. Moser, G. Camps-Valls, Multimodal classification of remote sensing images: a review and future directions, Proceedings of the IEEE 103 (9) (2015) 1560–1584.

[2] S. Chavez, J. Stuart, C. Sides, J. Anderson, Comparison of three different methods to merge multiresolution, multispectral data Landsat TM and SPOT panchromatic, Photogrammetric Engineering and Remote Sensing 57 (3) (1991) 259–303.

[3] P. Gamba, Image and data fusion in remote sensing of urban areas: status issues and research trends, International Journal of Image and Data Fusion 5 (1) (2014) 2–12.

[4] N. Joshi, M. Baumann, A. Ehammer, B. Waske, A review of the application of optical and radar remote sensing data fusion to land use mapping and monitoring, Remote Sensing 8 (1) (2016) 70.

[5] M.-T. Pham, G. Mercier, O. Regniers, J. Michel, Texture retrieval from VHR optical remote sensed images using the local extrema descriptor with application to vineyard parcel detection, Remote Sensing 8 (5) (2016) 368.

[6] C. Pohl, J. Van Genderen, Multi-sensor image fusion in remote sensing: concepts methods and applications, International Journal of Remote Sensing 19 (5) (1998) 823–854.

[7] M. Schmitt, X. Zhu, Data fusion and remote sensing: an ever-growing relationship, IEEE Geoscience and Remote Sensing Magazine 4 (4) (2016) 6–23.

[8] J.A. Benediktsson, G. Cavallaro, N. Falco, I. Hedhli, V.A. Krylov, G. Moser, S.B. Serpico, J. Zerubia, Remote sensing data fusion: Markov models and mathematical morphology for multisensor, multiresolution, and multiscale image classification, Springer International Publishing, 2018, pp. 277–323.

[9] W. Liao, J. Chanussot, W. Philips, Remote sensing data fusion: guided filter-based hyperspectral pansharpening and graph-based feature-level fusion, Springer International Publishing, 2018, pp. 243–275.

[10] H. Ghassemian, A review of remote sensing image fusion methods, Information Fusion 32 (2016) 75–89.

[11] W. Carper, T. Lillesand, R. Kiefer, The use of intensity-hue-saturation transform for merging SPOT panchromatic and multispectral image data, Photogrammetric Engineering and Remote Sensing 56 (4) (1990) 459–467.
[12] S. Yang, M. Wang, L. Jiao, Fusion of multispectral and panchromatic images based on support value transform and adaptive principal component analysis, Information Fusion 13 (3) (2012) 177–184.
[13] R. Gharbia, A. Azar, A. El Baz, A. Hassanien, Image fusion techniques in remote sensing, arXiv preprint, arXiv:1403.5473.
[14] L. Loncan, L. Almeida, J. Bioucas-Dias, W. Liao, X. Briottet, J. Chanussot, N. Dobigeon, S. Fabre, W. Liao, G. Licciardi, M. Simoes, J.-Y. Tourneret, M. Veganzones, G. Vivone, Q. Wei, N. Yokoya, Hyperspectral pansharpening: a review, IEEE Geoscience and Remote Sensing Magazine 3 (3) (2015) 27–46.
[15] Y. Yuan, X. Zheng, X. Lu, Hyperspectral image superresolution by transfer learning, IEEE Journal of Selected Topics in Applied Earth Observations and Remote Sensing 10 (5) (2017) 1963–1974.
[16] A. Gressin, C. Mallet, M. Paget, C. Barbanson, P.L. Frison, J.P. Rudant, N. Paparoditis, N. Vincent, Unsensored very high resolution land-cover mapping, in: Proc. of the IEEE International Geoscience and Remote Sensing Symposium, IGARSS, 2015, pp. 2939–2942.
[17] M. Fauvel, J. Benediktsson, J. Sveinsson, J. Chanussot, Spectral and spatial classification of hyperspectral data using SVMs and morphological profiles, in: Proc. of the IEEE International Geoscience and Remote Sensing Symposium, IGARSS, 2007, pp. 4834–4837.
[18] J.D. Wegner, R. Hänsch, A. Thiele, U. Soergel, Building detection from one orthophoto and high-resolution InSAR data using conditional random fields, IEEE Journal of Selected Topics in Applied Earth Observations and Remote Sensing 4 (2011) 83–91.
[19] Y. Ban, A. Jacob, Object-based fusion of multitemporal multiangle ENVISAT ASAR and HJ-1B multispectral data for urban land-cover mapping, IEEE Transactions on Geoscience and Remote Sensing 51 (4) (2013) 1998–2006.
[20] C. Dechesne, C. Mallet, A. Le Bris, V. Gouet-Brunet, Semantic segmentation of forest stands of pure specie as a global optimisation problem, ISPRS Annals of Photogrammetry, Remote Sensing and Spatial Information Sciences 4(1/W1) (2017) 141–148.
[21] C. Dechesne, C. Mallet, A. Le Bris, V. Gouet-Brunet, Semantic segmentation of forest stands of pure species combining airborne lidar data and very high resolution multispectral imagery, ISPRS Journal of Photogrammetry and Remote Sensing 126 (2017) 129–145.
[22] J. Xia, Z. Ming, A. Iwasaki, Multiple sources data fusion via deep forest, in: Proc. of the IEEE International Geoscience and Remote Sensing Symposium, IGARSS, 2018, pp. 1722–1725.
[23] N. Audebert, B. Le Saux, S. Lefèvre, Fusion of heterogeneous data in convolutional networks for urban semantic labeling, in: Proc. of the Joint Urban Remote Sensing Event, JURSE, 2017.
[24] X.X. Zhu, D. Tuia, L. Mou, G. Xia, L. Zhang, F. Xu, F. Fraundorfer, Deep learning in remote sensing: a comprehensive review and list of resources, IEEE Geoscience and Remote Sensing Magazine 5 (4) (2017) 8–36.
[25] C. Berger, M. Voltersen, R. Eckardt, J. Eberle, T. Heyer, N. Salepci, S. Hese, C. Schmullius, J. Tao, S. Auer, R. Bamler, K. Ewald, M. Gartley, J. Jacobson, A. Buswell, Q. Du, F. Pacifici, Multi-modal and multi-temporal data fusion: outcome of the 2012 GRSS Data Fusion Contest, IEEE Journal of Selected Topics in Applied Earth Observations and Remote Sensing 6 (3) (2013) 1324–1340.
[26] W. Liao, X. Huang, F.V. Coillie, S. Gautama, A. Pižurica, W. Philips, H. Liu, T. Zhu, M. Shimoni, G. Moser, D. Tuia, Processing of multiresolution thermal hyperspectral and digital color data: outcome of the 2014 IEEE GRSS Data Fusion Contest, IEEE Journal of Selected Topics in Applied Earth Observations and Remote Sensing 8 (6) (2015) 2984–2996.
[27] N. Yokoya, P. Ghamisi, J. Xia, S. Sukhanov, R. Heremans, I. Tankoyeu, B. Bechtel, B. Le Saux, G. Moser, D. Tuia, Open data for global multimodal land use classification: outcome of the 2017 IEEE GRSS Data Fusion Contest, IEEE Journal of Selected Topics in Applied Earth Observations and Remote Sensing 11 (5) (2018) 1363–1377.

[28] D. Dubois, H. Prade, Combination of fuzzy information in the framework of possibility theory, in: M.A. Abidi, R.C. Gonzalez (Eds.), Data Fusion in Robotics and Machine Intelligence, Academic Press, New York, 1992, pp. 481–505.
[29] D. Dubois, H. Prade, Possibility theory and data fusion in poorly informed environments, Control Engineering Practice 2 (5) (1994) 811–823.
[30] L. Zadeh, Fuzzy sets, Information and Control 8 (3) (1965) 338–353.
[31] N.R. Pal, J.C. Bezdek, Measuring fuzzy uncertainty, IEEE Transactions on Fuzzy Systems 2 (2) (1994) 107–118.
[32] M. Fauvel, J. Chanussot, J. Benediktsson, Decision fusion for the classification of urban remote sensing images, IEEE Transactions on Geoscience and Remote Sensing 44 (10) (2006) 2828–2838.
[33] G. Shafer, A Mathematical Theory of Evidence, Princeton University Press, 1976.
[34] V. Poulain, J. Inglada, M. Spigai, J.Y. Tourneret, P. Marthon, High-resolution optical and SAR image fusion for building database updating, IEEE Transactions on Geoscience and Remote Sensing 49 (8) (2011) 2900–2910.
[35] F. Rottensteiner, J. Trinder, S. Clode, K. Kubik, Building detection by fusion of airborne laser scanner data and multi-spectral images: performance evaluation and sensitivity analysis, ISPRS Journal of Photogrammetry and Remote Sensing 62 (2) (2007) 135–149.
[36] J. Tian, J. Dezert, Fusion of multispectral imagery and dsms for building change detection using belief functions and reliabilities, International Journal of Image and Data Fusion 10 (1) (2019) 1–27.
[37] R.O. Chavez-Garcia, O. Aycard, Multiple sensor fusion and classification for moving object detection and tracking, IEEE Transactions on Intelligent Transportation Systems 17 (2) (2016) 525–534.
[38] A. Lefebvre, C. Sannier, T. Corpetti, Monitoring urban areas with Sentinel-2A data: application to the update of the Copernicus high resolution layer imperviousness degree, Remote Sensing 8 (7) (2016) 606.
[39] S. Le Hégarat-Mascle, R. Seltz, Automatic change detection by evidential fusion of change indices, Remote Sensing of Environment 91 (3) (2004) 390–404.
[40] S. Le Hégarat-Mascle, R. Seltz, L. Hubert-Moy, S. Corgne, N. Stach, Performance of change detection using remotely sensed data and evidential fusion: comparison of three cases of application, International Journal of Remote Sensing 27 (16) (2006) 3515–3532.
[41] S.I. Oh, H.B. Kang, Object detection and classification by decision-level fusion for intelligent vehicle systems, Sensors 17 (1) (2017) 207.
[42] J. Dezert, Fondations pour une nouvelle théorie du raisonnement plausible et paradoxal: application à'la fusion d'informations incertaines et conflictuelles, Technical Report 1/06769 DTIM, Tech. rep., ONERA, 2003.
[43] A. Martin, C. Osswald, Une nouvelle règle de combinaison répartissant le conflit – applications en imagerie sonar et classification radar, Traitement du Signal 24 (2007) 71–82.
[44] A. Martin, Modélisation et gestion du conflit dans la théorie des fonctions de croyance, Habilitation à Diriger des Recherches de l'Université de Bretagne Occidentale, France, 2009.
[45] A. Appriou, Probabilités et incertitude en fusion de données multi-senseurs, Revue Scientifique Et Technique de la Défense 11 (1991) 27–40.
[46] Y. Cao, H. Lee, H. Kwon, Enhanced object detection via fusion with prior beliefs from image classification, arXiv preprint, arXiv:1610.06907.
[47] H. Lee, H. Kwon, R.M. Robinson, W.D. Nothwang, A.M. Marathe, Dynamic belief fusion for object detection, in: Proc. of the 2016 IEEE Winter Conference on Applications of Computer Vision, WACV, 2016.
[48] Z. Tu, X. Bai, Auto-context and its application to high-level vision tasks and 3D brain image segmentation, IEEE Transactions on Pattern Analysis and Machine Intelligence 32 (10) (2010) 1744–1757.
[49] C. Wendl, A. Le Bris, N. Chehata, A. Puissant, T. Postadjian, Decision fusion of SPOT-6 and multitemporal Sentinel-2 images for urban area detection, in: Proc. of the IEEE International Geoscience and Remote Sensing Symposium, IGARSS, 2018, pp. 1734–1737.
[50] U. Knauer, U. Seiffert, A comparison of late fusion methods for object detection, in: Proc. of IEEE International Conference on Image Processing, ICIP, 2013, pp. 3297–3301.

[51] B. Waske, J. Benediktsson, Fusion of support vector machines for classification of multisensor data, IEEE Transactions on Geoscience and Remote Sensing 45 (2007) 3858–3866.
[52] X. Ceamanos, B. Waske, J. Benediktsson, J. Chanussot, M. Fauvel, J. Sveinsson, A classifier ensemble based on fusion of support vector machines for classifying hyperspectral data, International Journal of Image and Data Fusion 1 (4) (2010) 293–307.
[53] J. Benediktsson, I. Kanellopoulos, Decision fusion methods in classification of multisource and hyperdimensional data, IEEE Transactions on Geoscience and Remote Sensing 37 (3) (1999) 1367–1377.
[54] F. Tupin, I. Bloch, H. Maitre, A first step toward automatic interpretation of SAR images using evidential fusion of several structure detectors, IEEE Transactions on Geoscience and Remote Sensing 37 (3) (1999) 1327–1343.
[55] B. Jeon, Decision fusion approach for multitemporal classification, IEEE Transactions on Geoscience and Remote Sensing 37 (3) (1999) 1227–1233.
[56] A. Mohammad-Djafari, A bayesian approach for data and image fusion, AIP Conference Proceedings 659 (2003) 386–408.
[57] A. Le Bris, D. Boldo, Extraction of land cover themes from aerial ortho-images in mountainous areas using external information, Photogrammetric Record 23 (124) (2008) 387–404.
[58] A. Le Bris, N. Chehata, Change detection in a topographic building database using submetric satellite images, The International Archives of the Photogrammetry, Remote Sensing and Spatial Information Sciences 38(3/W22) (2011) 25–30.
[59] M. Aitkenhead, I. Aalders, Automating land cover mapping of Scotland using expert system and knowledge integration methods, Remote Sensing of Environment 115 (5) (2011) 1285–1295.
[60] X. Huang, L. Zhang, A multilevel decision fusion approach for urban mapping using very high-resolution multi-hyper-spectral imagery, International Journal of Remote Sensing 33 (11) (2012) 3354–3372.
[61] S. Paisitkriangkrai, J. Sherrah, P. Janney, A. Van-Den Hengel, Effective semantic pixel labelling with convolutional networks and conditional random fields, in: Proc. of the IEEE Conference on Computer Vision and Pattern Recognition Workshop, CVPR, Boston, USA, 2015, pp. 36–43.
[62] P. Perona, J. Malik, Scale-space and edge detection using anisotropic diffusion, IEEE Transactions on Pattern Analysis and Machine Intelligence 12 (7) (1990) 629–639.
[63] P. Gong, P. Howarth, Performance analyses of probabilistic relaxation methods for land-cover classification, Remote Sensing of Environment 30 (1) (1989) 33–42.
[64] K. Schindler, An overview and comparison of smooth labeling methods for land-cover classification, IEEE Transactions on Geoscience and Remote Sensing 50 (11) (2012) 4534–4545.
[65] G. Moser, S. Serpico, J. Benediktsson, Land-cover mapping by Markov modeling of spatial contextual information in very-high-resolution remote sensing images, Proceedings of the IEEE 101 (3) (2013) 631–651.
[66] A. Hervieu, A. Le Bris, C. Mallet, Fusion of hyperspectral and vhr multispectral image classifications in urban areas, ISPRS Annals of Photogrammetry, Remote Sensing and Spatial Information Sciences III-3 (2016) 457–464.
[67] V. Andrejchenko, R. Heylen, W. Liao, W. Philips, P. Sheunders, MRF-based decision fusion for hyperspectral image classification, in: Proc. of the IEEE International Geoscience and Remote Sensing Symposium, IGARSS, Valencia, Spain, 2018, pp. 8070–8073.
[68] D. Dubois, H. Prade, Possibility theory and data fusion in poorly informed environment, Control Engineering Practice 2 (1997) 811–823.
M. Fauvel, J. Chanussot, J. Benediktsson, Decision fusion for the
[69] W. Ouerghemmi, A. Le Bris, N. Chehata, C. Mallet, A two-step decision fusion strategy: application to hyperspectral and multispectral images for urban classification, The International Archives of the Photogrammetry, Remote Sensing and Spatial Information Sciences XLII-1-W1 (2017) 167–174.
[70] I. Bloch, Information combination operators for data fusion: a comparative review, IEEE Transactions on Systems, Man and Cybernetics 26 (1) (1996) 52–67.

[71] L. Breiman, Random forests, Machine Learning 45 (1) (2001) 5–32.
[72] V.N. Vapnik, An overview of statistical learning theory, IEEE Transactions on Neural Networks 10 (1999) 988–999.
[73] V. Kolmogorov, R. Zabih, What energy functions can be minimized via graph cuts?, IEEE Transactions on Pattern Analysis and Machine Intelligence 26 (2) (2004) 65–81.
[74] C. Rother, V. Kolmogorov, A. Blake, "GrabCut": interactive foreground extraction using iterated graph cuts, ACM Transactions on Graphics 23 (3) (2004) 309–314.
[75] V. Kolmogorov, C. Rother, Minimizing non-submodular functions with graph cuts – a review, IEEE Transactions on Pattern Analysis and Machine Intelligence 29 (7) (2007) 1274–1279.
[76] C. Rother, V. Kolmogorov, V. Lempitsky, M. Szummer, Optimizing binary MRFs via extended roof duality, in: Conference on Computer Vision and Pattern Recognition, CVPR, 2007.
[77] W. Heldens, U. Heiden, T. Esch, E. Stein, A. Muller, Can the future EnMAP mission contribute to urban applications? A literature survey, Remote Sensing 3 (2011) 1817–1846.
[78] H. Shafri, E. Taherzadeh, S. Mansor, R. Ashurov, Hyperspectral remote sensing of urban areas: an overview of techniques and applications, Research Journal of Applied Sciences, Engineering and Technology 4 (11) (2012) 1557–1565.
[79] M. Herold, X. Liu, K. Clarke, Spatial metrics and image texture for mapping urban land-use, Photogrammetric Engineering and Remote Sensing 69 (9) (2003) 991–1001.
[80] C. Cleve, M. Kelly, F. Kearns, M. Moritz, Classification of the wildland–urban interface: a comparison of pixel- and object-based classifications using high-resolution aerial photography, Computers, Environment and Urban Systems 32 (4) (2008) 317–326.
[81] N. Thomas, C. Hendrix, R. Congalton, A comparison of urban mapping methods using high-resolution digital imagery, Photogrammetric Engineering and Remote Sensing 69 (9) (2003) 963–972.
[82] A.P. Carleer, O. Debeir, E. Wolff, Assessment of very high spatial resolution satellite image segmentations, Photogrammetric Engineering and Remote Sensing 71 (11) (2005) 1285–1294.
[83] Q. Yu, P. Gong, N. Clinton, G. Biging, M. Kelly, D. Schirokauer, Object-based detailed vegetation classification with airborne high resolution remote sensing imagery, Photogrammetric Engineering and Remote Sensing 72 (7) (2006) 799–811.
[84] A. Le Bris, N. Chehata, X. Briottet, N. Paparoditis, Spectral band selection for urban material classification using hyperspectral libraries, ISPRS Annals of Photogrammetry, Remote Sensing and Spatial Information Sciences 3 (7) (2016) 33–40.
[85] J. Platt, Probabilistic outputs for support vector machines and comparison to regularized likelihood methods, in: Advances in Large Margin Classifiers, MIT Press, 2000, pp. 61–74.
[86] J. Bioucas-Dias, A. Plaza, N. Dobigeon, M. Parente, Q. Du, P. Gader, J. Chanussot, Hyperspectral unmixing overview: geometrical, statistical, and sparse regression-based approaches, IEEE Journal of Selected Topics in Applied Earth Observations and Remote Sensing 5 (2) (2012) 354–379.
[87] K. Adeline, A. Le Bris, F. Coubard, X. Briottet, N. Paparoditis, F. Viallefont, N. Rivière, J.-P. Papelard, P. Deliot, J. Duffaut, S. Airault, N. David, G. Maillet, L. Poutier, P.-Y. Foucher, V. Achard, J.-P. Souchon, C. Thom, Description de la campagne aéroportée UMBRA: étude de l'impact anthropique sur les écosystèmes urbains et naturels avec des images THR multispectrales et hyperspectrales, Revue Française de Photogrammétrie et de Télédétection 202 (2013) 79–92.
[88] S. Michel, M.-J. Lefèvre-Fonollosa, S. Hosford, HYPXIM – a hyperspectral satellite defined for science, security and defence users, in: Proc. of the 3rd Workshop on Hyperspectral Image and Signal Processing: Evolution in Remote Sensing, WHISPERS, Lisbon, Portugal, 2011.
[89] C. Kurtz, N. Passat, P. Gançarski, A. Puissant, Extraction of complex patterns from multiresolution remote sensing images: a hierarchical top-down methodology, Pattern Recognition 45 (2) (2012) 685–706.
[90] C. Wemmert, A. Puissant, G. Forestier, P. Gancarski, Multiresolution remote sensing image clustering, IEEE Geoscience and Remote Sensing Letters 6 (3) (2009) 533–537.
[91] J. Inglada, A. Vincent, M. Arias, B. Tardy, D. Morin, I. Rodes, Operational high resolution land cover map production at the country scale using satellite image time series, Remote Sensing 9 (1) (2017) 95.

[92] M. Pesaresi, C. Corbane, A. Julea, V. Florczyk, A. Syrris, P. Soille, Assessment of the added value of Sentinel-2 for detecting built-up areas, Remote Sensing 8 (4) (2016) 299.

[93] M. Li, A. Stein, W. Bijker, Q. Zhan, Urban land use extraction from Very High Resolution remote sensing imagery using a Bayesian network, ISPRS Journal of Photogrammetry and Remote Sensing 122 (2016) 192–205.

[94] M.K. Ridd, Exploring a V-I-S (vegetation-impervious surface-soil) model for urban ecosystem analysis through remote sensing: comparative anatomy for cities, International Journal of Remote Sensing 16 (1995) 2165–2185.

[95] Q. Weng, Remote sensing of impervious surfaces in the urban areas: requirements, methods, and trends, Remote Sensing of Environment 117 (2012) 34–49.

[96] A. Puissant, S. Rougier, A. Stumpf, Object-oriented mapping of urban trees using random forest classifiers, International Journal of Applied Earth Observation and Geoinformation 26 (2014) 235–245.

[97] E. Maggiori, Y. Tarabalka, G. Charpiat, P. Alliez, Convolutional neural networks for large-scale remote-sensing image classification, IEEE Transactions on Geoscience and Remote Sensing 55 (2) (2017) 645–657.

[98] T. Postadjian, A. Le Bris, H. Sahbi, C. Mallet, Investigating the potential of deep neural networks for large-scale classification of very high resolution satellite images, ISPRS Annals of Photogrammetry, Remote Sensing and Spatial Information Sciences IV-1-W1 (2017) 183–190.

[99] M. Volpi, D. Tuia, Dense semantic labeling of sub-decimeter resolution images with convolutional neural networks, IEEE Transactions on Geoscience and Remote Sensing 55 (2) (2017) 881–893.

[100] F. Sabo, C. Corbane, S. Ferri, Inter-sensor comparison of built-up derived from Landsat, Sentinel-1, Sentinel-2 and SPOT5/SPOT6 over selected cities, Tech. rep., JRC, 2017.

CHAPTER 12

Cross-modal Learning by Hallucinating Missing Modalities in RGB-D Vision

Nuno C. Garcia[*,†], **Pietro Morerio**[*], **Vittorio Murino**[*,‡]

[*]Pattern Analysis & Computer Vision (PAVIS), Istituto Italiano di Tecnologia (IIT), Genova, Italy
[†]Universita' degli Studi di Genova, Genova, Italy [‡]Universita' degli Studi di Verona, Verona, Italy

Contents

12.1 Introduction

Depth perception refers to the interesting ability to reason about the 3D world from visual information captured by the retinal surface. It is vital for the survival of many different animal species which use it for hunting or escaping, and an important skill for humans to understand

Multimodal Scene Understanding
https://doi.org/10.1016/B978-0-12-817358-9.00018-4

and interact with the surrounding environment. Humans start to develop depth perception very early, when babies start to crawl [1]. There have been identified several mechanisms that jointly contribute in different ways to the sense of relative and absolute position of objects, usually called depth cues. These can be divided in ocular motor cues (*e.g.* the movement of eyes converging or diverging) and, more interestingly in this scope, visual cues, which can be binocular or monocular. Binocular cues are related to stereovision and how the brain calculates depth based on the disparity of the left and right eyes' images. On the other hand, monocular cues refers to *a priori* visual assumptions derived from 2D single images, often related with physical factors such as shadows, perspective, motion parallax, texture gradient, occlusion, and others. For example, the assumption that an object looks blurrier the further it is, or that an object must be closer if it occludes another one, are signals that we can acquire with one eye only, and that our brain uses to reason about relative depth [2]. Although using only monocular vision affects object distance estimation [3], we are still able to perform most of our vision-related tasks efficiently with one eye only, and most importantly, to extract at least some depth information from 2D images without stereo mechanisms, using monocular cues, and make use it to navigate the 3D world.

Similarly, depth information is often of paramount importance for many computer vision tasks related to robotics, autonomous driving, scene understanding, to name a few. The emergence of cheap depth sensors and the need for big data to train bigger models led to big multimodal datasets containing RGB, depth, infrared, and joints sequences [4], which stimulated multimodal deep learning approaches. Traditional computer vision tasks like action recognition, object detection, or instance segmentation have been shown to benefit performance gains if the model considers other modalities, namely depth, instead of RGB-only [5–8].

Even though depth information brings improvements over RGB-only approaches, it is unrealistic to expect total availability of such data modality when the model is deployed in the real world. RGB cameras are still much more ubiquitous than depth sensors which may be difficult to install everywhere, and, moreover, good quality depth data might be difficult to acquire due to far-distance or reflectance issues, not to mention sensor or communications failure, or other unpredictable events.

Considering this limitation, we would like to answer the following question: what is the best way of using all data available at training time, in order to learn robust representations, knowing that there are missing (or noisy) modalities at test time? In other words, is there any added value in training a model by exploiting multimodal data, even if only one modality is available at test time? Unsurprisingly, the simplest and most commonly adopted solution consists in training the model using only the modality in which it will be tested. However, a more interesting alternative is trying to exploit the potential of the available data and train the model using all modalities, being, however, aware of the fact that not all of them will be accessible at

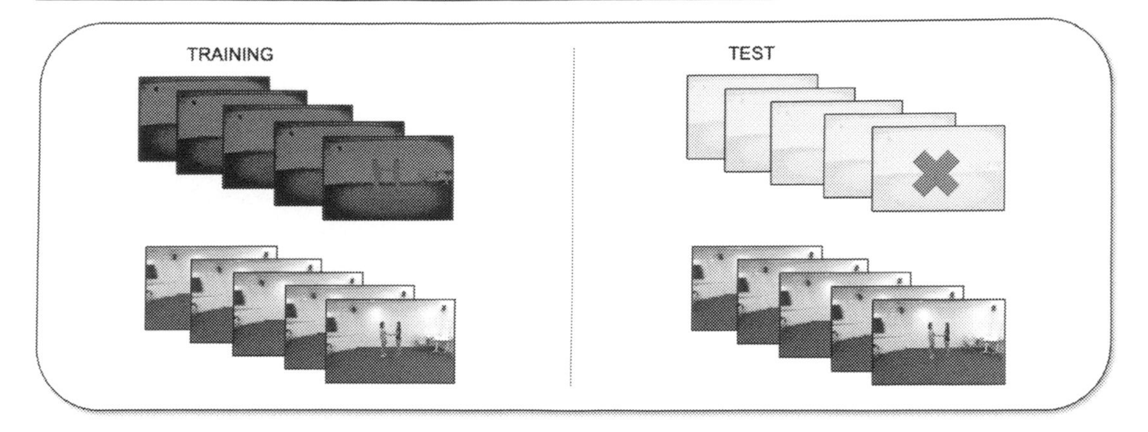

Figure 12.1 : What is the best way of using all data available at training time, considering a missing (or noisy) modality at test time?

test time (Fig. 12.1). This learning paradigm, i.e., when the model is trained using extra information, is generally known as *learning with privileged information* [9] or *learning with side information* [10].

In this chapter, we present a multimodal stream framework that learns from different data modalities and can be deployed and tested on a subset of these [11]. We design a model able to learn from RGB *and* depth video sequences, but due to its general structure, it can also be used to manage whatever combination of other modalities as well. To show its potential, we evaluate the performance on the task of video action recognition. The goal of the new learning paradigm, depicted in Fig. 12.2, is to *distill* the information conveyed by depth into a *hallucination* network, which is meant to "mimic" the missing stream at test time. Distillation [12] [13] refers to any training procedure where knowledge is transferred from a previously trained complex model to a simpler one. Our learning procedure introduces a new loss function, which is inspired by the *generalized distillation* framework [14], that formally unifies distillation and privileged information learning theories.

Our model is inspired by the two-stream network introduced by Simonyan and Zisserman [15], which has been notably successful in the traditional setting for video action recognition task [16] [17]. Differently from these works, we are interested in a multimodal setting, where we train one stream for each modality (RGB and depth in our case) and use these in the framework of privileged information. The most related work to ours is the inspiring method of Hoffman et al. [10], which proposes a hallucination network to learn with side information. We build on this idea, extending it by devising a new mechanism to *learn* and *use* such hallucination stream through a more general loss function and inter-stream connections.

This book chapter is strongly based on our recent conference paper [11], which main contributions and ideas are the following: (1) a multimodal stream network architecture able to exploit multiple data modalities at training while using only one at test time; (2) a new training procedure to learn a hallucination network within a novel two-stream model; (3) a more general loss function, based on the generalized distillation framework; (4) we report results comparable to/or achieving state of art – in the privileged information scenario – on the largest multimodal dataset for video action recognition, the NTU RGB+D [18], and on two other smaller ones, the UWA3DII [19] and the Northwestern-UCLA [20].

The rest of the chapter is organized as follows. Sect. 12.2 reviews similar approaches and discusses how they relate to the present work. Sect. 12.3 details the proposed architecture and the novel learning paradigm. Sect. 12.4 reports the results obtained on the various datasets, including a detailed ablation study performed on the NTU RGB+D dataset and a comparative performance with respect to the state of the art. Finally, we draw conclusions and future research directions in Sect. 12.5.

12.2 Related Work

Our work is at the intersection of three topics: privileged information [9], network distillation [12] [13], and multimodal video action recognition. However, Lopez et al. [14] noted that privileged information and network distillation are instances of the same more inclusive theory, called generalized distillation.

12.2.1 Generalized Distillation

Within the generalized distillation framework, our model is both related to the privileged information theory [9], considering that the extra modality (depth, in this case) is only used at training time, and, mostly, to the distillation framework. Indeed the core mechanism that our model uses to learn the hallucination network is derived from a distillation loss. More specifically, the supervision information provided by the teacher network (in this case, the network processing the depth data stream) is distilled into the hallucination network leveraging teacher's soft predictions and hard ground-truth labels in the loss function.

In this context, the closest works to our proposal are [21] and [10]. Luo et al. [21] addressed a similar problem to ours, where the model is first trained on several modalities (RGB, depth, joints and infrared), but tested only in one. The authors propose a graph-based distillation method that is able to distill information from all modalities at training time, while also passing through a validation phase on a subset of modalities. This showed to reach state-of-the-art results in action recognition and action detection tasks. Our work substantially differs from [21] since we benefit from a hallucination mechanism, consisting in an auxiliary network trained using the guidance distilled by the *teacher* network (that processes the depth data stream in our case). This mechanism allows the model to learn to emulate the presence of the missing modality at test time.

The work of Hoffman et al. [10] introduced a model to hallucinate depth features from RGB input for object detection task. While the idea of using a hallucination stream is similar to the one thereby presented, the mechanism used to learn it is different. In [10], the authors use a Euclidean loss between depth and hallucinated feature maps that is part of the total loss along with more than ten classification and localization losses, which makes its effectiveness very dependent on hyperparameter tuning to balance the different values, as the model is trained jointly in one step by optimizing the aforementioned composite loss. Differently, we propose a loss inspired to the distillation framework that not only uses the Euclidean distance between feature maps, and the one-hot labels, but also leverages soft predictions from the depth network. Moreover, we encourage the hallucination learning by design, by using cross-stream connections (see Sect. 12.3). This showed to largely improve the performance of our model with respect to the one-step learning process proposed in [10].

12.2.2 Multimodal Video Action Recognition

Video action recognition has a long and rich field of literature, spanning from classification methods using handcrafted features [22] [23] [24] [25] to modern deep learning approaches [26] [27] [28] [16], using either RGB-only or various multimodal data. Here, we focus on some of the more relevant works in multimodal video action recognition, including state-of-the-art methods considering the NTU RGB+D dataset, as well as architectures related to our proposed model.

The two-stream model introduced by Simonyan and Zisserman [15] is a landmark on video analysis, and since then has inspired a series of variants that achieved state-of-the-art performance on diverse datasets. This architecture is composed of an RGB and an optical flow stream, which are trained separately, and then fused at the prediction layer. The current state of the art in video action recognition [16] is inspired by such model, featuring 3D convolutions to deal with the temporal dimension, instead of the original 2D ones. In [17], a further variation of the two-stream approach is proposed, which models spatiotemporal features by injecting the motion stream's signal into the residual unit of the appearance stream. The idea of combining the two streams have also been explored previously by the same authors in [29].

Instead, in [5], the authors explore the complementary properties of RGB and depth data, taking the NTU RGB+D dataset as testbed. This work designed a deep autoencoder architecture and a structured sparsity learning machine, and showed to achieve state-of-the-art results for action recognition. Liu et al. [6] also use RGB and depth complementary information to devise a method for viewpoint invariant action recognition. Here, dense trajectories from RGB data are first extracted, which are then encoded in viewpoint invariant deep features. The RGB and depth features are then used as a dictionary to predict the test label.

All these previous methods exploited the rich information conveyed by the multimodal data to improve recognition. Our work, instead, proposes a fully convolutional model that exploits RGB and depth data at training time only, and uses exclusively RGB data as input at test time, reaching performance comparable to those utilizing the complete set of modalities in both stages.

12.3 Generalized Distillation with Multiple Stream Networks

This section describes our approach in terms of its architecture, the losses used to learn the different networks, and the training procedure.

12.3.1 Cross-stream Multiplier Networks

Typically in two-stream architectures, the two streams are trained separately and the predictions are fused with a late fusion mechanism [15] [17]. Such models use as input appearance (RGB) and motion (optical flow) data, which are fed separately into each stream, both in training and testing. Instead, in this work we use RGB and depth frames as inputs for training, but only RGB at test time, as already discussed (Fig. 12.2).

We use the ResNet-50-based [30] [31] model proposed in [17] as baseline architecture for each stream block of our model. In that paper, Feichtenhofer et al. proposed to connect the appearance and motion streams with multiplicative connections at several layers, as opposed to previous models which would only interact at the prediction layer. Such connections are depicted in Fig. 12.2 with the blue arrows. Fig. 12.3 illustrates this mechanism at a given layer of the multiple stream architecture, while it is actually implemented at the four convolutional layers of the Resnet-50 model. The underlying intuition is that these connections enable the model to learn better spatiotemporal representations, and help to distinguish between identical actions that require the combination of appearance and motion features. Originally, the cross-stream connections consisted of the injection of the motion stream signal into the other stream's residual unit, without affecting the skip path. ResNet's residual units are formally expressed as:

$$\mathbf{x}_{l+1} = f(h(\mathbf{x}_l) + F(\mathbf{x}_l, \mathcal{W}_l)),$$

where $\mathbf{x}_l$ and $\mathbf{x}_{l+1}$ are lth layer's input and output, respectively, F represents the residual convolutional layers defined by weights $\mathcal{W}_l$, $h(\mathbf{x}_l)$ is an identity mapping and f is a ReLU

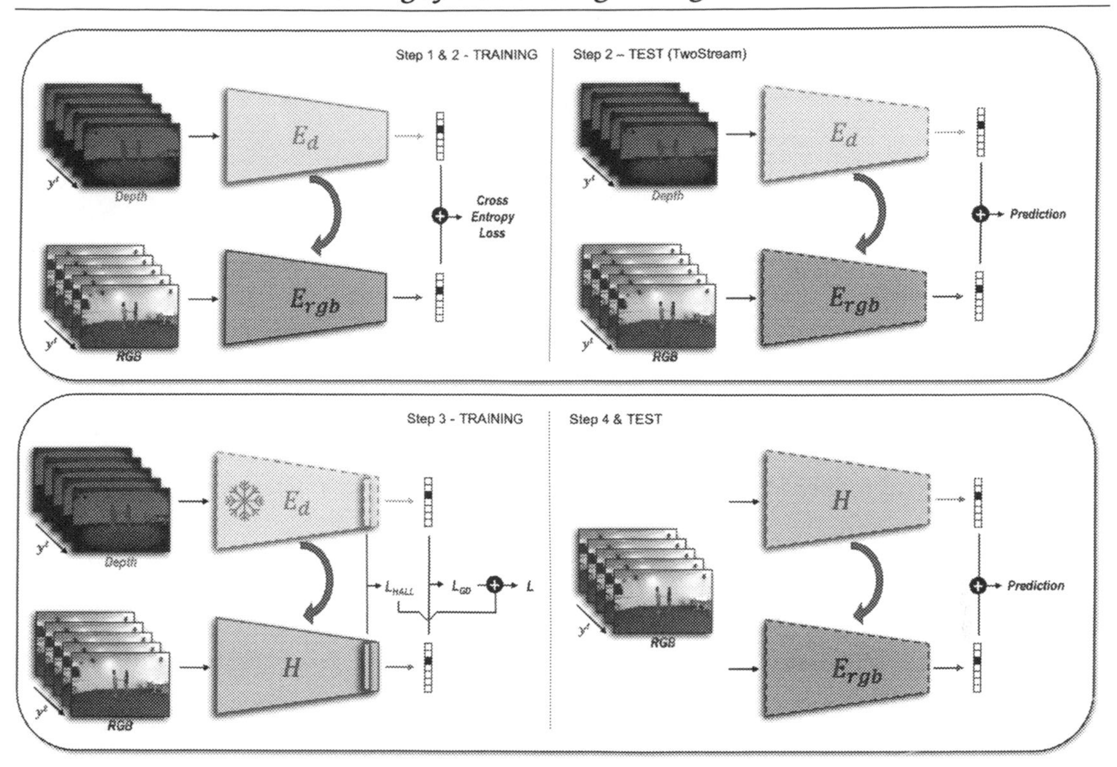

Figure 12.2 : Training procedure described in Sect. 12.3.3 (see also text therein). The first step refers to the separate (pre-)training of depth and RGB streams with standard cross-entropy classification loss, with both streams initialized with ImageNet weights. The second step, depicted in the scheme, represents the learning of the teacher network; both streams are initialized with the respective weights from step 1, and trained jointly with a cross-entropy loss as a traditional two-stream model, using RGB and depth data. The third step represents the learning of the student network: both streams are initialized with the depth stream weights from the previous step, but the actual depth stream is frozen; importantly, the input for the hallucination stream is RGB data; the model is trained using the loss proposed in Eq. (12.5). The fourth and last step refers to a fine-tuning step and also the test setup of our model, represented in the scheme; the hallucination stream is initialized from the respective weights from previous step, and the RGB stream with the respective weights from the second step; this model is fine-tuned using a cross-entropy loss, and importantly, using only RGB data as input for both streams.

non-linearity. The cross-streams connections are then defined as

$$\mathbf{x}_{l+1}^a = f(\mathbf{x}_l^a) + F(\mathbf{x}_l^a \odot f(\mathbf{x}_l^m), \mathcal{W}_l),$$

where $\mathbf{x}^a$ and $\mathbf{x}^m$ are the appearance and motion streams, respectively, and $\odot$ is the element-wise multiplication operation. Such mechanism implies a spatial alignment between both

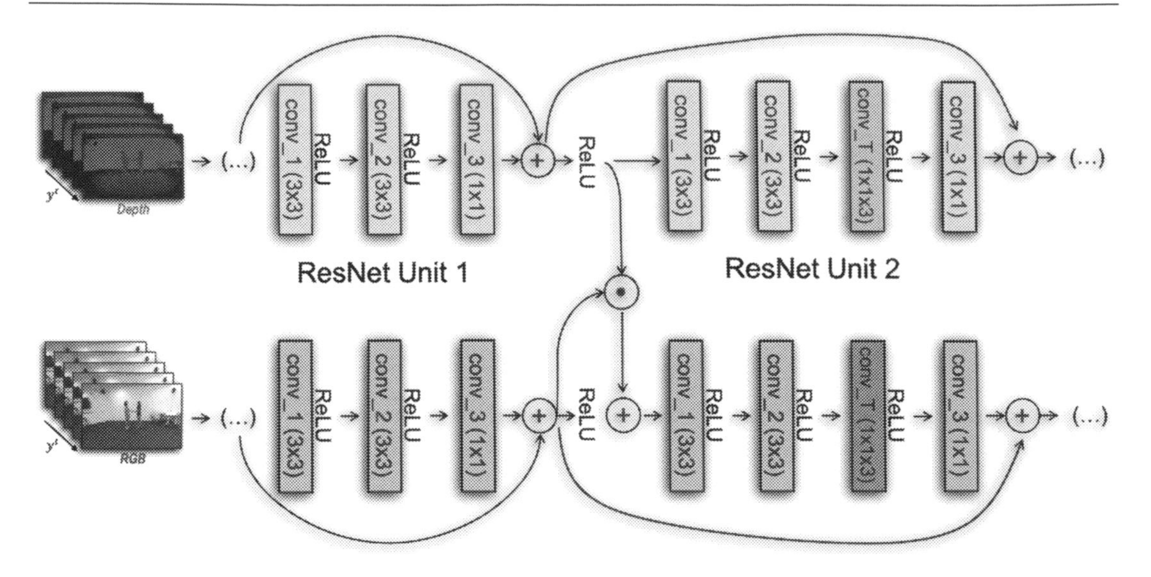

Figure 12.3 : Detail of the ResNet residual unit, showing the multiplicative connections and temporal convolutions [17]. In our architecture, the signal injection occurs before the second residual unit of each of the four ResNet blocks.

feature maps, and therefore between both modalities. This alignment comes for free when using RGB and optical flow, since the latter is computed from the former in a way that spatial arrangement is preserved. However, this is an assumption we cannot generally make. For instance, depth and RGB are often captured from different sensors, likely resulting in spatially misaligned frames. The alignment procedure, described in Sect. 12.4.2, is part of the pre-processing phase and it refers uniquely to the NTU RGB+D dataset.

Temporal convolutions. In order to augment the model temporal support, we implement 1D temporal convolutions in the second residual unit of each ResNet layer (as in [17]), as illustrated in Fig. 12.3. The weights $W_l \in \mathbb{R}^{1\times 1\times 3\times C_l\times C_l}$ are convolutional filters initialized as identity mappings at feature level, and centered in time, and C_l is the number of channels in layer l.

Very recently in [32], the authors explored various network configurations using temporal convolutions, comparing several different combinations for the task of video classification. This work suggests that decoupling 3D convolutions into 2D (spatial) and 1D (temporal) filters is the best setup in action recognition tasks, producing best accuracies. The intuition for the latter setup is that factorizing spatial and temporal convolutions in two consecutive convolutional layers eases training of the spatial and temporal tasks (also in line with [33]).

12.3.2 Hallucination Stream

We also introduce and learn a hallucination network [10], using a new learning paradigm, loss function and interaction mechanism. The hallucination stream network has the same architecture as the appearance and depth stream models.

This network receives RGB as input, and is trained to "imitate" the depth stream at different levels, *i.e.* at feature and prediction layers. In this work, we explore several ways to implement such learning paradigm, including both the training procedure and the loss, and how they affect the overall performance of the model.

In [10] it is proposed a regression loss between the hallucination and depth feature maps, defined as:

$$L_{hall}(l) = \lambda_l \|\sigma(A_l^d) - \sigma(A_l^h)\|_2^2, \tag{12.1}$$

where σ is the sigmoid function, and A_l^d and A_l^h are the l-th layer activations of depth and hallucination network. This Euclidean loss forces both activation maps to be similar. In [10], this loss is weighted along with another ten classification and localization loss terms, making it hard to balance the total loss. One of the main motivations behind the proposed new staged learning paradigm, described in Sect. 12.3.3, is to avoid the inefficient, heuristic-based tweaking of so many loss weights, a.k.a. hyperparameter tuning.

Instead, we adopt an approach inspired by the generalized distillation framework [14], in which a *student* model $f_s \in \mathcal{F}_s$ distills the representation $f_t \in \mathcal{F}_t$ learned by the *teacher* model. This is formalized as

$$f_s = \arg\min_{f \in \mathcal{F}_s} \frac{1}{n} \sum_{i=1}^{n} L_{GD}(i), n = 1, ..., N \tag{12.2}$$

where N is the number of examples in the dataset. The generalized distillation loss is so defined as:

$$L_{GD}(i) = (1 - \lambda)\ell(y_i, \varsigma(f(x_i))) + \lambda\ell(s_i, \varsigma(f(x_i))), \ \lambda \in [0, 1] \ f_s \in \mathcal{F}_s, \tag{12.3}$$

ς is the *softmax* operator and s_i is the soft prediction from the teacher network:

$$s_i = \varsigma(f_t(x_i)/T), \ T > 0. \tag{12.4}$$

The parameter λ in Eq. (12.3) allows one to tune the loss by giving more importance either to imitating ground-truth hard or soft teacher targets, y_i and s_i, respectively. This mechanism indeed allows the transfer of information from the depth (teacher) to the hallucination (student) network. The temperature parameter T in Eq. (12.4) allows one to smooth the probability vector predicted by the teacher network. The intuition is that such smoothing may expose relations between classes that would not easily be revealed in raw predictions, further facilitating the distillation by the student network f_s.

We suggest that both Euclidean and generalized distillation losses are indeed useful in the learning process. In fact, by encouraging the network to decrease the distance between hallucinated and true depth feature maps, it can help to distill depth information encoded in the generalized distillation loss. Thus, we formalize our final loss function as follows:

$$L = (1-\alpha)L_{GD} + \alpha L_{hall}, \ \alpha \in [0, 1], \tag{12.5}$$

where α is a parameter balancing the contributions of the two loss terms during training. The parameters λ, α and T are estimated by utilizing a validation set and discussed in Sect. 12.4.2.

In summary, the generalized distillation framework proposes to use the student-teacher framework introduced in the distillation theory to extract knowledge from the privileged information source. We explore this idea by proposing a new learning paradigm to train a hallucination network using privileged information, which we will describe in the next section. In addition to the loss functions introduced above, we also allow the teacher network to share information with the student network by design, through the cross-stream multiplicative connections. We test how all these possibilities affect the model's performance in the experimental section through an extensive ablation study.

12.3.3 Training Paradigm

In general, the proposed training paradigm, illustrated in Fig. 12.2, is divided in two core parts: the first part (steps 1 and 2 in the figure) focuses on learning the teacher network f_t, leveraging RGB and depth data (the privileged information in this case); the second part (steps 3 and 4 in the figure) focuses on learning the hallucination network, referred to as student network f_s in the distillation framework, using the general hallucination loss defined in Eq. (12.5).

The *first training step* consists in training both streams separately, which is a common practice in two-stream architectures. Both depth and appearance streams are trained minimizing cross-entropy, after being initialized with a pre-trained ImageNet model for all experiments. Temporal kernels are initialized as $[0, 1, 0]$, *i.e.* only information on the central frame is used at the beginning—this eventually changes as the training continues. As in [34], depth frames are encoded into color images using a jet colormap.

The *second training step* is still focused on further training the teacher model. Since the model trained in this step has the architecture and capacity of the final one, and *has access to both modalities*, its performance represents an upper bound for the task we are addressing. This is one of the major differences between our approach and the one used in [10]: by decoupling the teacher learning phase with the hallucination learning, we are able to both learn a better teacher *and* a better student, as we will show in the experimental section.

In the *third training step*, we focus on learning the hallucination network from the teacher model, *i.e.*, the depth stream network just trained. Here, the weights of the depth network are frozen, while receiving in input depth data. Instead, the hallucination network, receiving in input RGB data, is trained with the loss defined in Eq. (12.5), while also receiving feedback from the cross-stream connections from the depth network. We found that this helps the learning process.

In the *fourth and last step*, we carry out fine tuning of the whole model, composed of the RGB and the hallucination streams. This step uses RGB-only as input, and it also precisely resembles the setup used at test time. The cross-stream connections inject the hallucinated signal into the appearance RGB stream network, resulting in the multiplication of the hallucinated feature maps and the RGB feature maps. The intuition is that the hallucination network has learned to inform the RGB model where the action is taking place, similarly to what the depth model would do with real depth data.

12.4 Experiments

12.4.1 Datasets

We evaluate our method on three datasets, while the ablation study is performed only on the NTU RGB+D dataset. Our model is initialized with ImageNet pre-trained weights and trained and evaluated on the NTU RGB+D dataset. We later fine-tune this model on each of the two smaller datasets for the corresponding evaluation experiments.

NTU RGB+D [18]. This is the largest public dataset for multimodal video action recognition. It is composed of 56,880 videos, available in four modalities: RGB videos, depth sequences, infrared frames, and 3D skeleton data of 25 joints. It was acquired with a Kinect v2 sensor in 80 different viewpoints, and includes 40 subjects performing 60 distinct actions. We follow the two evaluation protocols originally proposed in [18], which are cross-subject and cross-view. As in the original paper, we use about 5% of the training data as validation set for both protocols, in order to select the parameters λ, α and T. In this work, we use only RGB and depth data. The masked depth maps are converted to a three channel map via a jet mapping, as in [34].

UWA3DII [19]. This dataset consists on 1075 samples of RGB, depth and skeleton sequences. It features 10 subjects performing 30 actions captured in 5 different views.

Northwestern-UCLA [20]. Similarly to the other datasets, it provides RGB, depth and skeleton sequences for 1475 samples. It features 10 subjects performing 10 actions captured in 3 different views.

12.4.2 Pre-processing and Alignment of RGB and Depth Frames

The multiplicative cross-stream connections require both RGB and depth frames to be spatially aligned, since they are element-wise operations over the feature maps. The NTU RGB+D dataset is acquired using a Kinect sensor, which result in different dimensions and aspect ratios for RGB and depth frames. Fortunately, this dataset provides the joints' spatial coordinates in every RGB and depth frames, $rgb_{x,y}$ and $depth_{x,y}$, respectively, which we use to align both modalities. Because the other two smaller datasets do not provide such information, this alignment procedure is only applicable to the NTU RGB+D dataset. Consequently, experiments using the other datasets refer to the model with no cross-stream connections.

For every frame of a given video, we first compute the ratio $ratio_x^{A,B} = (rgb_x^A - rgb_x^B)/(depth_x^A - depth_x^B)\ \forall A, B \in S$, using all depth and RGB x coordinates from the frame's well-tracked joints set S, and similarly for the y dimension. The video aspect ratio is then calculated as the mean between the median aspect ratio for x and the median aspect ratio for y dimensions. The RGB frames of a given video are scaled according to this ratio. Finally, both RGB and depth frames are overlaid by aligning both skeletons, and the intersection is cropped on both modalities. The cropped sections are then re-scaled according to the network's input dimension, in this case 224×224.

Similarly to what was done in [17], we sample 5 frames evenly spaced in time for each video, both for training and testing. Sampling random frames for training results in similar accuracy values. For training, we also flip horizontally the video frames with probability $P = 0.5$.

12.4.3 Hyperparameters and Validation Set

After validation, we have selected the following set of hyperparameters: $\alpha = 0.5$, $\lambda = 0.5$, $T = 10$; slightly different values do not show a significant change in performance. The networks are optimized using ADAM [35] with default learning rate, except for the fine-tuning steps where the learning rate was decreased by a factor of 10.

Regarding the NTU RGB+D dataset, the validation set is not defined in the original paper where the dataset is presented [18]. For the sake of experiments reproducibility, we explain here how we defined the validation set. For the cross-subject protocol, we choose the subject #1 (from the training set), which corresponds to around 5% of the training set. For the cross-view protocol, we do the following: 1) create a dictionary of sorted videos for each

Table 12.1: Ablation study. A full set of experiments is provided for the NTU cross-subject evaluation protocol. For the cross-view protocol, only the most important results are reported.

#	Method	Test Modality	Loss	Cross-Subject	Cross-View
1	Ours - step 1, depth stream	Depth	x-entr	70.44%	75.16%
2	Ours - step 1, RGB stream	RGB	x-entr	66.52%	71.39%
3	Hoffman [10] w/o conn.	RGB	Eq. (12.1)	64.64%	-
4	Hoffman [10] w/o conn.	RGB	Eq. (12.3)	68.60%	-
5	Hoffman [10] w/o conn.	RGB	Eq. (12.5)	70.70%	-
6	Ours - step 2, depth stream	Depth	x-entr	71.09%	77.30%
7	Ours - step 2, RGB stream	RGB	x-entr	66.68%	56.26%
8	Ours - step 2	RGB & Depth	x-entr	**79.73%**	81.43%
9	Ours - step 2 w/o conn.	RGB & Depth	x-entr	78.27%	**82.11%**
10	Ours - step 3 w/o conn.	RGB (*hall*)	Eq. (12.1)	69.93%	70.64%
11	Ours - step 3 w/ conn.	RGB (*hall*)	Eq. (12.1)	70.47%	-
12	Ours - step 3 w/ conn.	RGB (*hall*)	Eq. (12.3)	71.52%	-
13	Ours - step 3 w/ conn.	RGB (*hall*)	Eq. (12.5)	71.93%	74.10%
14	Ours - step 3 w/o conn.	RGB (*hall*)	Eq. (12.5)	71.10%	-
15	**Ours - step 4**	**RGB**	x-entr	**73.42%**	**77.21%**

key=action (from the training set); 2) set numpy random seed equal to 0; 3) sample 31 videos using `numpy.random.choice` for each action, which in the end will correspond to around 5% of the training set.

12.4.4 Ablation Study

In this section we discuss the results of the experiments carried out to understand the contribution of each part of the model and of the training procedure. Table 12.1 reports performances at the several training steps, different losses and model configurations.

Rows #1 and #2 refer to the first training step, where depth and RGB streams are trained separately. We note that the depth stream network provides better performance with respect to the RGB one, as expected.

The second part of the table (Rows #3–5) shows the results using Hoffman et al.'s method [10] – *i.e.*, adopting a model initialized with the pre-trained networks from the first training step, and the hallucination network initialized using the depth network. Row #3 refers to the original paper [10] (i.e., using the loss L_{hall}, Eq. (12.1)), and rows #4 and #5 refer to the training using the proposed losses L_{GD} and L, in Eqs. (12.3) and (12.5), respectively. It can be noticed that the accuracies achieved using the proposed loss functions overcome that obtained in [10] by a significant margin (about 6% in the case of the total loss L).

The third part of the table (rows #6–9) reports performances after the training step 2. Rows #6 and #7 refer to the accuracy provided by depth and RGB stream networks belonging to the model of row #8, taken individually. The final model constitutes the upper bound for our hallucination model, since it uses RGB and depth for training and testing. Performances obtained by the model in row #8 and #9, with and without cross-stream connections, respectively, are the highest in absolute since using both modalities (around 78–79% for the cross-subject and 81–82% for the cross-view protocols, respectively), largely outperforming the accuracies obtained using only one modality (in rows #6 and #7).

The *fourth* part of the table (rows #10–14) shows results for our hallucination network after the several variations of learning processes, different losses and with and without cross-stream connections.

Finally, the last row, #15, reports results after the last fine-tuning step which further narrows the gap with the upper bound.

12.4.4.1 Contribution of the cross-stream connections

We claim that the signal injection provided by the cross-stream connections helps the learning of a better hallucination network. Row #13 and #14 show the performances for the hallucination network learning process, starting from the same point and using the same loss. The hallucination network that is learned using multiplicative connections performs better than its counterpart, where depth and RGB frames are properly aligned. It is important to note though that this is not observed in the other two smaller datasets, due to the spatial misalignment of modalities, and consequently between feature maps.

12.4.4.2 Contributions of the proposed distillation loss (Eq. (12.5))

The distillation and Euclidean losses have complementary contributions to the learning of the hallucination network. This is observed by looking at the performances reported in rows #3, #4 and #5, and also #11, #12 and #13. In both the training procedure proposed by Hoffman et al. [10] and our staged training process, the distillation loss improves over the Euclidean loss, and the combination of both improves over the rest. This suggests that both Euclidean and distillation losses have its own share and act differently to align the hallucination (student) and depth (teacher) feature maps and outputs' distributions.

12.4.4.3 Contributions of the proposed training procedure

The intuition behind the staged training procedure proposed in this work can be ascribed to the *divide et impera* (divide-and-conquer) strategy. In our case, it means breaking the problem in two parts: learning the actual task we aim to solve and learning the student network to face test-time limitations. Row #5 reports accuracy for the architecture proposed by Hoffman et al., and rows #15 report the performance for our model with connections. Both use the same loss

Table 12.2: Accuracy of the model tested with clean RGB and noisy depth data. Accuracy of the proposed hallucination model, i.e. with *no depth* at test time, is 77.21%.

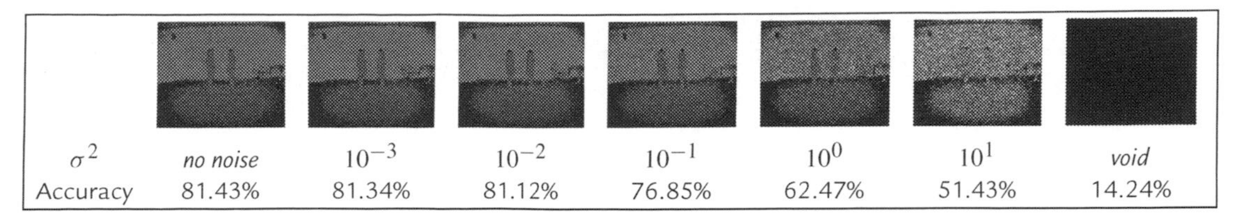

σ^2	*no noise*	10^{-3}	10^{-2}	10^{-1}	10^{0}	10^{1}	*void*
Accuracy	81.43%	81.34%	81.12%	76.85%	62.47%	51.43%	14.24%

to learn the hallucination network, and both start from the same initialization. We observe that our method outperform the one in row #5, which justifies the proposed staged training procedure.

12.4.5 Inference with Noisy Depth

Suppose that in a real test scenario we can only access unreliable sensors which produce noisy depth data. The question we now address is: to which extent can we trust such noisy data? In other words, at which level of noise does it become favorable to hallucinate the depth modality with respect to using the full teacher model (step 2) with noisy depth data?

The depth sensor used in the NTU dataset (Kinect), is an IR emitter coupled with an IR camera, and has very complex noise characterization comprising at least six different sources [36]. It is beyond the scope of this work to investigate noise models affecting the depth channel, so, for our analysis we choose the most common one, i.e., the multiplicative speckle noise. Hence, we inject Gaussian noise in the depth images I in order to simulate speckle noise: $I = I * n, n \sim \mathcal{N}(1, \sigma)$. Table 12.2 shows how performances of the network degrade when depth is corrupted with such Gaussian noise with increasing variance (NTU cross-view protocol only). Results show that accuracy significantly decreases w.r.t. to the one guaranteed by our hallucination model (77.21% – row #15 in Table 12.1), even with low noise variance. This means, in conclusion, that *training a hallucination network is an effective way not only to obviate to the problem of a missing modality, but also to deal with noise affecting the input data channel.*

12.4.6 Comparison with Other Methods

Table 12.3 compares performances of different methods on the various datasets. The standard performance measure used for this task and datasets is classification accuracy, estimated according to the protocols (training and testing splits) reported in the respective works we are comparing with.

Table 12.3: Classification accuracies and comparisons with the state of the art. Performances referred to the several steps of our approach (ours) are highlighted in bold. x refers to comparisons with unsupervised learning methods. Δ refers to supervised methods: here train and test modalities coincide. □ refers to privileged information methods: here training exploits RGB+D data, while test relies on RGB data only. The second column refers to the modalities used at test time: R-RGB, D-Depth, and J-Joints. The third column refers to cross-subject and the fourth to the cross-view evaluation protocols on the NTU dataset. The results reported on the other two datasets are for the cross-view protocol.

Method	Test Mods.	NTU (p1)	NTU (p2)	UWA3DII	NW-UCLA	
Luo [37]	D	66.2%	–	–	–	
Luo [37]	R	56.0%	–	–	–	×
Rahmani [38]	R	–	–	67.4%	78.1%	
HOG-2 [39]	D	32.4%	22.3%	–	–	
Action Tube [40]	R	–	–	37.0%	61.5%	
Ours – depth step 1	**D**	**70.44%**	**75.16%**	**75.28%**	**72.38%**	
Ours – RGB step 1	**R**	**66.52%**	**71.39%**	**63.67%**	**85.22%**	
Deep RNN [18]	J	56.3%	64.1%	–	–	△
Deep LSTM [18]	J	60.7%	67.3%	–	–	
Sharoudy [18]	J	62.93%	70.27%	–	–	
Kim [41]	J	74.3%	83.1%	–	–	
Sharoudy [5]	R+D	74.86%	–	–	–	
Liu [6]	R+D	77.5%	84.5%	–	–	
Rahmani [42]	D+J	75.2	83.1	84.2%	–	
Ours – step 2	**R+D**	**79.73%**	**81.43%**	**79.66%**	**88.87%**	
Hoffman et al. [10]	R	64.64%	–	66.67%	83.30%	
ADMD [43]	R	73.11%	81.50%	–	91.64%	□
Ours – step 3	**R**	**71.93%**	**74.10%**	**71.54%**	**76.30%**	
Ours – step 4	**R**	**73.42%**	**77.21%**	**73.23%**	**86.72%**	

The first part of the table (indicated by × symbol) refers to unsupervised methods, which achieve surprisingly high results even without relying on labels in learning representations.

The second part refers to supervised methods (indicated by △), divided according to the modalities used for training and testing. Here, we list the performance of the separate RGB and depth streams trained in step 1, as a reference. We expect our final model to perform better than the one trained on RGB-only, whose accuracy constitutes a lower bound for our student network. The values reported for *our step 1* models for UWA3DII and NW-UCLA datasets refer to the fine-tuning of our NTU model. We have experimented training using pre-trained ImageNet weights, which led from 20% to 30% less accuracy. We also propose our baseline, consisting in the teacher model trained in step 2. Its accuracy represents an upper bound for the final model, which will not rely on depth data at test time.

Table 12.4: RGB distillation (NTU RGB-D, cross-view protocol).

#	Method	Test Modality	Loss	Cross-View
13a	Ours – step 3	Depth (*hall*)	Eq. (12.5)	76.12%
15a	Ours – step 4	Depth	x-entr	76.41%

The last part of the table (indicated by □) reports our model's performances at two different stages together with the other privileged information methods [10] [11]. For all datasets and protocols, we can see that our privileged information approach outperforms [10], which is the only fair *direct* comparison we can make (same training & test data). Besides, as expected, our final model performs better than "Ours – RGB model, step 1" since it exploits more data at training time, and worse than " Ours – step 2", since it exploits less data at test time. Other RGB+D methods perform better (which is comprehensible since they rely on RGB+D in both training and test) but not by a large margin. The method by Garcia et al. [11] is similar to these two in the sense that is also uses a hallucination network to cope with the missing depth modality. However, it takes a different look to it, by learning the hallucination stream through an adversarial strategy.

12.4.7 Inverting Modalities – RGB Distillation

The results presented in Table 12.4 address the opposite case of what is studied in the rest of the chapter, *i.e.*, the case when RGB data is missing. In this case, the hallucination stream distills knowledge from the RGB stream in step 3 (Fig. 12.2).

We observe that the performance of the final model degrades by almost 1%, 76.41% vs. 77.21% (cf. line 15 of Table 12.2). A more consistent setting would be to modify the model, inverting the cross-stream connections in steps 3 and 4, thus having information flowing again from depth to RGB.

12.5 Conclusions and Future Work

This chapter addresses the task of video action recognition in the context of privileged information. We describe a new learning paradigm to teach a hallucination network to mimic the depth stream. Our model outperforms many of the supervised methods recently evaluated on the NTU RGB+D dataset, as well as the original hallucination model proposed in [10]. We conducted an extensive ablation study to verify how the several parts composing our learning paradigm contribute to the model performance. As a future work, we would like to extend this approach to dealing with additional modalities that may be available at training time, such as skeleton joints data or infrared sequences. Finally, the current model cannot be applied to still images due to the presence of temporal convolutions. In principle, we could remove them and apply our method to still images and other tasks such as object detection.

References

[1] E.J. Gibson, R.D. Walk, The "visual cliff", Scientific American 202 (4) (1960) 64–71.

[2] M.R. Watson, J.T. Enns, Encyclopedia of Human Behavior, Elsevier, 2012, pp. 690–696, Ch. Depth Perception.

[3] P. Servos, Distance estimation in the visual and visuomotor systems, Experimental Brain Research 130 (1) (2000) 35–47.

[4] M. Firman, Rgbd datasets: past, present and future, in: Proceedings of the IEEE Conference on Computer Vision and Pattern Recognition Workshops, 2016, pp. 19–31.

[5] A. Shahroudy, T.-T. Ng, Y. Gong, G. Wang, Deep multimodal feature analysis for action recognition in rgb+d videos, IEEE Transactions on Pattern Analysis and Machine Intelligence 40 (5) (2018) 1045–1058.

[6] J. Liu, N. Akhtar, A. Mian, Viewpoint invariant action recognition using rgb-d videos, IEEE Access (2017), https://doi.org/10.1109/DICTA.2017.8227505.

[7] S. Gupta, R. Girshick, P. Arbeláez, J. Malik, Learning rich features from rgb-d images for object detection and segmentation, in: European Conference on Computer Vision, 2014, pp. 345–360.

[8] C. Hazirbas, L. Ma, C. Domokos, D. Cremers, Fusenet: incorporating depth into semantic segmentation via fusion-based cnn architecture, in: Asian Conference on Computer Vision, Springer, 2016, pp. 213–228.

[9] V. Vapnik, A. Vashist, A new learning paradigm: learning using privileged information, Neural Networks 22 (5) (2009) 544–557.

[10] J. Hoffman, S. Gupta, T. Darrell, Learning with side information through modality hallucination, in: Proceedings of the IEEE Conference on Computer Vision and Pattern Recognition, 2016, pp. 826–834.

[11] N.C. Garcia, P. Morerio, V. Murino, Modality distillation with multiple stream networks for action recognition, in: The European Conference on Computer Vision, ECCV, 2018.

[12] G. Hinton, O. Vinyals, J. Dean, Distilling the knowledge in a neural network, in: NIPS Deep Learning and Representation Learning Workshop, 2014.

[13] J. Ba, R. Caruana, Do deep nets really need to be deep?, in: Advances in Neural Information Processing Systems, 2014, pp. 2654–2662.

[14] D. Lopez-Paz, B. Schölkopf, L. Bottou, V. Vapnik, Unifying distillation and privileged information, in: International Conference on Learning Representations, ICLR, 2016.

[15] K. Simonyan, A. Zisserman, Two-stream convolutional networks for action recognition in videos, in: Advances in Neural Information Processing Systems, 2014, pp. 568–576.

[16] J. Carreira, A. Zisserman, Quo vadis, action recognition? A new model and the kinetics dataset, in: Computer Vision and Pattern Recognition, 2017 IEEE Conference on, CVPR, IEEE, 2017, pp. 4724–4733.

[17] C. Feichtenhofer, A. Pinz, R.P. Wildes, Spatiotemporal multiplier networks for video action recognition, in: Proceedings of the IEEE Conference on Computer Vision and Pattern Recognition, 2017, pp. 4768–4777.

[18] A. Shahroudy, J. Liu, T.-T. Ng, G. Wang, Ntu rgb+d: a large scale dataset for 3d human activity analysis, in: Proceedings of the IEEE Conference on Computer Vision and Pattern Recognition, 2016, pp. 1010–1019.

[19] H. Rahmani, A. Mahmood, D. Huynh, A. Mian, Histogram of oriented principal components for cross-view action recognition, IEEE Transactions on Pattern Analysis and Machine Intelligence 38 (12) (2016) 2430–2443.

[20] J. Wang, X. Nie, Y. Xia, Y. Wu, S.-C. Zhu, Cross-view action modeling, learning and recognition, in: Proceedings of the IEEE Conference on Computer Vision and Pattern Recognition, 2014, pp. 2649–2656.

[21] Z. Luo, J.-T. Hsieh, L. Jiang, J. Carlos Niebles, L. Fei-Fei, Graph distillation for action detection with privileged modalities, in: The European Conference on Computer Vision, ECCV, 2018.

[22] N. Dalal, B. Triggs, Histograms of oriented gradients for human detection, in: Computer Vision and Pattern Recognition, 2005, IEEE Computer Society Conference on, vol. 1, CVPR 2005, IEEE, 2005, pp. 886–893.
[23] H. Wang, A. Kläser, C. Schmid, C.-L. Liu, Action recognition by dense trajectories, in: Computer Vision and Pattern Recognition, 2011 IEEE Conference on, CVPR, IEEE, 2011, pp. 3169–3176.
[24] H. Wang, C. Schmid, Action recognition with improved trajectories, in: Proceedings of the IEEE International Conference on Computer Vision, 2013, pp. 3551–3558.
[25] I. Laptev, M. Marszalek, C. Schmid, B. Rozenfeld, Learning realistic human actions from movies, in: Computer Vision and Pattern Recognition, 2008, IEEE Conference on, CVPR 2008, IEEE, 2008, pp. 1–8.
[26] A. Karpathy, G. Toderici, S. Shetty, T. Leung, R. Sukthankar, L. Fei-Fei, Large-scale video classification with convolutional neural networks, in: Proceedings of the IEEE Conference on Computer Vision and Pattern Recognition, 2014, pp. 1725–1732.
[27] D. Tran, L. Bourdev, R. Fergus, L. Torresani, M. Paluri, Learning spatiotemporal features with 3d convolutional networks, in: Proceedings of the IEEE International Conference on Computer Vision, 2015, pp. 4489–4497.
[28] X. Wang, R. Girshick, A. Gupta, K. He, Non-local neural networks, in: The IEEE Conference on Computer Vision and Pattern Recognition, vol. 1, CVPR, 2018, p. 4.
[29] C. Feichtenhofer, A. Pinz, A. Zisserman, Convolutional two-stream network fusion for video action recognition, in: Proceedings of the IEEE Conference on Computer Vision and Pattern Recognition, 2016, pp. 1933–1941.
[30] K. He, X. Zhang, S. Ren, J. Sun, Deep residual learning for image recognition, in: Proceedings of the IEEE Conference on Computer Vision and Pattern Recognition, 2016, pp. 770–778.
[31] K. He, X. Zhang, S. Ren, J. Sun, Identity mappings in deep residual networks, in: European Conference on Computer Vision, Springer, 2016, pp. 630–645.
[32] D. Tran, H. Wang, L. Torresani, J. Ray, Y. LeCun, M. Paluri, A closer look at spatiotemporal convolutions for action recognition, in: Proceedings of the IEEE International Conference on Computer Vision, 2018.
[33] L. Sun, K. Jia, D.-Y. Yeung, B.E. Shi, Human action recognition using factorized spatio-temporal convolutional networks, in: Proceedings of the IEEE International Conference on Computer Vision, 2015, pp. 4597–4605.
[34] A. Eitel, J.T. Springenberg, L. Spinello, M. Riedmiller, W. Burgard, Multimodal deep learning for robust rgb-d object recognition, in: Intelligent Robots and Systems, 2015 IEEE/RSJ International Conference on, IROS, IEEE, 2015, pp. 681–687.
[35] D.P. Kingma, J. Ba, Adam: a method for stochastic optimization, in: International Conference on Learning Representations, ICLR, 2015.
[36] T. Mallick, P.P. Das, A.K. Majumdar, Characterizations of noise in Kinect depth images: a review, IEEE Sensors Journal 14 (6) (2014) 1731–1740.
[37] Z. Luo, B. Peng, D.-A. Huang, A. Alahi, L. Fei-Fei, Unsupervised learning of long-term motion dynamics for videos, in: Proceedings of the IEEE Conference on Computer Vision and Pattern Recognition, 2017, pp. 2203–2212.
[38] H. Rahmani, A. Mian, M. Shah, Learning a deep model for human action recognition from novel viewpoints, IEEE Transactions on Pattern Analysis and Machine Intelligence 40 (3) (2018) 667–681.
[39] E. Ohn-Bar, M.M. Trivedi, Joint angles similarities and hog2 for action recognition, in: Computer Vision and Pattern Recognition Workshops, 2013 IEEE Conference on, CVPRW, IEEE, 2013, pp. 465–470.
[40] G. Gkioxari, J. Malik, Finding action tubes, in: Computer Vision and Pattern Recognition, 2015 IEEE Conference on, CVPR, IEEE, 2015, pp. 759–768.
[41] T. Soo Kim, A. Reiter, Interpretable 3d human action analysis with temporal convolutional networks, in: Proceedings of the IEEE Conference on Computer Vision and Pattern Recognition Workshops, 2017, pp. 20–28.
[42] H. Rahmani, M. Bennamoun, Learning action recognition model from depth and skeleton videos, in: Proceedings of the IEEE Conference on Computer Vision and Pattern Recognition, 2017, pp. 5832–5841.
[43] N.C. Garcia, P. Morerio, V. Murino, Learning with privileged information via adversarial discriminative modality distillation, arXiv preprint, arXiv:1810.08437.

Index

D

N

O

P

Q

R

Printed in the United States
By Bookmasters